Editorial Board Members' Collection Series: Gastrointestinal and Hepatic Diseases

Editorial Board Members' Collection Series: Gastrointestinal and Hepatic Diseases

Guest Editors

Marcello Candelli
Ludovico Abenavoli

MDPI

Basel • Beijing • Wuhan • Barcelona • Belgrade • Novi Sad • Cluj • Manchester

Guest Editors
Marcello Candelli
Emergency, Reanimation and Anesthesiology Sciences
Department of Fondazione Policlinico Agostino Gemelli—IRCCS
Catholic University of Sacred Heart
Rome
Italy

Ludovico Abenavoli
Department of Health Sciences
University "Magna Græcia"
Catanzaro
Italy

Editorial Office
MDPI AG
Grosspeteranlage 5
4052 Basel, Switzerland

This is a reprint of the Special Issue, published open access by the journal *Medicina* (ISSN 1648-9144), freely accessible at: https://www.mdpi.com/journal/medicina/special_issues/ZE7W54BEW9.

For citation purposes, cite each article independently as indicated on the article page online and as indicated below:

Lastname, A.A.; Lastname, B.B. Article Title. *Journal Name* **Year**, *Volume Number*, Page Range.

ISBN 978-3-7258-4331-2 (Hbk)
ISBN 978-3-7258-4332-9 (PDF)
https://doi.org/10.3390/books978-3-7258-4332-9

Cover image courtesy of Ludovico Abenavoli

Contents

medicina

MDPI

Editorial

Editorial Board Members' Collection Series: Gastrointestinal and Hepatic Diseases

Ludovico Abenavoli [1,*] and Marcello Candelli [2]

1 Department of Health Sciences, University "Magna Graecia", Viale Europa, 88100 Catanzaro, Italy
2 Emergency, Anesthesiological and Reanimation Sciences Department of Fondazione Policlinico Agostino Gemelli—IRCCS, Catholic University of Rome, Largo Agostino Gemelli 8, 00168 Rome, Italy; marcello.candelli@policlinicogemelli.it
* Correspondence: l.abenavoli@unicz.it; Tel.: +39-0961-3694387

The Special Issue titled "Editorial Board Members' Collection Series: Gastrointestinal and Hepatic Diseases" is a collection of papers from our Editorial Board Members and researchers invited by them. The aim is to provide a venue for networking and communication between *Medicina* and scholars in the field of gastrointestinal and hepatic diseases. Recent studies have highlighted significant advancements in understanding various medical conditions through innovative biomarkers, novel therapeutic strategies, and epidemiological insights. One of the key findings involves hemogram-derived ratios, such as platelet-to-lymphocyte and platelet-to-neutrophil ratios, which have emerged as potential tools for differentiating between idiopathic and secondary pulmonary fibrosis. This approach, along with hepatic biomarkers, could refine diagnostic accuracy and improve disease management [1]. The interplay between thyroid function and liver cirrhosis has also gained attention, with research suggesting that thyroid-stimulating hormone levels may correlate with disease severity and prognosis. Similarly, the impact of Coronavirus Disease-19 on acute cholangitis has been explored, revealing extended hospital stays, a higher prevalence of malignant causes, and notable shifts in microbial infections. In oncology, the increasing incidence of malignant polyps in younger populations has prompted a reconsideration of colorectal cancer screening guidelines [2]. The growing recognition of shifting tumor biology underscores the need for earlier interventions. Meanwhile, treatment strategies for inflammatory bowel diseases (IBD) continue to evolve, with subcutaneous vedolizumab proving effective in maintaining remission after intravenous therapy [3]. The relationship between metabolic disorders and gastrointestinal diseases remains a critical area of research. In this regard, non-alcoholic fatty liver disease appears increasingly prevalent among individuals with IBD, particularly those with ulcerative colitis. Given the close link between metabolic dysfunction and liver disease, experts emphasize the importance of comprehensive management strategies targeting cardiometabolic risk factors [4]. Beyond metabolic disorders, new perspectives are emerging on the potential connections between *Helicobacter pylori* infection and coronary artery disease, highlighting the role of chronic inflammation in cardiovascular health. Dietary habits, particularly those influenced by the Western diet and food additives, are also being scrutinized for their contribution to metabolic dysfunction-associated steatotic liver disease [5]. Advances in gastrointestinal research extend to conditions such as irritable bowel syndrome and slow transit constipation, both of which significantly affect quality of life [6]. The gut–brain axis, microbiome imbalances, and potential therapeutic interventions, including probiotics and fecal microbiota transplantation, are shaping new treatment paradigms. Similarly, the management of hepatic hemangiomas has shifted toward minimally invasive procedures, reflecting broader trends in interventional medicine [7]. Ongoing research into biofilm formation in biliary stents and the challenges of cirrhotic cardiomyopathy further illustrate the complexity of modern medical science. As diagnostic and therapeutic approaches continue to advance, integrating these findings into clinical practice will be crucial for improving patient outcomes across a wide range of diseases [8]. Gastroenterology is a rapidly advancing field that comprises both pre-clinical and clinical areas. The development of new treatments, cutting-edge technologies, and a deeper understanding of disease mechanisms are key factors driving this progress. In this regard, this Research Topic provides scholars with a comprehensive and current perspective, emphasizing the latest innovations and developments in the pursuit of global excellence in gastroenterology.

Author Contributions: Conceptualization, L.A. and M.C.; visualization, L.A. and M.C. All authors have read and agreed to the published version of the manuscript.

Acknowledgments: We would like to thank Simone Scarlata for his critical review of the English language.

Conflicts of Interest: The authors declare no conflicts of interest.

References

1. Paluschinski, M.; Loosen, S.; Kordes, C.; Keitel, V.; Kuebart, A.; Brandenburger, T.; Schöler, D.; Wammers, M.; Neumann, U.P.; Luedde, T.; et al. Extracellular Vesicles as Markers of Liver Function: Optimized Workflow for Biomarker Identification in Liver Disease. *Int. J. Mol. Sci.* **2023**, *24*, 9631. [CrossRef] [PubMed]
2. Tanadi, C.; Tandarto, K.; Stella, M.M.; Sutanto, K.W.; Steffanus, M.; Tenggara, R.; Bestari, M.B. Colorectal cancer screening guidelines for average-risk and high-risk individuals: A systematic review. *Rom. J. Intern. Med.* **2023**, *62*, 101–123. [CrossRef] [PubMed]
3. Crooks, B.; Barnes, T.; Limdi, J.K. Vedolizumab in the treatment of inflammatory bowel disease: Evolving paradigms. *Drugs Context* **2020**, *9*, 2019-10-2. [CrossRef] [PubMed]
4. Abenavoli, L.; Spagnuolo, R.; Scarlata, G.G.M.; Scarpellini, E.; Boccuto, L.; Luzza, F. Ultrasound Prevalence and Clinical Features of Nonalcoholic Fatty Liver Disease in Patients with Inflammatory Bowel Diseases: A Real-Life Cross-Sectional Study. *Medicina* **2023**, *59*, 1935. [CrossRef] [PubMed]
5. Armandi, A.; Bugianesi, E. Dietary and pharmacological treatment in patients with metabolic-dysfunction associated steatotic liver disease. *Eur. J. Intern. Med.* **2024**, *122*, 20–27. [CrossRef] [PubMed]
6. Cassar, G.E.; Youssef, G.J.; Knowles, S.; Moulding, R.; Austin, D.W. Health-Related Quality of Life in Irritable Bowel Syndrome: A Systematic Review and Meta-analysis. *Gastroenterol. Nurs.* **2020**, *43*, E102–E122. [CrossRef] [PubMed]
7. Loh, J.S.; Mak, W.Q.; Tan, L.K.S.; Ng, C.X.; Chan, H.H.; Yeow, S.H.; Foo, J.B.; Ong, Y.S.; How, C.W.; Khaw, K.Y. Microbiota-gut-brain axis and its therapeutic applications in neurodegenerative diseases. *Signal Transduct. Target. Ther.* **2024**, *9*, 37. [CrossRef] [PubMed]
8. Prasad, S.S.; Potter, M.; Keely, S.; Talley, N.J.; Walker, M.M.; Kairuz, T. Roles of healthcare professionals in the management of chronic gastrointestinal diseases with a focus on primary care: A systematic review. *JGH Open* **2019**, *4*, 221–229. [CrossRef] [PubMed]

medicina

MDPI

Article

Exploring the Role of Hemogram-Derived Ratios and Liver Fibrosis Scores in Pulmonary Fibrosis

Vera Ciornolutchii [1], Victoria Maria Ruta [2], Adina Milena Man [3], Nicoleta Stefania Motoc [3], Stefan-Lucian Popa [1], Dan L. Dumitrascu [1], Abdulrahman Ismaiel [1,*,†] and Daniel-Corneliu Leucuta [4,†]

1 2nd Department of Internal Medicine, "Iuliu Hatieganu" University of Medicine and Pharmacy, 400006 Cluj-Napoca, Romania; veraciornolutchii11@gmail.com (V.C.); popa.stefan@umfcluj.ro (S.-L.P.); ddumitrascu@umfcluj.ro (D.L.D.)
2 Department of Pulmonology, "Leon Daniello" Clinical Hospital of Pulmonology, 400371 Cluj-Napoca, Romania; victoria.suteu@yahoo.com
3 Department of Pneumology, "Iuliu Hatieganu" University of Medicine and Pharmacy, 400000 Cluj-Napoca, Romania; manmilena50@yahoo.com (A.M.M.); motoc_nicoleta@yahoo.com (N.S.M.)
4 Department of Medical Informatics and Biostatistics, "Iuliu Hatieganu" University of Medicine and Pharmacy, 400349 Cluj-Napoca, Romania; dleucuta@umfcluj.ro
* Correspondence: abdulrahman.ismaiel@yahoo.com
† These authors contributed equally to this work.

Abstract: *Background and Objectives:* Pulmonary fibrosis, including idiopathic pulmonary fibrosis (IPF) and secondary pulmonary fibrosis (SPF), is a progressive lung disease that significantly impairs respiratory function. Accurate differentiation between IPF and SPF is crucial for effective management. This study explores the association between pulmonary fibrosis and hepatic conditions, evaluating the utility of various hemogram-derived ratios and hepatic fibrosis scores in distinguishing between IPF and SPF. *Materials and Methods:* We conducted a retrospective study involving patients diagnosed with IPF or SPF at the "Leon Daniello" Clinical Hospital of Pneumology in Cluj-Napoca, Romania. Pulmonary fibrosis was confirmed via imaging techniques, and hepatic steatosis and fibrosis were assessed using non-invasive scores. We analyzed clinical, laboratory, and pulmonary function data, focusing on hemogram-derived ratios and hepatic scores. Statistical analyses, including ROC curves, were used to evaluate the effectiveness of these biomarkers in differentiating IPF from SPF. *Results:* We included a total of 38 patients with IPF and 28 patients with SPF. Our findings revealed that IPF patients had a significantly higher FIB-4 score compared to SPF patients, suggesting increased hepatic fibrosis risk in IPF, as well as an increased RDW/PLT ratio. Conversely, SPF patients exhibited elevated PLR, PNR, and SII, reflecting a more pronounced inflammatory profile. PLR and PNR demonstrated the highest discriminatory ability between IPF and SPF, while traditional hepatic fibrosis scores showed limited differentiation capabilities. No significant differences in pulmonary function tests were observed across hepatic fibrosis risk categories. *Conclusions:* The study highlights the value of biomarkers like PLR and PNR in differentiating between IPF and SPF, offering additional diagnostic insights beyond traditional imaging. Integrating hepatic assessments into the management of pulmonary fibrosis could improve diagnostic accuracy and patient care.

Keywords: pulmonary fibrosis; idiopathic pulmonary fibrosis (IPF); secondary pulmonary fibrosis (SPF); hepatic steatosis; biomarkers; hemogram-derived ratios; liver fibrosis

1. Introduction

Pulmonary fibrosis is a chronic and progressive interstitial lung disease characterized by the thickening and scarring of lung tissue, which leads to a decline in respiratory function [1]. Idiopathic Pulmonary Fibrosis (IPF) represents the most severe form of this condition, with a median survival rate of only 3 to 5 years following diagnosis [2]. It predominantly affects older adults and is slightly more common in men than women [3].

While IPF remains a disease of unknown etiology, Secondary Pulmonary Fibrosis (SPF) arises from known causes, including environmental exposures, autoimmune diseases, and certain medications [4]. Both IPF and SPF present significant clinical challenges, due to their progressive nature and the limited efficacy of current treatment options [5].

Differentiating between IPF and SPF is critical for patient management and prognosis. Current assessment methods primarily rely on high-resolution computed tomography (HRCT) to identify characteristic patterns of lung damage, such as the usual interstitial pneumonia (UIP) pattern, which is highly suggestive of IPF [6]. However, in cases where the HRCT findings are inconclusive or where the lung biopsy is not feasible, distinguishing between IPF and SPF can be challenging [7]. In these instances, additional diagnostic tools and biomarkers are often necessary to accurately categorize the type of pulmonary fibrosis and guide appropriate treatment strategies [8,9].

Patients with pulmonary fibrosis frequently present with multiple comorbidities that complicate their clinical management [10]. Cardiovascular diseases, including pulmonary hypertension, are common, as are conditions like gastroesophageal reflux disease (GERD) and sleep apnea [11–13]. The presence of these comorbidities often worsens the patient's overall prognosis and contributes to the complexity of their care [14]. Understanding the full spectrum of associated conditions is essential for the holistic management of patients with pulmonary fibrosis.

Emerging evidence suggests a potential link between pulmonary fibrosis and hepatic disorders, particularly hepatic steatosis and liver fibrosis [15]. These conditions may share common pathophysiological mechanisms, including chronic inflammation and fibrogenesis, which could contribute to their co-occurrence [16,17]. Non-invasive scores and biomarkers have been developed to assess hepatic steatosis and fibrosis, offering a valuable tool for investigating their prevalence and impact in patients with pulmonary fibrosis [18]. This association may have implications for the severity of pulmonary involvement and overall disease progression [19].

The current study aims to evaluate the association between pulmonary fibrosis, both idiopathic and secondary, diagnosed using pulmonary CT, and hepatic steatosis and fibrosis assessed through non-invasive scores and biomarkers. We also seek to explore the utility of hemogram-derived ratios and other biomarkers in distinguishing between IPF and SPF. Furthermore, we examined the relationship between hepatic conditions and pulmonary function tests to assess whether hepatic involvement predicts more severe pulmonary disease.

2. Methods

2.1. Study Participants

This retrospective study was conducted at the "Leon Daniello" Clinical Hospital of Pneumology in Cluj-Napoca, Romania. The study included patients diagnosed with either IPF or SPF during their hospitalization. IPF was confirmed based on a typical pattern of UIP identified via native CT or HRCT, either with or without lung biopsy, provided there were no other identifiable causes of pulmonary fibrosis. SPF was confirmed by native CT or HRCT in patients who had identifiable underlying causes, including sarcoidosis or collagen vascular diseases. Patients with conditions such as tumors, hepatitis B or C, alcoholic liver disease, autoimmune hepatitis, drug-induced liver injury (DILI)/herb-induced liver injury (HILI), acute hemolytic diseases, acute inflammatory pathologies, and any liver disease were excluded. Additionally, patients with unclear pulmonary fibrosis diagnoses or missing data were not included in the study.

2.2. Pulmonary Fibrosis Assessment

IPF was defined as a specific form of chronic, progressive fibrosing interstitial pneumonia of unknown cause, occurring primarily in older adults. It was characterized by the histopathologic and/or radiologic pattern of UIP. Diagnosis was confirmed either by native CT or HRCT showing a typical UIP pattern, or through lung biopsy when imaging was inconclusive [20,21]. The UIP pattern included reticular opacities, often associated with honeycombing and traction bronchiectasis, typically with a subpleural and basal predominance. Exclusion criteria for IPF included any other identifiable cause of pulmonary fibrosis, such as heavy-metal exposure, drug-induced reactions, pulmonary irradiation, hypersensitivity pneumonitis, sarcoidosis, bronchiolitis obliterans, HIV infection, viral hepatitis, cancer, or collagen vascular diseases (e.g., scleroderma, polymyositis/dermatomyositis, systemic lupus erythematosus, rheumatoid arthritis).

SPF was defined as pulmonary fibrosis that occurs secondary to an identifiable cause, such as autoimmune diseases, chronic hypersensitivity pneumonitis, or exposure to environmental or occupational agents [22]. Diagnosis was confirmed by native CT or HRCT, revealing fibrotic changes consistent with known etiologies of pulmonary fibrosis. Underlying conditions that could lead to SPF, such as sarcoidosis or collagen vascular diseases (e.g., scleroderma, polymyositis/dermatomyositis, systemic lupus erythematosus, rheumatoid arthritis), were thoroughly investigated to establish the diagnosis.

2.3. Data Collection

Data were retrospectively collected from hospital records, covering several clinical and laboratory parameters.

2.3.1. Laboratory Analysis

All laboratory analyses were performed using standardized methods in the hospital's central laboratory. Blood samples were collected during routine clinical assessments, and the following tests were performed: Complete Blood Count (CBC), including measurements of white blood cell count, red blood cell count, hemoglobin levels, hematocrit, platelet count, and differential leukocyte count. These values were used to calculate hemogram-derived ratios. Liver Function Tests (LFTs) including aspartate aminotransferase (AST), alanine aminotransferase (ALT), alkaline phosphatase (ALP), and gamma-glutamyl transferase (GGT), were measured using automated biochemical analyzers. These enzymes were critical in calculating hepatic steatosis and fibrosis scores. Lipid-profile levels of total cholesterol and triglycerides were measured to assess metabolic status and to contribute to the calculation of certain hepatic indices. Renal function tests including creatinine and urea were assessed, and estimated Glomerular Filtration Rate (eGFR) was calculated. Fasting plasma sugar levels were measured to evaluate metabolic conditions which may be associated with hepatic steatosis. These laboratory data were essential for calculating non-invasive hepatic scores and assessing systemic inflammation, providing a basis for investigating their potential association with pulmonary fibrosis severity.

2.3.2. Pulmonary Function Tests (PFTs)

Pulmonary function was assessed using standard spirometry, which measured several parameters including Forced Vital Capacity (FVC), Forced Expiratory Volume in one second (FEV1), FEV1/FVC, Maximal Expiratory Flow (MEF) 75%, MEF 50%, and MEF 25%, as well as Diffusing Capacity of the Lung for Carbon Monoxide (DLCO) [23]. These tests provided insight into the severity of lung impairment in patients with pulmonary fibrosis. The results were expressed according to the predicted normal values, adjusted for age, gender, height, and ethnicity.

2.3.3. Hepatic Steatosis and Liver Fibrosis Scores

Hepatic steatosis and liver fibrosis were evaluated using non-invasive scores including the Fibrosis-4 Index (FIB-4), AST to Platelet Ratio Index (APRI), aspartate aminotransferase to alanine aminotransferase (AST/ALT) ratio, body mass index (BMI), AST/ALT ratio, and diabetes (BARD), age, bilirubin, INR and serum creatinine level (ABIC) score, King's score for liver fibrosis, Logarithm of Odds for Lok Index (LogOddsLok), Lok Index for Liver Fibrosis (Lok index), and triglyceride to glucose (TyG) ratio [23,24]. These scores were calculated based on routinely available laboratory data and were used to assess the degree of hepatic involvement in the study population. Advanced fibrosis probability was calculated according to the current recommendations for each score, individually.

2.3.4. Hemogram-Derived Ratios and Biomarkers

Several hemogram-derived ratios and systemic inflammation markers were assessed including Neutrophil-to-Lymphocyte Ratio (NLR), Derived Neutrophil-to-Lymphocyte Ratio (dNLR), Platelet-to-Lymphocyte Ratio (PLR), Lymphocyte-to-Monocyte Ratio (LMR), Eosinophil-to-Lymphocyte Ratio (ELR), Basophil-to-Lymphocyte Ratio (BLR), red cell distribution width (RDW-CV)-to Platelet ratio, and Systemic Immune–Inflammation Index (SII), a composite index calculated as (Platelet count × Neutrophil count)/Lymphocyte count, reflecting the balance between inflammation and immune response.

2.4. Statistical Analysis

Data were presented as means with standard deviations (SDs) for normally distributed quantitative variables, medians with interquartile ranges (IQRs) for non-normally distributed quantitative data, and as numbers with percentages for categorical variables. Clinical characteristics were compared between groups using appropriate statistical tests: a *t*-test for independent samples for normally distributed quantitative variables, the Wilcoxon rank-sum test for non-normally distributed quantitative variables, and $\chi 2$ tests or Fisher's exact tests for categorical data. ROC analysis was employed to assess the accuracy of hepatic steatosis and liver fibrosis scores, as well as hemogram-derived ratios, in differentiating between IPF and SPF. A *p*-value of <0.05 was considered statistically significant. All analyses were performed using R software version 4.1.2 (R Foundation for Statistical Computing).

3. Results

3.1. General Characteristics

In this study, a comparative analysis between patients with IPF and SPF revealed several significant findings, as outlined in Table 1. The median age of patients with IPF was significantly higher, at 73 years, compared to 68 years in the SPF group (*p*-value = 0.021). A significantly higher proportion of males were present in the IPF group (60.53%) compared to the SPF group (32.14%) (*p*-value = 0.023). However, other variables, such as BMI and smoking history, did not show significant differences between the two groups. The inflammatory marker C-reactive protein (CRP) and liver function tests, including AST and ALT, also showed no significant differences. Notably, total bilirubin levels were higher in the IPF group compared to the SPF group, with a median of 0.74 mg/dL versus 0.52 mg/dL, respectively (*p*-value = 0.011). Other biochemical markers, such as ALP, GGT, and renal function tests, including urea, creatinine, and eGFR, showed no significant differences between the two groups. Additionally, lipid profiles and RDW-CV were comparable, indicating similar metabolic and hematological profiles between the IPF and SPF patients.

Most etiologies of SPF included rheumatoid arthritis (*n* = 9), as well as rheumatoid arthritis associated with antisynthetase syndrome (*n* = 1), COVID-19 infection (*n* = 1), RS3PE syndrome (*n* = 1), Sjogren syndrome and antisynthetase syndrome (*n* = 1), Sjogren syndrome (*n* = 1), and Sjogren syndrome and dermatomyositis (*n* = 1), followed by sarcoidosis (*n* = 4), scleroderma (*n* = 4), mixed connective tissue disease (*n* = 3), systemic lupus erythematosus (*n* = 1), mixed connective tissue disease and COVID-19 infection (*n* = 1).

Table 1. General characteristics and laboratory tests of included participants.

Variables	IPF (*n* = 38)	SPF (*n* = 28)	*p*-Value
Age (years), median (IQR)	73 (67.5–75.75)	68 (62.25–74)	0.021
Sex (Male), nr (%)	23 (60.53)	9 (32.14)	0.023
BMI (kg/m^2), median (IQR)	27.87 (25.59–30.31)	28.65 (26.31–29.96)	0.825

Table 1. *Cont.*

Variables	IPF (n = 38)	SPF (n = 28)	p-Value
BMI interpretation, nr (%)	Normal: 6 (15.79) Obesity: 10 (26.32) Overweight: 22 (57.89)	Normal: 6 (21.43) Obesity: 7 (25) Overweight: 15 (53.57)	0.841
Smoking history, nr (%)	Ex-smoker: 15 (39.47) Non-smoker: 21 (55.26) Smoker: 2 (5.26)	Ex-smoker: 7 (25) Non-smoker: 18 (64.29) Smoker: 3 (10.71)	0.363
Smoking (pack/years), median (IQR)	20 (17–29)	30 (20–30)	0.45
Professional exposure (Yes), nr (%)	10 (26.32)	8 (28.57)	0.839
Atopic risk (Yes), nr (%)	5 (13.16)	8 (28.57)	0.12
CRP (mg/L), median (IQR)	6 (1.25–22.67)	7.5 (3–16.62)	0.617
AST (UI/L), median (IQR)	22.5 (19–27.25)	22.5 (17.75–27.5)	0.697
ALT (UI/L), median (IQR)	17.5 (13.25–26)	16 (11–20.5)	0.32
Total bilirubin (mg/dL), median (IQR)	0.74 (0.46–1.11)	0.52 (0.31–0.71)	0.011
ALP (UI/L), median (IQR)	97.5 (69.25–137.75)	85 (69.5–103.5)	0.392
GGT (UI/L), median (IQR)	29.5 (17.25–50.5)	22 (14.5–30.25)	0.099
INR, median (IQR)	1.12 (1.04–1.24)	1.08 (1.04–1.27)	0.455
PT (s), median (IQR)	12.7 (11.4–14)	12 (11.55–13.9)	0.632
Urea (mg/dL), median (IQR)	37 (28.5–45.5)	32 (25–42.5)	0.192
Creatinine (mg/dL), median (IQR)	0.9 (0.72–1.04)	0.81 (0.68–1)	0.452
eGFR, mean (SD)	76.9 (21.85)	77.42 (25.02)	0.929
Total cholesterol (mg/dL), mean (SD)	168.43 (48.95)	175.54 (44.57)	0.547
Triglycerides (mg/dL), median (IQR)	113.5 (84–149.5)	120 (94.5–164.5)	0.523
RDW-CV (%), median (IQR)	13.8 (12.83–14.67)	14.45 (13.28–15.33)	0.109

ALP: Alkaline Phosphatase; ALT: Alanine Aminotransferase; AST: Aspartate Aminotransferase; BMI: Body Mass Index; CRP: C-reactive Protein; eGFR: Estimated Glomerular Filtration Rate; GGT: Gamma-Glutamyl Transferase; INR: International Normalized Ratio; IPF: Idiopathic Pulmonary Fibrosis; IQR: Interquartile Range; PT: Prothrombin Time; RDW-CV: Red Cell Distribution Width–Coefficient of Variation; SD: Standard Deviation; SPF: Secondary Pulmonary Fibrosis; UI/L: Units per Liter.

3.2. Associated Comorbidities and Treatment

In the analysis of associated comorbidities between patients with IPF and SPF, several notable differences were observed, as demonstrated in Table 2. The exacerbation rate was also higher in the IPF group, with 15.79% experiencing exacerbations, whereas no exacerbations were reported in the SPF group (p-value = 0.035). Additionally, ischemic heart disease was more prevalent among IPF patients (42.11%) compared to those with SPF (17.86%) (p-value = 0.037). There were no significant differences in the presence of chronic obstructive pulmonary disease (COPD), diabetes, and hypertension between the groups. While the use of antifibrotic therapy, specifically Nintedanib, was significantly more common in the IPF group (63.16% vs. 17.86%; p-value < 0.001), the use of Pirfenidone did not differ significantly between the groups.

Table 2. Associated comorbidities and administered treatment in IPF and SPF.

Variables	IPF (n = 38)	SPF (n = 28)	p-Value
Comorbidities			
Exacerbation (Yes), nr (%)	6 (15.79)	0 (0)	0.035
COPD (Yes), nr (%)	5 (13.16)	5 (17.86)	0.732
COPD exacerbation (Yes), nr (%)	1 (2.63)	1 (3.57)	1
Asthma (Yes), nr (%)	4 (10.53)	4 (14.29)	0.714
Bronchiectasis (Yes), nr (%)	20 (52.63)	14 (50)	0.833
Home oxygen therapy (Yes), nr (%)	18 (47.37)	7 (25)	0.064
Obstructive sleep apnea (Yes), nr (%)	1 (2.63)	2 (7.14)	0.57
COVID-19 (Yes), nr (%)	8 (21.05)	7 (25)	0.705
Diabetes (Yes), nr (%)	11 (28.95)	5 (17.86)	0.299
Hypertension (Yes), nr (%)	29 (76.32)	18 (64.29)	0.286

Table 2. *Cont.*

Variables	IPF (n = 38)	SPF (n = 28)	p-Value
Comorbidities			
Heart failure (Yes), nr (%)	14 (36.84)	10 (35.71)	0.925
Ischemic heart disease (Yes), nr (%)	16 (42.11)	5 (17.86)	0.037
Previous MI (Yes), nr (%)	2 (5.26)	3 (10.71)	0.643
Pulmonary hypertension (Yes), nr (%)	21 (55.26)	13 (46.43)	0.478
Cerebrovascular accident (Yes), nr (%)	0 (0)	2 (7.14)	0.176
Carotid atherosclerosis (Yes), nr (%)	7 (18.42)	7 (25)	0.518
Treatment			
Corticosteroids (Yes), nr (%)	2 (5.26)	3 (10.71)	0.643
Biological treatment (Yes), nr (%)	0 (0)	2 (7.14)	0.176
Biological treatment type (RITUXIMAB), nr (%)	0 (0)	2 (7.14)	0.176
Nintedanib, nr (%)	Initiated: 6 (15.79) Interrupted: 3 (7.89) No: 5 (13.16) Yes: 24 (63.16)	Initiated: 2 (7.14) Interrupted: 0 (0) No: 21 (75) Yes: 5 (17.86)	<0.001
Pirfenidone, nr (%)	Initiated: 1 (2.63) No: 34 (89.47) Yes: 3 (7.89)	Initiated: 0 (0) No: 28 (100) Yes: 0 (0)	0.256

COPD: Chronic Obstructive Pulmonary Disease; COVID-19: Coronavirus disease; IPF: Idiopathic Pulmonary Fibrosis; MI: Myocardial Infarction; SPF: Secondary Pulmonary Fibrosis.

3.3. *Hepatic Steatosis, Liver Fibrosis, and Hemogram-Derived Ratios*

In the comparative analysis of hemogram-derived ratios, hepatic steatosis, and liver fibrosis scores between patients with IPF and SPF, several significant differences were identified, as mentioned in Table 3. The FIB-4 score was significantly higher in the IPF group (median: 1.66) compared to the SPF group (median: 1.42) (p-value = 0.049). Additionally, the RDW/PLT was also significantly higher in the IPF group (median: 0.07) compared to the SPF group (median: 0.06) (p-value = 0.037). The PLR and PNR were significantly higher in the SPF group with a median of 160.52, compared to 107.24 in the IPF group (p-value = 0.005), and a median of 160.52 in the SPF group, compared to 107.24 in the IPF group, with a p-value of 0.005. Similarly, the SII was elevated in the SPF group (median: 957.35) versus the IPF group (median: 560.88) (p-value = 0.011). The LMR was significantly lower in the SPF group, with a median of 2.52 compared to 3.5 in the IPF group (p-value = 0.044).

Table 3. Evaluated hemogram-derived ratios, hepatic steatosis and liver fibrosis scores.

Variables	IPF (n = 38)	SPF (n = 28)	p-Value
Hemogram-derived ratios			
NLR, median (IQR)	2.75 (2.09–4.06)	3.65 (2.57–4.99)	0.091
dNLR, median (IQR)	14.16 (7.92–25.46)	19.07 (12.37–33.89)	0.153
PLR, median (IQR)	107.24 (78.59–163.32)	160.52 (119.64–265.17)	0.005
LMR, median (IQR)	3.5 (2.54–5.05)	2.52 (1.93–4)	0.044
SII, median (IQR)	560.88 (386.72–999.12)	957.35 (669.49–1308.34)	0.011
PNR, median (IQR)	107.24 (78.59–163.32)	160.52 (119.64–265.17)	0.005
ELR, median (IQR)	0.12 (0.05–0.2)	0.11 (0.08–0.18)	0.756
BLR, median (IQR)	0.01 (0–0.03)	0.02 (0.01–0.03)	0.081
RDW TO PLT RATIO, median (IQR)	0.07 (0.05–0.08)	0.06 (0.04–0.07)	0.037
Hepatic steatosis and liver fibrosis scores			
FIB-4, median (IQR)	1.66 (1.35–2.23)	1.42 (1.07–1.98)	0.049
APRI, median (IQR)	0.33 (0.2–0.43)	0.23 (0.17–0.35)	0.05
BARD, median (IQR)	3 (2–3)	3 (2–3)	0.55
AST/ALT ratio, median (IQR)	1.27 (0.95–1.57)	1.39 (1.16–1.64)	0.246
ABIC, median (IQR)	8.67 (8.19–8.76)	8.25 (7.33–8.76)	0.272

Table 3. *Cont.*

Variables	IPF (*n* = 38)	SPF (*n* = 28)	*p*-Value
Hepatic steatosis and liver fibrosis scores			
King score, median (IQR)	9.34 (6.34–11.34)	7.02 (4.91–9.9)	0.083
LogOddsLok, median (IQR)	0.29 (−0.74–1)	−0.46 (−1.22–0.3)	0.101
Lok index, mean (SD)	0.55 (0.25)	0.43 (0.26)	0.117
TyG ratio, mean (SD)	3.78 (0.21)	3.8 (0.27)	0.793

ALT: Alanine Aminotransferase; AST: Aspartate Aminotransferase; ABIC: Age-Bilirubin-INR-Creatinine Score; APRI: Aspartate Aminotransferase-to-Platelet Ratio Index; BARD: Bilirubin, Age, AST/ALT Ratio, and Diabetes Score; BLR: Basophil-to-Lymphocyte Ratio; ELR: Eosinophil-to-Lymphocyte Ratio; FIB-4: Fibrosis-4 Index; IPF: Idiopathic Pulmonary Fibrosis; IQR: Interquartile Range; King score: King's Score for Liver Fibrosis; LMR: Lymphocyte-to-Monocyte Ratio; LogOddsLok: Log Odds of Lok Index; NLR: Neutrophil-to-Lymphocyte Ratio; PLR: Platelet-to-Lymphocyte Ratio; PNR: Platelet-to-Neutrophil Ratio; RDW TO PLT RATIO: Red Cell Distribution Width-to-Platelet Ratio; SII: Systemic Immune–Inflammation Index; SD: Standard Deviation; SPF: Secondary Pulmonary Fibrosis; TyG ratio: Triglyceride Glucose Index.

3.4. Pulmonary Function Tests

A comparison of pulmonary function parameters between patients with IPF and SPF is outlined in Table 4, revealing a significant difference in the MEF25%. Specifically, the MEF25% was significantly lower in the SPF group (median: 44) compared to the IPF group (median: 61.5) with a *p*-value of 0.042. However, other pulmonary function metrics, including FVC%, FEV1%, FEV1/FVC, MEF75%, MEF50%, and DLCO, did not show significant differences between the two groups.

Table 4. Pulmonary function tests according to pulmonary fibrosis type.

Variables	IPF (*n* = 38)	SPF (*n* = 28)	*p*-Value
FVC%, mean (SD)	76.08 (23.21)	76.11 (25)	0.996
FEV1%, mean (SD)	79.47 (22.33)	76.81 (27.09)	0.667
FEV1/FVC, median (IQR)	109 (105–114)	105 (98.5–111)	0.164
MEF75%, mean (SD)	78.24 (28.19)	83.33 (33.44)	0.509
MEF50%, median (IQR)	78.5 (58–93.5)	68 (46.5–92)	0.405
MEF25%, median (IQR)	61.5 (48–82.5)	44 (27–71.5)	0.042
MEF75% ≥ 80% (Yes), nr (%)	16 (42.11)	16 (59.26)	0.173
MEF50% ≥ 80% (Yes), nr (%)	19 (50)	12 (44.44)	0.659
MEF25% ≥ 80% (Yes), nr (%)	11 (28.95)	6 (22.22)	0.543
DLCO, mean (SD)	46.61 (16.96)	56.5 (26.77)	0.093
DLCO category, nr (%)	Mild: 10 (26.32) Moderate: 12 (31.58) Normal: 2 (5.26) Severe: 14 (36.84)	Mild: 7 (25) Moderate: 8 (28.57) Normal: 5 (17.86) Severe: 8 (28.57)	0.466

DLCO: Diffusing Capacity for Carbon Monoxide; FEV1%: Forced Expiratory Volume in 1 s percentage; FEV1/FVC: Ratio of FEV1 to FVC; FVC%: Forced Vital Capacity percentage; IPF: Idiopathic Pulmonary Fibrosis; IQR: Interquartile Range; MEF25%: Mid-Expiratory Flow at 25% of FVC; MEF50%: Mid-Expiratory Flow at 50% of FVC; MEF75%: Mid-Expiratory Flow at 75% of FVC; SD: Standard Deviation; SPF: Secondary Pulmonary Fibrosis.

3.5. Pulmonary Function Tests in Relation to Hemogram-Derived Ratios, Hepatic Steatosis and Liver Fibrosis Scores

3.5.1. Mean Expiratory Flow

As outlined in Supplementary Materials Table S1, the comparison of various biomarkers and scores based on whether the MEF75% was ≥80% revealed the following results. NLR, dNLR, PLR, LMR, SII, PNR, ELR, BLR, FIB-4, APRI, BARD, AST/ALT ratio, ABIC, LogOddsLok, RDW to PLT Ratio, FEV1/FVC, and MEF50% did not show significant differences between the groups with MEF75% ≥ 80% and those with MEF75% < 80%. Nevertheless, King Score was significantly higher in the group with MEF75% < 80% (median: 9.38) compared to the group with MEF75% ≥ 80% (median: 6.67) with a *p*-value of 0.04. Moreover, Lok Index and TyG Ratio showed trends, but did not reach statistical significance, with *p*-values of 0.103 and 0.197, respectively.

The analysis comparing various biomarkers and scores based on whether the MEF50% is ≥80%, is demonstrated in Supplementary Materials Table S2, yielding the following results. PLR was significantly higher in the group with MEF50% < 80% (median: 155.37) compared to the group with MEF50% ≥ 80% (median: 100.84) with a *p*-value of 0.031. SII was significantly higher in the group with MEF50% < 80% (median: 857.27) compared to the group with MEF50% ≥ 80% (median: 576.11) with a *p*-value of 0.045. AST/ALT ratio was significantly higher in the group with MEF50% ≥ 80% (median: 1.44) compared to the group with MEF50% < 80% (median: 1.23) with a *p*-value of 0.042. No significant differences were observed for NLR, dNLR, LMR, PNR, ELR, BLR, FIB-4, APRI, BARD, ABIC, King Score, LogOddsLok, RDW to PLT Ratio, Lok Index, or TyG Ratio between the two groups.

As mentioned in Supplementary Materials Table S3, the analysis of biomarkers and scores in relation to MEF25% being ≥80% revealed the following. ABIC was significantly higher in the group with MEF25% < 80% (median: 8.24) compared to the group with MEF25% ≥ 80% (median: 8.73) with a *p*-value of 0.031. No significant differences were observed for NLR, dNLR, PLR, LMR, SII, PNR, ELR, BLR, FIB-4, APRI, BARD, AST/ALT ratio, King Score, LogOddsLok, RDW to PLT Ratio, Lok Index, or TyG Ratio between the two groups.

3.5.2. DLCO

Supplementary Materials Table S4 summarizes the analysis of biomarkers and scores by DLCO categories, revealing that most indicators, including NLR, dNLR, PLR, LMR, SII, ELR, BLR, FIB-4, APRI, BARD, and AST/ALT ratio, showed no significant differences between mild, moderate, normal, and severe categories. Key measures like the ABIC score, King Score, and RDW-to-PLT Ratio also did not vary significantly. SII and ELR trends were observed, but differences were not statistically significant.

3.6. Correlations Between Pulmonary Function Tests and Assessed Biomarkers/Scores

The analysis of correlations between hemogram-derived ratios (Figure 1) and liver fibrosis scores (Figure 2) with lung function parameters revealed several significant findings (Table 5). NLR and dNLR exhibited notable negative correlations with FVC% (*p*-value = 0.022 and *p* = 0.006, respectively) and FEV1% (*p*-value = 0.014 and *p*-value = 0.01, respectively). In contrast, LMR showed a significant positive correlation with FVC% (*p*-value = 0.03). PLR correlates positively with FEV1/FVC (*p*-value = 0.033) and negatively with MEF50% (*p*-value = 0.009). ABIC also shows a significant correlation with FEV1/FVC (*p*-value = 0.026).

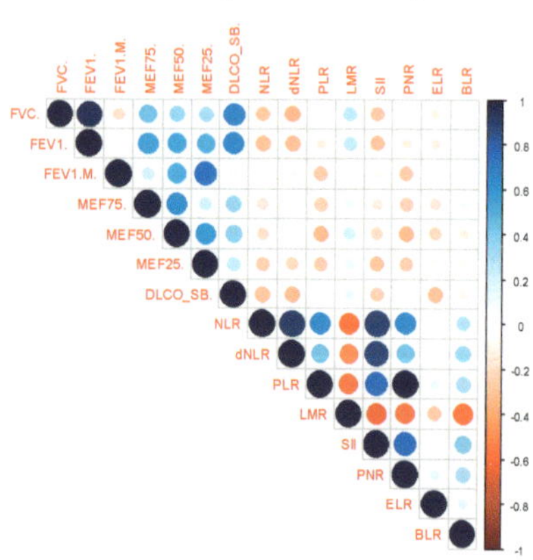

Figure 1. Correlation matrix evaluating pulmonary function tests (FVC%, FEV1%, FEV1/FVC, MEF75%, MEF50%, MEF25%, DLCO) and hemogram-derived ratios (NLR, dNLR, PLR, LMR, SII, PNR, ELR, BLR).

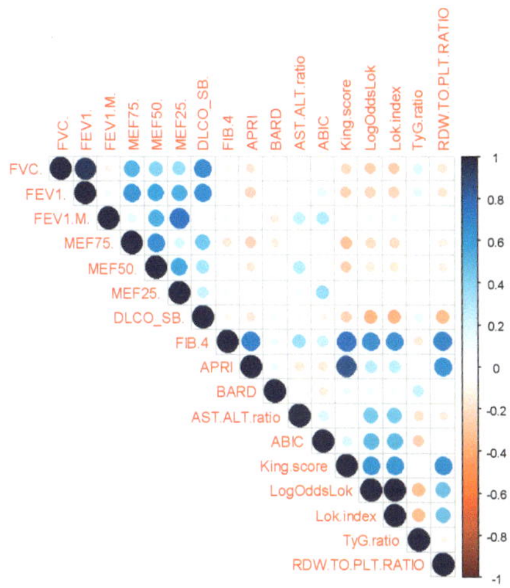

Figure 2. Correlation matrix displaying relationships between pulmonary function tests (FVC%, FEV1%, FEV1/FVC, MEF75%, MEF50%, MEF25%, DLCO), various liver fibrosis scores (FIB-4, APRI, BARD, AST/ALT ratio, ABIC, King score, LogOddsLok, Lok index, TyG ratio), and RDW-to-PLT ratio.

For MEF75%, there was a significant negative correlation with LogOddsLok (p-value = 0.027). MEF50% was notably negatively correlated with PLR and dNLR (p-value = 0.009 for both). MEF25% had significant negative correlations with NLR (p-value = 0.038) and PLR (p-value = 0.04), along with a positive correlation with King score (p-value = 0.005).

DLCO displayed significant negative correlations with NLR (p-value = 0.035) and dNLR (p-value = 0.012), and a positive correlation with LogOddsLok (p-value = 0.049).

Overall, NLR and dNLR were consistently negatively correlated with various lung function measures, while other biomarkers demonstrated less consistent and variable associations.

3.7. AUROC to Differentiate Between IPF and SPF

To differentiate between IPF and SPF, various biomarkers were evaluated for their diagnostic performance as outlined in Table 6. Among them, PLR and PNR demonstrated the highest discriminatory ability, with an AUC of 0.702, reflecting good sensitivity (71.43%) and specificity (65.79%). The RDW-to-PLT Ratio also showed strong performance, with an AUC of 0.651 and perfect sensitivity (100%), but it lacked specificity (0%). The SII had an AUC of 0.684, providing high sensitivity (85.71%) and moderate specificity (57.89%). In contrast, FIB-4 and APRI were highly sensitive (100%) but did not effectively differentiate PF, due to their zero specificity. The NLR and dNLR had moderate AUCs (0.623 and 0.604), with balanced sensitivity and specificity. The LMR had high specificity (94.74%) but very low sensitivity (7.14%). Other markers like ABIC and King Score showed lower AUCs (0.591 and 0.642), indicating less reliability. LogOddsLok and Lok Index also had high specificity (96.55%) but very low sensitivity (8.7%). Overall, while FIB-4 and APRI are highly sensitive, PLR and PNR offer a balanced approach, making them more effective for distinguishing pulmonary fibrosis.

Table 5. The association between hemogram-derived ratios, hepatic steatosis, and liver fibrosis scores with pulmonary function tests.

Variable	NLR	dNLR	PLR	LMR	SII	PNR	ELR	BLR	RDW to PLT Ratio	FIB-4	APRI	BARD	AST/ALT Ratio	ABIC	King Score	Log-Odds-Lok	Lok Index	TyG Ratio
FVC%	−0.28 (0.022)	−0.34 (0.006)	−0.05 (0.68)	0.27 (0.03)	−0.26 (0.033)	−0.05 (0.68)	−0.12 (0.358)	−0.06 (0.641)	−0.02 (0.888)	0.05 (0.689)	−0.02 (0.844)	0.02 (0.863)	−0.05 (0.698)	0.14 (0.343)	−0.16 (0.272)	−0.2 (0.161)	−0.2 (0.161)	0.15 (0.233)
FEV1%	−0.3 (0.014)	−0.32 (0.01)	−0.13 (0.305)	0.24 (0.053)	−0.3 (0.017)	−0.13 (0.305)	−0.13 (0.308)	−0.02 (0.901)	−0.01 (0.93)	0.08 (0.525)	−0.04 (0.742)	−0.02 (0.88)	0.02 (0.864)	0.22 (0.129)	−0.2 (0.158)	−0.18 (0.205)	−0.18 (0.205)	0.11 (0.382)
FEV1/FVC	−0.01 (0.966)	0.08 (0.526)	−0.26 (0.033)	−0.02 (0.857)	−0.09 (0.494)	−0.26 (0.033)	−0.07 (0.564)	0.01 (0.912)	−0.01 (0.965)	0.1 (0.441)	−0.02 (0.851)	−0.11 (0.38)	0.28 (0.026)	0.3 (0.032)	−0.03 (0.828)	0.11 (0.431)	0.11 (0.431)	−0.09 (0.47)
MEF75%	−0.15 (0.235)	−0.06 (0.653)	−0.24 (0.058)	0.11 (0.376)	−0.14 (0.252)	−0.24 (0.058)	−0.16 (0.198)	0.1 (0.448)	−0.11 (0.388)	−0.08 (0.531)	−0.13 (0.291)	−0.1 (0.434)	0.08 (0.55)	−0.03 (0.831)	−0.31 (0.027)	−0.21 (0.131)	−0.21 (0.131)	0.1 (0.421)
MEF50%	−0.2 (0.116)	−0.08 (0.549)	−0.32 (0.009)	0.2 (0.106)	−0.21 (0.095)	−0.32 (0.009)	−0.22 (0.085)	−0.12 (0.323)	−0.07 (0.554)	0.04 (0.769)	−0.06 (0.645)	0 (0.994)	0.18 (0.141)	0.08 (0.579)	−0.23 (0.11)	−0.12 (0.396)	−0.12 (0.396)	0.07 (0.583)
MEF25%	−0.26 (0.038)	−0.21 (0.087)	−0.26 (0.04)	0.13 (0.284)	−0.3 (0.017)	−0.26 (0.04)	−0.07 (0.558)	−0.09 (0.467)	0.01 (0.919)	0.12 (0.357)	0.03 (0.784)	−0.04 (0.749)	0.11 (0.381)	0.39 (0.005)	0.04 (0.8)	0.03 (0.814)	0.03 (0.814)	−0.09 (0.486)
DLCO	−0.26 (0.035)	−0.31 (0.012)	−0.02 (0.875)	0.11 (0.399)	−0.21 (0.087)	−0.02 (0.875)	−0.26 (0.033)	−0.13 (0.297)	−0.19 (0.118)	−0.01 (0.912)	−0.06 (0.639)	0.09 (0.47)	0.01 (0.917)	−0.03 (0.858)	−0.17 (0.234)	−0.27 (0.049)	−0.27 (0.049)	0.17 (0.188)

ABIC: Age-Bilirubin-INR-Creatinine Score; APRI: AST-to-Platelet Ratio Index; AST/ALT Ratio: Aspartate Aminotransferase-to-Alanine Aminotransferase Ratio; BARD: BMI, AST/ALT Ratio, Diabetes Score; BLR: Basophil-to-Lymphocyte Ratio; DLCO: Diffusing Capacity of the Lungs for Carbon Monoxide; dNLR: Derived Neutrophil-to-Lymphocyte Ratio; ELR: Eosinophil-to-Lymphocyte Ratio; FEV1%: Forced Expiratory Volume in One-Second Percent Predicted; FEV1/FVC: Ratio of FEV1 to FVC; FIB-4: Fibrosis-4 Index; FVC%: Forced Vital Capacity Percent Predicted; LMR: Lymphocyte-to-Monocyte Ratio; Lok index: Lok Index; LogOddsLok: Log Odds of Lok Model; MEF25%: Maximal Expiratory Flow at 25% of FVC; MEF50%: Maximal Expiratory Flow at 50% of FVC; MEF75%: Maximal Expiratory Flow at 75% of FVC; NLR: Neutrophil-to-Lymphocyte Ratio; PNR: Platelet-to-Neutrophil Ratio; PLR: Platelet-to-Lymphocyte Ratio; RDW-to-PLT Ratio: Red Cell Distribution-Width-to-Platelet Ratio; SII: Systemic Immune–Inflammation Index; TyG Ratio: Triglyceride Glucose Index Ratio.

Table 6. AUROC of several hemogram-derived ratios, hepatic steatosis, and liver fibrosis scores for differentiating IPF from SPF.

Variable	AUC (95% CI)	Se	Sp	Cut-Off
RDW-CV (%)	0.617 (0.48–0.75)	50	71.05	14.4
NLR	0.623 (0.481–0.758)	60.71	65.79	3.283261803
dNLR	0.604 (0.461–0.743)	67.86	60.53	17.36452174
PLR	0.702 (0.569–0.824)	71.43	65.79	135.042735
LMR	0.646 (0.507–0.784)	7.14	94.74	7.114285714
SII	0.684 (0.548–0.812)	85.71	57.89	607.1794872
PNR	0.702 (0.565–0.827)	71.43	65.79	135.042735
ELR	0.477 (0.338–0.619)	78.57	36.84	0.070422535
BLR	0.626 (0.483–0.761)	85.71	39.47	0.00862
FIB-4	0.643 (0.496–0.773)	100	0	−Inf
APRI	0.642 (0.508–0.776)	100	0	−Inf
BARD	0.541 (0.403–0.672)	100	7.89	0
AST/ALT ratio	0.585 (0.451–0.723)	60.71	57.89	1.285714286

APRI: AST-to-Platelet Ratio Index; ALT: Alanine Aminotransferase; AST: Aspartate Aminotransferase; AUROC: Area Under the Receiver Operating Characteristic Curve; BARD: Bilirubin, Age, AST/ALT Ratio, and Diabetes Score; BLR: Basophil-to-Lymphocyte Ratio; dNLR: Derived Neutrophil-to-Lymphocyte Ratio; ELR: Eosinophil-to-Lymphocyte Ratio; FIB-4: Fibrosis-4 Index; LMR: Lymphocyte-to-Monocyte Ratio; NLR: Neutrophil-to-Lymphocyte Ratio; PLR: Platelet-to-Lymphocyte Ratio; PNR: Platelet-to-Neutrophil Ratio; RDW-CV: Red Cell Distribution-Width–Coefficient of Variation; SII: Systemic Immune–Inflammation Index; Se: Sensitivity; Sp: Specificity.

3.8. Risk of Advanced Hepatic Fibrosis and Pulmonary Function Tests

In assessing advanced hepatic fibrosis risk using FIB-4, spirometric measures showed varying results, as demonstrated in Supplementary Materials Table S5. FVC% and FEV1% did not differ significantly between high, indeterminate, and low fibrosis-risk categories. However, FEV1/FVC was notably higher in the high-risk group. Other measures like MEF75%, MEF50%, MEF25%, and DLCO showed no significant differences across the risk categories.

For advanced fibrosis risk assessed by APRI, spirometric and DLCO measures showed no significant differences between high- and low-risk groups- as reported in Supplementary Materials Table S6. Median values for FVC%, FEV1%, and other spirometric measures like MEF75%, MEF50%, and MEF25% were similar across groups. DLCO also did not differ significantly, indicating limited variability in these parameters relative to APRI-defined fibrosis risk.

When comparing advanced fibrosis risk based on the BARD score, spirometric measures and DLCO did not show significant differences between high- and low-risk groups, as mentioned in Supplementary Materials Table S7. Median values for FVC%, FEV1%, FEV1%M%, MEF75%, MEF50%, MEF25%, and DLCO were similar across both risk categories, with no statistically significant differences found.

3.9. Comparative Analysis of Advanced Fibrosis Risk in IPF vs. SPF

In the context of pulmonary fibrosis, IPF cases had a significantly higher proportion of high FIB-4 advanced-fibrosis risk compared to SPF cases (18.42% vs. 0%, *p*-value= 0.015), as outlined in Table 7. APRI advanced-fibrosis risk showed no significant difference between IPF and SPF (5.26% vs. 0%, *p*-value= 0.504). For BARD advanced-fibrosis risk, the proportion of high-risk cases was similar between the two groups (89.47% in IDP vs. 96.43% in SPF, *p*-value = 0.385).

Table 7. Type of pulmonary fibrosis and associated risk of advanced liver fibrosis assessed using FIB-4, APRI, and BARD.

Variable	IPF (n = 38)	SPF (n = 28)	p-Value
FIB-4 Advanced-fibrosis risk, nr (%)	High: 7 (18.42) Indeterminate: 23 (60.53) Low: 8 (21.05)	High: 0 (0) Indeterminate: 16 (57.14) Low: 12 (42.86)	0.015
APRI Advanced-fibrosis risk (High), nr (%)	2 (5.26)	0 (0)	0.504
BARD Advanced-fibrosis risk (High), nr (%)	34 (89.47)	27 (96.43)	0.385

APRI: AST-to-Platelet Ratio Index; BARD: Bilirubin, Age, AST/ALT Ratio, and Diabetes Score; FIB-4: Fibrosis-4 Index; IPF: Idiopathic Pulmonary Fibrosis; SPF: Secondary Pulmonary Fibrosis.

4. Discussion

This study investigated the relationship between pulmonary fibrosis, hepatic conditions, and various biomarkers, providing new insights into differentiating IPF from SPF. Our main findings reveal that IPF patients exhibit higher FIB-4 scores and RDW/PLT ratios compared to SPF patients, suggesting increased hepatic fibrosis risk. Conversely, SPF patients show elevated PLR and SII, reflecting a more pronounced inflammatory profile. Notably, PLR and PNR demonstrated the highest discriminatory ability between IPF and SPF, while traditional hepatic fibrosis scores such as FIB-4 and APRI had limited differentiation capabilities. Our analysis also found no significant differences in pulmonary function tests across hepatic fibrosis-risk categories, indicating that hepatic fibrosis risk may not directly impact pulmonary function.

In our study, we observed a significant association between pulmonary fibrosis and hepatic conditions, with IPF patients showing elevated FIB-4 scores and RDW/PLT ratios compared to SPF patients. The FIB-4 score, which combines age, AST, ALT, and platelet count, is a validated non-invasive marker of hepatic fibrosis. Higher FIB-4 scores in IPF align with findings from Cocconcelli et al., who reported a notable overall survival risk related to liver fibrosis risk in IPF patients [15]. This association may be attributed to shared pathogenic mechanisms such as chronic inflammation and oxidative stress, which drive both pulmonary and hepatic fibrosis.

Conversely, SPF patients exhibited higher PLR and SII. Elevated PLR and SII in SPF could reflect the inflammatory and immune dysregulation often observed in secondary causes of pulmonary fibrosis, such as autoimmune diseases or occupational exposures. The heightened inflammatory profile in SPF may contribute to distinct clinical and histopathological features compared to IPF. Our analysis identified PLR and PNR as particularly effective in differentiating between IPF and SPF, with PLR showing the highest discriminatory power among the evaluated biomarkers. PLR has been increasingly recognized for its role in various inflammatory and fibrotic diseases. Achaiah et al. demonstrated that blood neutrophil and lymphocyte counts are more reliable than monocytes for predicting disease progression in individuals with established IPF [25]. Chen et al. found that higher levels of NLR expression correlate with reduced overall survival in patients with IPF, regardless of other prognostic factors. This suggests that NLR could serve as a dependable prognostic biomarker for individuals with IPF [26].

However, FIB-4 and APRI, despite their high sensitivity, did not effectively distinguish between IPF and SPF. The discrepancy may stem from differences in patient populations or the specific characteristics of pulmonary fibrosis. Our findings suggest that while these scores are useful for assessing hepatic fibrosis, they may not fully capture the complexities of pulmonary fibrosis. The lack of significant differences in pulmonary function tests across hepatic fibrosis-risk categories (assessed by FIB-4, APRI, and BARD) is noteworthy. Our results suggest that while hepatic fibrosis risk may influence lung function, the relationship might be less direct or more complex than previously thought [27]. Factors such as concurrent comorbidities, disease duration, and treatment effects could contribute to these nuanced findings. Our results indicate a higher proportion of advanced fibrosis risk (as assessed by FIB-4) in IPF compared to SPF, corroborating earlier studies, which found a strong link between IPF and liver fibrosis. The similar proportions of high-risk cases for APRI and BARD between IPF and SPF suggest that these scores may not be as effective in distinguishing between the two conditions. This could be due to the overlap in the underlying mechanisms of fibrosis and the influence of various confounding factors.

Our study underscores the utility of biomarkers such as the PLR and PNR in differentiating IPF from SPF. These biomarkers offer a valuable complement to traditional diagnostic methods, which primarily rely on HRCT and sometimes lung biopsy. While HRCT is effective in identifying the UIP pattern, it may not always clearly distinguish between IPF and SPF, especially in complex cases [28]. PLR and PNR reflect systemic inflammation and immune responses, providing additional diagnostic insights that can enhance accuracy. Integrating these biomarkers into clinical practice can improve diagnostic precision and patient management by offering a quantitative measure of inflammation that supports a more comprehensive and nuanced approach to distinguishing between IPF and SPF.

It is important to mention that antifibrotic medications, including Nintedanib and Pirfenidone, play a critical role in managing IPF by slowing disease progression and improving patient outcomes. However, these medications can pose risks of hepatotoxicity [29]. Nintedanib, a tyrosine kinase inhibitor, has been associated with mild and generally reversible elevations in liver enzymes, necessitating regular monitoring of liver function throughout treatment [30]. Similarly, Pirfenidone, an anti-inflammatory and antifibrotic agent, can cause increases in transaminases and bilirubin levels, though it is usually well-tolerated when managed appropriately [31]. Both drugs have been linked to cases of DILI, underscoring the importance of close surveillance for hepatic adverse effects. The risk of hepatotoxicity may require dose adjustments or discontinuation of therapy, highlighting the need to carefully balance the therapeutic benefits of these medications against potential liver-related risks. Effective management of these hepatic side effects is essential for optimizing patient care and minimizing severe liver complications associated with IPF treatment [5].

Our findings underscore the importance of integrating hepatic assessments into the management of pulmonary fibrosis. Given the high prevalence of hepatic involvement in IPF and the distinct inflammatory profiles in SPF, clinicians should consider a holistic approach to evaluating patients with pulmonary fibrosis. The use of biomarkers such as PLR and PNR can enhance diagnostic accuracy and guide personalized treatment strategies. Future research should focus on validating these findings in larger, multi-center cohorts and exploring the underlying mechanisms linking hepatic and pulmonary fibrosis. Longitudinal studies assessing the impact of hepatic involvement on disease progression and treatment response in pulmonary fibrosis are also warranted. Such research will help refine diagnostic tools and therapeutic approaches, ultimately improving patient outcomes in this challenging field.

Our study has several limitations that should be considered when interpreting the results. First, its retrospective design may introduce selection bias, as it relies on existing patient records and clinical data, which can limit the generalizability of our findings. Additionally, the sample size, while adequate for preliminary analyses, may not be large enough to detect more subtle differences between IPF and SPF. The lack of a longitudinal follow-up means that we could not assess the long-term outcomes or progression of disease in relation to the biomarkers studied. Furthermore, although we evaluated various biomarkers and hepatic fibrosis scores, the absence of a gold standard for diagnosing hepatic fibrosis limits our ability to definitively validate the efficacy of these measures. Lastly, the study excluded patients with certain comorbid conditions and incomplete data, which might have affected the representativeness of the patient cohort.

Despite these limitations, our study presents several strengths that enhance its contribution to the field. The inclusion of a comprehensive set of biomarkers and hepatic fibrosis scores provides a robust analysis of their potential utility in differentiating between IPF and SPF. By incorporating both traditional diagnostic methods and novel biomarkers, the study offers a holistic approach to understanding pulmonary fibrosis and its associated hepatic involvement. The use of well-defined criteria for diagnosing IPF and SPF, based on HRCT and clinical evaluation, adds to the accuracy of our classifications. Additionally, the significant findings regarding the utility of PLR and PNR in distinguishing between IPF and SPF provide valuable insights for clinical practice and suggest directions for future research. Overall, the study's thorough examination of these biomarkers contributes valuable information to the ongoing efforts to refine diagnostic and prognostic tools for pulmonary fibrosis.

5. Conclusions

IPF patients have higher FIB-4 scores, indicating greater hepatic fibrosis risk, and RDW/PLT ratio, while SPF patients exhibit higher PLR and SII, reflecting a stronger inflammatory response. PLR and PNR were the most effective at distinguishing IPF from SPF, whereas traditional fibrosis scores like FIB-4 and APRI were less discriminative. Additionally, hepatic fibrosis risk did not significantly affect pulmonary function test results.

Supplementary Materials: The following supporting information can be downloaded at https://www.mdpi.com/article/10.3390/medicina60101702/s1, Table S1: Hemogram-derived ratios, hepatic steatosis, and liver fibrosis scores according to MEF75% ≥ 80%; Table S2. Hemogram-derived ratios, hepatic steatosis, and liver fibrosis scores according to MEF50% ≥ 80%; Table S3. Hemogram-derived ratios, hepatic steatosis, and liver fibrosis scores according to MEF25% ≥ 80%; Table S4. Hemogram-derived ratios, hepatic steatosis, and liver fibrosis scores according to DLCO categories (Normal: >75% of predicted, up to 140%; Mild: 60–75%; Moderate: 40–60%; Severe: <40%); Table S5. The association between pulmonary function tests and risk of advanced liver fibrosis assessed using FIB-4 score; Table S6. The association between pulmonary function tests and risk of advanced liver fibrosis assessed using APRI score; Table S7. The association between pulmonary function tests and risk of advanced liver fibrosis assessed using BARD score.

Author Contributions: Conceptualization, V.C. and A.I.; methodology, V.C., A.I. and D.-C.L.; software, V.C., A.I. and D.-C.L.; formal analysis, A.I. and D.-C.L.; investigation, V.C., A.I., V.M.R., A.M.M. and S.-L.P.; resources, V.M.R., N.S.M. and A.M.M.; data curation, V.C. and A.I.; writing—original draft preparation, V.C. and A.I.; writing—review and editing, A.I., V.M.R., A.M.M., N.S.M., D.-C.L., S.-L.P. and D.L.D.; visualization, A.I.; supervision, A.I., D.-C.L. and D.L.D.; project administration, A.I. and D.L.D. All authors have read and agreed to the published version of the manuscript.

Funding: This research received no external funding.

Institutional Review Board Statement: The study was conducted according to the guidelines of the Declaration of Helsinki and approved by the Ethics Committee of "Leon Daniello" Clinical Hospital of Pulmonology, Cluj-Napoca, Romania (no. 4406/23 July 2024).

Informed Consent Statement: Patient consent was waived, due to the retrospective study design.

Data Availability Statement: The datasets from the current study can be made available by the corresponding author upon reasonable request.

Conflicts of Interest: The authors declare no conflicts of interest.

References

1. Wilson, M.S.; Wynn, T.A. Pulmonary fibrosis: Pathogenesis, etiology and regulation. *Mucosal Immunol.* **2009**, *2*, 103–121. [CrossRef] [PubMed]
2. Tzilas, V.; Tzouvelekis, A.; Chrysikos, S.; Papiris, S.; Bouros, D. Diagnosis of Idiopathic Pulmonary Fibrosis "Pragmatic Challenges in Clinical Practice". *Front. Med.* **2017**, *4*, 151. [CrossRef] [PubMed]
3. Meltzer, E.B.; Noble, P.W. Idiopathic pulmonary fibrosis. *Orphanet J. Rare Dis.* **2008**, *3*, 8. [CrossRef] [PubMed]
4. Kreuter, M.; Ladner, U.M.; Costabel, U.; Jonigk, D.; Heussel, C.P. The Diagnosis and Treatment of Pulmonary Fibrosis. *Dtsch. Arztebl. Int.* **2021**, *118*, 152–162. [CrossRef]
5. Mackintosh, J.A.; Keir, G.; Troy, L.K.; Holland, A.E.; Grainge, C.; Chambers, D.C.; Sandford, D.; Jo, H.E.; Glaspole, I.; Wilsher, M.; et al. Treatment of idiopathic pulmonary fibrosis and progressive pulmonary fibrosis: A position statement from the Thoracic Society of Australia and New Zealand 2023 revision. *Respirology* **2024**, *29*, 105–135. [CrossRef]

6. Lederer, C.; Storman, M.; Tarnoki, A.D.; Tarnoki, D.L.; Margaritopoulos, G.A.; Prosch, H. Imaging in the diagnosis and management of fibrosing interstitial lung diseases. *Breathe* **2024**, *20*, 240006. [CrossRef]
7. Fell, C.D.; Martinez, F.J.; Liu, L.X.; Murray, S.; Han, M.K.; Kazerooni, E.A.; Gross, B.H.; Myers, J.; Travis, W.D.; Colby, T.V.; et al. Clinical predictors of a diagnosis of idiopathic pulmonary fibrosis. *Am. J. Respir. Crit. Care Med.* **2010**, *181*, 832–837. [CrossRef]
8. Cabrera Cesar, E.; Lopez-Lopez, L.; Lara, E.; Hidalgo-San Juan, M.V.; Parrado Romero, C.; Palencia, J.L.R.S.; Martín-Montañez, E.; Garcia-Fernandez, M. Serum Biomarkers in Differential Diagnosis of Idiopathic Pulmonary Fibrosis and Connective Tissue Disease-Associated Interstitial Lung Disease. *J. Clin. Med.* **2021**, *10*, 3167. [CrossRef]
9. Zheng, Z.; Peng, F.; Zhou, Y. Biomarkers in idiopathic pulmonary fibrosis: Current insight and future direction. *Chin. Med. J. Pulm. Crit. Care Med.* **2024**, *2*, 72–79. [CrossRef]
10. Aoki, A.; Hara, Y.; Fujii, H.; Murohashi, K.; Nagasawa, R.; Tagami, Y.; Enomoto, T.; Matsumoto, Y.; Masuda, M.; Watanabe, K.; et al. The clinical impact of comorbidities among patients with idiopathic pulmonary fibrosis undergoing anti-fibrotic treatment: A multicenter retrospective observational study. *PLoS ONE* **2023**, *18*, e0291489. [CrossRef]
11. Karuga, F.F.; Kaczmarski, P.; Szmyd, B.; Białasiewicz, P.; Sochal, M.; Gabryelska, A. The Association between Idiopathic Pulmonary Fibrosis and Obstructive Sleep Apnea: A Systematic Review and Meta-Analysis. *J. Clin. Med.* **2022**, *11*, 5008. [CrossRef] [PubMed]
12. van Cleemput, J.; Sonaglioni, A.; Wuyts, W.A.; Bengus, M.; Stauffer, J.L.; Harari, S. Idiopathic Pulmonary Fibrosis for Cardiologists: Differential Diagnosis, Cardiovascular Comorbidities, and Patient Management. *Adv. Ther.* **2019**, *36*, 298–317. [CrossRef]
13. Ruaro, B.; Pozzan, R.; Confalonieri, P.; Tavano, S.; Hughes, M.; Matucci Cerinic, M.; Baratella, E.; Zanatta, E.; Lerda, S.; Geri, P.; et al. Gastroesophageal Reflux Disease in Idiopathic Pulmonary Fibrosis: Viewer or Actor? To Treat or Not to Treat? *Pharmaceuticals* **2022**, *15*, 1033. [CrossRef] [PubMed]
14. Alhamad, E.H.; Cal, J.G.; Alrajhi, N.N.; Aharbi, W.M.; AlRikabi, A.C.; AlBoukai, A.A. Clinical characteristics, comorbidities, and outcomes in patients with idiopathic pulmonary fibrosis. *Ann. Thorac. Med.* **2020**, *15*, 208–214. [CrossRef]
15. Cocconcelli, E.; Tonelli, R.; Abbati, G.; Marchioni, A.; Castaniere, I.; Pelizzaro, F.; Russo, F.P.; Vegetti, A.; Balestro, E.; Pietrangelo, A.; et al. Subclinical liver fibrosis in patients with idiopathic pulmonary fibrosis. *Intern. Emerg. Med.* **2021**, *16*, 349–357. [CrossRef] [PubMed]
16. Makarev, E.; Izumchenko, E.; Aihara, F.; Wysocki, P.T.; Zhu, Q.; Buzdin, A.; Sidransky, D.; Zhavoronkov, A.; Atala, A. Common pathway signature in lung and liver fibrosis. *Cell Cycle* **2016**, *15*, 1667–1673. [CrossRef]
17. Antar, S.A.; Ashour, N.A.; Marawan, M.E.; Al-Karmalawy, A.A. Fibrosis: Types, Effects, Markers, Mechanisms for Disease Progression, and Its Relation with Oxidative Stress, Immunity, and Inflammation. *Int. J. Mol. Sci.* **2023**, *24*, 4004. [CrossRef]
18. Ismaiel, A.; Leucuta, D.C.; Popa, S.L.; Fagoonee, S.; Pellicano, R.; Abenavoli, L.; Dumitrascu, D.L. Noninvasive biomarkers in predicting nonalcoholic steatohepatitis and assessing liver fibrosis: Systematic review and meta-analysis. *Panminerva Med.* **2021**, *63*, 508–518. [CrossRef]
19. Ruta, V.M.; Man, A.M.; Alexescu, T.G.; Motoc, N.S.; Tarmure, S.; Ungur, R.A.; Todea, D.A.; Coste, S.C.; Valean, D.; Pop, M.C. Neutrophil-To-Lymphocyte Ratio and Systemic Immune-Inflammation Index-Biomarkers in Interstitial Lung Disease. *Medicina* **2020**, *56*, 381. [CrossRef]
20. Sankari, A.; Chapman, K.; Ullah, S. Idiopathic Pulmonary Fibrosis. In *StatPearls*; StatPearls Publishing: Treasure Island, FL, USA, 2024.
21. Raghu, G.; Remy-Jardin, M.; Myers, J.L.; Richeldi, L.; Ryerson, C.J.; Lederer, D.J.; Behr, J.; Cottin, V.; Danoff, S.K.; Morell, F.; et al. Diagnosis of Idiopathic Pulmonary Fibrosis. An Official ATS/ERS/JRS/ALAT Clinical Practice Guideline. *Am. J. Respir. Crit. Care Med.* **2018**, *198*, e44–e68. [CrossRef]
22. Wells, C.L. Chapter 47—Pulmonary diseases. In *Geriatric Rehabilitation Manual*, 2nd ed.; Kauffman, T.L., Barr, J.O., Moran, M., Eds.; Churchill Livingstone: Edinburgh, UK, 2007; pp. 297–304.
23. Ponce, M.C.; Sankari, A.; Sharma, S. Pulmonary Function Tests. In *StatPearls*; StatPearls Publishing: Treasure Island, FL, USA, 2024.
24. European Association for the Study of the Liver (EASL); European Association for the Study of Diabetes (EASD); European Association for the Study of Obesity (EASO). EASL-EASD-EASO Clinical Practice Guidelines on the management of metabolic dysfunction-associated steatotic liver disease (MASLD). *J. Hepatol.* **2024**, *81*, 492–542. [CrossRef] [PubMed]
25. Achaiah, A.; Rathnapala, A.; Pereira, A.; Bothwell, H.; Dwivedi, K.; Barker, R.; Iotchkova, V.; Benamore, R.; Hoyles, R.K.; Ho, L.-P. Neutrophil lymphocyte ratio as an indicator for disease progression in Idiopathic Pulmonary Fibrosis. *BMJ Open Respir. Res.* **2022**, *9*, e001202. [CrossRef]
26. Chen, Y.; Cai, J.; Zhang, M.; Yan, X. Prognostic Role of NLR, PLR and MHR in Patients with Idiopathic Pulmonary Fibrosis. *Front. Immunol.* **2022**, *13*, 882217. [CrossRef] [PubMed]
27. Song, J.-U.; Jang, Y.; Lim, S.-Y.; Ryu, S.; Song, W.J.; Byrne, C.D.; Sung, K.C. Decreased lung function is associated with risk of developing non-alcoholic fatty liver disease: A longitudinal cohort study. *PLoS ONE* **2019**, *14*, e0208736. [CrossRef]
28. Chung, J.H.; Goldin, J.G. Interpretation of HRCT Scans in the Diagnosis of IPF: Improving Communication between Pulmonologists and Radiologists. *Lung* **2018**, *196*, 561–567. [CrossRef]
29. Liao, K.M.; Chen, C.Y. Risk of potential hepatotoxicity from pirfenidone or nintedanib in patients with idiopathic pulmonary fibrosis: Results of a retrospective analysis of a large insurance database in Taiwan. *Front. Pharmacol.* **2024**, *15*, 1309712. [CrossRef] [PubMed]

30. Bendstrup, E.; Wuyts, W.; Alfaro, T.; Chaudhuri, N.; Cornelissen, R.; Kreuter, M.; Nielsen, K.M.; Münster, A.-M.B.; Myllärniemi, M.; Ravaglia, C.; et al. Nintedanib in Idiopathic Pulmonary Fibrosis: Practical Management Recommendations for Potential Adverse Events. *Respiration* **2018**, *97*, 173–184. [CrossRef]
31. Costabel, U.; Bendstrup, E.; Cottin, V.; Dewint, P.; Egan, J.J.J.; Ferguson, J.; Groves, R.; Hellström, P.M.; Kreuter, M.; Maher, T.M.; et al. Pirfenidone in Idiopathic Pulmonary Fibrosis: Expert Panel Discussion on the Management of Drug-Related Adverse Events. *Adv. Ther.* **2014**, *31*, 375–391. [CrossRef]

medicina

MDPI

Article

Evaluation of MELD Scores and Thyroid Hormones as Prognostic Factors of Liver Cirrhosis

Anca M. Belu [1,2], Alina D. Nicoara [1,2,*], Daniela M. Belu [1] and Eduard Circo [1,2]

[1] Faculty of Medicine, Ovidius University of Constanta, 1 Universitatii Alley, 900470 Constanta, Romania; ancabelu@doctor.com (A.M.B.); belu.d@yahoo.com (D.M.B.); eduard_circo@yahoo.com (E.C.)

[2] "St. Apostol Andrew" Emergency County Hospital, 145 Tomis Blvd., 900591 Constanta, Romania

* Correspondence: alinadnicoara@yahoo.com

Abstract: *Background and Objectives*: Hepatic cirrhosis is a disease with an increasing frequency globally, but its mechanisms of disease development are not yet completely known. The aim of this study was to evaluate the relationship between thyroid hormone levels (T3, fT4, and TSH) and survival in patients with chronic liver disease. *Materials and Methods*: A total of 419 patients diagnosed with liver cirrhosis were included in the study. The MELD score was computed, and TSH, T3, and fT4 were collected from each patient using the ELISA procedure. Signs and symptoms of liver failure and portal hypertension confirmed the clinical diagnosis of liver cirrhosis, and biological tests and imaging methods confirmed the diagnosis. *Results*: The MELD score was positively associated with TSH on admission and TSH on discharge and negatively associated with T3 at discharge. TSH levels were higher in non-survivors than in survivors. The values of T3 and fT4 present no significant changes to be considered as prognostic factors. *Conclusions*: Although the differences between the median TSH values of the patients who died and those who survived are not very large, the statistical significance of the data obtained demonstrates that there are changes in metabolism of the thyroid hormones during the progression of liver cirrhosis. It is possible that TSH is the one which maintains the normal balance of thyroid activity for patients with liver cirrhosis, so it can be considered as an important marker of evolution of these patients.

Keywords: evolution of cirrhosis; TSH; T3; fT4; encephalopathy; liver

1. Introduction

Liver cirrhosis has become an increasingly public health concern globally and in Europe. The Global Burden of Disease (GBD) study (2017) estimated that 112 million people worldwide have been diagnosed with compensated liver cirrhosis [1]. According to WHO, 2.4% of all deaths worldwide are due to liver cirrhosis [2].

The main causes of liver cirrhosis are infection with hepatitis C virus (HCV), hepatitis B virus (HBV), alcohol-associated liver disease, and non-alcoholic fatty liver disease (NAFLD) [3].

There is a tendency for viral liver infections to decrease as the cause of liver cirrhosis. This fact is due to the vaccination against virus B and the implementation of treatment programs for HBV, but also the successful treatment of HCV infection with direct acting antiviral (DAA) therapy [4].

Alcohol is the substance most often abused throughout the world and is still a main cause in the etiology of liver cirrhosis [5].

Alcohol consumption is an important factor in the occurrence and evolution of liver cirrhosis. The systemic effect of alcohol consumption causes complications of pre-existing pathologies, but also the unfavorable evolution of liver pathology. The association between viral liver infection and alcohol consumption causes repeated decompensations of liver disease and forms with unfavorable evolution [6].

End-stage chronic liver disease is characterized by the replacement of normal liver tissue with fibrotic tissue. Liver function loss has serious repercussions and is a cause of morbidity and mortality [7].

The management of chronic liver diseases is challenging because of the complexity of liver functions and the metabolic correlations in which they play a decisive role [8].

The survival of patients with liver cirrhosis depends on the way in which the triggering cause of cirrhosis is kept under control, but also the complications and the effect on other associated pathologies.

Liver cirrhosis develops over time and can have permanent, potentially fatal consequences. Monitoring the clinical course and treatment is challenging because of the lack of prognostic and developmental markers. The evolution of patients with end-stage chronic liver disease is completely unpredictable because of the lack of balance between cell destruction and regeneration and the absence of a clear link between numerous biological constants and disease progression [9]. Challenging research topics include the prevention of these problems and the determination of the links between disease progression and metabolic changes, including hormonal changes [10]. Recent research has traced thyroid dysfunction to liver disease. Changes in thyroid-stimulating hormone (TSH) and thyroid hormone values are associated with liver disease complications, including mortality [11,12].

Thyroid hormone profile was strongly associated with worse outcomes in patients with cirrhosis and might represent a promising prognostic tool that can be incorporated in clinical practice [13].

The thyroid hormones triiodothyronine (T3) and thyroxine (T4) are produced under the control of an endocrine feedback loop. Both hormones are bound to the bloodstream by transport proteins called albumin, transthyretin, and thyroid-binding globulin (TBG) [14]. They play a part in the proper growth, development, and operation of organs. Moreover, they influence liver function by regulating the basal metabolic rate of all cells, including hepatocytes. On the other hand, the liver plays a significant role in thyroid hormone metabolism, including conjugation, thyroglobulin-related synthesis, and peripheral deionization [15]. Triiodothyronine is the main regulator of thyroid function in various target organs. Most of the T3 hormone is produced by enzymatic deiodination at the $5'$ position of T4, mainly in the liver. Thus, T3 reflects the functional status of peripheral tissue rather than synthetic thyroid activity. Serum T3 concentration decreases as the conversion of T4 to T3 decreases [16].

The activity of the thyroid gland is directly connected with that of the liver. Thyroid hormones regulate the rate of basal metabolism of hepatocytes and dysthyroidism can produce alterations in liver metabolism and circulation at this level [17]. Thyroid hormones elicit non-genomic effects that usually begin at the plasma membrane and are mediated primarily by integrin $\alpha v\beta 3$, although other receptors such as TRα and TRβ are also capable of eliciting non-genomic responses [18].

There is a complex relationship between thyroid and liver pathology. Under normal conditions, the liver plays an essential physiological role in thyroid hormone activation and inactivation, transport, and metabolism, while thyroid hormones are involved in hepatocyte activity and liver metabolism [19]. In hypothyroidism, changes in liver enzymes can appear, which can be attributed to the impairment of lipid and protein metabolism. Also, severe hypothyroidism can develop with hyperammonemia and ascites, mimicking liver failure [20].

Thyroid hormones participate actively or inactively in all physiological processes in the body [21]. The homeostasis of the thyroid can affect the evolution of chronic liver diseases; there is even the idea that use of thyroid hormone-analog can be used in the treatment of liver disease [22].

Patients with liver cirrhosis may be clinically euthyroid, but the determination of thyroid hormone and TSH values may record changes most likely determined by metabolic alteration, especially related to dysproteinemia, with a decrease in total proteins and albumin [23].

The MELD score is commonly used to assess disease severity and it is based on paraclinical measurements: bilirubin, creatinine, and INR (International Normalized Ratio). We studied the possibility that calculating and tracking the MELD score and detecting changes in the level of thyroid hormones may improve the evolution of the patients with liver cirrhosis [24,25].

The main aim of this study was to investigate the possibility of using combined thyroid hormones and the MELD score to more accurately evaluate the mortality of patients with chronic liver disease. In this regard, as a specific objective, we aimed to evaluate the relationship between thyroid hormone levels (T3, fT4, and TSH) and the survival of patients with chronic liver disease.

2. Materials and Methods

Survival was assessed by recording death during hospitalization in patients with chronic liver disease. MELD (Model for End-Stage Liver Disease) is a reliable indicator of short-term survival in patients with end-stage liver disease and was designed based on bilirubin, creatinine, and INR. The lowest MELD score was 6, and the highest score was 40.

2.1. Samples

Thyroid Stimulating Hormone (TSH, thyrotropin): collection was performed under fasting conditions, but not after a recent thyroid biopsy or thyroid surgery. Venous blood was sampled using a collection container with a vacutainer without anticoagulant, with or without separating gel. Serum storage conditions: 20 °C or 2–8 °C. The samples were centrifuged to separate it. At least 0.5 mL of serum was extracted, and electrochemiluminescence detection immunochemistry (ELISA) was used as the analytical method. We excluded any analytical or drug interference and evaluated the range between 0.27 and 4.2 IU/mL as normal values, taking into account the adult age of the patients in the research group [26].

Thyroxine (T4): Blood was collected from venous blood in a vacutainer with or without separating gel and without anticoagulant. The storage conditions were 20 °C or 2–8 °C. Centrifugation was used to separate the serum. The serum sample should contain 0.5 mL. The electrochemiluminescence detection immunochemical assay (ELISA) was performed. Interferences in drug and kit components were excluded. Reference values ranged from 12.0 to 22.0 pmol/L [27].

Triiodothyronine (T3) is a thyroid hormone that circulates bound to the transporter protein but has a 10-fold lower affinity for the protein transporter than T4. The collection and determination method was similar to that for fT4 determination, and the reference range is 1.3–3.1 nmol/L [16].

All serological samples were processed during hospitalization and in the same laboratory.

2.2. Inclusion Criteria

The study included all patients over 18 years of age who provided written informed consent and were diagnosed with liver cirrhosis after clinical, paraclinical-biologic, and imaging examinations.

2.3. Exclusion Criteria

Patients who refused to complete the informed consent form were excluded. Patients who were taking drugs that might have affected thyroid hormone metabolism were also excluded. These included iodide contrast agents and amiodarone, which inhibit the conversion of T4 to triiodothyronine T3, as well as other classes of drugs, such as glucocorticoids and dopamine, which increase TSH secretion and, therefore, decrease T3 [28].

2.4. Participants

The study involved 419 patients diagnosed with liver cirrhosis who were hospitalized between March 2022 and March 2023 in Constanța County Hospital, Romania. The patients were followed during hospitalization, until discharge, with an interval between 1 and 41 days (M = 7.21, SD = 6.73). There were three categories of discharges—clinical and biological stabilization with resolution of the decompensation episode, discharge on request, and exitus.

2.5. Data Analysis Methods

All analysis were performed using R [29] and the R packages.

Initially, outliers, missing values, and univariate descriptive analyses, including analysis of compliance with the assumption of univariate normality for continuous data, were performed using the Shapiro–Wilk test statistic [30,31], and indicators of skewness and kurtosis were computed. Extreme univariate values (located beyond 3 standard deviations to the left or above 3 standard deviations to the right of the mean) were replaced using missing values, and imputation was performed using the K-nearest neighbor method [32]. Recorded hormone values as they resulted from the analysis (continuous variables, higher power) were used instead of transformation to categorical values (categorical variables, lower power), and statistical methods were used to compare rank means.

For the analysis of the associations between the main continuous variables, depending on whether the assumption of univariate normality was met, a Bravais–Pearson r correlation matrix was used if the assumption was met, or a Spearman ρ correlation matrix was used if the assumption was not met.

To test the hypotheses, depending on the fulfillment of the assumptions, either the two-sample *t*-test, comparing the means of two independent populations, or its nonparametric equivalent (Wilcoxon sum rank test) was used.

Given that the survival achievement variable is dichotomous, its prediction as a function of thyroid hormone levels was performed using receiver operator characteristic (ROC) curve and area under curve (AUC) analysis. The true positive rate (TPR) was calculated as the proportion of correctly predicted survival cases out of all survival cases, and the false positive rate (FPR or specificity) as the proportion of incorrectly predicted cases out of all deaths. The ROC curve displays the ratio between sensitivity and specificity, and closeness to the upper-left corner indicates very good classification performance, whereas the equality of sensitivity and specificity (TPR = FPR) indicates lack of concrete classification of predictions (random assignment), where the ROC curve follows the diagonal. AUC (Area Under Curve) and confidence interval were also calculated as measures of classification accuracy. The control level for the ROC curve analysis was death (the "Yes" variant).

The diagnosis of liver cirrhosis was supported clinically by the presence of portal hypertension and symptoms of liver failure and paraclinical findings on biological and imaging tests. Each patient was assigned a MELD score and a Child–Pugh score, depending on the course and severity. We also classified the patients with hepatic encephalopathy into stage categories, using the traditional West Haven Classification (formulated by Harold Conn). This classification divides patients with hepatic encephalopathy into four stages [33]. According to the West Haven criteria, stage 1 includes changes in attention, euphoria, or anxiety, and reduced intellectual performance; stage 2 is characterized by lethargy or apathy, minimal temporo-spatial disorientation, personality changes, and inappropriate behavior; stage 3 includes drowsiness up to semi-stupor, but remaining responsive to verbal stimuli, confusion, and severe temporal-spatial disorientation; and stage 4 is represented by coma [34].

Kaplan–Meier survival analysis was also conducted, in which participants were right-censored according to the number of hospitalization days. As an event of interest, the death of patients was monitored, and as a group variable, continuous hormone values were discretized, as follows: (a) for TSH, values between 0.27 and 4.2 were considered normal, and values lower or higher than these limits were considered abnormal; (b) for T3, values between 1.3 and 3.1 were considered normal, and values lower or higher than these limits were considered abnormal; and (c) for fT4, values between 12.0 and 22.0 were considered normal, and values lower or higher than these limits were considered abnormal.

3. Results

3.1. Socio-Demographic Data

The age of the participants ranged from 32 to 91 years (M = 63.26, SD = 9.57), and 68.97% were male. After diagnosis, patients were hospitalized between 0 and 96 months (M = 20.35, SD = 20.22), and the duration of hospitalization ranged from 0 to 41 days (M = 7.21, SD = 6.73).

In terms of etiology, most cases had alcoholic etiology (50.60%), followed by HVC (18.62%) and HVB (17.90%) etiology, as well as mixed etiology (12.89%). In terms of encephalopathy score, most patients were in the group determined by score 0 (67.78%), followed by those in the group determined by a score of 2 (17.66%), then those in the group with score 1 (6.68%) and 3 (5.01%), and lastly those with the highest score, 5 (2.86%). Moreover, (7.88%) are currently deceased.

3.2. Univariate Descriptive Analysis

Values > 10.4 for TSH at discharge, >5.4 for T3 at admission, and >4.3 for T3 at discharge were considered univariate extreme and were discarded, with a new K-nearest neighbor imputation. For TSH, only one missing value was found at admission and three missing values at discharge. For T3, five missing values were observed, the same participants at admission and discharge. For FT4, two missing values were found at admission and three missing values at discharge

The hospitalization data showed that the MELD score, TSH, T3, and fT4 are positively skewed, postulating the existence of highly, although not extremely, emphasized values. The MELD score and TSH have a leptokurtic distribution, with low variability around the mean, and the other variables had a mesokurtic distribution (Tables 1 and 2), so the assumption of univariate normal distribution is not fulfilled.

Table 1. Univariate descriptive analysis: data collected at admission.

Variable	N	Mean	Ab. Std	Median	Min	Max	Skew (ES)	Kurt (ES)	Normal
MELD Score	419	14.08	4.32	14	5	31	0.72 (0.12)	0.92 (0.24)	
TSH (mUI/L)	419	5.89	1.92	5.7	2.4	14.4	0.70 (0.12)	1.03 (0.24)	0.4–4.0
T3 (pmol/L)	419	0.84	0.50	0.77	0.05	2.5	0.51 (0.12)	−0.24 (0.24)	1.2–3.0
fT4 (pmol/L)	419	14.13	2.62	13.5	8.8	23	0.83 (0.12)	0.43 (0.24)	12.0–22.0

Table 2. Univariate descriptive analysis: data collected at discharge.

Variable	N	Mean	Ab. Std	Median	Min	Max	Skew (ES)	Kurt (ES)	Normal
TSH (mUI/L)	419	3.89	1.26	3.55	1.35	8.15	0.68 (0.12)	0.04 (0.24)	0.4–4.0
T3 (pmol/L)	419	0.78	0.52	0.65	0.00	2.25	0.64 (0.12)	−0.64 (0.24)	1.2–3.0
fT4 (pmol/L)	419	12.76	3.04	12.4	8.8	23.4	0.86 (0.12)	0.26 (0.24)	12.0–22.0

In the case of the discharge data, all distributions are positively skewed, and the distribution for T3 is platykurtic (see Table 1).

3.3. Bivariate Correlation Analysis

The assumption of univariate normality was not met, so the variables were correlated using Spearman's ρ correlation coefficient (Table 3) after first converting the dichotomous variable "non-surviving patients" to the corresponding numerical values.

Table 3. Spearman ρ correlation matrix.

	1	2	3	4	5	6	7	8	9
(1) MELD	-								
(2) Non-surviving patients	−0.09	-							
(3) Hospitalization (days)	−0.03	−0.02	-						
(4) TSH—Admission	0.27 ***	−0.19 ***	0.03	-					
(5) TSH—Discharge	0.21 ***	−0.14 **	0.00	0.75 ***	-				
(6) T3—Admission	−0.10	0.04	0.01	−0.35 ***	−0.28 ***	-			
(7) T3—Discharge	−0.17 ***	−0.02	−0.01	−0.23 ***	−0.14 **	0.46 ***	-		
(8) fT4—Admission	−0.02	0.01	0.06	−0.18 ***	−0.07	0.23 ***	0.13 **	-	
(9) fT4—Discharge	0.08	0.02	0.07	0.05	0.20 ***	0.09	0.51 ***	0.42 ***	-
Media	14.08	1.92	7.21	5.89	3.89	0.84	0.78	14.13	12.76
Standard deviations	4.32	0.27	6.73	1.92	1.26	0.50	0.52	2.62	3.04

*** $p < 0.001$; ** $p < 0.01$.

The MELD score was positively associated with TSH on admission (ρ = 0.27, $p < 0.001$), and TSH on discharge (ρ = 0.21, $p < 0.001$), negatively associated with T3 at discharge (ρ = −0.17, $p < 0.001$), negatively marginally associated with non-survival (ρ = −0.09, $p = 0.074$) and T3 on admission (ρ = −0.10, $p = 0.053$), and not associated with hospitalization days (ρ = −0.03, $p = 0.541$), fT4 at hospitalization (ρ = −0.02, $p = 0.629$), and fT4 at discharge (ρ = 0.08, $p = 0.109$).

Non-survivor status at discharge was negatively associated with TSH at hospitalization (ρ = −0.19, $p < 0.001$), and TSH at discharge (ρ = −0.14, $p = 0.004$), and not associated with hospitalization days (ρ = −0.02, $p = 0.702$), T3 at hospitalization (ρ = 0.04, $p = 0.0355$), T3 at discharge (ρ = −0.02, $p = 0.750$), fT4 at hospitalization (ρ = 0.02, $p = 0.745$), and fT4 at discharge (ρ = 0.02, $p = 0.745$).

The number of hospitalization days was not associated with any variable.

TSH on admission was positively associated with TSH on discharge (ρ = 0.75, $p < 0.001$), negatively associated with T3 on admission (ρ = −0.35, $p < 0.001$), T3 on discharge (ρ = −0.23, $p < 0.001$), and fT4 on admission (ρ = −0.18, $p < 0.001$), and not associated with fT4 on discharge (ρ = 0.05, $p = 0.315$).

TSH at discharge was positively associated with fT4 at discharge (ρ = 0.20, $p < 0.001$), negatively associated with T3 at hospitalization (ρ = −0.28, $p < 0.001$), and T3 at discharge (ρ = −0.11, $p = 0.005$), and not associated with fT4 at hospitalization (ρ = −0.07, $p = 0.147$).

T3 on admission was positively associated with T3 on discharge (ρ = 0.46, $p < 0.001$), fT4 on admission (ρ = 0.23, $p < 0.001$), and marginally positively associated with fT4 on discharge (ρ = 0.09, $p = 0.08$).

T3 at discharge was positively associated with fT4 at hospitalization (ρ = 0.13, $p = 0.008$) and fT4 at discharge (ρ = 0.50, $p < 0.001$), and fT4 at hospitalization was positively associated with fT4 at discharge (ρ = 0.42, $p < 0.001$).

3.4. Data Analysis

The non-parametric Wilcoxon rank sum test was used for two independent populations from which the samples were drawn, also known as the Mann–Whitney test, because the assumption of univariate normality was not met.

We found that the mean value for TSH on admission was 7.14 (SD = 1.79, median = 6.7) in decedents and 5.78 (SD = 1.9, median = 6.60) in survivors, with normal values ranging from 0.4 to 4.0 mIU/L (Figure 1). Thus, the decedents had statistically significantly higher TSH levels on admission (W = 8930, $p < 0.001$, es = 0.19), and the effect size was small.

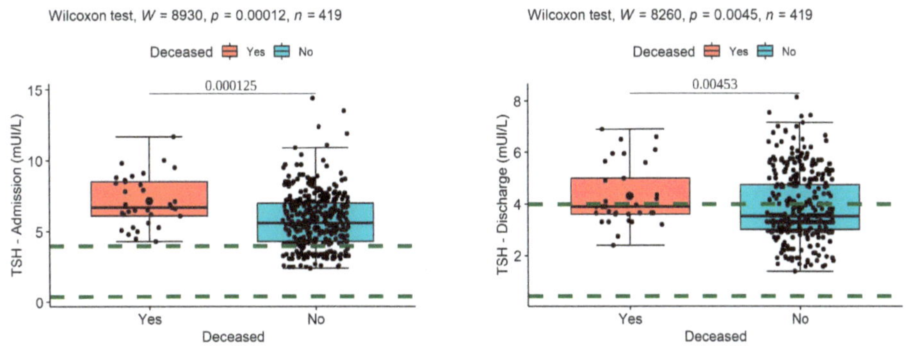

Figure 1. Comparisons of mean ranks of deceased and survivors' TSH levels on admission (**left**) and discharge (**right**).

The space between the dotted lines is the space of normal range of values. Similar results were also observed for TSH at discharge, the mean rank of survivors' values being 3.85 (SD = 1.26, median = 3.55), lower compared to that of the deceased 4.31 (SD = 1.19, median = 3.9), and this difference had statistical significance (W = 8260, p = 0.005, es = 0.14) with a small effect size.

ROC analysis showed an accuracy of survival classification based on TSH hormone of 70.33% (95% CI [61.61, 78.05]) at admission, and (62.78%, 95% CI [52.86, 71.27]) at discharge (Figure 2), so it can be considered an acceptable classifier of survival in patients with cirrhosis, even though the effects have been shown to be very small but statistically significant.

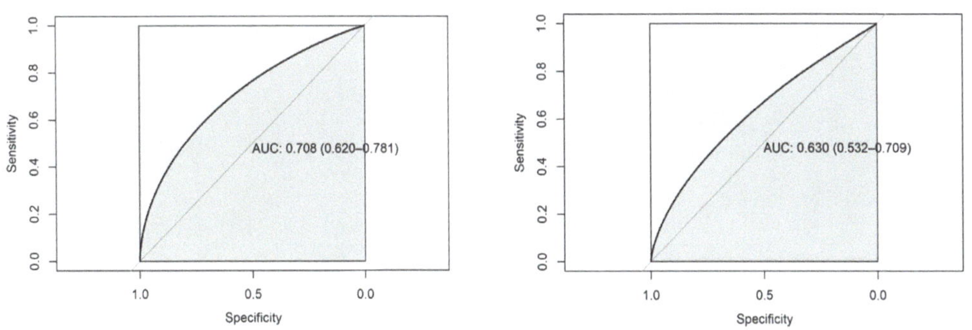

Figure 2. ROC curve analysis of survival classification according to TSH on admission (**left**) and discharge (**right**).

In the case of T3 on admission, the values of the decedents were (M = 0.74, SD = 0.38, median = 0.76), and of the survivors were (M = 0.85, SD = 0.50, median = 0.77), with no statistically significant differences (W = 5750.5, p = 0.354, es = 0.05) between the two means (Figure 3).

The space between the dotted lines is corresponding to the normal values of T3. At discharge, those who subsequently died had a mean of 0.77 (SD = 0.47, median = 0.80), and the mean of those who survived was 0.78 (SD = 0.53, median = 0.66), with at most marginally significant differences between the two means (W = 6581.5, p = 0.75, es = 0.02).

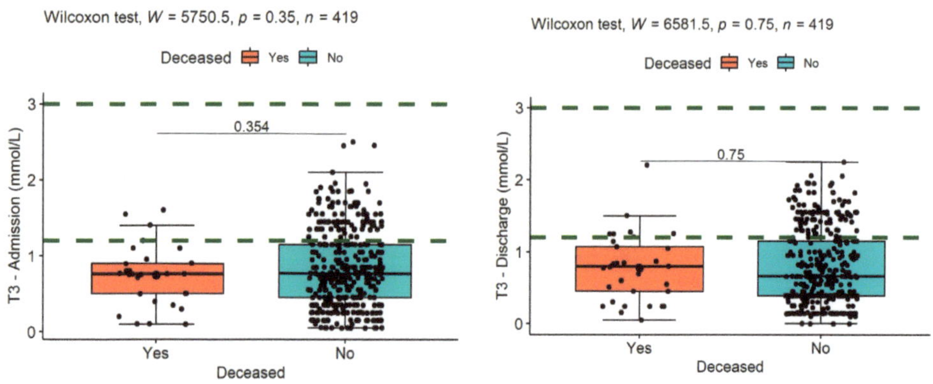

Figure 3. Comparisons of mean ranks of deceased and survivor T3 values on admission (**left**) and discharge (**right**).

Survival classification based on T3 hormone had a very low accuracy both at admission 55.95% (95% CI [46.41, 64.74]) and at discharge, with a specificity of 50.59% (95% CI [41.18, 59.87]) at discharge (Figure 4), and T3 cannot be considered an acceptable classifier of survival in patients with liver cirrhosis.

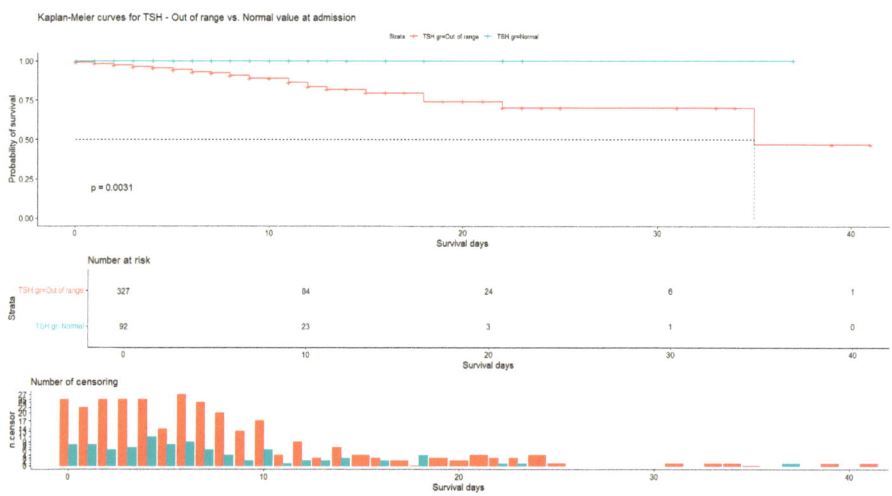

Figure 4. Kaplan–Meier curve for TSH at admission. Normal values vs. out-of-range values.

For fT4, there were no statistically significant differences between survivors and those subsequently deceased, neither at admission (W = 6229.5, p = 0.835, es = 0.01) nor at discharge (W = 6155.5, p = 0.745, es = 0.02).

Classification of survival based on fT4 had a very low accuracy, both at admission 52.54% (95% CI [41.77, 60.94]) and at discharge (52.07%, 95% CI [41.08, 64.29]). Therefore, fT4 cannot be considered an acceptable classifier of survival in patients with liver cirrhosis.

The mean MELD values were 15.54 (SD = 5.20, median = 15) for decedents and 13.95 (SD = 4.22, median = 14) for survivors. Thus, deceased patients have marginally significantly higher MELD score values (W = 7555.5, p = 0.075, es = 0.09), but the effect size was small.

Kaplan–Meier analysis was performed only for admission data. In the case of TSH, the average number of days of hospitalization was 7.34 (N = 327, SD = 6.88) for patients with abnormal hormone values and 6.74 (N = 92, SD = 6.16) for patients with normal values, with patient censoring observed almost every day. Our data show that all patients with normal TSH values survived, while for patients with abnormal TSH values the probability of survival was 88.8% (SE = 2.39%, 95% CI [84.2%, 93.6%]) at 9 days and 46.6% (SE = 19.52%, 95% CI [20.5%, 100.0%]) at 35 days. The log-rank test indicates a statistically significant difference ($p = 0.003$). The median survival time of patients in the group with abnormal TSH values was 35 days, which is significantly shorter than the survival time of patients in the group with normal TSH values (41 days, see Figure 6).

In the case of T3 values, the average number of days of hospitalization was 7.29 (N = 328, SD = 6.77) for patients with abnormal hormone values and 6.90 (N = 91, SD = 6.61) for patients with normal values, with patient censoring observed almost every day. Our data show that for patients with abnormal T3 values the probability of survival was 89.0% (SE = 24.16%, 95% CI [84.3%, 93.9%]) at 9 days and 71.5% (SE = 6.46%, 95% CI [59.9%, 85.4%]) at 22 days, whereas for patients with normal T3 values it was 92.3% (SE = 6.45%, 95% CI [80.5%, 100.0%]) at 12 days (Figure 5). The log-rank test indicated no statistically significant difference ($p = 0.10$) between the median survival time of patients with abnormal T3 values and that of patients with normal T3 values.

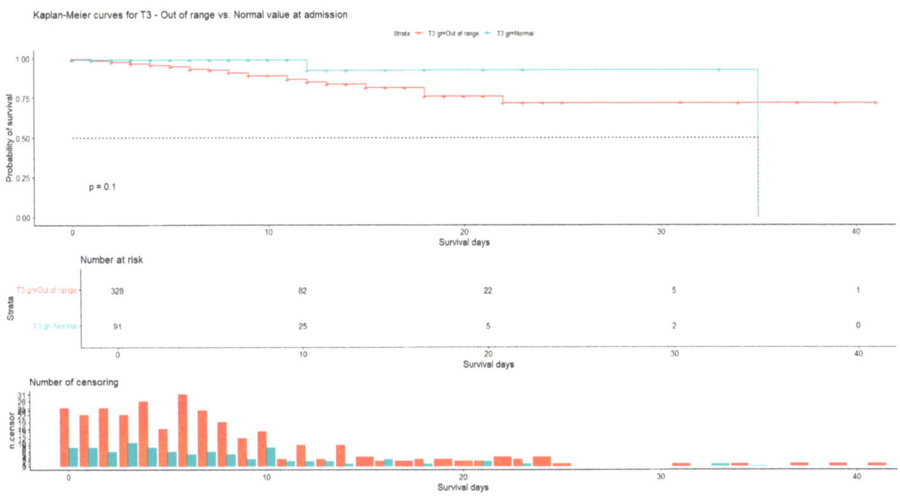

Figure 5. Kaplan–Meier curve for T3 at admission. Normal values vs. out-of-range values.

The same conclusions can be drawn for fT4. The average number of days of hospitalization was 7.78 (N = 97, SD = 7.40) for patients with abnormal hormone values and 7.03 (N = 322, SD = 6.52) for patients with normal values, with patient censoring observed almost every day. For patients with abnormal fT4 values, the probability of survival was 90.40% (SE = 5.03%, 95% CI [81.0%, 100.0%]) at 12 days and 73.8% (SE = 11.84%, 95% CI [53.9%, 100.0%]) at 22 days, whereas for patients with normal fT4 values it was 85.3% (SE = 3.39%, 95% CI [79.0%, 92.3%]) at 12 days and 77.5% (SE = 5.37%, 95% CI [67.7%, 88.8%]) at 18 days (Figure 6). The log-rank test indicated no significant difference ($p = 0.93$) between the median survival time of patients with abnormal fT4 values and that of patients with normal fT4 values.

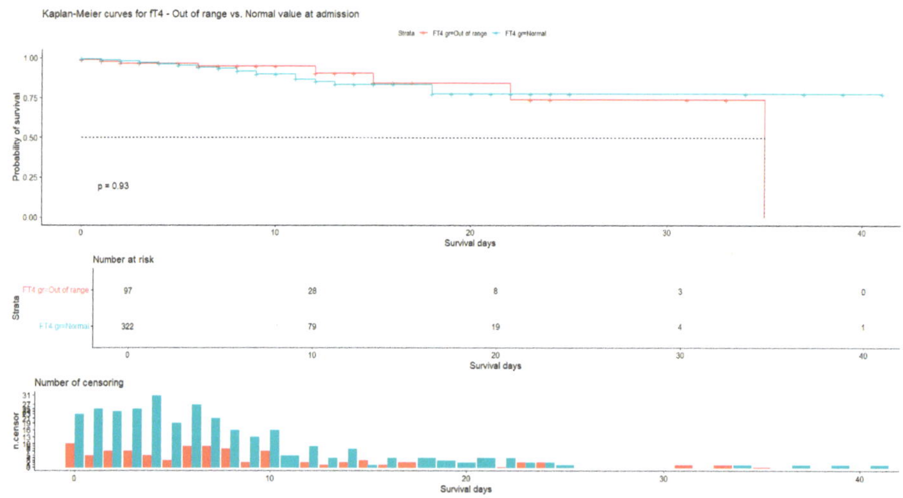

Figure 6. Kaplan–Meier curve for fT4 at admission. Normal values vs. out-of-range values.

4. Discussion

MELD score was positively associated with TSH at admission and at discharge and negatively associated with T3 at discharge. This finding corresponds with data published by Punekar, but refuted by studies by Vincken [11,35].

Non-survival was negatively associated with TSH at admission and at discharge and not associated with the number of hospitalizations, not associated with T3 at admission and at discharge, and not associated with fT4 at admission and at discharge. The number of days of hospitalization was not associated with any of the variables.

TSH at admission was positively associated with TSH at discharge, negatively associated with T3 at admission and at discharge and with fT4 at admission, and not associated with fT4 at discharge. The same situation is presented in the study of Vinken [35].

T3 at discharge was positively associated with fT4 at admission and at discharge. The deceased had statistically significantly higher TSH levels on admission and also at discharge, so it can be considered an acceptable classifier of survival of patients with liver cirrhosis. These results are consistent with the data obtained by Fei Ye, who considered the association between increased TSH and mortality to be statistically significant [36].

There are no statistically significant differences between T3 on admission to the hospital of the survivors and of the decedents, so T3 cannot be considered as an acceptable classifier of survival in liver cirrhosis. The same thing can be said about fT4, and the result is in agreement with studies by Trajkovic and Vincken who reported that they observed no associations between fT4 values and cirrhotic patients who died [35,37].

The Kaplan–Meier analysis shows that all the patients with normal TSH survived and for those with abnormal TSH values the probability of survival was 88.8% at 9 days of hospitalization and 46.6% at 35 days. The log-rank test shows no statistically significant difference in T3 and fT4 in survivors and in non-survivors. A similar situation was reported in the studies of Fei Ye [36]. Although the differences between the median TSH values of the patients who died and those who survived are not very large, the statistical significance of the data obtained demonstrates that there are changes in metabolism of the thyroid hormones during the progression of liver cirrhosis.

For fT4, there were no statistically significant differences between survivors and those who subsequently died, and the result is in agreement with studies by Trajkovic and Vincken who reported that they observed the same situation [35,37].

5. Conclusions

The findings of this study highlight the importance of careful monitoring of hormonal markers in patients with liver cirrhosis. The MELD score was positively associated with TSH on admission and TSH on discharge, and negatively associated with T3 at discharge. Increased TSH levels in patients with cirrhosis during hospitalization are associated with mortality. TSH may be a prognostic factor of mortality in patients with liver cirrhosis. Monitoring TSH may not only improve our understanding of disease progression, but significantly contribute to patient survival. The MELD score provides important prognostic and evolution data. These data can be correlated with the TSH variation and can be used to prevent unfavorable evolution towards exitus. The T3 and fT4 changes that occur during the decompensation of patients with liver cirrhosis can be considered transitory and not part of the permanent damage to the thyroid gland. The role of TSH is major, correcting thyroid hormone disorders that can occur in patients with liver cirrhosis. Even the conversion from T4 to T3 may not occur in proper conditions due to impaired liver function; it seems that the body has the ability to regulate the secretion of thyroid hormones so that there are no significant changes. This is probably achieved through the effect of TSH, which shows increased values along with the evolution towards death of patients with liver cirrhosis.

Author Contributions: Conceptualization, A.M.B. and E.C.; methodology, A.M.B.; software, D.M.B.; validation, A.D.N.; formal analysis, D.M.B.; investigation, A.M.B.; resources, A.M.B.; data curation, D.M.B.; writing—original draft preparation, A.M.B.; writing—review and editing, A.D.N.; visualization, E.C.; supervision, E.C. All authors have read and agreed to the published version of the manuscript.

Funding: This research received no external funding.

Institutional Review Board Statement: The study was conducted in accordance with the Declaration of Helsinki, and approved by the Institutional Review Board (or Ethics Committee) of University Ovidius of Constanta, protocol code 9944 and date of approval: 31 July 2022.

Informed Consent Statement: Informed consent was obtained from all subjects involved in the study.

Data Availability Statement: Data are available on request.

Conflicts of Interest: The authors declare no conflicts of interest.

References

1. WHO. The Global Health Observatory. Global Health Estimates: Leading Causes of Death. Available online: https://Www.Who.Int/Data/Gho/Data/Themes/Mortality-and-Global-Health-Estimates/Ghe-Leading-Causes-of-Death (accessed on 20 May 2023).
2. Huang, D.Q.; Terrault, N.A.; Tacke, F.; Gluud, L.L.; Arrese, M.; Bugianesi, E.; Loomba, R. Global Epidemiology of Cirrhosis—Aetiology, Trends and Predictions. *Nat. Rev. Gastroenterol. Hepatol.* **2023**, *20*, 388–398. [CrossRef] [PubMed]
3. Moon, A.M.; Singal, A.G.; Tapper, E.B. Contemporary Epidemiology of Chronic Liver Disease and Cirrhosis. *Clin. Gastroenterol. Hepatol. Off. Clin. Pract. J. Am. Gastroenterol. Assoc.* **2020**, *18*, 2650–2666. [CrossRef] [PubMed]
4. Jalilova, A.S. The spread of cirrhosis of the liver by etiological factors. *Orient. Renaiss. Innov. Educ. Nat. Soc. Sci.* **2022**, *2*, 253–257.
5. Patel, R.; Mueller, M. Alcoholic Liver Disease. In *StatPearls*; StatPearls Publishing: Treasure Island, FL, USA, 2024.
6. Llamosas-Falcón, L.; Probst, C.; Buckley, C.; Jiang, H.; Lasserre, A.M.; Puka, K.; Tran, A.; Zhu, Y.; Rehm, J. How Does Alcohol Use Impact Morbidity and Mortality of Liver Cirrhosis? A Systematic Review and Dose–Response Meta-Analysis. *Hepatol. Int.* **2024**, *18*, 216–224. [CrossRef]
7. GinËs, P.; Krag, A.; Abraldes, J.G. Liver Cirrhosis. *Lancet* **2021**, *398*, 1359–1376. [CrossRef] [PubMed]
8. Cheemerla, S.; Balakrishnan, M. Epidemiologia Globală a Bolii Hepatice Cronice. *Boală Hepatică Clin.* **2021**, *17*, 365–370.
9. Devarbhavi, H.; Asrani, S.K.; Arab, J.P.; Nartey, Y.A.; Pose, E.; Kamath, P.S. Global Burden of Liver Disease: 2023 Update. *J. Hepatol.* **2023**, *79*, 516–537. [CrossRef]
10. Aguirre-Villarreal, D.; Servin-Rojas, M.; Sánchez-Cedillo, A.; Chávez-Villa, M.; Hernandez-Alejandro, R.; Arab, J.P.; Ruiz, I.; Avendaño-Castro, K.P.; Matamoros, M.A.; Adames-Almengor, E.; et al. Liver Transplantation in Latin America: Reality and Challenges. *Lancet Reg. Health–Am.* **2023**, *28*, 100633. [CrossRef]
11. Punekar, P.; Sharma, A.K.; Jain, A. A Study of Thyroid Dysfunction in Cirrhosis of Liver and Correlation with Severity of Liver Disease. *Indian J. Endocrinol. Metab.* **2018**, *22*, 645–650. [CrossRef]

12. Gangadharam, D.Y.; Veeranarayana, D.G.; Kanha, D.M.M. A Study on Thyroid Function Tests as a Biomarker to Differentiate Acute from Chronic Liver Disease. *J. Cardiovasc. Dis. Res.* **2023**, *14*, 1–11.
13. Nardin, G.D.; Colombo, B.d.S.; Ronsoni, M.F.; Silva, P.E.S.e; Fayad, L.; Wildner, L.M.; Bazzo, M.L.; Dantas-Correa, E.B.; Narciso-Schiavon, J.L.; Schiavon, L.d.L. Thyroid Hormone Profile Is Related to Prognosis in Acute Decompensation of Cirrhosis. *Arch. Endocrinol. Metab.* **2024**, *68*, e230249. [CrossRef]
14. Kasper, D.L.; Fauci, A.S.; Hauser, S.L.; Longo, D.L.; Jameson, J.L.; Loscalzo, J. Thyroid Gland Disorders. In *Harrison's Manual of Medicine*; McGraw-Hill Education: New York, NY, USA, 2016.
15. Kharb, S.; Garg, M.K.; Puri, P.; Brar, K.S.; Pandit, A.; Srivastava, S. Assessment of Thyroid and Gonadal Function in Liver Diseases. *Indian J. Endocrinol. Metab.* **2015**, *19*, 89–94. [CrossRef] [PubMed]
16. Ritter, M.J.; Amano, I.; Hollenberg, A.N. Thyroid Hormone Signaling and the Liver. *Hepatology* **2020**, *72*, 742. [CrossRef]
17. Kim, H.J. Importance of Thyroid-Stimulating Hormone Levels in Liver Disease. *J. Pediatr. Endocrinol. Metab. JPEM* **2020**, *33*, 1133–1137. [CrossRef] [PubMed]
18. Gionfra, F.; De Vito, P.; Pallottini, V.; Lin, H.-Y.; Davis, P.J.; Pedersen, J.Z.; Incerpi, S. The Role of Thyroid Hormones in Hepatocyte Proliferation and Liver Cancer. *Front. Endocrinol.* **2019**, *10*, 532. [CrossRef]
19. Sinha, R.A.; Singh, B.K.; Yen, P.M. Direct Effects of Thyroid Hormones on Hepatic Lipid Metabolism. *Nat. Rev. Endocrinol.* **2018**, *14*, 259–269. [CrossRef]
20. Piantanida, E.; Ippolito, S.; Gallo, D.; Masiello, E.; Premoli, P.; Cusini, C.; Rosetti, S.; Sabatino, J.; Segato, S.; Trimarchi, F.; et al. The Interplay between Thyroid and Liver: Implications for Clinical Practice. *J. Endocrinol. Investig.* **2020**, *43*, 885–899. [CrossRef]
21. Mullur, R.; Liu, Y.-Y.; Brent, G.A. Thyroid Hormone Regulation of Metabolism. *Physiol. Rev.* **2014**, *94*, 355–382. [CrossRef]
22. Marino, L.; Kim, A.; Ni, B.; Celi, F.S. Thyroid Hormone Action and Liver Disease, a Complex Interplay. *Hepatology* **2023**, 10-1097. [CrossRef] [PubMed]
23. Verma, S.; Kumar, V.; Tiwari, P.; Joge, N.; Misra, R. Thyroid Profile in Patients of Cirrhosis of Liver: A Cross-Sectional Study. *J. Clin. Diagn. Res.* **2017**, *11*, 6. [CrossRef]
24. Ruf, A.; Dirchwolf, M.; Freeman, R.B. From Child-Pugh to MELD Score and beyond: Taking a Walk down Memory Lane. *Ann. Hepatol.* **2022**, *27*, 100535. [CrossRef] [PubMed]
25. Kim, W.R.; Mannalithara, A.; Heimbach, J.K.; Kamath, P.S.; Asrani, S.K.; Biggins, S.W.; Wood, N.L.; Gentry, S.E.; Kwong, A.J. MELD 3.0: The Model for End-Stage Liver Disease Updated for the Modern Era. *Gastroenterology* **2021**, *161*, 1887–1895.e4. [CrossRef]
26. Nallagangula, K.S.; Nagaraj, S.K.; Venkataswamy, L.; Chandrappa, M. Liver Fibrosis: A Compilation on the Biomarkers Status and Their Significance During Disease Progression. *Future Sci. OA* **2018**, *4*, FSO250. [CrossRef]
27. Kumar, A.; Ahuja, V.; Kaur, I.; Pandov, V.; Singh, A.; Sibia, R.P. Prevalence of Thyroid Dysfunction in Patients of Cirrhosis of Liver and Its Correlation with Severity of Cirrhosis. *Int. J. Adv. Res.* **2020**, *8*, 91–95. [CrossRef]
28. Zucchi, R. Thyroid Hormone Analogues: An Update. *Thyroid* **2020**, *30*, 1099–1105. [CrossRef]
29. R Core Team. *R: A Language and Environment for Statistical Computing*; R Foundation for Statistical Computing: Vienna, Austria, 2024.
30. Royston, J.P. An Extension of Shapiro and Wilk's W Test for Normality to Large Samples. *J. R. Stat. Soc. Ser. C Appl. Stat.* **1982**, *31*, 115–124. [CrossRef]
31. Royston, P. Remark AS R94: A Remark on Algorithm AS 181: The W-Test for Normality. *J. R. Stat. Soc. Ser. C Appl. Stat.* **1995**, *44*, 547–551. [CrossRef]
32. Templ, M.; Kowarik, A.; Alfons, A.; de Cillia, G.; Prantner, B.; Rannetbauer, W. VIM: Visualization and Imputation of Missing Values. *R Package Version* **2022**, *2*, 2.
33. West Haven Criteria for Hepatic Encephalopathy—GPnotebook. Available online: https://gpnotebook.com/pages/surgery/west-haven-criteria-for-hepatic-encephalopathy (accessed on 26 August 2024).
34. Mandiga, P.; Kommu, S.; Bollu, P.C. Hepatic Encephalopathy. In *StatPearls*; StatPearls Publishing: Treasure Island, FL, USA, 2024.
35. Vincken, S.; Reynaert, H.; Schiettecatte, J.; Kaufman, L.; Velkeniers, B. Liver Cirrhosis and Thyroid Function: Friend or Foe? *Acta Clin. Belg.* **2017**, *72*, 85–90. [CrossRef] [PubMed]
36. Ye, F.; Zhai, M.; Long, J.; Gong, Y.; Ren, C.; Zhang, D.; Lin, X.; Liu, S. The Burden of Liver Cirrhosis in Mortality: Results from the Global Burden of Disease Study. *Front. Public Health* **2022**, *10*, 909455. [CrossRef]
37. Trajkovic, M.; Visser, T.J.; Mittag, J.; Horn, S.; Lukas, J.; Darras, V.M.; Raivich, G.; Bauer, K.; Heuer, H. Abnormal Thyroid Hormone Metabolism in Mice Lacking the Monocarboxylate Transporter 8. *J. Clin. Investig.* **2007**, *117*, 627–635. [CrossRef] [PubMed] [PubMed Central]

medicina MDPI

Article

Outcomes in COVID-19 Patients with Acute Cholangitis: A Single-Center Retrospective Analysis

Deiana Vuletici [1,2], Bogdan Miutescu [1,2,*], Calin Burciu [1,2,3], Iulia Ratiu [1,2], Tudor Moga [1,2], Eyad Gadour [4,5], Alexandru Catalin Motofelea [6], Oana Koppandi [3], Roxana Sirli [1,2] and Alina Popescu [1,2]

1 Department of Gastroenterology and Hepatology, "Victor Babes" University of Medicine and Pharmacy Timisoara, Eftimie Murgu Square 2, 300041 Timisoara, Romania; deiana.vuletici@umft.ro (D.V.); calin.burciu@umft.ro (C.B.); ratiu.iulia@umft.ro (I.R.); moga.tudor@umft.ro (T.M.); sirli.roxana@umft.ro (R.S.); popescu.alina@umft.ro (A.P.)
2 Advanced Regional Research Center in Gastroenterology and Hepatology, "Victor Babes" University of Medicine and Pharmacy Timisoara, 300041 Timisoara, Romania
3 Department of Gastroenterology, Faculty of Medicine, Pharmacy and Dental Medicine, "Vasile Goldis" West University of Arad, 310414 Arad, Romania; oanakoppandi@yahoo.com
4 Department of Gastroenterology, King Abdulaziz Hospital-National Guard Health Affairs, Al Ahsa 31982, Saudi Arabia; eyadgadour@doctors.org.uk
5 Department of Medicine, Zamzam University College, Khartoum 11113, Sudan
6 Department of Internal Medicine, Faculty of Medicine, "Victor Babeș" University of Medicine and Pharmacy Timisoara, Eftimie Murgu Square 2, 300041 Timisoara, Romania; alexandru.motofelea@umft.ro
* Correspondence: miutescu.bogdan@umft.ro

Abstract: *Background and Objectives*: This study aimed to assess the impact of coronavirus disease 2019 (COVID-19) on patients with acute cholangitis (AC) by comparing outcomes, complications, and hospital stays in a tertiary Gastroenterology department. *Materials and Methods*: This retrospective observational cohort study was conducted in a tertiary gastroenterology department, collecting data from all AC and AC + COVID-19 patients between April 2020 and February 2022. Data included clinical and demographic information, COVID-19-specific details, acute cholangitis presentation, medical records, laboratory results, and interventions. AC was diagnosed using Tokyo Guidelines 2018 (TG18) criteria, with all patients undergoing bile culture sampling. *Results*: The study included 241 patients, 30 in the COVID group and 211 in the non-COVID group. The COVID group's mean age was significantly higher (74.3 vs. 67.3 years, $p < 0.009$). Abdominal pain was more common in the COVID group (90% vs. 70.6%, $p < 0.025$). Length of hospital stay was longer for COVID patients (13.5 vs. 7.9 days, $p < 0.001$). COVID patients had higher incidences of malignant causes of AC, with pancreatic cancer being the most common (30%). *Pseudomonas* spp. was significantly more prevalent in COVID patients (16.7% vs. 5.7%, $p = 0.028$). *Conclusions*: Our study results show that COVID-19 affected the duration of hospitalization for patients with AC. Furthermore, this study presents observations regarding the impact of COVID-19 on AC, revealing differences in microbial profiles.

Keywords: acute cholangitis; COVID-19; biliary drainage; microorganisms

1. Introduction

Acute cholangitis (AC) is a potentially fatal condition requiring prompt detection and treatment [1]. This clinical syndrome arises when bacterial infections invade the normally sterile biliary system, typically in the context of a bile duct obstruction caused by choledocholithiasis, although it can also occur in patients with neoplasms and strictures [2]. Management of acute cholangitis hinges on the severity of the condition, with biliary drainage and antibiotics being the primary treatment modalities [3].

The Tokyo Guidelines 2018 (TG18) provide criteria for diagnosing AC, which include systemic inflammation, cholestasis, and imaging evidence of bile-duct abnormalities [4]. These guidelines also offer recommendations for the appropriate use of antimicrobial agents [5].

The emergence of coronavirus disease 2019 (COVID-19), caused by the SARS-CoV-2 virus, has profoundly impacted medical practices worldwide, including gastroenterology [6]. COVID-19's rapid global spread has led to varied clinical presentations, from asymptomatic cases to severe illness, complicating the management of pre-existing conditions like acute cholangitis [7,8]. The need for emergency endoscopic procedures, such as endoscopic retrograde cholangiopancreatography (ERCP), persisted during the pandemic, despite overall reductions in gastrointestinal endoscopy volumes [9].

Notably, the pandemic did not significantly decrease the diagnosis of pancreaticobiliary cancers, nor did it alter the approach to ERCP for malignant and benign conditions [10]. However, patients with chronic liver disease who contracted COVID-19 faced higher risks of severe complications and mortality [11].

The liver's susceptibility to SARS-CoV-2 infection due to the high expression of ACE2 receptors in cholangiocytes further complicates the clinical picture [12].

While liver function abnormalities are common in COVID-19 patients, significant liver impairment is rare [13]. Severe cases may experience liver injury due to immune-mediated inflammation, including cytokine storms and hypoxia-associated pneumonia [14].

This study aims to evaluate the impact of COVID-19 on patients with AC, comparing outcomes, complications, and length of hospital stay within a tertiary gastroenterology department. Additionally, the primary objective of this study was to characterize the microbiological profiles of bile aspirates collected during ERCP in patients with acute cholangitis. By exploring these interactions, this study seeks to enhance the understanding and management of AC in the context of the recent pandemic.

2. Materials and Methods

2.1. Study Design and Participants

Ethical approval was obtained from the Internal Review Board of "Pius Brinzeu" Emergency County Hospital of Timisoara, Romania, and patient confidentiality and data security were strictly maintained. This study was designed as a retrospective observational cohort single-center study conducted in a tertiary gastroenterology department to investigate the intersection of COVID-19 and acute cholangitis. It aims to elucidate the clinical characteristics, therapeutic interventions, and outcomes in patients with both conditions. Data were collected from patients with acute cholangitis, with or without COVID-19, between April 2020 and February 2022. Clinical and demographic data were systematically and retrospectively collected, including COVID-19-specific information, cholangitis presentation, medical records, laboratory results, radiological findings, and medical interventions. Additionally, patient-reported outcomes and complications were documented.

The study aims to characterize the microbiological profiles of bile aspirates from patients undergoing ERCP. The PICO elements are P (Population)—patients diagnosed with AC and COVID-19; I (Intervention)—comparison of clinical outcomes and microbiological analysis of bile; C (Comparison)—outcomes in patients with AC with and without COVID-19; and O (Outcomes)—clinical outcomes (e.g., hospital stay, complications), and microbiological profiles.

2.2. Inclusion Criteria

Patient inclusion criteria included testing for COVID-19 diagnosis through RT-PCR on nasopharyngeal swabs and clinical and imaging evidence of AC based on the TG18. Additionally, participants were required to be above 18 years of age and willing to provide informed consent. Patients excluded from the study were those with inadequate medical records or incomplete clinical data, inability to provide informed consent due to medical or psychiatric conditions, age below 18 years, antibiotics treatment for other medical conditions at the time of acute cholangitis diagnosis, post-ERCP perforation, cholangitis secondary to ERCP, or percutaneous or surgical drainage.

2.3. Diagnosis of AC

The TG18 criteria determined the AC diagnosis. Based on the TG18 criteria, diagnosing acute cholangitis (AC) relies on three essential factors: systemic inflammation, cholestasis, and imaging-detected bile-duct abnormalities. Systemic inflammation is a mandatory criterion typically identified by fever or elevated inflammatory markers, such as increased leukocyte count or elevated C-reactive protein levels. Despite significant advancements in diagnostic imaging techniques, direct imaging-based diagnosis of AC remains challenging, requiring a continued dependence on clinical and laboratory findings to confirm the disease [4]. Following admission, all patients were administered antibiotics according to the TG18 recommendations for the grade specified in the diagnosis of AC [5]. Culture media were used to identify microorganisms in bile samples.

Various diagnostic techniques were applied to address the cause of obstruction at admission. B-mode ultrasonography was initially performed, and when the diagnosis remained uncertain, further methods, such as endoscopic ultrasound (EUS), contrast-enhanced Computer Tomography (CE-CT), or magnetic resonance (CE-MRI), were used to assist in diagnosing and staging malignancies. Additional diagnostic confirmation was achieved by evaluating tumor markers and reviewing histopathological data derived from ERCP or EUS biopsies.

2.4. Therapeutic Approach

ERCP was performed using a therapeutic duodenoscope provided by Olympus Corp., Tokyo, Japan, to access the common bile duct through a guidewire. Under careful sedation management, a specialized anesthesia and intensive care team expertly administered a blend of midazolam, propofol, and fentanyl, following their internal protocols during ERCP procedures. The timing for ERCP was determined by evaluating the severity of the condition and adhering to the guidelines set forth by the endoscopists as outlined in the Tokyo Guidelines.

2.5. Data Acquisition and Study Variables

Meticulously documenting a broad range of variables enabled a thorough analysis throughout the study. The collected data encompassed various aspects, including patient demographics (gender, age), clinical observations (symptoms such as abdominal pain, jaundice, fever, chills), laboratory analysis, duration of hospitalization, severity according to the Tokyo guidelines, and microbial cultures of bile samples.

Additionally, we included COVID-19 patients with pneumonia, those who received antiviral treatment for COVID-19, as well as those with comorbidities such as cardiac pathology, type 2 diabetes mellitus, and chronic kidney disease in both groups of patients.

2.6. Statistic Analysis

Continuous variables were assessed for normality using the Shapiro-Wilk test. Normally distributed data were presented as means $\pm$ standard deviations (SD), while non-normally distributed data were summarized using medians with interquartile ranges (IQRs; 25th to 75th percentiles). Categorical variables were described as counts and percentages. Differences between groups for normally distributed continuous variables were assessed using Welch's t-test for comparisons between two groups and one-way ANOVA for multiple groups, incorporating post-hoc tests (e.g., Tukey's HSD) to pinpoint specific group differences. For non-parametric continuous data, the Mann-Whitney U and Kruskal-Wallis tests were applied for two and multiple groups, respectively, with subsequent Dunn's post-hoc analyses as necessary. Categorical data comparisons were conducted using Chi-square tests or Fisher's exact test when expected cell counts were below five. We used binomial logistic regression to identify independent predictors of study outcomes, carefully considering potential confounding factors. Before the main analysis, we checked for multicollinearity among the predictors to ensure the integrity of our regression model. The logistic regression model quantified the association between independent variables and the outcome variable through regression coefficients. These coefficients provided insight into the direction and importance of the effect of each predictor, with statistical significance determined by a p-value of less than 0.05. We applied a pseudo-R-squared measure alongside the Hosmer-Lemeshow goodness-of-fit check to evaluate our logistic regression model's overall performance and suitability. Sample size calculations were conducted a priori to achieve a confidence level of 95% and a statistical power of 80% based on anticipated effect sizes and variance estimates derived from preliminary data. All statistical analyses were performed in R (version 3.6.3), leveraging the capabilities of several comprehensive packages within the Tidyverse for data manipulation and visualization, Finalfit for regression analyses, and other specialized packages (MCGV, Stringdist, Janitor, Hmisc) for various data processing needs.

3. Results

A total of 241 patients were included in this study. The etiology of AC is detailed in Table 1. Most patients in the benign group were diagnosed with choledochal lithiasis (43.6%). Malignant pathology was diagnosed in 53.3% of patients (n = 129/241), with pancreatic cancer (29%) being the most common cause. The mean age between the four groups showed statistical differences ($p < 0.009$). Patients in the COVID group had a mean age of 74.3 years (SD = 10.6), while patients in the non-COVID group had a mean age of 67.3 years (SD = 14.1). No gender differences were found between COVID and non-COVID patients ($p = 0.539$), as shown in Table 2. Abdominal pain was more common in the COVID group than in the non-COVID group (90% vs. 70.6%, $p < 0.025$). No differences were observed between the two groups in the incidence of fever ($p = 0.246$). Patients who tested positive for COVID-19 had an extended hospital stay of 13.5 days (SD = 6.6), compared to an average of 7.9 days (SD = 5.4) for patients not diagnosed with COVID-19.

A combination of correlation analysis, multiple regression analysis, and ANOVA was used to evaluate the relationships between hospitalization duration and factors such as sex, age, comorbidities (including cardiac pathology, Type 2 diabetes, and chronic kidney disease), pneumonia severity, and the probability of receiving COVID-19 treatment for patients with and without COVID-19.

The dataset included 30 COVID-19 patients, 15 of whom (50%) had pneumonia. Of those with pneumonia, 5 patients (16.7%) had minimal involvement, 7 patients (23.3%) had moderate involvement, and 3 patients (10%) had severe involvement. Additionally, 11 patients (36.7%) received COVID-19 treatment, while 19 (63.3%) did not.

Pearson correlation was used to explore the relationships between hospitalization duration and factors such as sex, age, and comorbidities in COVID-19 and non-COVID-19 groups. For the COVID-19 group, age ($r = -0.131$, $p = 0.490$) showed weak correlations with hospitalization duration. In the non-COVID-19 group, age ($r = 0.155$, $p = 0.024$) showed a weak but statistically significant positive correlation with hospitalization duration.

Multiple linear regression further quantified these relationships. In the COVID-19 group, Type 2 diabetes ($\beta = 3.59$, $p = 0.142$), gender ($\beta = 1.38$, $p = 0.530$), and age ($\beta = -0.009$, $p = 0.937$) were not statistically significant predictors of hospitalization duration. In the non-COVID-19 group, chronic kidney disease significantly affected hospitalization duration ($\beta = 4.02$, $p = 0.063$), and age also approached statistical significance ($\beta = 0.053$, $p = 0.058$). However, the model's explanatory power was low, with an R-squared value of 0.039, indicating that these variables explained only 3.9% of the variance in hospitalization duration.

An ANOVA test was conducted to assess whether there were significant differences in hospitalization duration based on pneumonia severity for COVID-19 patients. The results indicated no statistically significant differences in average hospitalization duration across different levels of pneumonia severity ($F = 0.192$, $p = 0.901$). A binomial logistic regression was also performed to evaluate the association between hospitalization duration and the probability of receiving COVID-19 treatment. The regression analysis showed that hospitalization duration (coefficient = 0.0664, $p = 0.265$) was not significantly associated with the probability of receiving COVID-19 treatment.

Table 1. Distribution of acute cholangitis etiologies among COVID-19 positive and negative patients.

Condition	COVID (n = 30)	Without COVID (n = 211)	p-Value
Benign	13.0 (43.3%)	99.0 (46.9%)	
Choledocholithiasis	13.0 (43.3%)	92.0 (43.6%)	0.978 [1]
Benign vaterian ampulloma	0.0 (0.0%)	2.0 (0.9%)	0.590 [1]
Benign coledochal stenosis	0.0 (0.0%)	4.0 (1.9%)	0.447 [1]
Liver abscess	0.0 (0.0%)	1.0 (0.5%)	0.706 [1]
Malignant	17.0 (56.7%)	112.0 (53.1%)	
Pancreatic cancer	9.0 (30.0%)	61.0 (28.9%)	0.902 [1]
Cholangiocarcinoma	6.0 (20.0%)	31.0 (14.7%)	0.450 [1]
Malignant vaterian ampulloma	2.0 (6.7%)	13.0 (6.2%)	0.915 [1]
Malignant extrinsic compression	0.0 (0.0%)	6.0 (2.8%)	0.350 [1]
Gallbladder cancer	0.0 (0.0%)	1.0 (0.5%)	0.706 [1]

n—number of patients; [1] Proportions are evaluated with a chi-square test.

Table 2. Clinical characteristics of the study population with biliary obstruction stratified by COVID-19 infection.

	COVID (n = 30)	Without COVID (n = 211)	p Value
Gender			0.539 [1]
F	18.0 (60.0%)	114.0 (54.0%)	
M	12.0 (40.0%)	97.0 (46.0%)	
Age			**0.009** [2]
Mean (SD)	74.3 (10.6)	67.3 (14.1)	
Range	52.0–93.0	19.0–96.0	
Jaundice			0.918 [1]
Yes	27.0 (90.0%)	192.0 (91%)	
No	3.0 (10.0%)	19.0 (9.0%)	
Abdominal pain			**0.025** [1]
Yes	27.0 (90.0%)	149.0 (70.6%)	
No	3.0 (10.0%)	62.0 (29.4%)	
Fever			0.264 [1]
Yes	6.0 (20.0%)	63.0 (29.9%)	
No	24.0 (80.0%)	148.0 (70.1%)	
CRP (mg/L)			0.476 [2]
Mean (SD)	119.65 (97.83)	105.70 (107.56)	
Range	11.0–322.8	2.14–545.9	
WBC ($\times 10^3/\mu L$)			0.881 [2]
Mean (SD)	11.73 (7.35)	11.52 (5.85)	
Range	3.61–40.0	2.81–41.9	
Total Bilirubin (mg/dL)			0.698 [2]
Mean (SD)	10.77 (7.12)	10.22 (7.29)	
Range	1.6–30.2	0.5–36.3	
Platelets ($\times 10^3/\mu L$)			0.959 [2]
Mean (SD)	261.84 (120.52)	263.05 (109.51)	
Range	32.0–501.8	24.0–777.0	
INR			0.219 [2]
Mean (SD)	1.52 (0.39)	1.42 (0.59)	
Range	0.96–2.77	0.91–5.05	
Cardiac pathology			0.058 [1]
Yes	25 (83.3%)	135 (63.9%)	
No	5 (16.6%)	76 (36%)	

Table 2. *Cont.*

	COVID (n = 30)	Without COVID (n = 211)	p Value
Type 2 Diabetes			0.727 [1]
Yes	9 (30%)	53 (25.1%)	
No	21 (70%)	158 (74.9%)	
Chronic Kidney Disease			0.757 [1]
Yes	0 (0%)	6 (2.8%)	
No	30 (100%)	205 (97.2%)	
Smoking status			
Ex-smoker	8 (30%)	64 (30.3%)	0.934 [1]
Smoker	11 (36.7%)	63 (29.9%)	0.635 [1]
Non-smoker	10 (33.3%)	84 (39.8%)	0.723 [1]
Previous stent			0.362 [1]
Yes	7.0 (23.3%)	35.0 (16.6%)	
No	23.0 (76.7%)	176.0 (83.4%)	
Hospitalization days			**<0.001** [2]
Mean (SD)	13.5 (6.6)	7.9 (5.4)	
Range	4.0–26.0	1.0–35.0	
Weekend admission			0.550 [1]
Yes	8.0 (26.7%)	46.0 (21.8%)	
No	22.0 (73.3%)	165.0 (78.2%)	
Tokyo severity score			0.103 [1]
Grade I	10.0 (33.3%)	85.0 (40.3%)	
Grade II	6.0 (20.0%)	67.0 (31.8%)	
Grade III	14.0 (46.7%)	59.0 (28.0%)	

n—number of patients; [1] Proportions are evaluated with a chi-square test; [2] Linear Model ANOVA; CRP—C-reactive protein; WBC-white blood cells; SD—Standard Deviation, INR—international normalized ratio.

The results presented in Table 3 show bile culture results associated with Tokyo severities and COVID-19 status, revealing different patterns of bacterial growth and culture sterility across various severity degrees. For patients with COVID-19, the highest proportion of sterile cultures was observed in mild Tokyo severity (60%), followed by severe cases (40%). No sterile cultures were identified in moderate-severity cases. In contrast, for patients without COVID-19 infection, sterile cultures were more evenly distributed, with a significant proportion observed in mild cases (40%), followed by moderate (33.3%) and severe grades (26.7%).

Table 3. Bacterial presence in bile specimens: Insights from the Tokyo Guidelines Severity Grades and COVID-19 Status.

	Sterile			1 Bacterium			2 Bacteria			>3 Bacteria		
	COVID	Non-COVID	p	COVID	Non-COVID	p	COVID	Non-COVID	p	COVID	Non-COVID	p
Tokyo Grade I	3 (60%)	30 (40%)		4 (25%)	35 (41.7%)		2 (28.6%)	18 (40.90%)		1 (50%)	2 (25%)	
Tokyo Grade II	0	25 (33.3%)	0.298 [1]	4 (25%)	27 (32.1%)	0.156 [1]	2 (28.6%)	12 (27.30%)	0.795 [1]	0	3 (37.5%)	0.429 [1]
Tokyo Grade III	2 (40%)	20 (26.7%)		8 (50%)	22 (26.2%)		3 (42.9%)	14 (31.80%)		1 (50%)	3 (37.5%)	

[1] Proportions are evaluated with a chi-square test.

Grade I (mild) acute cholangitis had a prevalence of monomicrobial growth in patients with COVID-19 (25%) compared to those without COVID-19 (41.7%). For grade III (severe) acute cholangitis, a higher prevalence of COVID-19 was observed (50%) compared to non-COVID-19 cases (26.2%). Interestingly, for grade II (moderate) acute cholangitis, the prevalence was relatively constant for both COVID-19 (25%) and non-COVID-19 (32.1%) patients. For those with two bacterial types identified, the occurrence was almost equally distributed across all three types of Tokyo classification for both COVID-19 and non-COVID-19 categories, with a slightly higher incidence in severe cases (42.9% for COVID-19 and 31.8% for non-COVID-19). Although less common, cultures with three or more bacteria showed a 50% prevalence for both mild and severe Tokyo disease in the COVID-19 group versus a lower prevalence in the non-COVID-19 group.

We utilized binomial logistic regression to assess the link between various bacterial infections and COVID-19 status among patients with AC. The model showed that approximately 26.5% of the variance in COVID-19 status could be attributed to the differential bacterial profiles, as evidenced by an R-squared value of 0.26. This highlights the substantial role that bacterial infections play in the context of COVID-19 among this patient cohort. Among the various bacteria examined, *Pseudomonas* spp. was the only microorganism that showed a statistically significant relationship with the COVID-19 status of the patients. The results revealed an increasing odds ratio for *Pseudomonas* spp., suggesting that the presence of this bacteria is associated with a fourfold increase in the likelihood of being a COVID-19-positive case after adjusting for other bacterial infections. This finding highlights the importance of *Pseudomonas* spp. as a significant indicator of COVID-19 status in patients with cholangitis, highlighting its potential role in the pathophysiology of COVID-19-related complications in such populations.

Tables 4 and 5 present a comprehensive analysis comparing the presence of various microorganisms in bile cultures of patients with COVID-19 and those without COVID-19. This comparison aims to identify significant differences in bacterial prevalence, which could inform treatment and management strategies for patients with AC during the pandemic.

Table 4. Binominal Logistic Regression Analysis of Bacterial Profiles at patients with acute cholangitis based on COVID-19 status.

						95% Confidence Interval	
Predictor	**Estimate**	**SE**	**Z**	***p***	**Odds Ratio**	**Lower**	**Upper**
Intercept	−2.494	0.339	7.356	**<0.001**	0.0826	0.0425	0.160
Other bile germs:							
Yes—No	0.999	0.627	1.593	0.111	2.7147	0.7945	9.275
Acinetobacter spp.							
Yes—No	0.531	1.181	0.450	0.653	1.7014	0.1682	17.205
Citrobacter spp.							
Yes—No	0.816	0.828	0.985	0.324	2.2608	0.4462	11.454
Pseudomonas spp.							
Yes—No	1.457	0.615	2.369	**0.018**	4.2923	1.2857	14.329
Enterococcus spp.							
Yes—No	0.320	0.488	0.655	0.512	1.3771	0.5289	3.586
Klebsiella spp.							
Yes—No	0.261	0.514	0.509	0.611	1.2988	0.4742	3.558
Escherichia coli:							
Yes—No	0.337	0.431	0.782	0.434	1.4008	0.6021	3.259

Note: Estimates represent the log odds of "Group = COVID" vs. "Group = Without COVID" (dependent variable). The predictors listed are considered independent variables. The model was not adjusted for gender and age. Each estimate is accompanied by its standard error (SE) and Z-score (Z), providing insight into each predictor's statistical significance and stability within the model. 'Yes vs. No' denotes the presence vs. absence of specific bacterial species.

Our findings indicate that *Pseudomonas* spp. exhibited a significant difference in occurrence rates between the two groups. Among patients with COVID-19, *Pseudomonas* spp. was identified in 16.7% of cases, compared to 5.7% in patients without COVID-19, with a *p*-value of 0.0281. This suggests a markedly higher prevalence of *Pseudomonas* spp. infections among patients suffering from COVID-19, highlighting a potential area of concern for managing secondary infections in these patients.

Table 5. Impact of COVID-19 on the prevalence of microbial germs in bile cultures.

	COVID (*n* = 30)	Without COVID (*n* = 211)	*p*-Value
Other bile germs			0.237 [1]
No	26.0 (86.7%)	196.0 (92.9%)	
Yes	4.0 (13.3%)	15.0 (7.1%)	
Acinetobacter spp.			0.605 [1]
No	29.0 (96.7%)	207.0 (98.1%)	
Yes	1.0 (3.3%)	4.0 (1.9%)	
Citrobacter spp.			0.555 [1]
No	28.0 (93.3%)	202.0 (95.7%)	
Yes	2.0 (6.7%)	9.0 (4.3%)	
Pseudomonas spp.			0.028 [1]
No	25.0 (83.3%)	199.0 (94.3%)	
Yes	5.0 (16.7%)	12.0 (5.7%)	
Enterobacter spp.			0.200 [1]
No	30.0 (100.0%)	200.0 (94.8%)	
Yes	0.0 (0.0%)	11.0 (5.2%)	
Enterococcus spp.			0.662 [1]
No	23.0 (76.7%)	169.0 (80.1%)	
Yes	7.0 (23.3%)	42.0 (19.9%)	
Klebsiella spp.			0.450 [1]
No	24.0 (80.0%)	180.0 (85.3%)	
Yes	6.0 (20.0%)	31.0 (14.7%)	
Streptococcus spp.			0.350 [1]
No	30.0 (100.0%)	205.0 (97.2%)	
Yes	0.0 (0.0%)	6.0 (2.8%)	
Escherichia coli			0.666 [1]
No	19.0 (63.3%)	142.0 (67.3%)	
Yes	11.0 (36.7%)	69.0 (32.7%)	

n—number of patients; [1] Proportions are evaluated with a chi-square test.

Other Gram-negative bacteria, such as *Escherichia coli* and *Klebsiella* spp., showed no statistically significant difference in prevalence between the two groups. *E. coli* was present in 36.7% of COVID-19 patients and 32.7% of non-COVID-19 patients, while *Klebsiella* spp. was found in 20.0% of COVID-19 patients compared to 14.7% in the control group.

Similarly, Gram-positive bacteria like *Enterococcus* spp. did not demonstrate significant differences in prevalence. *Enterococcus* spp. was observed in 23.3% of COVID-19 patients and 19.9% of non-COVID-19 patients.

4. Discussion

The present study offers a detailed analysis of AC's etiology, clinical characteristics, and microbial profiles in COVID-19 infection. Our research highlights notable differences in the presentation and outcomes of AC patients with and without COVID-19, providing informative observations into the pandemic's impact on the management and prognosis of biliary tract infections. The difference in mean age between COVID-19-positive and negative cohorts was statistically significant, with COVID-19 patients being notably older. This finding aligns with existing literature indicating that advanced age is a significant risk factor for severe COVID-19 outcomes [15–17]. Notably, gender distribution did not exhibit significant differences between the two groups, suggesting a lack of gender-specific predisposition to COVID-19 infection, consistent with previous findings [15,17].

Abdominal pain emerged as a significant symptom within the COVID-19-positive cohort, supporting emerging evidence that gastrointestinal manifestations, including abdominal pain, can be indicators of COVID-19 infection [16,18,19]. The presence of abdominal pain in COVID-19 patients warrants attention as a possible clinical marker for early detection and effective management strategies.

In contrast, no statistically significant difference was observed between the cohorts regarding fever. This finding contrasts with prior research emphasizing fever as a prevalent symptom in COVID-19 patients [16,17]. It underscores the heterogeneous nature of COVID-19's clinical presentation and emphasizes the importance of comprehensive symptom evaluation in the diagnostic process [20].

In our study, patients who tested positive for COVID-19 experienced a significantly longer hospital stay, averaging 13.5 (6.6) days compared to 7.9 (5.4) days without COVID-19. This extended hospitalization period highlights the additional healthcare burden imposed by COVID-19. A recent study focusing on individuals with decompensated liver cirrhosis and COVID-19 revealed that the significant influence of COVID-19 on patients with LC, particularly concerning organ failure, associated infections, hospitalization, and mortality, was expected to some extent [11]. The severity of symptoms does not just determine extended hospitalization in COVID-19 cases; it is influenced by several factors, such as compromised functional status, referrals from other hospitals, specific admission criteria, chronic health conditions, and the emergence of complications during the hospital stay. Prolonged hospital stays are associated with COVID-19 and not exclusively with AC progression, as observed with other diseases [21].

Regarding comorbidities in the COVID-19 group, Type 2 diabetes, gender, and age were not significant predictors of hospitalization duration. In our analysis within the COVID-19 group, type 2 diabetes showed a weak positive correlation with hospitalization duration. However, another study revealed that COVID-19 patients with diabetes had a substantially longer hospital stay and a markedly higher incidence of ICU admissions compared to non-diabetic patients [22]. This discrepancy may be because the sample size in our research is fairly limited.

To accentuate the impact of COVID-19 on other pathologies, a recent study on acute pancreatitis during the pandemic reported that patients had a threefold increase in relative death risk compared to those before the pandemic [23]. These findings highlight the severe impact of the pandemic on patient outcomes and the increased risks associated with concurrent SARS-CoV-2 infection. A notable aspect of the study was the analysis of bile cultures, revealing distinct bacterial profiles in COVID-19 patients compared to their non-COVID counterparts. Sterile cultures were more prevalent among COVID-19 patients with mild Tokyo severity, while non-COVID patients with severe cases had a higher proportion of sterile cultures. This differential pattern suggests COVID-19 may influence the biliary microbial environment, potentially through immune modulation or direct viral effects on biliary tissue.

During the examination of various bacterial strains, *Pseudomonas* spp. emerged as significantly associated with COVID-19 status, showing a fourfold increased likelihood of presence in 16.7% of cases of COVID-19 patients (Odds Ratio = 4.2923, $p = 0.018$) compared to 5.7% in patients without COVID-19. This association highlights the need for increased vigilance and possibly specialized antimicrobial strategies in managing cholangitis in COVID-19 patients. The presence of *Pseudomonas* spp. as a significant indicator of COVID-19 status may reflect the opportunistic nature of this pathogen in immunocompromised or critically ill patients, a category in which COVID-19 patients often fall. However, the literature review did not yield data corresponding to these findings regarding AC.

Most studies examining bacterial coinfections in COVID-19 patients faced limitations due to insufficient sample sizes, hindering the ability to detect outcome differences between those with and without bacterial coinfection. However, a recent comprehensive analysis focusing on bacterial coinfections upon admission found that patients with bacterial coinfection had longer hospital stays and increased in-hospital mortality compared to those without. While not frequently detected upon admission, bacterial infections often emerged during the prolonged hospitalization of patients, with prevalent pathogens including *Pseudomonas aeruginosa*, *Klebsiella* spp., and *S. aureus* [24]. Numerous international reports have documented a slight uptick in the incidence of *Pseudomonas aeruginosa* bacteremia during the COVID-19 pandemic [25].

Other bacteria, including *E. coli* and *Klebsiella* spp., did not show significant differences between the groups, suggesting that while these pathogens are common in AC, their prevalence is not necessarily influenced by COVID-19. The lack of significant differences in these common pathogens may indicate that standard prophylactic and therapeutic measures remain effective for these bacteria, regardless of COVID-19 status.

In a comprehensive multicenter observational study conducted by Gomi et al. in 2017 focusing on patients with acute cholangitis, *E. coli* emerged as the predominant organism detected in bile cultures [26]. Consistent with these findings, review studies have also highlighted the prevalence of coliform organisms, including *Escherichia coli* (25–50%), *Klebsiella* spp. (15–20%), and *Enterobacter* species (5–10%) as commonly identified bacteria in AC cases [27–29].

Comparing our findings to the TG18 for AC reveals several deviations. While the guidelines delineate cholangitis severity based on clinical criteria such as systemic inflammation, cholestasis, and imaging findings, our study suggests that COVID-19 status may influence the microbial profile of acute cholangitis cases. This influence leads to varying patterns of bacterial growth and culture sterility across severity grades [4].

The findings of this study have several clinical implications. Firstly, the increased age and prolonged hospitalization of COVID-19-positive cholangitis patients necessitate special consideration for resource allocation and management strategies in healthcare settings. The higher incidence of abdominal pain and the distinct bacterial profiles, particularly the prevalence of *Pseudomonas* spp., emphasize the need for specific clinical protocols and potentially more aggressive management strategies for co-infected patients.

Moreover, the analysis of the microbial profiles suggests that routine bile culture and sensitivity testing should be emphasized in COVID-19-positive cholangitis patients to guide appropriate antimicrobial therapy. The distinct microbial landscape in these patients could impose more precise and effective treatment regimens, potentially improving outcomes and reducing the length of hospital stays. The global health landscape has been significantly impacted by the COVID-19 pandemic, primarily due to the prevalence of severe respiratory illness. Nonetheless, emerging data suggests that COVID-19 can also lead to secondary sclerosing cholangitis, commonly called post-COVID-19 cholangiopathy. This rare yet serious complication manifests as inflammation and damage to the bile ducts following a bout of COVID-19 infection [30].

While comprehensive, this study has several limitations that should be addressed in future research. The relatively small sample size of COVID-19-positive patients may limit the generalizability of the findings. Additionally, the study's retrospective nature could introduce biases related to the accuracy of medical records and the diagnostic criteria used. Due to the retrospective design, it is challenging to ensure that all participants in the COVID-19 and non-COVID-19 groups had no previous infections before enrollment in the study. Our data is limited to what is recorded in our hospital's database, and we acknowledge that this may not capture the entire infectious history of the patients. Many patients might have been diagnosed or treated for COVID-19 in other hospitals or by their primary care physicians, and these records are not readily available to us. Additionally, in this study, we did not have detailed information on the COVID-19 vaccination status of all participants. This lack of data on vaccination status is another limitation that could potentially affect the results and interpretation of our study.

5. Conclusions

Our study results show that COVID-19 affected the duration of hospitalization for patients with acute cholangitis. Furthermore, this study presents observations regarding the impact of COVID-19 on acute cholangitis, revealing differences in microbial profiles. The advanced age and prolonged hospitalization of these patients demand a more precise approach, while the increased incidence of abdominal pain and distinct bacterial profiles, particularly the prevalence of *Pseudomonas* spp., indicate the importance of developing specific clinical protocols. These observations suggest that a more intensive and individualized treatment strategy may be necessary to improve outcomes in this vulnerable patient population.

Author Contributions: Conceptualization, D.V. and B.M.; methodology, D.V., C.B. and B.M.; software, C.B.; validation, D.V. and C.B.; formal analysis, I.R. and T.M.; investigation, I.R., B.M. and T.M.; resources, E.G., O.K. and A.C.M.; data curation, B.M. and A.C.M.; writing—original draft preparation, D.V.; writing—review and editing, B.M., E.G. and R.S.; visualization, A.C.M. and A.P.; supervision, B.M. and A.P.; project administration, D.V., B.M. and A.P. All authors have read and agreed to the published version of the manuscript.

Funding: We would like to acknowledge Victor Babes University of Medicine and Pharmacy Timisoara for covering the publication costs for this research paper.

Institutional Review Board Statement: The study was conducted according to the guidelines of the Declaration of Helsinki and approved by the Ethics Committee on 14 October 2022 (code I-27098).

Informed Consent Statement: Informed consent was obtained from all subjects involved in the study.

Data Availability Statement: Data are available upon request.

Conflicts of Interest: The authors declare no conflicts of interest.

References

1. Gromski, M.A.; Gutta, A.; Lehman, G.A.; Tong, Y.; Fogel, E.L.; Watkins, J.L.; Easler, J.J.; Bick, B.L.; McHenry, L.; Beeler, C.; et al. Microbiology of bile aspirates obtained at ERCP in patients with suspected acute cholangitis. *Endoscopy* **2022**, *54*, 1045–1052. [CrossRef]
2. Khashab, M.A.; Tariq, A.; Tariq, U.; Kim, K.; Ponor, L.; Lennon, A.M.; Canto, M.I.; Gurakar, A.; Yu, Q.; Dunbar, K.; et al. Delayed and unsuccessful endoscopic retrograde cholangiopancreatography are associated with worse outcomes in patients with acute cholangitis. *Clin. Gastroenterol. Hepatol. Off. Clin. Pract. J. Am. Gastroenterol. Assoc.* **2012**, *10*, 1157–1161. [CrossRef] [PubMed]
3. Miura, F.; Okamoto, K.; Takada, T.; Strasberg, S.M.; Asbun, H.J.; Pitt, H.A.; Gomi, H.; Solomkin, J.S.; Schlossberg, D.; Han, H.S.; et al. Tokyo Guidelines 2018: Initial management of acute biliary infection and flowchart for acute cholangitis. *J. Hepato-Biliary-Pancreat. Sci.* **2018**, *25*, 31–40. [CrossRef]
4. Kiriyama, S.; Kozaka, K.; Takada, T.; Strasberg, S.M.; Pitt, H.A.; Gabata, T.; Hata, J.; Liau, K.H.; Miura, F.; Horiguchi, A.; et al. Tokyo Guidelines 2018: Diagnostic criteria and severity grading of acute cholangitis (with videos). *J. Hepato-Biliary-Pancreat. Sci.* **2018**, *25*, 17–30. [CrossRef]
5. Gomi, H.; Solomkin, J.S.; Schlossberg, D.; Okamoto, K.; Takada, T.; Strasberg, S.M.; Ukai, T.; Endo, I.; Iwashita, Y.; Hibi, T.; et al. Tokyo Guidelines 2018: Antimicrobial therapy for acute cholangitis and cholecystitis. *J. Hepato-Biliary-Pancreat. Sci.* **2018**, *25*, 3–16. [CrossRef] [PubMed]
6. Gralnek, I.M.; Hassan, C.; Beilenhoff, U.; Antonelli, G.; Ebigbo, A.; Pellisè, M.; Arvanitakis, M.; Bhandari, P.; Bisschops, R.; Van Hooft, J.E.; et al. ESGE and ESGENA Position Statement on gastrointestinal endoscopy and the COVID-19 pandemic. *Endoscopy* **2020**, *52*, 483–490. [CrossRef]
7. Lantinga, M.A.; Theunissen, F.; Ter Borg, P.C.J.; Bruno, M.J.; Ouwendijk, R.J.T.; Siersema, P.D. Impact of the COVID-19 pandemic on gastrointestinal endoscopy in the Netherlands: Analysis of a prospective endoscopy database. *Endoscopy* **2021**, *53*, 166–170. [CrossRef]
8. Guan, W.J.; Ni, Z.Y.; Hu, Y.; Liang, W.H.; Ou, C.Q.; He, J.X.; Liu, L.; Shan, H.; Lei, C.L.; Hui, D.S.C.; et al. Clinical Characteristics of Coronavirus Disease 2019 in China. *N. Engl. J. Med.* **2020**, *382*, 1708–1720. [CrossRef]
9. Donato, G.; Forti, E.; Mutignani, M.; Laterra, M.A.; Arese, D.; Coppola, F.; Zaccari, P.; Mariani, A.; Arcidiacono, P.G.; Pigò, F.; et al. A multicenter survey on endoscopic retrograde cholangiopancreatography during the COVID-19 pandemic in northern and central Italy. *Endosc. Int. Open* **2021**, *9*, E629–E634. [CrossRef] [PubMed] [PubMed Central]
10. Ikemura, M.; Tomishima, K.; Ushio, M.; Takahashi, S.; Yamagata, W.; Takasaki, Y.; Suzuki, A.; Ito, K.; Ochiai, K.; Ishii, S.; et al. Impact of the Coronavirus Disease-2019 Pandemic on Pancreaticobiliary Disease Detection and Treatment. *J. Clin. Med.* **2021**, *10*, 4177. [CrossRef] [PubMed] [PubMed Central]
11. Moga, T.V.; Foncea, C.; Bende, R.; Popescu, A.; Burdan, A.; Heredea, D.; Danilă, M.; Miutescu, B.; Ratiu, I.; Bizerea-Moga, T.O.; et al. Impact of COVID-19 on Patients with Decompensated Liver Cirrhosis. *Diagnostics* **2023**, *13*, 600. [CrossRef] [PubMed] [PubMed Central]
12. Spearman, C.W.; Aghemo, A.; Valenti, L.; Sonderup, M.W. COVID-19 and the liver: A 2021 update. *Liver Int. Off. J. Int. Assoc. Study Liver* **2021**, *41*, 1988–1998. [CrossRef]
13. Zhang, Y.; Zheng, L.; Liu, L.; Zhao, M.; Xiao, J.; Zhao, Q. Liver impairment in COVID-19 patients: A retrospective analysis of 115 cases from a single centre in Wuhan city, China. *Liver Int. Off. J. Int. Assoc. Study Liver* **2020**, *40*, 2095–2103. [CrossRef]
14. Zhang, C.; Shi, L.; Wang, F.S. Liver injury in COVID-19: Management and challenges. *Lancet Gastroenterol. Hepatol.* **2020**, *5*, 428–430. [CrossRef]

15. Sharma, A.; Jaiswal, P.; Kerakhan, Y.; Saravanan, L.; Murtaza, Z.; Zergham, A.; Honganur, N.S.; Akbar, A.; Deol, A.; Francis, B.; et al. Liver disease and outcomes among COVID-19 hospitalized patients—A systematic review and meta-analysis. *Ann. Hepatol.* **2021**, *21*, 100273. [CrossRef]
16. Lin, L.; Jiang, X.; Zhang, Z.; Huang, S.; Zhang, Z.; Fang, Z.; Gu, Z.; Gao, L.; Shi, H.; Mai, L.; et al. Gastrointestinal symptoms of 95 cases with SARS-CoV-2 infection. *Gut* **2020**, *69*, 997–1001. [CrossRef]
17. Bastug, A.; Bodur, H.; Erdogan, S.; Gokcinar, D.; Kazancioglu, S.; Kosovali, B.D.; Ozbay, B.O.; Gok, G.; Turan, I.O.; Yilmaz, G.; et al. Clinical and laboratory features of COVID-19: Predictors of severe prognosis. *Int. Immunopharmacol.* **2020**, *88*, 106950. [CrossRef] [PubMed] [PubMed Central]
18. Cha, M.H.; Regueiro, M.; Sandhu, D.S. Gastrointestinal and hepatic manifestations of COVID-19: A comprehensive review. *World J. Gastroenterol.* **2020**, *26*, 2323–2332. [CrossRef]
19. Pan, L.; Mu, M.; Yang, P.; Sun, Y.; Wang, R.; Yan, J.; Li, P.; Hu, B.; Wang, J.; Hu, C.; et al. Clinical Characteristics of COVID-19 Patients With Digestive Symptoms in Hubei, China: A Descriptive, Cross-Sectional, Multicenter Study. *Am. J. Gastroenterol.* **2020**, *115*, 766–773. [CrossRef]
20. Zhai, L.L.; Xiang, F.; Wang, W.; Wu, L.; Ye, L.; Yao, L.C.; Tang, Z.G. Atypical presentations of coronavirus disease 2019 in a patient with acute obstructive suppurative cholangitis. *Clin. Res. Hepatol. Gastroenterol.* **2020**, *44*, e135–e140. [CrossRef] [PubMed] [PubMed Central]
21. Lucijanic, M.; Marelic, D.; Stojic, J.; Markovic, I.; Sedlic, F.; Kralj, I.; Rucevic, D.; Busic, N.; Javor, P.; Lucijanic, T.; et al. Predictors of prolonged hospitalization of COVID-19 patients. *Eur. Geriatr. Med.* **2023**, *14*, 511–516. [CrossRef]
22. Mohamed, Y.S.; Mukhtar, M.; Elmalti, A.; Kheirallah, K.; Panigrahi, D.; Abu-Rish, E.Y.; Bani, I.; Nasor, E.M.; Ahmed, W.; Alzoubi, A. Hospital Mortality and Morbidity in Diabetic Patients with COVID-19: A Retrospective Analysis from the UAE. *Int. J. Environ. Res. Public Health* **2024**, *21*, 697. [CrossRef] [PubMed]
23. Rădulescu, P.M.; Căluianu, E.I.; Traşcă, E.T.; Mercuţ, D.; Georgescu, I.; Georgescu, E.F.; Ciupeanu-Călugăru, E.D.; Mercuţ, M.F.; Mercuţ, R.; Padureanu, V.; et al. The Impact of the COVID-19 Pandemic on Outcomes in Acute Pancreatitis: A Propensity Score Matched Study Comparing before and during the Pandemic. *Diagnostics* **2023**, *13*, 2446. [CrossRef]
24. Westblade, L.F.; Simon, M.S.; Satlin, M.J. Bacterial Coinfections in Coronavirus Disease 2019. *Trends Microbiol.* **2021**, *29*, 930–941. [CrossRef]
25. Ng, Q.X.; Ong, N.Y.; Lee, D.Y.X.; Yau, C.E.; Lim, Y.L.; Kwa, A.L.H.; Tan, B.H. Trends in *Pseudomonas aeruginosa* (*P. aeruginosa*) Bacteremia during the COVID-19 Pandemic: A Systematic Review. *Antibiotics* **2023**, *12*, 409. [CrossRef] [PubMed] [PubMed Central]
26. Gomi, H.; Takada, T.; Hwang, T.L.; Akazawa, K.; Mori, R.; Endo, I.; Miura, F.; Kiriyama, S.; Matsunaga, N.; Itoi, T.; et al. Updated comprehensive epidemiology, microbiology, and outcomes among patients with acute cholangitis. *J. Hepato-Biliary-Pancreat. Sci.* **2017**, *24*, 310–318. [CrossRef]
27. Lee, C.C.; Chang, I.J.; Lai, Y.C.; Chen, S.Y.; Chen, S.C. Epidemiology and prognostic determinants of patients with bacteremic cholecystitis or cholangitis. *Am. J. Gastroenterol.* **2007**, *102*, 563–569. [CrossRef]
28. Rhodes, A.; Evans, L.E.; Alhazzani, W.; Levy, M.M.; Antonelli, M.; Ferrer, R.; Kumar, A.; Sevransky, J.E.; Sprung, C.L.; Nunnally, M.E.; et al. Surviving Sepsis Campaign: International Guidelines for Management of Sepsis and Septic Shock: 2016. *Intensive Care Med.* **2017**, *43*, 304–377. [CrossRef] [PubMed]
29. Miuţescu, B.; Vuletici, D.; Burciu, C.; Turcu-Stiolica, A.; Bende, F.; Raţiu, I.; Moga, T.; Sabuni, O.; Anjary, A.; Dalati, S.; et al. Identification of Microbial Species and Analysis of Antimicrobial Resistance Patterns in Acute Cholangitis Patients with Malignant and Benign Biliary Obstructions: A Comparative Study. *Medicina* **2023**, *59*, 721. [CrossRef] [PubMed]
30. Tafreshi, S.; Whiteside, I.; Levine, I.; D'Agostino, C. A case of secondary sclerosing cholangitis due to COVID-19. *Clin. Imaging* **2021**, *80*, 239–242. [CrossRef]

medicina

MDPI

Article

National Trends in the Incidence of Sporadic Malignant Colorectal Polyps in Young Patients (20–49 Years): An 18-Year SEER Database Analysis

Mark M. Aloysius [1], Tejas Nikumbh [2,*], Lekha Yadukumar [2], Udit Asija [2], Niraj J. Shah [3], Ganesh Aswath [1], Savio John [1] and Hemant Goyal [4,*]

1 Division of Gastroenterology, Department of Medicine, State University of New York Upstate Syracuse, New York, NY 13210, USA; vamedicalresident@gmail.com (M.M.A.)
2 Department of Internal Medicine, The Wright Center for Graduate Medical Education, Scranton, PA 18505, USA; yadukumarl@thewrightcenter.org (L.Y.); asijau@thewrightcenter.org (U.A.)
3 Division of Gastroenterology, University of Mississippi Medical Center, Jackson, MS 39216, USA
4 Advanced Endoscopy, Borland Groover Owntown Office, Jacksonville, FL 32207, USA
* Correspondence: drtejasnikumbh@gmail.com (T.N.); doc.hemant@yahoo.com (H.G.)

Abstract: *Background and Objectives*: Conflicting guidelines exist for initiating average-risk colorectal cancer screening at the age of 45 years. The United States Preventive Services Task Force (USPSTF) changed its guidelines in 2021 to recommend initiating screening at 45 years due to an increasing incidence of young-onset colorectal cancer. However, the American College of Physicians (ACP) recently recommended not screening average-risk individuals between 45 and 49 years old. We aim to study the national trends in the incidence of sporadic malignant polyps (SMP) in patients from 20 to 49 years old. *Materials and Methods*: We analyzed the Surveillance, Epidemiology, and End Results database (2000–2017) on patients aged 20–49 years who underwent diagnostic colonoscopy with at least a single malignant sporadic colorectal polyp. *Results*: Of the 10,742 patients diagnosed with SMP, 42.9% were female. The mean age of incidence was 43.07 years (42.91–43.23, 95% CI). Approximately 50% of malignant polyps were diagnosed between 45 and 49 years of age, followed by 25–30% between 40 and 45. There was an upward trend in malignant polyps, with a decreased incidence of malignant villous adenomas and a rise in malignant adenomas and tubulovillous adenomas. *Conclusions*: Our findings suggest that almost half of the SMPs under 50 years occurred in individuals under age 45, younger than the current screening threshold recommended by the ACP. There has been an upward trend in malignant polyps in the last two decades. This reflects changes in tumor biology, and necessitates further research and support in the USPSTF guidelines to start screening at the age of 45 years.

Keywords: young-onset colorectal cancer; sporadic malignant polyps; adenomatous adenomas; malignant tubulovillous adenomas; SEER database

1. Introduction

Colorectal carcinoma (CRC) is the third most common cancer worldwide and the second most common cause of cancer death. More than 1.9 million new colorectal cancer cases and 935,000 deaths were estimated to occur in 2020, representing about 1 in 10 cancer cases and deaths [1]. CRC is commonly diagnosed in older people, and the median age of diagnosis is 67 years [2]. Early-onset CRC, also known as young-onset colorectal carcinoma (YoCRC), is defined as CRC diagnosed in individuals younger than the age of 50 years who did not previously meet the traditional age criteria for average-risk screening in the United States [3,4]. In 2023, approximately 153,020 people will be diagnosed with CRC, and 52,550 will die from the disease, with 19,550 cases and 3750 deaths occurring in people under the age of 50 [5]. Over the last ten years, the incidence of CRC has been steadily increasing, and there is an increased annual incidence of YoCRC. The incidence of YoCRC has nearly doubled since 1990, with an increase of 2% per year, mainly in Western countries [6–8]. Based on current trends, it is anticipated that the incidence rates of cancers of the colon and rectum will rise by 90% and 124.2%, respectively, among people in the 20–34 age group, while they will increase by 27.7% and 46.0%, respectively, among patients in the 35–39 age group [6,9]. YoCRC is now the second and fourth most common cause of cancer in men and women under 50 years in the US. The death rate from YoCRC has been progressively increasing, with an estimated 12 deaths per 100,000 as of 2019 [10].

Young-onset colorectal cancer is associated with a number of risk factors, such as obesity, processed meat consumption, alcohol intake, family history, inflammatory bowel disease, genetic predisposition syndromes, and disruption of the gut microbiome [3,11]. Despite one-third of YoCRC being associated with familial risk factors, the majority of YoCRCs do not have associated hereditary syndromes and are linked with microsatellite instability. YoCRCs are more likely to present as advanced-stage III/IV carcinomas, and more frequently display aggressive histological characteristics such as poor differentiation and perineural and blood vessel invasion [12–14]. Patients with YoCRC also have a significantly longer median time to diagnosis, symptom duration, and time of evaluation. The time to diagnosis is 1.4 times longer for younger than older patients [14]. The definitive pathogenesis and molecular profile of YoCRC are not well understood, and there are few studies that have addressed this.

The increase in the incidence of YoCRC, together with the significant proportion of cases with sporadic characteristics within this subgroup of CRC, have led multiple societies to recommend starting regular screening by a stool-based test or colonoscopy at age 45 for people at an average risk for CRC. However, conflicting guidelines exist in terms of initiating average-risk colorectal cancer screening at age 45 years. The US Preventive Services Task Force changed its guidelines in 2021 to recommend initiating screening at 45 years due to the increasing incidence of young-onset colorectal cancer [15]. However, the American College of Physicians (ACP) recently recommended not to screen average-risk individuals between 45 and 49 years [16]. The American Cancer Society (ACS) published recommendations for CRC screening for average-risk adults starting at age 45 in May 2018 [17]. The US Preventive Services Task Force (USPSTF) expanded its recommendations for CRC screening guidelines in 2021 to include adults aged 45–49 [15,18]. This was followed by the American College of Gastroenterology (ACG) amending their CRC screening guidelines to begin from 45 years for average-risk individuals [19].

The initiation of screening at age 45 instead of 50 years added 19 million average-risk people to the screening pool and dropped national CRC screening rates for those 50 and older from 68% to 59% [5]. No specific changes in screening recommendations were made for people at higher-than-average risk by the ACS, USPSTF, or ACG. The USPSTF recommends beginning screening for those with a family history of CRC 10 years before the age of the youngest affected relative's diagnosis or age 40, whichever is earlier [20]. The NCCN guidelines recommend genetic screening for patients with young-onset CRC [21]. It is of value to note that only half of young-onset CRC patients with germline mutations have a history of CRC in a first-degree relative, despite the fact that family history is frequently used to screen for elevated CRC risk [22].

Owing to the lack of screening in younger patients, the incidence of YoCRC reflects diagnostically detected CRCs in symptomatic patients or inherently high-risk patients. Data about the new case trends in sporadic malignant polyps (SMP) in younger patients are limited, given the lack of established screening in those at an average risk of CRC under the age of 45. The lack of uniformity in guidelines prevents definitive management and screening in the 45 to 49 years age group population. Hence, we aimed to study the trends in new cases of sporadic malignant polyps in patients between 20 and 49 years of age over an 18-year period from the National Cancer Institute's SEER database.

2. Materials and Methods

2.1. Study Design

The Surveillance, Epidemiology, and End Results (SEER) Program of the National Cancer Institute (NCI) is an authoritative source of information on cancer incidence and survival in the United States (U.S.). The SEER database contains cancer incidences dating back four decades, from 1975 [23]. It was launched on 1 January 1973 as a part of the National Cancer Act. Initially, 7 registries (SEER 7) with epidemiologically significant populations that consisted of racial and ethnic minorities were included, and this was gradually enlarged to the current 22 cancer registries (SEER 22). SEER collects demographic, clinical, and outcome information on all cancers diagnosed in representative geographic regions and subpopulations. Data are obtained for all primary invasive malignancies and some additional diagnoses, such as in situ carcinomas, and include the date of diagnosis, as well as demographic data such as age, gender, race/ethnicity, and county of residence [24]. Appropriate use of the SEER database can ensure that the correct research conclusions are drawn and maximize the benefits for clinicians and patients [25].

SEER is widely regarded as the gold standard for data quality in US and international cancer registries. For cancer registries worldwide, the SEER Program serves as a model due to its emphasis on quality control from the program's beginning, its long-standing commitment to representing all population segments, and its recent success in funding research advancements. Contractual arrangements with regional registries ensure quality, and SEER standards must be met before data are transferred [24]. Originally, there were only 9 tumor registries, and now there are 22 US geographic areas participating in the SEER program. Recently, the SEER Program has been moving toward more automation to improve its consistency and reduce delays in cancer reporting. It currently collects and publishes cancer incidence and survival data from population-based cancer registries, covering approximately 41.9 percent of the U.S. population as per the 2020 census, and has the largest geographic coverage available for survival. SEER 17 (accessed in November 2022) includes 17 cancer registries that collect cancer incidence data from various geographic regions of the U.S. SEER 17 contains 1 record for each of 9,208,295 tumors.

The American Community Survey (ACS) is an ongoing community-based survey that has been conducted by the US Census Bureau since 2005. The ACS is the primary source of high-resolution geographic data on the U.S. population. It provides vital demographic, housing, and socioeconomic information, including employment, migration, and disability information about the US population. These community-level indicators, which include occupation, income, and education, have been linked to a number of health outcomes, including life expectancy, self-reported health, chronic conditions, certain types of cancer, mental disorders, cardiovascular disease, obesity, and infant mortality [26]. In fact, studies have assessed the feasibility of linking patient data from HER to microdata from the ACS, with the goal of improving the understanding health disparities and social determinants of health in the population [27]. The ACS denominators generally perform comparably well and yield estimates with little bias [28]. We utilized information from the ACS as a measure for the current population to derive the incidence rates of malignant colorectal polyps. The incidence rates were calculated using a population specific to the particular year as a denominator, per 100,000 people. Because these are de-identified datasets, the study was exempted from review/approval by the Wright Center for Graduate Medical Education's institutional review board.

2.2. Patient Selection

All patients diagnosed with at least a single malignant colorectal polyp on colonoscopies performed for any indication or screening between 2000 and 2017 were eligible for the analysis. All these patients were stratified according to the 6th edition of AJCC, thus ensuring uniformity in staging. We included data from patients aged between 20 and 49 years, diagnosed over the 18-year time period. Malignant colorectal polyps found during colonoscopy in the entire colon, including the rectum, were included, however, the SEER database does not differentiate data between right vs. left colon or rectum. Most of these patients underwent diagnostic colonoscopies, since screening colonoscopy began at age 50 years prior to 2018, after which, the USPSTF guidelines changed. We excluded familial cancers, lesions with a high microsatellite instability (MSI-H), and Adenomatous polyposis coli (APC).

2.3. Statistical Analysis

We accessed the data using the SEER diagnostic codes 8210/2 and 3 for tubular adenomas, 8221/2 for serrated polyps, 8221/2 and 3 for multiple adenomatous polyps, 8261/3 and 8262/3 for villous adenomas, and 8263/2 and 3 for tubulovillous adenomas. We performed a descriptive statistical analysis using the SPSS v27 Macintosh(SPSS v27, IBM Corporation, Armonk, NY, USA). The incidence rates per 100,000 people were calculated using data from the American Community Survey [29]. CRC incidence was stratified based on gender and histology.

3. Results

Between 2000 and 2017, a total of 10,742 patients with biopsy-proven sporadic malignant colorectal polyps between the ages of 20 and 49 years were included. The cohort consisted of 57.1% men and 42.9% women (Table 1).

Table 1. Gender distribution and histological trend over time (2000–2017) of malignant colorectal polyps in patients aged 20–49 years.

	2000	2001	2002	2003	2004	2005	2006	2007	2008	2009	2010	2011	2012	2013	2014	2015	2016	2017
Female	226	235	242	242	223	235	282	307	293	329	355	315	346	332	342	315	332	340
Male	245	233	266	250	247	268	286	291	356	349	325	365	341	316	379	356	389	306
Adenocarcinoma in adenomatous polyp	169	152	191	166	184	179	239	237	245	270	247	283	281	276	309	298	280	275
Serrated adenocarcinoma	0	0	0	0	0	1	0	0	1	0	0	0	0	2	0	0	1	2
Adenocarcinoma in multiple adenomatous polyp	1	3	2	4	3	6	1	3	1	0	2	1	3	4	2	0	1	0
Adenocarcinoma in villous adenoma	108	112	101	101	91	95	93	101	101	96	91	79	72	61	64	59	49	46
Adenocarcinoma in tubulovillous adenoma	193	212	214	221	192	222	244	257	301	312	340	317	331	305	346	314	390	323

The average age of malignant colorectal polyp diagnosis was 43.07 years (average age range for CRC diagnosis: 42.92–43.23 years). The annual mean age (95% CI) of new-onset malignant polyps trending over time is visualized in Figure 1. The majority of sporadic malignant polyps were diagnosed in patients aged 45–49 years, with 50% diagnosed between the ages of 45 and 49 and 25–30% diagnosed between the ages of 40 and 44.

The trends of incidence of sporadic malignant polyps were compared with local-stage CRC cases, which is depicted in Figure 2. SEER defines local stage as a malignancy limited to the organ of origin; no spread beyond the organ of origin; and infiltration past the basement membrane of the epithelium into the stroma of the organ. This roughly corresponds to stage T1N0M0, as per the TNM cancer staging classification for CRC. The incidence was calculated using the number of cases of sporadic malignant polyps from the SEER database and divided by the population during the same year in the same age group (20–49 years) obtained from the ACS. Thus, the incidence rate was per 100,000 people for that particular year. Since 2013, there has been a noticeable rise in cases of local-stage CRC, while the incidence of sporadic malignant colorectal polyps remained constant from 2010 to 2019 (~0.5 cases/10,000 population). Table 1 lists the histological subtypes and gender distribution of sporadic malignant colorectal polyps. There has been a significant decrease in new cases of malignant villous adenomas. However, the incidence of malignant tubular adenomatous polyps and tubulovillous adenomas has significantly increased over time. Serrated adenocarcinomas in serrated polyps were reported as very few in the population studied.

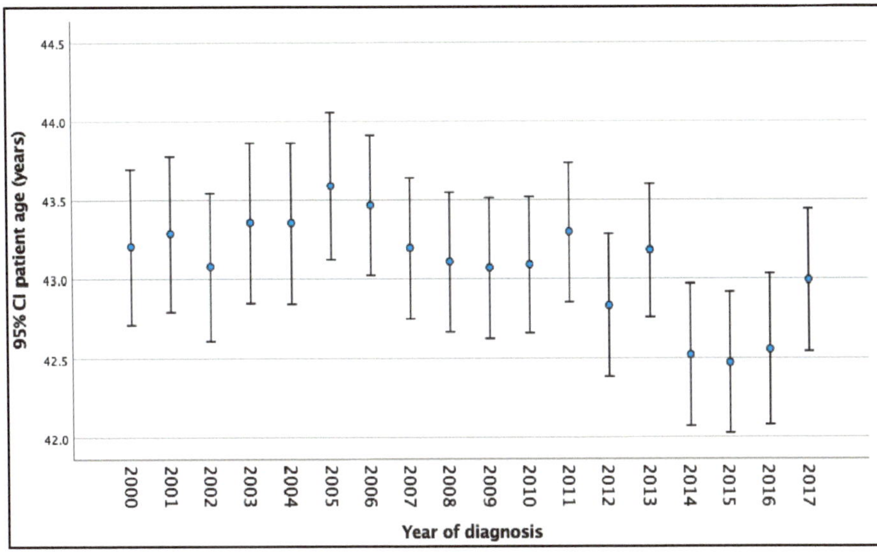

Figure 1. Age of malignant colorectal polyp incidence over time in 20- to 50-year age group.

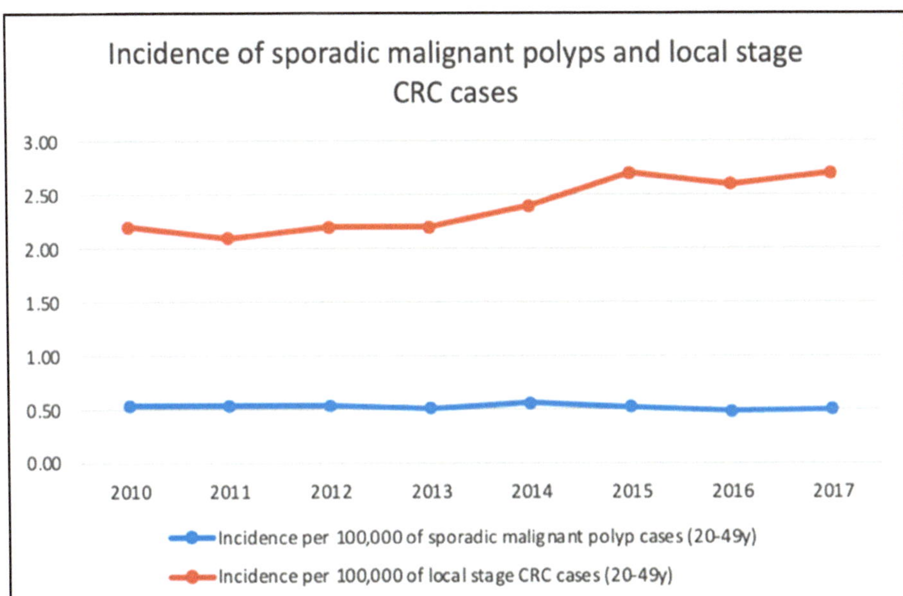

Figure 2. Observed incidence per 100,000 of sporadic malignant polyps and local-stage CRC cases.

4. Discussion

We looked at data from colonoscopies performed on symptomatic patients across the country. The objective of the present study was to analyze the incidence and trends of young-onset CRC between 2000 and 2017. To the best of our knowledge, this is the largest known cohort to date, where we analyzed 10,742 cases of sporadic young-onset CRC over a period of 18 years across the United States.

One in five patients diagnosed with CRC under 50 years of age has a genetic predisposition syndrome [11]. There are multi-society guidelines to begin early screening in these individuals [30,31]. However, the detection of cancer among the other 80 percent of patients poses a considerable challenge, since there is no family history to advocate for early screening in this group. Conflicting guidelines exist in terms of initiating average-risk colorectal cancer screening at age 45 years. The USPSTF changed its guidelines in 2021 to recommend initiating screening at 45 years due to the increasing incidence of young-onset colorectal cancer [15]. It expanded its recommendations for CRC screening guidelines in 2021 to include adults aged 45–49 [15,18]. The ACP recently recommended not to screen average-risk individuals between 45 and 49 years [16]. This ACP guideline was based on the absence of direct evidence that screening younger individuals reduces CRC incidence or mortality [16,32].

Of note is that early experiences with screening in the 45 to 49 years group show similar rates of neoplasia as those in the 50 to 55 years group of patients [33]. Lowering the starting age of population screening for sporadic CRC to 45 years also seems to be cost-effective. Our data support screening for CRC in the younger population. The average age of CRC diagnosis was 43.07 years (range: 42.92–43.23), which is well below the ACP recommendation to start screening at 50 years. In fact, it is also lower than the USPSTF recommendation to start CRC screening at 45 years. This lack of uniformity prevents having definitive guidelines for screening in the 45 to 49 years population.

According to studies conducted around the world, the incidence of CRC is increasing in people under 50 and at a slower rate in people over 50 [34–36]. The total incidence of YoCRC in the United States and the number of cases of advanced-stage colorectal carcinomas increased twofold between 1990 and 2013, according to a study analyzing the SEER database [37]. While older research has established that diabetes and obesity are risk factors in the younger population [38], Austin et al. studied mean body mass index (BMI) increases across age groups, which was unable to explain the selective increase in incidence amongst younger age groups compared to the older population [39].

Our study population comprised 42.9% females, and showed increased incidence in both genders. Although there was no significant gender preponderance in our study, it was seen that, during some years, females were diagnosed with a higher number of cases of YoCRC. Our results are consistent with those of Lall et al., who found that young females are more susceptible to CRCs [40]. In contrast, some cohorts revealed a marginally higher incidence of YoCRC in men compared to women, with women consistently having fewer cases than men, while other studies found no difference between the two genders over the past three decades [37,41]. It has been studied that gender and body mass index (BMI) are associated with CRC diagnosis at a younger age, and there is a linear relationship between BMI and YoCRC [42,43]. Men are more likely than women to be diagnosed with CRC according to global trends; this has been attributed to a number of factors, including the frequency of colonoscopies, race, socioeconomic status, and insurance coverage [41,44].

The mean age at presentation and diagnosis of sporadic malignant polyps in our study was 43 years (42.9–43.23, 95% CI). As shown in Figure 1, the average age at which malignant polyps is diagnosed has decreased since 2014. The majority of malignant polyps (50%) were discovered in patients between the ages of 45 and 49, with 25 to 30 percent discovered in patients between the ages of 40 and 44. The SEER study by Wang et al. found a similar pattern, with 27% of cases diagnosed between 40 and 45 and 47% between 45 and 49 [37]. According to Abualkhair et al., 95.1% of CRCs diagnosed between the ages of 45 and 50 are invasive, with 46% increased incidence rates in 1 year age transitions [45]. The patients in our study who underwent colonoscopies had symptoms that persisted and necessitated diagnostic scopes. These lesions would have been discovered earlier had screening been performed at an earlier age.

Our analysis revealed a stable trend in overall local-stage CRCs over time, however, there has been a steady increase in the subset of sporadic malignant polyps (Figure 2). This is an interesting trend which implies that, while the number of CRCs accounted for by sporadic malignant polyps has remained steady, there has been a rise in polyps that are going undetected and progressing to local-stage CRC at presentation. This could mean an increase in the incidence of adenomas in the younger age group which progress to CRC before they are detected. Patients in the age group of 20–49 years have not been subject to CRC screening until recently, when the minimum age for cancer screening was decreased from 50 to 45 years by the USPSTF in 2019 [46]. This could also mean that there are 'alternative' pathways other than the traditional adenoma–carcinoma sequence of CRC which are at play. Recently, there has been increasing interest in and evidence in favor of the serrated carcinogenesis pathway [47]. Serrated polyps are the second most common type of polyp identified during a colonoscopy (after conventional adenomas). Approximately 15–30% of CRCs arise from the serrated polyp pathway. Evidence suggests that serrated polyp subtypes, especially traditional serrated adenoma (TSA) and sessile serrated adenoma/polyp (SSA/P), can cause adenocarcinoma via the serrated route. Additionally, the data indicate that SSA/Ps are the precursors of CRC through microsatellite instability (MSI) and could rapidly progress to malignancy [48]. Recent data from surveillance colonoscopies after the development of YoCRC have shown that the absence or presence of polyps is an important prognostic factor. The development of polyps during surveillance shows that it is necessary to extend the follow-up time, even in cases with microsatellite-stable YoCRC [49].

Over the course of this 18-year analysis, histopathology trends showed an increase in malignant tubular and tubulovillous adenomas and a gradual decline in malignant villous adenomas. YoCRC patients are known to be more likely to develop poorly differentiating aggressive forms of CRC, which have worse prognoses [38,50–52]. According to a study by Abualkhair et al. that spanned over 15 years, the incidence of localized CRC increased by 75.9%, that of regional CRC increased by 30%, and that of advanced CRC increased by 15.7% between the ages of 49 and 50 [45]. Similar research found that 90% of polyps were tubular adenomas with low-grade dysplasia, with tubulovillous adenomas with low-grade dysplasia making up 8% of these polyps [40]. Seven serrated adenocarcinomas in serrated polyps were found in our study during this period. Serrated adenomas are frequently found in the proximal colon and are challenging to see without improved colonoscopic techniques [53]. Vogelstein et al. described the pathogenesis of YoCRC to be based on an adenoma to carcinoma sequence [54]. However, this cannot explain genetic etiologies of CRC, as these appear phenotypically different from old CRC [55]. Changes in genotypic driver mutations in the adenoma–carcinoma sequence have been linked to phenotypic profile changes found in YoCRC. Many genes identified to have caused mutations in YoCRC are not commonly seen in the presumed adenoma–carcinoma sequencing. Further research should focus on younger patients with sporadic malignant colorectal polyps [56,57].

It is important to consider the fact that gastrointestinal diseases are responsible for considerable healthcare use and expenditure. In 2018 alone, gastrointestinal healthcare expenditures totaled USD 119.6 billion in the United States [58]. Increasing awareness about early CRC detection through regular screening is an effective strategy to reduce the economic burden of CRC [59]. In the past 5 years, the USPTF, ACG, and ACS updated their screening recommendations for CRC from the age of 50 to start at the age of 45 years [17–19]. The biggest challenge in the near future will be to improve screening in the newly eligible, those overdue and unscreened, and reduce barriers to cancer care [32]. Since this modification to standard CRC screening, there have not been enough studies looking at the prevalence of YoCRC. Our study examined data collected prior to 2018, when screening recommendations changed and colonoscopies were performed in response to symptomatic presentations. We noticed a significant increase in the incidence of YoCRC in patients aged 45–49, as well as those aged 40–45 years. It is known that these polyps might have existed years before the age of screening; if these cases had been screened for earlier, the majority of diagnoses of YoCRC might have been avoided [60]. We also looked at changes in the histopathological trends of young-onset malignant polyps, which are probably the result of evolving genotypic mutations. Additionally, there is a need for more studies to better understand the pathogenesis and evolution of YoCRC.

Among our study's strengths is that SEER remains widely regarded as the gold standard for data quality among cancer registries in the United States and around the world. We leveraged the strengths of the SEER program with regard to representativeness and generalizability to the U.S. population, the lengthy period of data collection, the large numbers of cases, and the collection of cancer specific outcomes. The registry includes patients treated in a wide range of practice settings and currently represents 42% of the US population. SEER's stringent protocols and quality control ensure data reliability and accuracy. The SEER database was used to provide epidemiological trends in sporadic malignant polyps of young onset in our study. We calculated the incidence per 100,000 people in the United States to provide a standardized denominator for comparison.

Since this is a population-based retrospective study, it has some inherent limitations. The SEER database has an inclusion bias, with limited staging and metastatic data recorded. Because the database only collects malignant data, we were unable to include any premalignant lesions. SEER only reports first treatment interventions. As a result, there was no information on the subsequent recurrence of malignant polyps. However, recurrence data should not be used to calculate the incidence of primary sporadic malignant polyps. There is also a lack of information on comorbidities. Furthermore, detailed information regarding treatment and prognosis is lacking. Nonetheless, the study remains convincing, given the large demographics.

5. Conclusions

Malignant sporadic polyps account for one-quarter of all localized CRCs diagnosed in people under the age of 50. Significant upward trends in sporadic malignant polyps were observed over time, possibly reflecting changes in tumor biology, necessitating further research. The majority of these sporadic malignant polyps were diagnosed between 40 and 49 years. These findings may have potential implications for future CRC screening strategies in younger patients. Improving screening in the newly eligible population within a framework of health equity and reducing barriers to care remains important to further reduce the burden of CRC.

Author Contributions: Conceptualization, M.M.A. and H.G.; methodology, N.J.S. and H.G.; validation, M.M.A.; formal analysis, M.M.A.; data curation, M.M.A.; writing—original draft preparation, L.Y., U.A. and T.N.; writing—review and editing, T.N., S.J. and G.A. All authors have read and agreed to the published version of the manuscript.

Funding: This research received no external funding.

Institutional Review Board Statement: The study was conducted in accordance with the Declaration of Helsinki, and approved by the Institutional Review Board of The Wright Center for Graduate Medical Education. Approval code 1698777-1. Approval date: December 17, 2020.

Informed Consent Statement: Informed consent was obtained from all subjects involved in the study as per SEER policies.

Data Availability Statement: The data that support the findings of this study are openly available within the Surveillance Epidemiology and End Results (SEER) database at http://seer.cancer.gov/ (accessed on 21 October 2023). NIH. National Cancer Institute: Surveillance, Epidemiology and End results (SEER) program. Available online: https://seer.cancer.gov/ (accessed on 21 October 2023). The Registry of Research Data Repositories persistent identifier for this resource is RRID: nif-0000-21366.

Conflicts of Interest: The authors declare no conflicts of interest.

References

1. Sung, H.; Ferlay, J.; Siegel, R.L.; Laversanne, M.; Soerjomataram, I.; Jemal, A.; Bray, F. Global Cancer Statistics 2020: GLOBOCAN Estimates of Incidence and Mortality Worldwide for 36 Cancers in 185 Countries. *CA A Cancer J. Clin.* **2021**, *71*, 209–249. [CrossRef] [PubMed]
2. Siegel, R.L.; Miller, K.D.; Goding Sauer, A.; Fedewa, S.A.; Butterly, L.F.; Anderson, J.C.; Cercek, A.; Smith, R.A.; Jemal, A. Colorectal Cancer Statistics, 2020. *CA A Cancer J. Clin.* **2020**, *70*, 145–164. [CrossRef] [PubMed]
3. Siegel, R.L.; Torre, L.A.; Soerjomataram, I.; Hayes, R.B.; Bray, F.; Weber, T.K.; Jemal, A. Global Patterns and Trends in Colorectal Cancer Incidence in Young Adults. *Gut* **2019**, *68*, 2179–2185. [CrossRef] [PubMed]
4. Vuik, F.E.; Nieuwenburg, S.A.; Bardou, M.; Lansdorp-Vogelaar, I.; Dinis-Ribeiro, M.; Bento, M.J.; Zadnik, V.; Pellisé, M.; Esteban, L.; Kaminski, M.F.; et al. Increasing Incidence of Colorectal Cancer in Young Adults in Europe over the Last 25 Years. *Gut* **2019**, *68*, 1820–1826. [CrossRef] [PubMed]
5. Siegel, R.L.; Wagle, N.S.; Cercek, A.; Smith, R.A.; Jemal, A. Colorectal Cancer Statistics, 2023. *CA A Cancer J. Clin.* **2023**, *73*, 233–254. [CrossRef] [PubMed]
6. Bailey, C.E.; Hu, C.-Y.; You, Y.N.; Bednarski, B.K.; Rodriguez-Bigas, M.A.; Skibber, J.M.; Cantor, S.B.; Chang, G.J. Increasing Disparities in the Age-Related Incidences of Colon and Rectal Cancers in the United States, 1975–2010. *JAMA Surg.* **2015**, *150*, 17. [CrossRef] [PubMed]
7. Goyal, H.; Desai, R.; Aloysius, M.M.; Jecmenica, M.; Enders, G.H.; Bansal, P. Young-Onset Colorectal Cancer: Hospitalization Trends and Gender Disparities in the United States 2010–2014. *Int. J. Colorectal. Dis.* **2019**, *34*, 1611–1615. [CrossRef] [PubMed]
8. Mauri, G.; Sartore-Bianchi, A.; Russo, A.-G.; Marsoni, S.; Bardelli, A.; Siena, S. Early-Onset Colorectal Cancer in Young Individuals. *Mol. Oncol.* **2019**, *13*, 109–131. [CrossRef] [PubMed]
9. Ashktorab, H.; Kupfer, S.S.; Brim, H.; Carethers, J.M. Racial Disparity in Gastrointestinal Cancer Risk. *Gastroenterology* **2017**, *153*, 910–923. [CrossRef] [PubMed]
10. CDC. Colorectal Cancer Statistics. Available online: https://www.cdc.gov/cancer/colorectal/statistics/ (accessed on 27 November 2023).
11. O'Reilly, M.; Linehan, A.; Krstic, A.; Kolch, W.; Sheahan, K.; Winter, D.C.; Mc Dermott, R. Oncotherapeutic Strategies in Early Onset Colorectal Cancer. *Cancers* **2023**, *15*, 552. [CrossRef]
12. Sultan, I.; Rodriguez-Galindo, C.; El-Taani, H.; Pastore, G.; Casanova, M.; Gallino, G.; Ferrari, A. Distinct Features of Colorectal Cancer in Children and Adolescents: A Population-based Study of 159 Cases. *Cancer* **2010**, *116*, 758–765. [CrossRef] [PubMed]
13. Rodriguez, L.; Brennan, K.; Karim, S.; Nanji, S.; Patel, S.V.; Booth, C.M. Disease Characteristics, Clinical Management, and Outcomes of Young Patients With Colon Cancer: A Population-Based Study. *Clin. Color. Cancer* **2018**, *17*, e651–e661. [CrossRef] [PubMed]
14. Chen, F.W.; Sundaram, V.; Chew, T.A.; Ladabaum, U. Advanced-Stage Colorectal Cancer in Persons Younger Than 50 Years Not Associated With Longer Duration of Symptoms or Time to Diagnosis. *Clin. Gastroenterol. Hepatol.* **2017**, *15*, 728–737. [CrossRef] [PubMed]
15. Knudsen, A.B.; Rutter, C.M.; Peterse, E.F.P.; Lietz, A.P.; Seguin, C.L.; Meester, R.G.S.; Perdue, L.A.; Lin, J.S.; Siegel, R.L.; Doria-Rose, V.P.; et al. *Colorectal Cancer Screening: An Updated Decision Analysis for the U.S. Preventive Services Task Force*; U.S. Preventive Services Task Force Evidence Syntheses, Formerly Systematic Evidence Reviews; Agency for Healthcare Research and Quality (US): Rockville, MD, USA, 2021.
16. Qaseem, A.; Harrod, C.S.; Crandall, C.J.; Wilt, T.J.; Clinical Guidelines Committee of the American College of Physicians; Wilt, T.J.; Crandall, C.J.; Balk, E.M.; Cooney, T.G.; Cross, J.T.; et al. Screening for Colorectal Cancer in Asymptomatic Average-Risk Adults: A Guidance Statement from the American College of Physicians (Version 2). *Ann. Intern. Med.* **2023**, *176*, 1092–1100. [CrossRef] [PubMed]
17. ACS. American Cancer Society Guidelines for the Early Detection of Cancer. Available online: https://www.cancer.org/healthy/find-cancer-early/american-cancer-society-guidelines-for-the-early-detection-of-cancer.html (accessed on 27 December 2023).
18. USPSTF. Colorectal Cancer: Screening. Available online: https://www.uspreventiveservicestaskforce.org/uspstf/recommendation/colorectal-cancer-screening (accessed on 27 November 2023).
19. Shaukat, A.; Kahi, C.J.; Burke, C.A.; Rabeneck, L.; Sauer, B.G.; Rex, D.K. ACG Clinical Guidelines: Colorectal Cancer Screening 2021. *Am. J. Gastroenterol.* **2021**, *116*, 458–479. [CrossRef] [PubMed]
20. Rex, D.K.; Boland, C.R.; Dominitz, J.A.; Giardiello, F.M.; Johnson, D.A.; Kaltenbach, T.; Levin, T.R.; Lieberman, D.; Robertson, D.J. Colorectal Cancer Screening: Recommendations for Physicians and Patients from the U.S. Multi-Society Task Force on Colorectal Cancer. *Gastroenterology* **2017**, *153*, 307–323. [CrossRef] [PubMed]
21. Provenzale, D.; Ness, R.M.; Llor, X.; Weiss, J.M.; Abbadessa, B.; Cooper, G.; Early, D.S.; Friedman, M.; Giardiello, F.M.; Glaser, K.; et al. NCCN Guidelines Insights: Colorectal Cancer Screening, Version 2.2020: Featured Updates to the NCCN Guidelines. *J. Natl. Compr. Cancer Netw.* **2020**, *18*, 1312–1320. [CrossRef] [PubMed]
22. Stoffel, E.M.; Koeppe, E.; Everett, J.; Ulintz, P.; Kiel, M.; Osborne, J.; Williams, L.; Hanson, K.; Gruber, S.B.; Rozek, L.S. Germline Genetic Features of Young Individuals with Colorectal Cancer. *Gastroenterology* **2018**, *154*, 897–905.e1. [CrossRef] [PubMed]
23. NIH. National Cancer Institute: Surveillance, Epidemiology and End Results (SEER) Pro-Gram. Available online: https://seer.cancer.gov/ (accessed on 21 October 2023).

24. Park, H.S.; Lloyd, S.; Decker, R.H.; Wilson, L.D.; Yu, J.B. Overview of the Surveillance, Epidemiology, and End Results Database: Evolution, Data Variables, and Quality Assurance. *Curr. Probl. Cancer* **2012**, *36*, 183–190. [CrossRef] [PubMed]
25. Che, W.-Q.; Li, Y.-J.; Tsang, C.-K.; Wang, Y.-J.; Chen, Z.; Wang, X.-Y.; Xu, A.-D.; Lyu, J. How to Use the Surveillance, Epidemiology, and End Results (SEER) Data: Research Design and Methodology. *Mil. Med. Res.* **2023**, *10*, 50. [CrossRef] [PubMed]
26. Liang, Z.; Nau, C.; Xie, F.; Vogel, R.; Chen, W. The Application of Community-Based Information from the American Community Survey in a Large Integrated Health Care Organization. *Perm. J.* **2020**, *25*, 1–14. [CrossRef] [PubMed]
27. Udalova, V.; Carey, T.S.; Chelminski, P.R.; Dalzell, L.; Knoepp, P.; Motro, J.; Entwisle, B. Linking Electronic Health Records to the American Community Survey: Feasibility and Process. *Am. J. Public Health* **2022**, *112*, 923–930. [CrossRef] [PubMed]
28. Nethery, R.C.; Rushovich, T.; Peterson, E.; Chen, J.T.; Waterman, P.D.; Krieger, N.; Waller, L.; Coull, B.A. Comparing Denominator Sources for Real-Time Disease Incidence Modeling: American Community Survey and WorldPop. *SSM-Popul. Health* **2021**, *14*, 100786. [CrossRef] [PubMed]
29. United States Census Bureau. American Community Survey (ACS). Available online: https://www.census.gov/programs-surveys/acs (accessed on 21 October 2023).
30. Wilkinson, A.N.; Lieberman, D.; Leontiadis, G.I.; Tse, F.; Barkun, A.N.; Abou-Setta, A.; Marshall, J.K.; Samadder, J.; Singh, H.; Telford, J.J.; et al. Colorectal Cancer Screening for Patients with a Family History of Colorectal Cancer or Adenomas. *Can. Fam. Physician* **2019**, *65*, 784–789. [PubMed]
31. Wilkins, T.; McMechan, D.; Talukder, A.; Herline, A. Colorectal Cancer Screening and Surveillance in Individuals at Increased Risk. *Am. Fam. Physician* **2018**, *97*, 111–116. [PubMed]
32. Shaukat, A.; Crockett, S.D. Colorectal Cancer Screening: Time to Spring Forward. *Am. J. Gastroenterol.* **2024**, *119*, 395–396. [CrossRef]
33. Ladabaum, U.; Shepard, J.; Mannalithara, A. Adenoma and Sessile Serrated Lesion Detection Rates at Screening Colonoscopy for Ages 45–49 Years vs Older Ages Since the Introduction of New Colorectal Cancer Screening Guidelines. *Clin. Gastroenterol. Hepatol.* **2022**, *20*, 2895–2904.e4. [CrossRef] [PubMed]
34. Yong, K.K.; Kyaw, M.; Chadwick, G.; Sundaram, K. Increasing Incidence of Young-Onset Colorectal Cancers in the UK and Rising Mortality in Rectal Cancers. *Gut* **2020**, *69*, 2267–2268. [CrossRef] [PubMed]
35. Sung, J.J.Y.; Chiu, H.-M.; Jung, K.-W.; Jun, J.K.; Sekiguchi, M.; Matsuda, T.; Kyaw, M.H. Increasing Trend in Young-Onset Colorectal Cancer in Asia: More Cancers in Men and More Rectal Cancers. *Am. J. Gastroenterol.* **2019**, *114*, 322–329. [CrossRef]
36. Lui, R.N.; Tsoi, K.K.F.; Ho, J.M.W.; Lo, C.M.; Chan, F.C.H.; Kyaw, M.H.; Sung, J.J.Y. Global Increasing Incidence of Young-Onset Colorectal Cancer Across 5 Continents: A Joinpoint Regression Analysis of 1,922,167 Cases. *Cancer Epidemiol. Biomark. Prev.* **2019**, *28*, 1275–1282. [CrossRef]
37. Wang, W.; Chen, W.; Lin, J.; Shen, Q.; Zhou, X.; Lin, C. Incidence and Characteristics of Young-Onset Colorectal Cancer in the United States: An Analysis of SEER Data Collected from 1988 to 2013. *Clin. Res. Hepatol. Gastroenterol.* **2019**, *43*, 208–215. [CrossRef] [PubMed]
38. Ahnen, D.J.; Wade, S.W.; Jones, W.F.; Sifri, R.; Mendoza Silveiras, J.; Greenamyer, J.; Guiffre, S.; Axilbund, J.; Spiegel, A.; You, Y.N. The Increasing Incidence of Young-Onset Colorectal Cancer: A Call to Action. *Mayo Clin. Proc.* **2014**, *89*, 216–224. [CrossRef] [PubMed]
39. Austin, H.; Jane Henley, S.; King, J.; Richardson, L.C.; Eheman, C. Changes in Colorectal Cancer Incidence Rates in Young and Older Adults in the United States: What Does It Tell Us about Screening. *Cancer Causes Control.* **2014**, *25*, 191–201. [CrossRef] [PubMed]
40. Lall, V.; Ismail, A.G.M.; Ayonrinde, O.T. Disparate Age and Sex Distribution of Sessile Serrated Lesions and Conventional Adenomas in an Outpatient Colonoscopy Population–Implications for Colorectal Cancer Screening? *Int. J. Colorectal. Dis.* **2022**, *37*, 1569–1579. [CrossRef] [PubMed]
41. Shepherdson, M.; Leemaqz, S.; Singh, G.; Ryder, C.; Ullah, S.; Canuto, K.; Young, J.P.; Price, T.J.; McKinnon, R.A.; Pandol, S.J.; et al. Young-Onset Gastrointestinal Adenocarcinoma Incidence and Survival Trends in the Northern Territory, Australia, with Emphasis on Indigenous Peoples. *Cancers* **2022**, *14*, 2870. [CrossRef] [PubMed]
42. Shen, J.; Wu, Y.; Mo, M.; Feng, X.; Zhou, C.; Wang, Z.; Cai, G.; Zheng, Y. Risk Factors Associated with Early-Onset Colorectal Neoplasm in Chinese Youth: A Prospective Population-Based Study. *Front. Oncol.* **2021**, *11*, 702322. [CrossRef] [PubMed]
43. Abdel-Rahman, O.; Karachiwala, H.; Koski, S. Patterns of Colorectal Cancer Diagnosis among Younger Adults in a Real-World, Population-Based Cohort. *Future Oncol.* **2022**, *18*, 47–54. [CrossRef] [PubMed]
44. Siegel, R.L.; Miller, K.D.; Fedewa, S.A.; Ahnen, D.J.; Meester, R.G.S.; Barzi, A.; Jemal, A. Colorectal Cancer Statistics, 2017. *CA A Cancer J. Clin.* **2017**, *67*, 177–193. [CrossRef] [PubMed]
45. Abualkhair, W.H.; Zhou, M.; Ahnen, D.; Yu, Q.; Wu, X.-C.; Karlitz, J.J. Trends in Incidence of Early-Onset Colorectal Cancer in the United States Among Those Approaching Screening Age. *JAMA Netw. Open* **2020**, *3*, e1920407. [CrossRef] [PubMed]
46. US Preventive Services Task Force; Davidson, K.W.; Barry, M.J.; Mangione, C.M.; Cabana, M.; Caughey, A.B.; Davis, E.M.; Donahue, K.E.; Doubeni, C.A.; Krist, A.H.; et al. Screening for Colorectal Cancer: US Preventive Services Task Force Recommendation Statement. *JAMA* **2021**, *325*, 1965. [CrossRef] [PubMed]
47. Saldana, C.; Urman, J.M.; Gomez, M.; Basterra, M.; Irisarri, R.; Amorena, E.; Garaigorta, M.; Francisco, G.; Ruiz-Clavijo, D.; Gomez-Doronsorro, M.L.; et al. Association between Ser-Rated Polyps and the Risk of Synchronous Colorectal Cancer: A Community-Based Study. *Color. Cancer* **2021**, *7*, 8308.

48. Szylberg, Ł.; Janiczek, M.; Popiel, A.; Marszałek, A. Serrated Polyps and Their Alternative Pathway to the Colorectal Cancer: A Systematic Review. *Gastroenterol. Res. Pract.* **2015**, *2015*, 573814. [CrossRef] [PubMed]
49. Perea García, J.; Arribas, J.; Cañete, Á.; García, J.L.; Álvaro, E.; Tapial, S.; Narváez, C.; Vivas, A.; Brandáriz, L.; Hernández-Villafranca, S.; et al. Association of Polyps with Early-Onset Colorectal Cancer and Throughout Surveillance: Novel Clinical and Molecular Implications. *Cancers* **2019**, *11*, 1900. [CrossRef] [PubMed]
50. You, Y.N. Young-Onset Colorectal Cancer: Is It Time to Pay Attention? *Arch. Intern. Med.* **2012**, *172*, 287. [CrossRef] [PubMed]
51. Bhandari, A.; Woodhouse, M.; Gupta, S. Colorectal Cancer Is a Leading Cause of Cancer Incidence and Mortality among Adults Younger than 50 Years in the Usa: A Seer-Based Analysis with Comparison to Other Young-Onset Cancers. *J. Investig. Med.* **2017**, *65*, 311–315. [CrossRef] [PubMed]
52. Holowatyj, A.N.; Ruterbusch, J.J.; Rozek, L.S.; Cote, M.L.; Stoffel, E.M. Racial/Ethnic Disparities in Survival Among Patients With Young-Onset Colorectal Cancer. *JCO* **2016**, *34*, 2148–2156. [CrossRef] [PubMed]
53. Luisetto, M.; Ahmadabadi, B.N.; Nili-Ahmadabadi, H.; Rafa, A.Y.; Abdul, G. Comparison of Risk Factors and Molecular Analysis of Right-Sided Colon and Left Sided Colon Cancer. *Sci. World J. Cancer Sci. Ther.* **2019**, *1*, 1–24. [CrossRef]
54. Vogelstein, B.; Kinzler, K.W. Cancer Genes and the Pathways They Control. *Nat. Med.* **2004**, *10*, 789–799. [CrossRef] [PubMed]
55. Parsons, D.W.; Wang, T.-L.; Samuels, Y.; Bardelli, A.; Cummins, J.M.; DeLong, L.; Silliman, N.; Ptak, J.; Szabo, S.; Willson, J.K.V.; et al. Mutations in a Signalling Pathway. *Nature* **2005**, *436*, 792. [CrossRef] [PubMed]
56. Smit, W.L.; Spaan, C.N.; Johannes De Boer, R.; Ramesh, P.; Martins Garcia, T.; Meijer, B.J.; Vermeulen, J.L.M.; Lezzerini, M.; MacInnes, A.W.; Koster, J.; et al. Driver Mutations of the Adenoma-Carcinoma Sequence Govern the Intestinal Epithelial Global Translational Capacity. *Proc. Natl. Acad. Sci. USA* **2020**, *117*, 25560–25570. [CrossRef]
57. Carvalho, B.; Sillars-Hardebol, A.H.; Postma, C.; Mongera, S.; Droste, J.T.S.; Obulkasim, A.; Van De Wiel, M.; Van Criekinge, W.; Ylstra, B.; Fijneman, R.J.A.; et al. Colorectal Adenoma to Carcinoma Progression Is Accompanied by Changes in Gene Expression Associated with Ageing, Chromosomal Instability, and Fatty Acid Metabolism. *Cell. Oncol.* **2012**, *35*, 53–63. [CrossRef] [PubMed]
58. Peery, A.F.; Crockett, S.D.; Murphy, C.C.; Jensen, E.T.; Kim, H.P.; Egberg, M.D.; Lund, J.L.; Moon, A.M.; Pate, V.; Barnes, E.L.; et al. Burden and Cost of Gastrointestinal, Liver, and Pancreatic Diseases in the United States: Update 2021. *Gastroenterology* **2022**, *162*, 621–644. [CrossRef] [PubMed]
59. Schlueter, D.; DeGroff, A.; Soloe, C.; Arena, L.; Melillo, S.; Tangka, F.; Hoover, S.; Subramanian, S. Factors That Support Sustainability of Health Systems Change to Increase Colorectal Cancer Screening in Primary Care Clinics: A Longitudinal Qualitative Study. *Health Promot. Pract.* **2023**, *24*, 755–763. [CrossRef] [PubMed]
60. Bhat, S.K.; East, J.E. Colorectal Cancer: Prevention and Early Diagnosis. *Medicine* **2015**, *43*, 295–298. [CrossRef]

medicina

MDPI

Article

Remission Is Maintained after Switch from Dose-Optimised Intravenous Treatment to Subcutaneous Treatment with Vedolizumab in Inflammatory Bowel Disease

Špela Pintar [1,†], Jurij Hanžel [1,2,†], David Drobne [1,2], Matic Koželj [1], Tina Kurent [1], Nataša Smrekar [1] and Gregor Novak [1,2,*]

[1] Department of Gastroenterology, University Medical Centre Ljubljana, Japljeva 2, 1000 Ljubljana, Slovenia; spela.pintar@kclj.si (Š.P.); jurij.hanzel@kclj.si (J.H.); david.drobne@kclj.si (D.D.); matic.kozelj@kclj.si (M.K.); tina.kurent@kclj.si (T.K.); natasa.smrekar@kclj.si (N.S.)

[2] Faculty of Medicine, University of Ljubljana, Vrazov trg 2, 1000 Ljubljana, Slovenia

* Correspondence: gregor.novak.gastro@kclj.si; Tel.: +386-1-522-9462

† These authors contributed equally to this work.

Abstract: *Background and Objectives*: The subcutaneous (SC) formulation of vedolizumab has proven to be effective for the maintenance of remission after intravenous induction. Little is known about the efficacy of switching from intravenous maintenance treatment to SC. We aimed to assess the real-world efficacy of switching to SC treatment and to assess the impact of a baseline treatment regimen. *Materials and Methods*: In this observational cohort study, adult patients with inflammatory bowel disease who were switched to SC vedolizumab maintenance treatment were enrolled. Patients after intravenous induction and patients who switched from intravenous maintenance treatment (every 8 weeks or every 4 weeks) were included. The SC vedolizumab dosing was 108 mg every 2 weeks, regardless of the previous regimen. The clinical, biochemical, and endoscopic disease activity parameters and vedolizumab serum concentrations at the time of the switch and at the follow-up were assessed. *Results*: In total, 135 patients (38% Crohn's disease, 62% ulcerative colitis) were switched to SC vedolizumab treatment. The median time to the first follow-up (FU) was 14.5 weeks (IQR 12–26), and the median time to the second FU was 40 weeks (IQR 36–52). Nine patients (7%) discontinued SC vedolizumab treatment, with two-thirds of them discontinuing due to active disease. In all dosing regimens, there were no significant changes in the clinical scores and CRP at the baseline and first and second FUs. Clinical and biochemical remission appeared to be maintained irrespective of the previous dosing regimen. *Conclusions*: The results of this real-world study suggest that the maintenance of clinical and biomarker remission can be achieved in patients who switched from intravenous to SC vedolizumab. The baseline vedolizumab dosing regimen (every 4 weeks versus every 8 weeks) did not have an impact on outcomes.

Keywords: inflammatory bowel disease; Crohn's disease; ulcerative colitis; vedolizumab; subcutaneous formulation

1. Introduction

Inflammatory bowel disease (IBD) is a chronic inflammatory disorder of the gastrointestinal tract that most commonly manifests with bloody diarrhoea and abdominal pain. If the inflammation is uncontrolled, it can cause progressive functional and structural damage and impair patients' quality of life. Among the current therapeutic armamentarium, several small molecules and biological agents are available, including vedolizumab [1].

Vedolizumab is a monoclonal antibody that targets $\alpha4\beta7$ integrin, which is preferentially expressed on gut-homing lymphatic cells and prevents their trafficking into the inflamed gut. The gut-selective mechanism of action contributes to vedolizumab's favourable benefit–risk profile [2,3].

Vedolizumab has been registered as an intravenous (IV) induction and maintenance treatment for ulcerative colitis (UC) [4] and Crohn's disease (CD) [5]. The recommended dose regimen of vedolizumab is 300 mg administered via intravenous (IV) infusion at weeks 0, 2, and 6, followed by infusions every 8 weeks thereafter. In the case of a loss of response or an inadequate response, dose escalation by shortening the dosing interval to every 4 weeks was proven effective in approximately half of the patients [6,7]. Recently, a subcutaneous (SC) formulation of vedolizumab has been approved for maintenance treatment after showing efficacy and safety in the phase III clinical trials VISIBLE 1 [8] and VISIBLE 2 [9]. The serum levels of vedolizumab were higher with SC administration compared to IV. Both clinical trials evaluated SC maintenance treatment (108 mg every 2 weeks) in patients who responded at week 6 to an induction with two infusions of vedolizumab [8,9]. However, these clinical trials did not assess IBD patients who were treated with maintenance IV vedolizumab before transitioning to SC.

Switching from IV infusion to SC injections is an appealing option due to the potential for self-administration at home, which could enhance patient convenience [10]. Additional advantages include a shorter application time, a decreased incidence of infusion-related adverse events, an improved quality of life, a reduction in the time needed to travel to a healthcare institution, and a decrease in healthcare system costs [10,11]. Conversely, more frequent injection-site reactions are observed with SC applications compared to IV applications [8]. Nevertheless, most patients and healthcare professionals express a preference for SC application over IV [10,11]. There are limited data about switching from IV maintenance to SC treatment. Four prospective real-world studies from Europe reported that transition to SC maintenance treatment is feasible, effective, and safe [12–15]. Data regarding switching from IV maintenance to SC treatment in patients who were previously dose-optimised due to a loss of response or an inadequate response by shortening the IV dosing interval to every 4 weeks are even more deficient.

In our study, we aimed to assess the drug persistence, efficacy, and pharmacokinetics of switching to an SC vedolizumab formulation in a real-world cohort of IBD patients and to assess if the baseline IV regimen (maintenance every 8 weeks (q8), maintenance every 4 weeks (q4), or IV induction) impacts the outcomes after the transition.

2. Materials and Methods

2.1. Study Design and Population

This observational study was conducted in a tertiary referral IBD centre (University Medical Centre Ljubljana, Slovenia) following the rules of the Declaration of Helsinki of 1975, revised in 2013. Ethical permission was granted by the Slovenian National Ethics Committee (0120-013/2016-2). Consecutive prospectively followed adult patients with IBD who were switched to SC vedolizumab maintenance treatment from May 2021 onward were enrolled. Follow-up lasted until July 2022. Therefore, the duration of follow-up differs among enrolled patients. All patients being treated with vedolizumab (with response after IV induction or undergoing IV maintenance treatment) were offered the option of switching to SC formulations after an exact explanation of known data.

The included patients had to be ≥18 years of age with a confirmed diagnosis of IBD (UC, CD, or unclassified IBD) and undergoing treatment with SC vedolizumab—either after IV induction or after IV maintenance. There were no exclusion criteria. Demographics and baseline characteristics were extracted from medical files, including age, gender, diagnosis, disease duration, disease phenotype, smoking status, weight, height, previous biological therapy, previous exposure to corticosteroids, and duration of IV vedolizumab therapy before switching to SC.

The included patients were grouped based on the IV treatment regimen before switching to SC: IV maintenance treatment every 8 weeks (q8 cohort), optimised IV maintenance treatment every 4 or 6 weeks (q4 cohort), and IV induction (two or three IV infusions). After the switch to SC, all groups were treated with the same SC vedolizumab regimen (108 mg every 2 weeks), regardless of the previous IV regimen. Follow-up visits were individually scheduled by the treating physician; therefore, not all patients had visits at the same time point. Data on SC vedolizumab discontinuation, clinical and endoscopic scores, and biochemical parameters (including vedolizumab serum concentrations) were documented at baseline and throughout follow-up.

2.2. Study Endpoints and Definitions

The main outcome was the proportion of patients who discontinued SC vedolizumab. Drug discontinuation could be due to disease activity, intolerance or side effects, the need for IBD surgery or hospitalisation, patients' wishes, moving to another IBD centre, and being lost to follow-up. The discontinuation date was defined as the day of the last SC vedolizumab application.

Additional outcomes included clinical remission, biochemical remission, and endoscopic remission after switching to SC treatment. Vedolizumab serum concentrations were assessed at the time of switch and at follow-up visits.

Clinical disease activity was assessed using the Harvey Bradshaw index [16] (HBI) for CD or the partial Mayo score [17] (pMayo) for UC. Clinical remission was defined as HBI < 5 in CD and pMayo < 2 in UC.

Fecal calprotectin (FC), C-reactive protein (CRP), and vedolizumab serum concentration were measured at the time of switch and at follow-up visits. FC was measured using the Calprest assay (Eurospital, Trieste, Italy) and CRP was measured with the ADVIA 1800 Chemistry System (Siemens, Germany). Biochemical remission was defined as CRP < 5 mg/L and FC < 100 µg/g [18]. Vedolizumab concentrations were measured with the Conformité Européenne-marked apDia vedolizumab enzyme-linked immunosorbent assay (Turnhout, Belgium) with a measurement range between 1 and 50 µg/mL.

The endoscopies were scheduled by the treating physician. Disease activity was assessed using the endoscopic Mayo score [19] in UC and by the presence of ulcers in CD. Endoscopic remission was defined as a Mayo endoscopic score < 2 in UC and an absence of ulcers > 5 mm in CD.

2.3. Statistical Methods

All analyses were performed on a per-protocol basis. The continuous variables are presented as medians with interquartile ranges (IQRs). The vedolizumab serum concentrations before and after switching were compared using a paired Wilcoxon rank test. Values of $p < 0.05$ were considered statistically significant. All analyses were performed using IBM SPSS statistics for Windows, Version 21.0 (IBM Corp., Armonk, NY, USA).

3. Results

3.1. Patient Characteristics

In total, 135 patients were enrolled, 51 (37.8%) had CD, and 84 (62.2%) had UC (Table 1). The median time to the first follow-up visit was 14.5 weeks (IQR 12–26 weeks), and the median time to last follow-up was 40 weeks (IQR 36–52 weeks). No patients underwent dose escalation during the follow-up period.

3.2. Drug Survival

A total of 9/135 patients (6.7%) discontinued SC vedolizumab treatment until the end of follow-up: 6 (7.1%) had UC and 3 (5.9%) had CD. Reasons for discontinuation were active disease with the need for treatment escalation in six patients, dysplasia at surveillance colonoscopy requiring colectomy in one patient, discontinuation due to the patient's wishes in one patient, and loss to follow-up for one patient. The median time to treatment discontinuation was 22 weeks (IQR 14–45 weeks).

Out of the six patients who discontinued SC vedolizumab due to active disease, three were in the IV induction group and three were in the q4 group (Table 2). All were switched to another treatment.

Table 1. Patient demographics and baseline characteristics.

	q8		q4		IV Induction	
	CD (*n* = 26)	UC (*n* = 39)	CD (*n* = 17)	UC (*n* = 24)	CD (*n* = 8)	UC (*n* = 21)
Age at first SC dose, median in years (range)	55 (39–63)	47 (41–65)	51 (40–61)	41 (30–66)	54 (30–70)	43 (24–53)
Male sex, *n* (%)	17 (63)	25 (64)	8 (47)	9 (38)	5 (63)	12 (57)
Disease duration: years (IQR)	9 (1.6–17.1)	10 (4–19)	16 (6–21)	6 (3–10)	12 (9–22)	10 (4–13)
CD location, *n* (%)						
L1 (ileal)	6 (23)		3 (18)		1 (13)	
L2 (colonic)	11 (44)		4 (24)		5 (63)	
L3 (ileocolonic)	9 (34)		10 (59)		2 (25)	
L4+ (isolated upper gastrointestinal disease)	3 (11)		1 (6)		1 (13)	
CD behaviour, *n* (%)						
B1 (non-stricturing/ non-penetrating)	21 (81)		5 (30)		6 (75)	
B2 (stricturing)	3 (12)		6 (35)		1 (12.5)	
B3 (penetrating)	2 (7)		6 (35)		1 (12.5)	
Perianal disease	1 (4)		6 (35)		2 (25)	
UC extent, *n* (%)						
E1 (proctitis)		5 (13)		0		9 (43)
E2 (left-sided colitis)		14 (36)		8 (33)		4 (19)
E3 (pancolitis)		20 (51)		16 (67)		8 (38)
Smoking status, *n* (%)						
Current smoker	4 (15)	3 (8)	3 (18)	2 (8)	1 (13)	3 (14)
Previous smoker	5 (19)	8 (21)	4 (24)	3 (13)	0	1 (5)
Non-smoker	17 (66)	28 (71)	10 (59)	19 (79)	7 (87)	17 (81)
Previous therapy with biologic, *n* (%)	10 (39)	14 (36)	15 (88)	11 (46)	4 (50)	6 (29)
>1 prior biologic *n* (%)	5 (19)	6 (15)	8 (47)	1 (4)	3 (37)	0
Previous therapy with corticosteroids, *n* (%)	14 (54)	28 (72)	11 (65)	21 (88)	6 (75)	11 (52)
Duration of IV vedolizumab treatment, months (IQR)	28 (21–44)	18 (11–31)	25 (16–43)	22 (14–32)	/	/

Abbreviations: UC: ulcerative colitis; CD: Crohn's disease; q8: intravenous vedolizumab every 8 weeks; q4: intravenous vedolizumab every 4 weeks; IV intravenous; IQR: interquartile range.

Table 2. Patients who discontinued vedolizumab SC due to active disease.

Group	Patient	Duration of SC Vedolizumab	Reason for Discontinuation
q4	CD, after right hemicolectomy	10 months	Endoscopically active disease (Rutgeerts i2); asymptomatic
	CD, after ileo-caecal resection	6 months	Endoscopically active disease (Rutgeerts i4); asymptomatic
	UC	6 months	Clinically and endoscopically (endoscopic Mayo score 2) active disease

Table 2. *Cont.*

Group	Patient	Duration of SC Vedolizumab	Reason for Discontinuation
IV induction group	CD, after proctocolectomy	13 months	Endoscopically active disease (ulcers in the stomach and small bowel); asymptomatic
	UC	11 months	Endoscopically active disease (endoscopic Mayo score 3); asymptomatic
	UC	11 months	Clinically active disease

Abbreviations: UC: ulcerative colitis; CD: Crohn's disease; q4: intravenous vedolizumab every 4 weeks; SC: subcutaneous.

3.3. Clinical and Biochemical Disease Activity after Switch

The proportion of patients with CD in clinical remission was 17/18 (94%), 20/20 (100%), 19/11 (91%) in the q8 group, 7/11 (64%), 13/15 (87%), 5/7 (72%) in the q4 group, and 7/7 (100%), 7/7 (100%), 2/2 (100%) in the IV induction group at baseline and first and second follow-up, respectively. The proportion of patients with UC in clinical remission was 26/27 (96%), 25/28 (89%), 12/13 (92%) in the q8 group, 12/17 (71%), 15/17 (88%), 7/10 (70%) in the q4 group, and 13/18 (72%), 16/18 (89%), 9/9 (100%) in the IV induction group at baseline and first and second follow-up, respectively. In all groups of patients, there were no significant changes in clinical disease activity scores at baseline and first and second follow-up. Data for clinical disease activity are shown in Table 3.

Table 3. Clinical and biochemical disease activity.

	q8		q4		IV Induction	
	CD (*n* = 26)	**UC (*n* = 39)**	**CD (*n* = 17)**	**UC (*n* = 24)**	**CD (*n* = 8)**	**UC (*n* = 21)**
Median HBI (IQR; *n*)						
At baseline	1 (0–2; 18)		2 (0–6; 11)		1 (0–3; 7)	
At 1st FU	1 (0–2; 20)		1 (0–3; 15)		0 (0–1; 7)	
At 2nd FU	1 (0–2; 11)		3 (0–6;7)		1 (/;2)	
HBI < 5 (%)						
At baseline	17/18 (94)		7/11 (64)		7/7 (100)	
At 1st FU	20/20(100)		13/15 (87)		7/7 (100)	
At 2nd FU	10/11 (91)		5/7 (72)		2/2 (100)	
Median pMayo (IQR; *n*)						
At baseline		0 (0–1; 27)		1 (0–2; 17)		1 (0–2; 18)
At 1st FU		0 (0–1; 28)		0 (0–1; 17)		0 (0–1; 18)
At 2nd FU		0 (0–0; 13)		1 (1–3; 10)		1 (0–1; 9)
pMayo <2 (%)						
At baseline		26/27 (96)		12/17 (71)		13/18 (72)
At 1st FU		25/28 (89)		15/17 (88)		16/18 (89)
At 2nd FU		12/13 (92)		7/10 (70)		9/9 (100)
CRP, mg/L, median (IQR; *n*)						
At baseline	3 (3–9; 22)	3 (3–7; 30)	3 (3–6; 15)	3(3–11;23)	5 (3–17;7)	3 (3–3; 20)
At 1st FU	3 (3–4; 17)	3 (3–5; 30)	3 (3–3; 15)	3 (3–5; 17)	8 (3–8; 8)	3 (3–3; 18)
At 2nd FU	3 (3–5; 14)	3 (3–6; 12)	4 (3–6; 8)	4 (3–11;10)	7 (/; 2)	3 (3–3; 11)
CRP < 5 mg/L (%)						
At baseline	13/22 (59)	21/30 (70)	9/15 (60)	13/23 (57)	3/7 (37)	16/20 (80)
At 1st FU	13/17 (76)	22/30 (73)	13/15 (87)	11/17 (65)	3/8 (38)	15/18 (87)
At 2nd FU	8/14 (57)	9/12 (75)	4/8 (50)	7/10 (70)	1/2 (50)	8/11 (73)

Table 3. *Cont.*

	q8		q4		IV Induction	
	CD (*n* = 26)	UC (*n* = 39)	CD (*n* = 17)	UC (*n* = 24)	CD (*n* = 8)	UC (*n* = 21)
FC, mg/kg, median (IQR; *n*)						
At baseline	29 (16–61; 20)	16 (16–50; 24)	131 (16–252; 7)	174 (35–419; 14)	16 (16–324; 5)	156 (34–212; 15)
At 1st FU	39 (16–119; 9)	16 (16–16; 15)	48 (18–337; 8)	165 (53–330; 6)	125 (16–147; 3)	91 (16–221; 12)
At 2nd FU	49 (48–147; 7)	40 (27–135; 10)	274 (/; 2)	500 (/; 4)	/	27 (27–53; 6)
FC < 100 mg/kg (%)						
At baseline	16/22 (72)	20/24 (87)	3/17 (18)	5/14 (36)	3/5 (60)	5/15 (33)
At 1st FU	6/9 (67)	13/15 (87) 7/10 (70)	5/8 (63)	2/7 (28)	1/3 (33)	6/12 (50)
At 2nd FU	3/7 (50)		1/2 (50)	0/4 (0)	/	5/6 (83)

Abbreviations: UC: ulcerative colitis; CD: Crohn's disease; q8: intravenous vedolizumab every 8 weeks; q4: intravenous vedolizumab every 4 weeks; IV Induction: intravenous vedolizumab as an induction—2 or 3 infusions; HBI: Harvey Bradshaw index; pMayo: partial Mayo score; FU: follow up; CRP: C-reactive protein; FC faecal calprotectin; IQR: interquartile range; *n*: number of patients.

Similar trends were noted with biochemical markers of disease activity (Table 3). In all groups of patients, there were no significant changes in CRP at baseline and first and second follow-up. Due to missing data, calculations were not performed for FC.

Clinical proportions in biochemical remission rates in patients with CD and UC are presented in Figures 1 and 2.

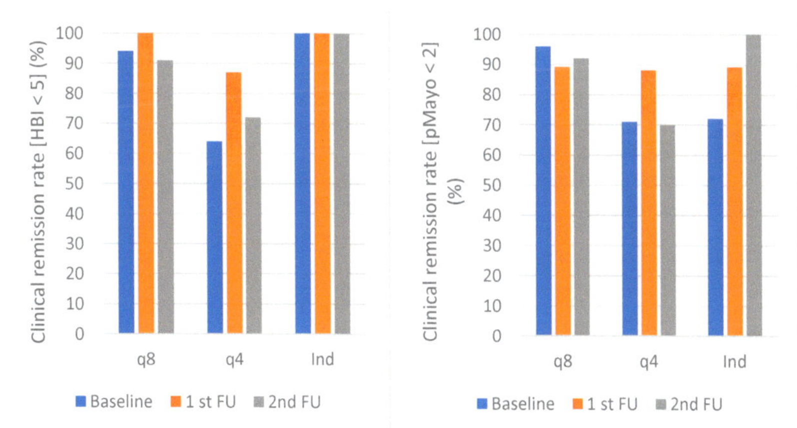

Figure 1. Clinical remission rates in CD (**left**) and UC (**right**) at baseline and first and second follow-up. Clinical remission was defined as HBI < 5 in CD and pMayo < 2 in UC. The median time to the first follow-up visit was 14.5 weeks (IQR 12–26 weeks), the median time to last follow-up was 40 weeks (IQR 36–52 weeks). Abbreviations: UC: ulcerative colitis; CD: Crohn's disease; q8: intravenous vedolizumab every 8 weeks; q4: intravenous vedolizumab every 4 weeks; Ind: intravenous vedolizumab induction—2 or 3 infusions; HBI: Harvey Bradshaw index; pMayo: partial Mayo score.

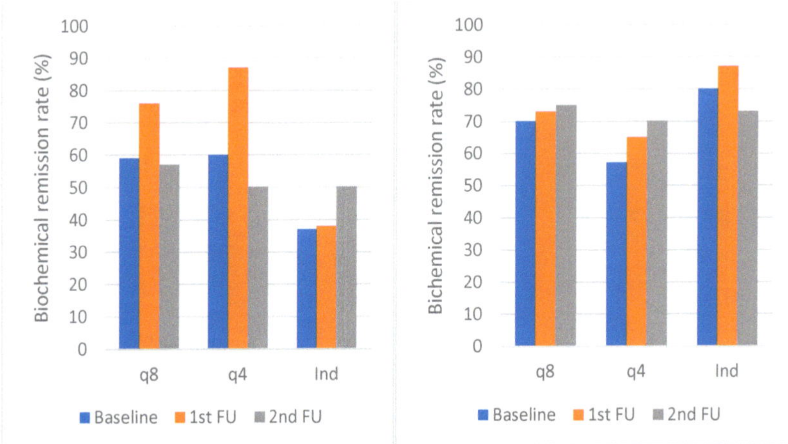

Figure 2. Biochemical remission rates in CD (**left**) and UC (**right**) at baseline and first and second FU. Biochemical remission was defined as CRP < 5 mg/L. The median time to the first FU visit was 14.5 weeks (IQR 12–26 weeks), the median time to last FU was 40 weeks (IQR 36–52 weeks). Abbreviations: UC: ulcerative colitis; CD: Crohn's disease; q8: intravenous vedolizumab every 8 weeks; q4: intravenous vedolizumab every 4 weeks; Ind: intravenous vedolizumab induction—2 or 3 infusions; CRP: C-reactive protein; FU: follow-up.

3.4. Endoscopic Disease Activity after Switch

Endoscopic disease activity after the switch to SC formulation was assessed after a median time of 24.5 weeks (IQR 17–42), 29 weeks (IQR 21–42), and 21 weeks (IQR 16–29) in the q8, q4, and IV induction groups, respectively. Endoscopic remission in UC was reached in 89%, 67%, and 75% of patients in the q8, q4, and IV induction groups, respectively.

The percentage of patients in endoscopic remission after the switch to SC formulation is presented in Table 4.

Table 4. Endoscopic disease activity.

	q8		**q4**		**IV Induction**	
	CD (*n* = 26)	**UC (*n* = 39)**	**CD (*n* = 17)**	**UC (*n* = 24)**	**CD (*n* = 8)**	**UC (*n* = 21)**
Median time to endoscopy after switch to SC in weeks (IQR)	24.5 (17–42)		29 (21–42)		21 (16–29)	
Endoscopic remission (percentage)	1/1 (100)	8/9 (89)	0/5 (0)	4/6 (67)	0/2 (0)	9/12 (75)

Endoscopic remission was defined as Mayo endoscopic score < 2 in UC and the absence of ulcers in CD. Abbreviations: UC: ulcerative colitis; CD: Crohn's disease; q8: intravenous vedolizumab every 8 weeks; q4: intravenous vedolizumab every 4 weeks; IV induction: intravenous vedolizumab as an induction—2 or 3 infusions; IQR: interquartile range.

3.5. Pharmacokinetics

The median vedolizumab serum concentration in each group at baseline and first and second follow-up is presented in Table 5. Vedolizumab serum concentration at first follow-up was significantly higher than at baseline in the q8 group ($p < 0.001$). However, no significant change in vedolizumab serum concentrations between baseline and first follow-up was observed in the q4 and IV induction groups.

Table 5. Vedolizumab serum concentration at baseline and 1st and 2nd follow up.

	q8		q4		IV Induction	
	CD (n = 26)	UC (n = 39)	CD (n = 17)	UC (n = 24)	CD (n = 8)	UC (n = 21)
Vedolizumab serum concentration µg/mL (IQR. n)						
At baseline	10.9 (7.2–16.6; 46)		28.5 (17.3–42.1; 32)		26.0 (17–37; 23)	
At first follow-up	28.6 * (20.8–34.8; 7)		22.7 ** (17.9–29.9; 9)		25.3 *** (24.0–34.3; 11)	
At second follow-up	29.7 (22.4–36.4; 6)		19.0 (15.4–28.0; 6)		27.0 (18.9–32.3; 6)	

* $p < 0.001$, ** $p = 0.575$, *** $p > 0.05$. Abbreviations: UC: ulcerative colitis; CD: Crohn's disease; q8: intravenous vedolizumab every 8 weeks; q4: intravenous vedolizumab every 4 weeks; IV induction: intravenous vedolizumab as an induction—2 or 3 infusions; IQR: interquartile range; n: number of patients.

4. Discussion

To the best of our knowledge, this is the largest reported cohort of patients who transitioned from escalated 4-week IV vedolizumab dosing to SC vedolizumab. Our findings confirm the feasibility of switching to regular SC maintenance treatment (108 mg every 2 weeks) even in patients with more refractory disease, receiving the optimised dosing of IV vedolizumab. Notably, clinical and biochemical remission appear to be maintained after switching to the SC formulation, regardless of the previous IV treatment regimen.

Over a median follow-up of 40 weeks, only 9 out of 135 (7%) patients discontinued SC treatment with vedolizumab, with a median time to treatment discontinuation of 22 weeks. This contrasts with discontinuation rates of 27% and 39% within 52 weeks of treatment reported in the registration trials VISIBLE 1 and VISIBLE 2 [8,9]. However, the results of our study cannot be compared to randomised controlled trials due to different patient populations, variations in IV treatment regimens (notably, only induction with two infusions in VISIBLE trials) and variable follow-up time. A more reliable comparison can be made with other real-world studies on transitioning to SC treatment. In a study from Netherlands, 11.9% of patients discontinued treatment after a median follow-up of 27 weeks [12]. An English cohort reported an 8% discontinuation rate at week 12 [14]. Similarly, in a Swedish cohort, discontinuation rates at 6 and 12 months were 4.5% and 11.5%, respectively [15]. A Norwegian study reported a 7.4% discontinuation rate by week 26 [13]. These findings indicate comparable discontinuation rates to our results.

On the other hand, our findings indicate lower discontinuation rates compared to other real-world cohorts receiving IV vedolizumab maintenance treatment. For instance, in a French cohort, 7.5% discontinued treatment with vedolizumab by week 14 and 43.5% by week 54 [20]. In the long-term follow-up of the same study, the 1-, 2- and 3-year persistence rates of vedolizumab in patients with CD were 48.5%, 31.4% and 26.3% and in patients with UC, 61.0%, 49.9% and 42.9%, respectively [21]. In a Danish cohort, the 12-week, 52-week, and 17-month drug continuation rates were 81%, 61%, and 58%, respectively [22]. Finally, in the Dutch cohort, the probability of continuing receiving vedolizumab treatment after 52 and 104 weeks was 54.0% and 38.4% for CD and 60.8% and 51.3% for UC, respectively [23]. From these data, it appears that transitioning to the SC formulation might be both feasible and, at the very least, similarly effective in preventing treatment discontinuation. However, low discontinuation rates in our study might be explained by the long median duration of IV vedolizumab treatment in the q4 and q8 groups. It is possible that many non-responsive patients were already discontinued before transitioning to SC.

The primary reason for treatment discontinuation in our study was active disease with the need for treatment escalation in two-thirds of patients. Interestingly, 4/6 of these patients were in clinical remission but had endoscopically active disease. In many of them, endoscopy was scheduled while still on IV maintenance treatment, meaning that disease could have been endoscopically active even before the switch to SC. Similar trends were observed in the Swedish cohort, where the majority discontinued due to active disease

(5 out of 9) [15]. Conversely, in the Dutch cohort, most patients discontinued treatment due to adverse events (9/16) and a fear of needles (3/16). Only four (25%) discontinued treatment due to a loss of response [12]. Similarly, in the English cohort, 8/10 (80%) patients discontinued treatment due to adverse events [14]. Interestingly, none of the patients in our study discontinued treatment due to adverse events or a fear of needles. This could be attributed to the shared decision-making process allowing patients to choose between continuing IV treatment or transitioning to SC treatment. Moreover, all patients underwent comprehensive injection training by a specialised IBD nurse.

The proportion of patients with UC and CD in clinical remission across the q8, q4, and IV induction groups appeared to be stable at the first and second follow-up compared to baseline. Notably, our study observed a high percentage of patients in clinical remission, ranging between 64% and 100% for CD and 70% and 100% for UC at various time points. A non-significant trend towards a lower proportion of patients in clinical remission was noticed in the q4 group. This population represents patients more refractory to therapy, which already needed to be optimised due to insufficient response. Another possible explanation might be the favourable safety profile of vedolizumab. Many elderly patients, patients with comorbidities or cancer, and patients wary of side effects tend to continue vedolizumab due to its selective mechanisms of action, despite only partial response to treatment. Similarly, in line with our findings, no statistically significant differences in clinical disease activity scores between baseline and follow-up were found in other studies [12–15]. For instance, in the Dutch study, the steroid-free clinical remission rates were 70%, 68%, and 39% for CD and 71%, 67%, and 44% for UC at baseline, week 12, and week 24, respectively [12]. Moreover, in the Swedish cohort, the clinical remission rates were 72%, 82%, and 73% in CD and 92%, 88%, and 92% in UC at switch, 3 months, and 6 months, respectively [13]. These percentages are comparable to our results.

No significant changes in CRP were observed at baseline and first and second follow-up across all patient groups. A similar trend was noted with FC, although our results were affected by missing data due to the retrospective nature of the study. Consistent with our findings, other real-world cohorts also reported no changes in biochemical parameters after switching to SC treatment [12–15]. However, a significant increase in FC after 12 weeks in the English cohort (from 31 μg/g to 47 μg/g) [14] and a significant decrease in FC after 6 months in the Swedish cohort (from 64 to 49 μg/g) were reported [15], both unlikely to be clinically relevant. Endoscopic remission in UC was achieved in the majority of patients: 89% in the q8, 67% in the q4, and 75% in the IV induction group. However, our analysis was again hampered by missing data due to the retrospective nature of the study and relatively short follow-up period.

After transitioning to the SC formulation, median serum vedolizumab trough levels increased significantly from baseline in the q8 group ($p < 0.001$). However, no significant changes in vedolizumab serum concentrations between baseline and the first follow-up were found in the q4 group. These results could have been expected since patients in the q4 group received a double dose of IV vedolizumab compared to patients in the q8 group before transitioning to the same standard SC dose. Additionally, the q4 group represents the most refractory population of patients with high inflammatory burden who had inadequate or loss of response to standard vedolizumab dose every 8 weeks. Lower vedolizumab serum concentrations could reflect higher drug clearance due to active disease [6,24]. Similarly to our findings, the VISIBLE 1 study reported higher vedolizumab serum concentrations in the SC vedolizumab treatment group compared to the IV vedolizumab treatment group (infusions every 8 weeks) [8]. Notably, IV and SC formulations have different pharmacokinetic profiles. SC administration leads to incomplete bioavailability, gradual absorption, and lower peak concentrations, whereas IV infusion results in immediate systemic drug exposure and a high peak concentration [12,25]. However, the average drug exposure between 108 mg SC vedolizumab every 2 weeks and 300 mg IV vedolizumab every 8 weeks should be similar [12]. In our study, vedolizumab levels ranged between 22.7 and 28.6 μg/mL at week 14. These concentrations are comparable to those observed in the

English (22.7 μg/mL at week 12) [14] and Swedish cohort (19.0 μg/mL at 6 months) [15] but lower compared to levels in the VISIBLE trials (34.6 μg/mL in the UC trial and 30.2 μg/mL in the CD trial) [8,9] and the Dutch cohort (31 μg/mL at week 12 and 37 μg/mL at week 24 in the Dutch cohort) [12].

While vedolizumab pharmacokinetics has been associated with clinical outcomes [4,5], the practical utility of measuring vedolizumab drug levels in clinical practice has been questioned. A vedolizumab serum concentration of 3 μg/mL induces a near-complete saturation of α4β7 on peripheral blood T cells [26]. Whether increasing vedolizumab concentrations (similar to concentrations with TNF antagonists) improves clinical outcomes remains unknown. After loss of response, approximately half of patients regain clinical response after dose escalation of IV vedolizumab by shortening the interval from 8 weeks to 4 weeks [6]. In real-world cohorts, more than half of patients on IV vedolizumab are optimised to infusions every 4 weeks [21]. However, Ungar at al. have argued against a pharmacokinetic basis for insufficient response to vedolizumab and questioned the need for dose escalation. In a retrospective study, they showed that lower vedolizumab levels before vedolizumab IV escalation were not predictive of success. On the contrary, higher pre-escalation vedolizumab levels were associated with better outcomes, possibly indicating lower inflammatory burden and higher probability of success [7,24]. Similarly, the endpoints of the ENTERPRET trial, comparing dose-optimisation strategy with vedolizumab to standard dosing in UC patients who had high drug clearance and exhibited primary nonresponse, were not met. The rates of endoscopic remission after 30 weeks were similar in the standard dosing arm (300 mg every 8 weeks) and optimised-dosing arm (300 mg or 600 mg every 4 weeks) [27]. The above findings are in line with our results. Patients optimised to 300 mg every 4 weeks due to insufficient response can be de-escalated to standard SC dosing (equivalent to 300 mg every 8 weeks) and maintain clinical and biochemical remission.

Our real-world study, assessing the effectiveness of switching from IV to SC vedolizumab maintenance treatment in IBD patients, has several strengths. Notably, it represents the largest real-world cohort to date consisting of patients previously optimised due to loss of response or inadequate response by shortening IV dosing interval to every 4 weeks. Furthermore, our study has a longer follow-up period compared to other published real-world cohorts [12,14,15]. Due to the real-world nature of the study without exclusion criteria, our results reflect everyday clinical practice with a very heterogenous IBD population. It is the only published real-world cohort of transitioning to vedolizumab SC with endoscopic data.

Our current study has several limitations. Firstly, its retrospective study design contributed to a substantial amount of missing data, posing challenges to the statistical analysis and comprehensiveness of our results. Secondly, follow-up visits were scheduled by the treating physician at varying time points after the switch. Consequently, clinical disease activity scores and laboratory testing were performed at different intervals following transition to SC, potentially influencing data consistency. Thirdly, our study did not have a comparator arm as it was an observational study. The results were compared to baseline parameters (IV treatment before switch). Fourthly, all enrolled patients had to be willing to switch to SC formulation, potentially introducing selection bias. More refractory patients may have been less prone to switch to SC vedolizumab. Fifthly, the scheduling of endoscopies by the treating physician, especially in patients with suspected active disease, might have influenced our endoscopic remission rates (which might explain low endoscopic remission rates in CD). Lastly, all patients were included from a tertiary centre, which may represent a more refractory IBD population, potentially limiting the generalizability of our findings.

5. Conclusions

The results of our real-world study suggest that transitioning patients established on IV vedolizumab to SC appears effective and safe. Transitioning to SC vedolizumab maintenance treatment is feasible also in the refractory patient population, which had to be optimised by shortening the IV interval to every 4 weeks due to insufficient response to standard dosing. Notably, clinical and biochemical remission in patients transitioning from IV to SC vedolizumab appears to be maintained, regardless of the baseline vedolizumab dosing regimen (every 4 weeks versus every 8 weeks). Further prospective studies with longer follow-up periods are needed to confirm these findings.

Author Contributions: Conceptualization, G.N. and Š.P.; methodology, Š.P., G.N., D.D. and J.H.; software and validation, G.N. and J.H.; formal analysis, J.H.; investigation, Š.P., G.N., J.H., M.K., T.K., N.S. and D.D.; resources; Š.P., G.N., J.H., M.K., T.K., N.S. and D.D.; data curation: Š.P.; writing-original draft preparation, G.N. and Š.P.; writing-review and editing: Š.P., G.N. and J.H.; visualization, G.N. and Š.P.; supervision, G.N.; project administration: G.N. All authors have read and agreed to the published version of the manuscript.

Funding: This research received no external funding.

Institutional Review Board Statement: Ethical permission was granted by the Slovenian national ethics committee (0120-013/2016-2), date: 4 February 2016.

Informed Consent Statement: Informed consent was obtained from all subjects involved in the study.

Data Availability Statement: Data and materials available on request from the corresponding authors.

Acknowledgments: We would like to thank Carmen Bobnar Sekulić and Tadeja Polanc for their administrative support.

Conflicts of Interest: Jurij Hanžel served as a speaker and/or consultant for Abbvie, Alimentiv Inc, Janssen, and Takeda outside the submitted work. Nataša Smrekar has served as a speaker, a consultant and/or an advisory board member for Abbvie, Takeda, Pfizer, Janssen, Biogen, Oktal Pharma, Novartis, Sobi, MSD and Amgen outside the submitted work. Matic Koželj reports lecture fees from Takeda, Amgen and Abbvie outside the submitted work. Tina Kurent reports lecture fees from Takeda, Ferring, Janssen, Abbvie and MSD outside the submitted work. David Drobne has served as a speaker, a consultant and/or an advisory board member for MSD, Abbvie, Takeda, Pfizer, Janssen, Krka, Eli Lilly, Oktal Pharma, Roche, Novartis, Amgen and Lek outside of the submitted work. Gregor Novak has served as a speaker, a consultant and/or an advisory board member for Abbvie, Takeda, Pfizer, Janssen, Sandoz, Krka, Lek, Oktal Pharma, Sobi and Biogen outside of the submitted work. For the remaining authors, there are no conflicts of interest.

References

1. Cai, Z.; Wang, S.; Li, J. Treatment of Inflammatory Bowel Disease: A Comprehensive Review. *Front. Med.* **2021**, *8*, 765474. [CrossRef]
2. Wyant, T.; Fedyk, E.; Abhyankar, B. An Overview of the Mechanism of Action of the Monoclonal Antibody Vedolizumab. *J. Crohn's Colitis* **2016**, *10*, 1437–1444. [CrossRef]
3. Novak, G.; Hindryckx, P.; Khanna, R.; Jairath, V.; Feagan, B.G. The safety of vedolizumab for the treatment of ulcerative colitis. *Expert. Opin. Drug Saf.* **2017**, *16*, 501–507. [CrossRef]
4. Feagan, B.G.; Rutgeerts, P.; Sands, B.E.; Hanauer, S.; Colombel, J.-F.; Sandborn, W.J.; Vn Assche, G.; Axler, J.; Kim, H.-J.; Danese, S.; et al. Vedolizumab as Induction and Maintenance Therapy for Ulcerative Colitis. *N. Engl. J. Med.* **2013**, *369*, 699–710. [CrossRef]
5. Sandborn, W.J.; Feagan, B.G.; Rutgeerts, P.; Hanauer, S.; Colombel, J.-F.; Sands, B.E.; Lukas, M.; Fedorak, R.N.; Lee, S.; Bressler, B.; et al. Vedolizumab as induction and maintenance therapy for Crohn's disease. *N. Engl. J. Med.* **2013**, *369*, 711–721. [CrossRef]
6. Outtier, A.; Wauters, L.; Rahier, J.; Bossuyt, P.; Colard, A.; Franchimont, D.; Lambrecht, G.; Macken, E.; Van Moerkercke, W.; Baert, F.; et al. Effect of vedolizumab dose intensification on serum drug concentrations and regain of response in inflammatory bowel disease patients with secondary loss of response. *GastroHep* **2021**, *3*, 63–71. [CrossRef]
7. Ungar, B.; Malickova, K.; Hanžel, J.; Abu Arisha, M.; Paul, S.; Rocha, C.; Ben Shatach, Z.; Abitbol, C.M.; Haj Natour, O.; Selinger, L.; et al. Dose optimisation for Loss of Response to Vedolizumab-Pharmacokinetics and Immune Mechanisms. *J. Crohn's Colitis* **2021**, *15*, 1707–1719. [CrossRef]
8. Sandborn, W.J.; Baert, F.; Danese, S.; Krznarić, Ž.; Kobayashi, T.; Yao, X.; Chen, J.; Rosario, M.; Bhatia, S.; Kisfalvi, K.; et al. Efficacy and Safety of Vedolizumab Subcutaneous Formulation in a Randomized Trial of Patients with Ulcerative Colitis. *Gastroenterology* **2020**, *158*, 562–572. [CrossRef]
9. Vermeire, S.; D'Haens, G.; Baert, F.; Danese, S.; Kobayashi, T.; Loftus, E.V.; Bhatia, S.; Agboton, C.; Rosario, M.; Chen, C.; et al. Efficacy and Safety of Subcutaneous Vedolizumab in Patients with Moderately to Severely Active Crohn's Disease: Results from the VISIBLE 2 Randomised Trial. *J. Crohn's Colitis* **2022**, *16*, 27–38. [CrossRef]

10. Buisson, A.; Serrero, M.; Orsat, L.; Nancey, S.; Rivière, P.; Altwegg, R.; Peyrin-Biroulet, L.; Nachury, M.; Hébuterne, X.; Gilletta, C.; et al. Comparative Acceptability of Therapeutic Maintenance Regimens in Patients with Inflammatory Bowel Disease: Results from the Nationwide ACCEPT2 Study. *Inflamm. Bowel Dis.* **2023**, *29*, 579–588. [CrossRef]
11. Jonaitis, L.; Marković, S.; Farkas, K.; Gheorghe, L.; Krznarić, Ž.; Salupere, R.; Mokricka, V.; Spassova, Z.; Gatev, D.; Grosu, I.; et al. Intravenous versus subcutaneous delivery of biotherapeutics in IBD: An expert's and patient's perspective. *BMC Proc.* **2021**, *15*, 25. [CrossRef]
12. Volkers, A.; Straatmijer, T.; Duijvestein, M.; Sales, A.; Levran, A.; van Schaik, F.; Maljaars, J.; Gecse, K.; Ponsioen, C.; Grootjans, J.; et al. Real-world experience of switching from intravenous to subcutaneous vedolizumab maintenance treatment for inflammatory bowel diseases. *Aliment. Pharmacol. Ther.* **2022**, *56*, 1044–1054. [CrossRef]
13. Wiken, T.H.; Høivik, M.L.; Buer, L.; Warren, D.J.; Bolstad, N.; Moum, B.A.; Anisdahl, K.; Småstuen, M.C.; Medhus, A.W. Switching from intravenous to subcutaneous vedolizumab maintenance treatment in patients with inflammatory bowel disease followed by therapeutic drug monitoring. *Scand. J. Gastroenterol.* **2023**, *58*, 1–11. [CrossRef]
14. Ventress, E.; Young, D.; Rahmany, S.; Harris, C.; Bettey, M.; Smith, T.; Moyses, H.; Lech, M.; Gwiggner, M.; Felwick, R.; et al. Transitioning from Intravenous to Subcutaneous Vedolizumab in Patients with Inflammatory Bowel Disease [TRAVELESS]. *J. Crohn's Colitis* **2022**, *16*, 911–921. [CrossRef]
15. Bergqvist, V.; Holmgren, J.; Klintman, D.; Marsal, J. Real-world data on switching from intravenous to subcutaneous vedolizumab treatment in patients with inflammatory bowel disease. *Aliment. Pharmacol. Ther.* **2022**, *55*, 1389–1401. [CrossRef]
16. Harvey, R.F.; Bradshaw, J.M. A simple index of Crohn's-disease activity. *Lancet* **1980**, *1*, 514. [CrossRef]
17. Lewis, J.D.; Chuai, S.; Nessel, L.; Lichtenstein, G.R.; Aberra, F.N.; Ellenberg, J.H. Use of the noninvasive components of the Mayo score to assess clinical response in ulcerative colitis. *Inflamm. Bowel Dis.* **2008**, *14*, 1660–1666. [CrossRef]
18. Labaere, D.; Smismans, A.; Van Olmen, A.; Christiaens, P.; D'Haens, G.; Moons, V.; Cuyle, P.-J.; Frans, J.; Bossuyt, P. Comparison of six different calprotectin assays for the assessment of inflammatory bowel disease. *United Eur. Gastroenterol. J.* **2014**, *2*, 30–37. [CrossRef]
19. Sharara, A.I.; Malaeb, M.; Lenfant, M.; Ferrante, M. Assessment of Endoscopic Disease Activity in Ulcerative Colitis: Is Simplicity the Ultimate Sophistication? *Inflamm. Intest. Dis.* **2022**, *7*, 7–12. [CrossRef]
20. Amiot, A.; Serrero, M.; Peyrin-Biroulet, L.; Filippi, J.; Pariente, B.; Roblin, X.; Buisson, A.; Stefanescu, C.; Trang-Poisson, C.; Altwegg, R.; et al. One-year effectiveness and safety of vedolizumab therapy for inflammatory bowel disease: A prospective multicentre cohort study. *Aliment. Pharmacol. Ther.* **2017**, *46*, 310–321. [CrossRef]
21. Amiot, A.; Serrero, M.; Peyrin-Biroulet, L.; Filippi, J.; Pariente, B.; Roblin, X.; Buisson, A.; Stefanescu, C.; Trang-Poisson, C.; Altwegg, R.; et al. Three-year effectiveness and safety of vedolizumab therapy for inflammatory bowel disease: A prospective multi-centre cohort study. *Aliment. Pharmacol. Ther.* **2019**, *50*, 40–53. [CrossRef]
22. Eriksson, C.; Marsal, J.; Bergemalm, D.; Vigren, L.; Björk, J.; Eberhardson, M.; Karling, P.; Söderman, C.; SWIBREG Vedolizumab Study Group; Myrelid, P.; et al. Long-term effectiveness of vedolizumab in inflammatory bowel disease: A national study based on the Swedish National Quality Registry for Inflammatory Bowel Disease (SWIBREG). *Scand. J. Gastroenterol.* **2017**, *52*, 722–729. [CrossRef]
23. Biemans, V.B.C.; van der Woude, C.J.; Dijkstra, G.; van der Meulen-de Jong, A.E.; Oldenburg, B.; de Boer, N.K.; Löwenberg, M.; Srivastava, N.; Bodelier, A.G.L.; West, R.L.; et al. Vedolizumab for Inflammatory Bowel Disease: Two-Year Results of the Initiative on Crohn and Colitis (ICC) Registry, A Nationwide Prospective Observational Cohort Study: ICC Registry—Vedolizumab. *Clin. Pharmacol. Ther.* **2020**, *107*, 1189–1199. [CrossRef]
24. Ungar, B.; Malickova, K.; Hanžel, J.; Abu-Arisha, M.; Paul, S.; Rocha, C.; Ben-Shatach, Z.; Haj-Natour, O.; Yavzori, M.; Fudim, E.; et al. P177 Lower vedolizumab trough levels before interval shortening are not predictive of success of the intervention. *J. Crohn's Colitis* **2020**, *14*, S226–S227. [CrossRef]
25. Bittner, B.; Richter, W.; Schmidt, J. Subcutaneous Administration of Biotherapeutics: An Overview of Current Challenges and Opportunities. *BioDrugs* **2018**, *32*, 425–440. [CrossRef]
26. Ungar, B.; Kopylov, U.; Yavzori, M.; Fudim, E.; Picard, O.; Lahat, A.; Coscas, D.; Waterman, M.; Haj-Natour, O.; Orbach-Zingboim, N.; et al. Association of Vedolizumab Level, Anti-Drug Antibodies, and α4β7 Occupancy With Response in Patients With Inflammatory Bowel Diseases. *Clin. Gastroenterol. Hepatol.* **2018**, *16*, 697–705.e7. [CrossRef]
27. Dosing, S. A Randomized Trial of Vedolizumab Dose Optimization in Patients with Moderate to Severe Ulcerative Colitis Who Have Early Nonresponse and High Drug Clearance: The ENTERPRET Trial. *Gastroenterol. Hepatol.* **2022**, *18*, 7–8.

medicina

MDPI

Review

Bacterial Biofilms—A Threat to Biliary Stents, Understanding Their Formation, Clinical Consequences and Management

Jolanta Gruszecka [1,2,*] and Rafał Filip [3,4]

1 Institute of Health Sciences, Medical College of Rzeszow University, 35-959 Rzeszow, Poland
2 Department of Clinical Microbiology, Clinical Hospital No. 2, 35-301 Rzeszow, Poland
3 Faculty of Medicine, University of Rzeszow, 35-959 Rzeszow, Poland
4 Department of Gastroenterology with IBD Unit, Clinical Hospital No. 2, 35-301 Rzeszow, Poland
* Correspondence: jagrusz@onet.pl

Abstract: A biofilm is a community of microbial cells which are enclosed in an external matrix and separated by a network of water channels attached to natural or artificial surfaces. Biofilms formed inside biliary stents consist of a mixed spectrum of bacterial communities, most of which usually originate from the intestines. The patency of biliary stents is the most important problem. Stent occlusion can threaten the health and even life of patients. The main cause of this phenomenon is bile sludge, which is an excellent environment for the multiplication and existence of microorganisms. Due to the great clinical importance of maintaining the patency of biliary stents, several methods have been developed to prevent the accumulation of sludge and the subsequent formation of biofilm; these include, among others, the use of anti-adhesive materials, coating the inner surface of stents with metal cations (silver, copper) or other antimicrobial substances, the implementation of biodegradable drug-eluting biliary stents and the development of a new stent design with an anti-reflux effect. This article presents the latest information on the formation of biofilms in biliary stents, as well as historical and future methods of prevention.

Keywords: biofilm; biliary tract stents; biofilm elimination

1. Introduction

In the 1940s, it was observed that the majority of microorganisms in the aquatic environment formed aggregates that adhered to objects immersed in water, exhibiting different properties from microorganisms that occur as single cells [1]. This specific form of existence of bacteria and fungi was called biofilm [2]. It has an advantage over planktonic forms (occurring as single, scattered cells, most often in an aquatic environment) in that it provides a greater chance of survival in a changing environment [3–6]. An important feature of a biofilm is its reduced sensitivity to physicochemical factors as well as stress [7]. In their natural habitat, more than 90% of bacteria occur in this form [3,8]. Biofilms are communities of microorganisms that adhere to each other and are embedded in an extracellular matrix with a diverse chemical and structural composition created by the microorganisms themselves [9,10]. The gradients that exist in the biofilm matrix allow for the formation of microniches, created by different microorganisms [9]. Anaerobic microorganisms and cells which are more sensitive to environmental stressors, such as hazardous chemicals, inappropriate pH, or physical damage, can live in deeper layers of biofilm [11]. The top layers of the biofilm, with an appropriate partial oxygen concentration and access to nutrients, enable microbial cells to carry out active metabolic processes with a high rate of division [12].

The goal of this review thesis was to present the issues related to the phenomenon of biofilm formation on medical devices, especially in biliary tract stents, as well as to present the methods of preventing and combating this very dangerous process for patients.

2. Biofilm Formation

Biofilm formation is a complex process. The biofilm life cycle consists of distinct stages: (1) initialization, (2) bacterial adhesion and aggregation, (3) microenvironment formation, (4) microenvironment maturation, (5) dispersion and (6) quorum sensing QS, as shown in Figure 1 [13–15].

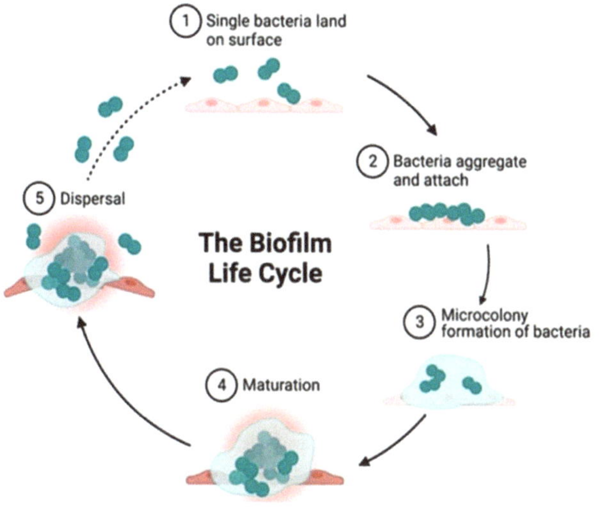

Figure 1. Stages of biofilm creation [14].

The first reversible stage of initialization occurs mainly due to physicochemical reactions between the colonized natural or artificial surface and the microbial cell [16]. Reversible binding most often occurs as a result of electrostatic, hydrophobic, van der Waals and surface tension interactions, or due to gravitational forces [17]. During the reversible attachment stage, microbial cells come into contact with the surface and begin to adhere, but can still be relatively easily removed. Reversible binding is generally mediated by the proteins found on the surface of the microorganisms. The rate of microbial adhesion is significantly dependent on the characteristics of the colonized surface, including hydrophobicity, topography and charge [18]. The chemical composition of the pathogen's cell wall and the roughness of the substrate surface are extremely important at this stage, because both factors affect the type of physicochemical interactions [7].

During the adhesion and aggregation phase of bacteria, irreversible attachment occurs; the cells completely bind to the surface and begin to produce an extracellular matrix that prevents their physical removal from the surface [2,13]. After attachment, the microorganisms change their profile from planktonic to sessile. The composition of the biofilm is different; it is a mixture of various secreted biomolecules: polysaccharides, proteins, lipids, teichoic acids and environmental DNA (eDNA) [11].

In the phase of microenvironment formation, biofilms grow and gain a three-dimensional structure due to cell proliferation, adhesion between microbial cells and the secretion of extracellular mucus [19]. Cells in the center of the biofilm, which have limited access to oxygen and nutrients, can often become dormant. These microorganisms are metabolically dormant, but not dead [20,21]. Anaerobic metabolic pathways become dominant among the microorganisms living deep in the biofilm due to their limited access to oxygen and nutrients [17,19,22].

During the maturation stage, changes and differentiation occur among the cells of the microorganisms forming the biofilm [7,16]. Differences in the metabolism of microorganisms can be observed depending on their location in the biofilm structures [23]. In the biofilm, there is increased diversity in activities of the cells forming it. There are dead cells, dormant cells and cells with aerobic and anaerobic metabolism [2].

In the dispersion stage, the mature biofilm reaches a critical size, bursts and disperses planktonic microorganisms [13]. Active detachment is triggered by various environmental signals such as changes in temperature, pH, nitric oxide, nutrient deficiency, oxygen deficiency and other stress factors [24,25]. The resulting chemical gradients experienced by the cells in the biofilm are believed to be the main causes of its dispersion [26].

An important factor in biofilm structures is the phenomenon of quorum sensing (QS). It represents chemical communication through signal substances or autoinducers (farnesol, tyrosol, dodecanol) which accumulate with increasing cell density, responding to changes in the external environment as well as to processes inside the biofilm. Microorganisms use autoinducers to regulate the course of physiological processes or the expression of pathogenicity factors in a controlled manner, depending on their number [27]. When the appropriate number of cells, i.e., the quorum, is reached, the concentration of the autoinducer exceeds the threshold value, and the controlled regulation of gene expression occurs, which enables the cooperation of a given population of microorganisms and may cause the simultaneous production of virulence factors. These factors affect, among others, sensitivity or resistance to biocides [2,23,28,29]. The metabolic diversity of microorganisms in individual layers of the biofilm may also lead to a differing sensitivity to antibiotics [12]. The QS system occurs both between cells of the same and different species and provides an opportunity for the coordinated regulation of important life processes in the entire population [2].

3. Biofilm Distribution

Biofilms are ubiquitous in almost every environment, affecting human health and industry [3,30]. They have been created on a variety of surfaces in different habitats, both natural and man-made, including in the hospital environment [30,31]. One of the first to recognize the importance of biofilms in medicine was Niels Høiby [32]. Since then, this phenomenon has been supported by numerous pieces of evidence [32]. Biofilms are involved in many different bacterial infections in the body. The National Institutes of Health (NIH) revealed that of all bacterial infections, 60–80% are associated with biofilm formation [15,33]. Biofilms are formed on various medical devices such as contact lenses, catheters, prostheses, biliary stents, valves and pacemakers, but also on various surfaces of the human body, including the skin or the mucous membranes of the respiratory and digestive tracts, constituting an important reservoir for the initiation of new infections [11]. Environmental biofilms in drinking water systems may be a source of the respiratory pathogen *Legionella pneumophila*, the causative factor of Legionnaires' disease, and opportunistic pathogens such as *Mycobacterium avium*, which poses a health risk, especially to immunocompromised patients. *Legionella* spp. often form biofilms, particularly in shower houses, which are believed to promote the persistence and resistance of the respiratory pathogen to chlorine [34]. Cholera, a waterborne diarrheal disease, is caused by *Vibrio cholerae*. This pathogen moves between the water body, where it forms biofilms on chitinous surfaces, and the human body, where it successfully colonizes the gastrointestinal tract. Studies with neonatal mice showed that both intact biofilms and dispersed sessile *V. cholerae* cells are more infectious than free-living planktonic cells [35]. From a clinical point of view, the most important features of a biofilm are its high resistance to antimicrobial agents and the immune system, as well as its strong ability to colonize patient tissues and biomedical materials [36]. Studies have shown that bacteria originating from biofilms are characterized by a higher resistance to antimicrobial compounds than their individual, planktonic counterparts [13]. Factors that cause higher antibiotic resistance in biofilm-associated infections include the following: metabolic changes in bacterial cells, antibiotic inactivation and reduced penetration through the extracellular matrix, inoculum effects related to the high density of bacterial cells in relation to the number of available antibiotic molecules and the increased exchange of resistance mechanisms between bacteria in close proximity to each other [11]. Biofilm-forming microorganisms may be dangerous for patients with predisposing factors, such as comorbidities or immunosuppression.

4. Biofilms in the Human Body and on Medical Equipment and Devices

Biofilms form on biomaterials, such as dental prostheses, catheters, endoprostheses, biliary tract stents, as well as on living tissues. Microorganisms within the biofilms are up to 1000 times more tolerant to antibiotic therapy than their planktonic counterparts, which allows them to evade elimination excellently [13]. Opportunistic biofilms readily colonize virtually any surface, especially those that are foreign to the body, such as implanted medical devices, used both in the short-term and for extended periods of time. As various medical devices are increasingly used in all branches of medicine, strategies to control biofilm formation in various environments are of great importance [37]. The spread of biofilms on medical implants is one of the main factors triggering persistent and chronic infections in clinical settings [38]. Biofilm formation and microbial colonization are encountered on a wide variety of implantable medical devices. Common examples include catheters, feeding tubes, cochlear implants, cardiac valves and pacemakers, urologic and breast implants, biliary stents, endoscopic tubes, contact lenses and neurosurgical and orthopedic implants. The abundance of microorganisms on various surfaces and sites in the body is observed depending on environmental characteristics, such as the presence of fluid flow and the surface properties of the implants, as well as the interplay between colonization and the human immune response [38]. The extracellular matrix protects microbial cells from drying out, constituting a barrier that impedes the penetration of antibiotics and antiseptics, impedes the interaction of the host's immune system (including impeding phagocytosis and inhibiting the penetration of antibodies), reduces the effective concentration of antibiotics reaching bacterial cells and creates optimal conditions for the formation of microbial colonies [7,9].

Multi-species biofilms in the human body can be both a positive and negative phenomenon. They are created by microbiota living in the oral cavity—mainly on the surface of teeth, in the intestines, in the vagina or on the skin [2]. There are over 700 different species of bacteria in the human oral cavity. They can initiate the formation of dental biofilms, also known as dental plaque. The exact composition of dental biofilms varies both from site to site in the mouth and from person to person. The core composition of the microbiome has been proposed to include species from the following genera: *Streptococcus*, *Veillonella*, *Granulicatella*, *Neisseria*, *Haemophilus*, *Corynebacterium*, *Rothia*, *Actinomyces*, *Prevotella*, *Capnocytophaga*, *Porphyromonas* and *Fusobacterium* [39]. Dental biofilm is a permanent reservoir of microorganisms, which can systematically spread throughout the body. Dental biofilm bacteria are also directly and indirectly associated with various systemic diseases, such as aspiration pneumonia, premature birth and low-birth-weight children, diabetes, circulatory system diseases, atherosclerosis and infective endocarditis [40]. For example, the caries production of *Streptococcus mutans* results from the adhesive properties (biofilm) of the extracellular polymeric substances secreted by it, the production of which is partially stimulated by the presence of fructose and the conversion of simple sugars into intracellular polysaccharides (mutan, dextran, levan) [41,42]. The final bacterial metabolites that make up dental plaque are organic acids that damage the enamel, allowing various cariogenic bacteria to begin the process of tooth destruction [42]. *S. mutans* can also cause bacterial endocarditis, especially the subacute clinical form in 50–70% of all cases of this disease entity. In people with risk factors for the development of the disease, which include congenital heart defects, rheumatic fever, heart surgery and damage to the oral mucosa, streptococci are allowed to enter the blood vessels. This can cause transient bacteremia with heart valve colonization and biofilm formation [42]. The intestinal biofilm, built by multi-species microorganisms, protects against chronic gastrointestinal diseases, retains water in the body, stimulates the host's immunity and participates in the production of vitamins (vitamin K, biotin) and the breakdown of food. The gastrointestinal microflora contains more than 1000 microbial species and the intestinal biofilm is formed by, among others, bacteria of the genera *Bacteroides*, *Bifidobacterium*, *Enterococcus* and *Streptococcus* [43]. The ability to form biofilms is also characteristic of the lactic acid bacteria of the *Lactobacillus* genus. Colonizing the vagina and intestines, they protect against infections of the digestive tract, urinary tract and sexually transmitted diseases. In the vagina, these bacteria participate in the protection of the mucous membrane against pathogens, secreting metabolites (organic acids, hydrogen peroxide, bacteriocins) with antimicrobial activity [44,45]. Over 60% of the microorganisms colonizing human skin are various bacteria that form biofilms. The dominant flora includes *Staphylococcus* spp., *Corynebacterium* spp. and *Propionibacterium* spp. [46]. The natural microflora on the surface of healthy skin performs a protective function; a biofilm is the predominant form of microbial life on its surface [2].

Modern medicine increasingly relies on surgical interventions and the placement of permanent medical devices in the patient's body. Both surgical procedures and medical devices can introduce foreign microorganisms into the body, which can serve as a permanent reservoir of infection and cause biofilm formation. Almost 80% of device-related infections are caused by biofilms formed by Gram-positive *Staphylococcus* spp. bacteria, primarily *Staphylococcus epidermidis* and *Staphylococcus aureus* [47]. *Staphylococcus* spp. are a commensal of the skin, but in favorable conditions they can cause infection. They can be introduced into the body via contaminated medical devices, from medical personnel or from patients themselves [21]. Medical devices are made of many materials, including metals, plastics and ceramics. Plastics are more easily colonized than metal surfaces, but bacterial biofilms can form on both surfaces [48]. The van der Waals and hydrophobic forces are the main factors influencing the adhesion of bacteria to the surfaces of medical devices [49]. Surface characteristics, including hydrophobicity, texture and electrostatic charge, can facilitate the attachment of microorganisms and influence which strains have an affinity for it [13]. The most frequently isolated bacterial strains associated with biofilms in medical devices used on the long-term and short-term are presented in Table 1.

Table 1. Most frequently isolated bacterial species on the surface of implanted medical devices.

Permanent Medical Device for Long-Term Use	Most Frequently Isolated Bacteria
Orthopedic implants [38,50–52]	*K. pneumoniae*
	A. baumannii
	S. epidermidis
	S. aureus

Table 1. *Cont.*

Permanent Medical Device for Long-Term Use	Most Frequently Isolated Bacteria
Stents [53]	*E. coli*
	Enterobacter spp.
	Klebsiella spp.
	P. aeruginosa
	E. faecalis
	Streptococcus spp.
	S. aureus
	S. epidermidis
Cochlear implants [54]	*P. aeruginosa*
	S. pyogenes
	S. epidermidis
	S. aureus
Breast implants [55]	*E. coli*
	Mycobacterium spp.
	S. epidermidis
	S. aureus
	Streptococcus spp.
	Bacillus spp.
Medical Device For Short-Term Use	**Most Frequently Isolated Bacteria**
Urinary catheter [56,57]	*E. coli*
	P. aeruginosa
	K. pneumoniae
	A. baumannii
	Enterobacter spp.
	S. epidermidis
	E. faecalis
Central line catheter [58]	*P. aeruginosa*
	K. pneumoniae
	S. epidermidis
	S. aureus
	E. faecalis
Endotracheal tube [59]	*P. aeruginosa*
	K. pneumoniae
	Acinetobacter spp.
	Enterobacter spp.
	S. aureus
	E. faecalis

Table 1. *Cont.*

Medical Device For Short-Term Use	Most Frequently Isolated Bacteria
Feeding tube [60]	*P. aeruginosa*
	Enterococcus spp.
	Bacillus spp.
	Staphylococcus spp.
Contact lenses [38,56,61]	*E. coli*
	P. aeruginosa
	S. aureus

5. Biofilms on the Inner Surface of Biliary Stents

Plastic stents in the biliary tract are often occluded by biliary sludge, which provides an excellent environment for microorganisms to adhere, multiply and thrive in. This is an additional factor contributing to biliary stent obstruction [16]. Stent patency is a major concern for patients, endoscopists and physicians, because it can affect both the life expectancy and treatment schedule of patients and depends on biliary tract injury and stent location. Biliary stent occlusion can occur due to several factors: biliary sludge causing the slowing of bile flow, bile viscosity, food exposure and the subsequent formation of a coating from dietary fibers. The reflux of intestinal contents into the bile duct allows for the easy adhesion, colonization and growth of bacteria on the inner surface of the stent, leading to an ascending bacterial infection. No ideal stent with permanent patency has been identified to date [62–64].

Microorganisms isolated from obstructed biliary stents (anaerobic and aerobic bacteria and fungi) secrete several types of proteins, such as fibronectin, vitronectin, laminin, fibrin and collagen, which increase their adhesion. It is believed that the biofilm on the inner surface of the stent causes it to be irregular, which further facilitates the accumulation of sediment and debris, precipitating the occurrence of obstruction and the recurrence of cholangitis [65]. Biofilms formed inside stents consist of a mixed spectrum of microorganisms [10]. Polymicrobial communities act synergistically on biofilm maturation, causing it to gradually become thicker [66]. Stent occlusion leads to jaundice and bacterial cholangitis with polymicrobial infections in 90% of patients, as shown in Figure 2 [16]. The inappropriate use of antimicrobial agents may lead to the emergence of antimicrobial resistance and, consequently, ineffective treatment of stent-related cholangitis [67].

Currently, more than 70% of patients with biliary jaundice are treated with the implantation of a biliary stent made of plastic or metal [68]. Plastic stents can be removed and replaced if necessary, which is their main advantage. Self-expanding metal stents are durable and have the advantages of a larger lumen and a longer period of patency [68]. In recent years, biodegradable biliary stents have also been developed for endoscopic applications [69]. Studies comparing the properties and safety of different types of stents for preoperative biliary drainage are limited, and no consensus has yet been reached on the optimal type [70].

A study conducted in Italy analyzed the composition of biofilms colonizing biliary stents. For this purpose, biliary stents were collected from 56 patients. The study participants were 32 to 89 years old (mean 67.30 $\pm$ 15.75) and had been wearing stents for 13 to 330 days (mean 70.21 $\pm$ 73.35). All stents were collected from patients who had not undergone antibiotic prophylaxis or chemotherapy. The time of stent patency ranged from 5 to 330 days [68]. The study used metal stents made of a braided nickel–titanium alloy (Nitinol) with a full-length silicone polymer lining, or plastic—made of polyethylene. The species associated with stents were usually anaerobic and Gram-positive bacteria, comprising 50% and 58.3%, respectively. The three species *Streptococcus anginosus*, *Escherichia coli* and *Enterococcus faecalis* were found in more than 80% of the samples (prevalence = 83.0%) [68].

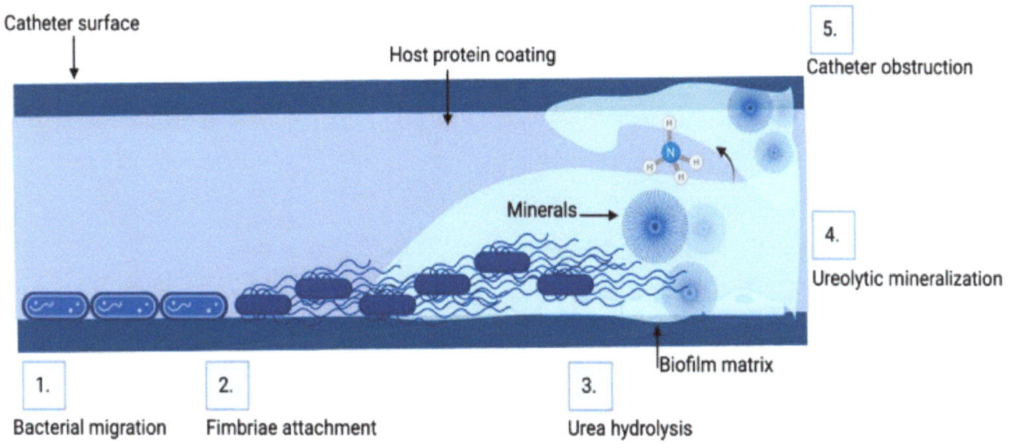

Figure 2. Biofilm formation on the inner surface of biliary tract stents [14].

In another study, a prospective microbiological analysis of biliary stent biofilm from all patients requiring elective or emergency stent replacement/removal was performed in northern India between April 2011 and March 2014. A total of 81 patients (41 males) aged 20–86 years were included in the study. The primary reasons for stent placement were gallstones (n = 46, 56.8%), benign stenosis (n = 29, 35.8%) and malignant stenosis (n = 6, 7.4%). All stents were made of polyethylene and were placed endoscopically. The median duration of stent placement was 65 days (range 5–1095 days). Cholangitis at the time of stent placement was present in 50 (61.7%) patients. A polybacterial biofilm was detected in most stents (n = 73, 90.1%), while single species were found in the remaining eight (9.9%) cases. The most common Gram-negative bacteria in the cited study were *Pseudomonas* spp. (n = 38), *Citrobacter* spp. (n = 23), *Klebsiella* spp. (n = 22), *Serratia* spp. (n = 16), *Escherichia coli* (n = 14), *Aeromonas* spp. (n = 12), *Proteus* spp. (n = 10) and *Enterobacter* spp. (n = 9). Among the Gram-positive bacteria, the most common were *Staphylococcus* spp. (n = 20), *Streptococcus* spp. (n = 13) and *Enterococcus* spp. (n = 13) [16].

In a prospective study that was conducted in Rome between July 2019 and February 2021 in patients requiring urgent biliary stent exchange/removal due to benign biliary stenosis, the mean duration from stent placement was 120 days [71]. A microbiological analysis of bile and stent samples taken from 22 patients was performed. The dominant species isolated in the bile and stent samples were *Lactobacillus* spp. (7.1 and 13.7% in bile and stent samples, respectively), *Enterococcus faecalis* (9.2% and 9.7%), followed by *Escherichia coli* (8.2 and 9.1%), *Klebsiella pneumoniae* (7.7% and 9.1%, respectively) and *Enterococcus faecium* (6.6% and 6.9%). Among the anaerobic Gram-positive bacteria, the most frequently isolated genus was *Clostridium* spp. (5.1% and 5.1%), especially *C. perfringens* (3.1% and 2.9%), while among the Gram-negative anaerobic bacteria, the most common genus was *Bacteroides* spp. (2.0% and 1.1%), with no differences between the individual species. The most commonly isolated yeast species were *Candida* spp. (11.7% and 8.0%), especially *Candida albicans* (8.7% and 7.4%) [71]. The biofilm which formed inside stents was an organized community of microorganisms enclosed in a self-produced exopolysaccharide matrix containing proteins and other polymers, which grew on a solid, synthetic surface, [16,66].

The experiment conducted in Japan aimed to investigate the antibacterial efficacy of polyurethane biliary stents coated with silver compared to polyurethane stents without this coating. Stent obstruction is caused mainly by the deposition of bile sediments, which consist of cholesterol crystals, calcium bilirubinate, calcium palmitate, bacteria and/or fungi, microbiological by-products, proteins, dietary fiber and glycoproteins. Silver ions proved to be a much stronger inhibitor of biofilm formation than many other antibacterial agents, and at lower concentrations. The observation of an almost complete lack of bacterial adhesion on the surfaces of silver-coated biliary stents after a longer period of time indicated the possibility of achieving the long-term patency of polyurethane stents coated with this metal [63].

6. Preventing and Combating Bacterial Biofilm

Because of the medical importance of bacterial biofilms, effective methods of preventing their formation and combating them are of great importance in clinical practice.

Bacterial cells in a biofilm are constantly growing and dispersing. These processes are regulated by complex signaling pathways. Therefore, the mass of the biofilm is constantly changing over days and weeks. The inhibition of biofilm formation cannot be achieved solely by preventing the adhesion of cells or proteins, as each bacterial species has its own surface characteristics that regulate its adhesion to a given surface [13].

The preventive action is aimed at changing the physical properties of the surface by modifying the self-assembled monolayer (SAM) that inhibits bacterial adhesion, disrupts biofilm formation or promotes its removal, in order to prevent the accumulation of a biofilm layer [72,73]. A self-assembled monolayer is a thin, single-layer film of small molecules that are attached to a surface in a highly ordered way [13]. SAMs are made of small molecules; they usually have a thickness in the range of 1–5 nm depending on the size of the molecule, and thus belong to the category of nanoscale materials. Compared to polymer films or metals, SAMs are resistant to release into the surrounding environment due to their strong interaction with the surface [74]. In addition to inhibiting bacterial biofilm formation, the ultimate functionality of SAMs can also be used to retain a biocidal agent on their surface. Most bactericidal SAMs kill by contact, using biocidal agents that act on the outside of the bacteria. In the case of these surfaces, bacteria are killed upon contact [13]. The covalent attachment between the surface and the bactericide is particularly important because it prevents its release, which could lead to the development of bacterial resistance. It also allows the use of relatively low concentrations of the active substance compared to the doses administered in vivo. If the bactericide requires internalization into the bacterial cell to act effectively, it can be used in an SAM, but then the mechanism of action is killing by release, meaning that the SAM releases the bactericide over time. In this case, there may be potential problems with controlling the concentration of the bactericide. This can lead to toxic effects if too much is released at the beginning of the application, and to inactivity when the coating loses its bactericide [13]. To inhibit biofilm formation, SAMs can form quaternary ammonium compounds on gold, titanium or silicone surfaces [75,76]. Another effective strategy for inhibiting the formation of biofilms of Gram-positive *S. aureus* and *S. epidermidis* bacteria is the covalent placement of the drug (vancomycin) on the surface of titanium and stainless steel alloys. Much lower overall concentrations of the antibiotic are used than the therapeutic, while maintaining the antibiotic at the biomedical implant site. As studies have shown, this effect can be maintained even after exposure to serum proteins [13,77]. A study was conducted to investigate the properties of an aminosilanized titanium surface onto which the antiseptic chlorhexidine was grafted using glutaraldehyde as a linker. The resulting surface inhibited the formation of an *S. aureus* biofilm in proportion to the concentration of chlorhexidine used [78]. Similarly, the applied salicylic acid was released into the substrate and showed an up to 90% inhibitory effect on the viability and growth of settled bacteria: *E. coli*, *S. aureus* and *S. epidermidis*. This solution may be suitable for implementation in situations with a limited exposure window, e.g., during the healing period after surgery, although these functionalized surfaces retain their antibiotic activity for a limited time [79]. Metal cations, particularly silver ions, which are known for their antibacterial properties, have also been grafted onto self-assembled monolayers on various surfaces to disrupt the biofilm formation process, in addition to antibiotics, which have also been grafted onto SAMs [80]. Silver has a broad spectrum of antimicrobial activity and, if used in small amounts due to toxicity concerns and so as to minimize costs, can be an effective bactericide [81,82]. Silver cations have been coordinated to titanium and stainless steel surfaces. Studies have confirmed its biocidal efficacy against *E. coli*, *S. aureus*, *S. epidermidis* and *P. aeruginosa*. A silver-coordinating SAM inhibited bacterial cell adhesion by three orders of magnitude and reduced the likelihood of biofilm formation by 80%. This SAM coating has been proven to kill bacteria and prevent their adhesion. The amount of silver required for this effect is less than 1 $nmol/cm^2$, which is less than many other antibacterial silver treatments [83]. Copper (cations of Cu^{2+}) also has antibacterial properties; its bactericidal effect has been assessed against *E. coli* and *S. aureus*. After five hours, almost 95% of bacterial cells were killed, and more than 99.9% were killed after 24 h [84]. Currently, the chemical composition of the biofilm matrix is known for most pathogenic microorganisms, so it would be realistic to disperse bacterial cells enclosed in biofilms by degrading the matrix. One of the main components of many bacterial biofilms is eDNA. Bacteria produce their own nucleases to digest eDNA, among other things, in order to disperse the biofilm matrix depending on the environmental conditions [85]. eDNA is a polymeric component of the matrix of many bacterial biofilms and most likely originates from cell lysis [86]. Nucleases can therefore become therapeutic agents by destroying the protective matrix and making bacteria sensitive to other treatments [11]. Many *Enterobacterales* produce extracellular amyloid fibrils, which are harmful because of their ability to adhere to surfaces and form and maintain biofilms. Specific bioactive compounds that inhibit the formation of these fibrils have been identified, effectively preventing biofilm formation and destabilizing the mature biofilms of pathogenic *E. coli* [87].

Stent obstruction is a serious problem in the treatment of biliary tract strictures; therefore, some modifications (i.e., design changes, special coatings and new biomaterials) have been proposed to prolong patency time, but there are no definitive data to support the introduction of these solutions into clinical practice [71].

7. Future Perspectives

The problem of the internal stent occlusion and the exact cause of its occurrence is not completely solved. The results of various studies have shown that the composition of the sediments from the biliary stent is not particularly dependent on the material from which it is made, the extraction procedure or any of the patient characteristics taken into account [68,88]. The main solution to this problem would be to develop innovative stents made of materials that have permanent antimicrobial properties, thus offering a promising solution to this long-standing problem [71]. Technical developments remain desirable to develop new stent materials and designs that minimize or eliminate the obstruction phenomenon [70]. The implementation of such improvements in stent design could significantly improve patient outcomes and reduce the risk of related complications [71]. Microscopic biofilms can cause serious infections and patient complications, especially in cases involving long-term medical devices. They can also be difficult or impossible to detect without removing the medical device. Further studies should focus on developing surfaces that prevent biofilm formation and facilitate their detection to provide clinicians with additional information and to enable the early identification of potential complications and the sources of these complications [89].

After analyzing the literature, the following main strategies for combating biofilms were identified:

- Development and creation of antiadhesive materials and substances with prolonged properties.
- Inhibition of the attachment of microorganisms to the substrate by using special compounds, and the destruction of biofilms early in their formation.
- Use of compounds that disrupt QS, causing the detachment of biofilms and the destruction of their vital activity.
- Use of physical destruction means (lasers, cold plasma, etc.).
- Development of drugs that destroy the biofilm matrix, facilitating cell access.
- Genetic engineering of phages.
- Use of antibacterials together with matrix-destroying factors.
- Drug-eluting biodegradable biliary stents. The drug administered in this way acts on a specific site, limiting the undesirable effects on the rest of the body, and the speed of its release can be controlled.
- Development of a new stent design with anti-reflux action [15,63,64,90–92].

The limitations in the above manuscript result from the small number of studies on this topic, and the fact that the published results are not methodologically consistent, which makes their comparison very difficult. Due to the small number of available data, the authors of this article describe the composition of biofilms to a negligible extent, focusing mainly on the types of isolated microorganisms. In the future, it would be reasonable to relate the pathogens found in biofilms to the duration the drain stays in the patient's body, the chemical composition of the biofilm and the medical indications for the insertion of a prosthesis.

Author Contributions: Conceptualization, J.G. and R.F.; writing—original draft preparation, J.G.; writing—review and editing, J.G. and R.F.; comprehensive review, R.F. All authors have read and agreed to the published version of the manuscript.

Funding: This research received no external funding.

Conflicts of Interest: The authors declare no conflicts of interest.

References

1. Valen, H.; Scheie, A.A. Biofilms and their properties. *Eur. J. Oral Sci.* **2018**, *126*, 13–18. [CrossRef] [PubMed]
2. Bigos, P.; Czerwińska, R.; Pajączkowska, M.; Nowicka, J. Mixed Oral Biofilm. *Postępy Mikrobiol.—Adv. Microbiol.* **2021**, *60*, 47–58. [CrossRef]
3. Boudarel, H.; Mathias, J.D.; Blaysat, B.; Grédiac, M. Towards standardized mechanical characterization of microbial biofilms: Analysis and critical review. *NPJ Biofilms Microb.* **2018**, *4*, 17. [CrossRef]
4. Kelsey, J.; Kielian, Y.T. Biofilm-Leukocyte Cross-Talk: Impact on Immune Polarization and Immunometabolism. *J. Innate Immun.* **2019**, *11*, 280–288.
5. Verderosa, A.D.; Totsika, M.; Fairfull-Smith, K.E. Bacterial Bio- film Eradication Agents: A Current Review. *Front. Chem.* **2019**, *7*, 824. [CrossRef]
6. Rumbaugh, K.P.; Sauer, K. Biofilm dispersion. *Nat. Rev. Microbiol.* **2020**, *18*, 571–586. [CrossRef] [PubMed]
7. Sulik-Tyszka, B.; Ceślik, J.; Swoboda-Kopeć, E. Impact of *Candida* biofilm on treatment fungal infections. *Forum Zakażeń* **2015**, *6*, 23–27. [CrossRef]
8. Wu, Y.; Cai, P.; Jing, X.; Niu, X.; Ji, D.; Ashry, N.M.; Gao, C.; Huang, Q. Soil biofilm creation enhances microbial community diversity and metabolic activity. *Environ. Int.* **2019**, *132*, 105116. [CrossRef] [PubMed]
9. Dzięgielewska, M.; Bartoszewicz, M.; Junka, A. Infections in ophthalmology complicated by a bacterial biofilm. *Forum Zakażeń* **2022**, *13*, 135–139. [CrossRef]

10. Kwon, C.I.; Lehman, G.A. Mechanisms of Biliary Plastic Stent Occlusion and Efforts at Prevention. *Clin. Endosc.* **2016**, *49*, 139–146. [CrossRef]
11. Schulze, A.; Mitterer, F.; Pombo, J.P.; Schild, S. Biofilms by bacterial human pathogens: Clinical relevance—Development, composition and regulation—Therapeutical strategies. *Microb. Cell* **2021**, *8*, 28–56. [CrossRef] [PubMed]
12. Bruinsma, G.M.; van der Mei, H.C.; Busscher, H.J. Bacterial adhesion to surface hydrophilic and hydrophobic contact lenses. *Biomaterials* **2001**, *22*, 3217–3224. [CrossRef]
13. Lundin, P.M.; Fiser, B.L.; Blackledge, M.S.; Pickett, H.L.; Copeland, A.L. Functionalized Self-Assembled Monolayers: Versatile Strategies to Combat Bacterial Biofilm Formation. *Pharmaceutics* **2022**, *14*, 1613. [CrossRef]
14. Sharma, S.; Mohler, J.; Mahajan, S.D.; Schwartz, S.A.; Bruggemann, L.; Aalinkeel, R. Microbial Biofilm: A Review on Formation, Infection, Antibiotic Resistance, Control Measures, and Innovative Treatment. *Microorganisms* **2023**, *11*, 1614. [CrossRef]
15. Guliy, O.I.; Evstigneeva, S.S. Bacterial Communities and Their Role in Bacterial Infections. *Front. Biosci. Elite Ed.* **2024**, *16*, 36. [CrossRef]
16. Vaishnavi, C.; Samanta, J.; Kochhar, R. Characterization of biofilms in biliary stents and potential factors involved in occlusion. *World J. Gastroenterol.* **2018**, *24*, 112–123. [CrossRef]
17. Khatoon, Z.; McTiernan, C.D.; Suuronen, E.J.; Mah, T.F.; Emilio, I.; Alarcon, E.I. Bacterial biofilm formation on implantable devices and approaches to its treatment and prevention. *Heliyon* **2018**, *4*, e01067. [CrossRef] [PubMed]
18. Wi, Y.M.; Patel, R. Understanding Biofilms and Novel Approaches to the Diagnosis, Prevention, and Treatment of Medical Device-Associated Infections. *Infect. Dis. Clin. N. Am.* **2018**, *32*, 915–929. [CrossRef] [PubMed]
19. Figueiredo, A.M.S.; Ferreira, F.A.; Beltrame, C.O.; Cortes, M.F. The role of biofilms in persistent infections and factors involved in ica-independent biofilm development and gene regulation in *Staphylococcus aureus*. *Crit. Rev. Microbiol.* **2017**, *43*, 602–620. [CrossRef]
20. Yan, J.; Bassler, B.L. Surviving as a Community: Antibiotic Tolerance and Persistence in Bacterial Biofilms. *Cell Host Microbe* **2019**, *26*, 15–21. [CrossRef]
21. Schilcher, K.; Horswill, A.R. *Staphylococcal* Biofilm Development: Structure, Regulation, and Treatment Strategies. *Microbiol. Mol. Biol. Rev.* **2020**, *84*, e00026-19. [CrossRef] [PubMed]
22. Achinas, S.; Charalampogiannis, N.; Euverink, G.J.W. A Brief Recap of Microbial Adhesion and Biofilms. *Appl. Sci.* **2019**, *9*, 2801. [CrossRef]
23. Rabin, N.; Zheng, Y.; Opoku-Temeng, C.; Du, Y.; Bonsu, E.; Sintim, H.O. Biofilm formation mechanisms and targets for developing antibiofilm agents. *Future Med. Chem.* **2015**, *7*, 493–512. [CrossRef] [PubMed]
24. McDougald, D.; Rice, S.A.; Barraud, N.; Steinberg, P.D.; Kjelleberg, S. Should we stay or should we go: Mechanisms and ecological consequences for biofilm dispersal. *Nat. Rev. Microbiol.* **2011**, *10*, 39–50. [CrossRef] [PubMed]
25. Pokrowiecki, R.; Tyski, S.; Zaleska, M. Problematyka zakażeń okołowszczepowych. *Post. Mikrobiol.* **2014**, *53*, 123–134.
26. Dean, S.N.; Chung, M.C.; van Hoek, M.L. *Burkholderia* Diffusible Signal Factor Signals to *Francisella novicida* To Disperse Biofilm and Increase Siderophore Production. *Appl. Environ. Microbiol.* **2015**, *81*, 7057–7066. [CrossRef]
27. Rutherford, S.T.; Bassler, B.L. Bacterial quorum sensing: Its role in virulence and possibilities for its control. *Cold Spring Harb. Perspect. Med.* **2012**, *2*, a012427. [CrossRef]
28. Solano, C.; Echeverz, M.; Lasa, I. Biofilm dispersion and quorum sensing. *Curr. Opin. Microbiol.* **2014**, *18*, 96–104. [CrossRef]
29. Nawrot, U. Patogeny grzybicze odpowiedzialne za zakażenia szpitalne w onkologii. In *Zakażenia Szpitalne w Onkologii*; Szawłowski, A., Ed.; PZWL (National Institute of Medical Publications): Warsaw, Poland, 2018; pp. 47–103.
30. Bisht, K.; Wakeman, C.A. Discovery and Therapeutic Targeting of Differentiated Biofilm Subpopulations. *Front. Microbiol.* **2019**, *10*, 1908. [CrossRef]
31. Furtak, A.; Cześnikiewicz-Guzik, M. Skuteczna walka z biofilmem bakteryjnym—Kluczowy element profilaktyki chorób jamy ustnej. *Med. Prakt. Stomatol.* **2015**, *2*, 32–46.
32. Høiby, N. A Personal History of Research on Microbial Biofilms and Biofilm Infections. *Pathog. Dis.* **2014**, *70*, 205–211. [CrossRef]
33. Jamal, M.; Ahmad, W.; Andleeb, S.; Jalil, F.; Imran, M.; Nawaz, M.A.; Hussain, T.; Ali, M.; Rafiq, M.; Kamil, M.A. Bacterial biofilm and associated infections. *J. Chin. Med. Assoc.* **2018**, *81*, 7–11. [CrossRef] [PubMed]
34. Proctor, C.R.; Reimann, M.; Vriens, B.; Hammes, F. Biofilms in shower hoses. *Water Res.* **2018**, *131*, 274–286. [CrossRef] [PubMed]
35. Seper, A.; Fengler, V.H.; Roier, S.; Wolinski, H.; Kohlwein, S.D.; Bishop, A.L.; Camilli, A.; Reidl, J.; Schild, S. Extracellular nucleases and extracellular DNA play important roles in *Vibrio cholerae* biofilm formation. *Mol. Microbiol.* **2011**, *82*, 1015–1037. [CrossRef]
36. Junka, A.; Żywicka, A.; Chodaczek, G.; Dziadas, M.; Czajkowska, J.; Duda-Madej, A.; Bartoszewicz, M.; Mikołajewicz, K.; Krasowski, G.; Szymczyk, P.; et al. Potential of biocellulose carrier impregnated with essential oils to fight against biofilms formed on hydroxyapatite. *Sci. Rep.* **2019**, *9*, 1256. [CrossRef] [PubMed]
37. Bekmurzayeva, A.; Duncanson, W.J.; Azevedo, H.S.; Kanayeva, D. Surface modification of stainless steel for biomedical applications: Revisiting a century-old material. *Mater. Sci. Eng. C* **2018**, *93*, 1073–1089. [CrossRef]

38. Caldara, M.; Belgiovine, C.; Secchi, E.; Rusconi, R. Environmental, Microbiological, and Immunological Features of Bacterial Biofilms Associated with Implanted Medical Devices. *Clin. Microbiol. Rev.* **2022**, *35*, e0022120. [CrossRef]
39. Larsen, T.; Fiehn, N.E. Dental biofilm infections—An update. *APMIS* **2017**, *125*, 376–384. [CrossRef]
40. Dubey, S.; Dubey, S.; Gupta, A.; Sharma, V. Biofilm-Mediated Dental Diseases. In *Biofilms in Human Diseases: Treatment and Control*; Kumar, S., Chandra, N., Singh, L., Hashmi, M.Z., Varma, A., Eds.; Springer International Publishing: Cham, Switzerland, 2019; pp. 91–116.
41. Forssten, S.D.; Björklund, M.; Ouwehand, A.C. *Streptococcus mutans*, caries and simulation models. *Nutrients* **2010**, *2*, 290–298. [CrossRef]
42. Bulanda, M. Ziarenkowce Gram-dodatnie. In *Medical Microbiology*; Heczko, P.B., Wróblewska, M., Pietrzyk, A., Eds.; PZWL (National Institute of Medical Publications): Warsaw, Poland, 2022; pp. 100–114.
43. Hussain, A.; Ansari, A.Z.; Ahmad, R. Microbial biofilms: Human mucosa and intestinal microbiota. In *New and Future Developments in Microbial Biotechnology and Bioengineering Microbial Biofilms: Current Research and Future Trends*; Yadav, M.K., Singh, B.K., Eds.; Elsevier: Amsterdam, The Netherlands, 2019; pp. 47–60.
44. Campisciano, G.; Zanotta, N.; Petix, V.; Corich, L.; De Seta, F.; Comar, M. Vaginal microbiota dysmicrobism and role of biofilm-forming bacteria. *Front. Biosci. Elite Ed.* **2018**, *10*, 528–536.
45. Sanchez, B.; Delgado, S.; Blanco-Mıguez, A.; Lourenco, A.; Gueimonde, M.; Margolles, A. Probiotics gut microbiota and their influence on host health and disease. *Mol. Nutr. Food Res.* **2017**, *61*, 1600240. [CrossRef] [PubMed]
46. Singh, A.K.; Gaur, V.; Singh, S.K. Biofilm-Mediated Skin Infections. In *Biofilms in Human Diseases: Treatment and Control*; Kumar, S., Chandra, N., Singh, L., Hashmi, M.Z., Varma, A., Eds.; Springer International Publishing: Berlin/Heidelberg, Germany, 2019; pp. 215–231.
47. Oliveira, W.F.; Silva, P.M.S.; Silva, R.C.S.; Silva, G.M.M.; Machado, G.; Coelho, L.; Correia, M.T.S. *Staphylococcus aureus* and *Staphylococcus epidermidis* infections on implants. *J. Hosp. Infect.* **2018**, *98*, 111–117. [CrossRef] [PubMed]
48. Lazar, V.; Chifiriuc, M.C. Medical significance and new therapeutical strategies for biofilm associated infections. *Roum. Arch. Microbiol. Immunol.* **2010**, *69*, 125–138.
49. Stoica, P.; Chifiriuc, M.C.; Rapa, M.; Lazăr, V. 1-Overview of biofilm-related problems in medical devices. In *Biofilms and Implantable Medical Devices*; Deng, Y., Lv, W., Eds.; Woodhead Publishing: Sawston, UK, 2017; pp. 3–23.
50. National Guideline Centre. *Evidence Review for Ultra-Clean Air: Joint Replacement (Primary): Hip, Knee and Shoulder*; Evidence Review I; NICE Evidence Reviews Collection: London, UK, 2020.
51. Cano, E.J.; Caflisch, K.M.; Bollyky, P.L.; Van Belleghem, J.D.; Patel, R.; Fackler, J.; Brownstein, M.J.; Horne, B.; Biswas, B.; Henry, M.; et al. Phage Therapy for Limb-threatening Prosthetic Knee *Klebsiella pneumoniae* Infection: Case Report and In Vitro Characterization of Anti-biofilm Activity. *Clin. Infect. Dis.* **2021**, *73*, e144–e151. [CrossRef]
52. DaSilva, R.B.; Araujo, R.O.; Salles, M.J. Non-elective and revision arthroplasty are independently associated with hip and knee prosthetic joint infection caused by *Acinetobacter baumannii*: A Brazilian single center observational cohort study of 98 patients. *BMC Musculoskelet Disord.* **2021**, *22*, 511.
53. Khoddami, S.; Chew, B.H.; Lange, D. Problems and solutions of stent biofilm and encrustations: A review of literature. *Turk. J. Urol.* **2020**, *46*, S11–S18. [CrossRef] [PubMed]
54. Gheorghe, D.C.; Ilie, A.; Niculescu, A.G.; Grumezescu, A.M. Preventing Biofilm Formation and Development on Ear, Nose and Throat Medical Devices. *Biomedicines* **2021**, *9*, 1025. [CrossRef]
55. Virden, C.P.; Dobke, M.K.; Paul, S.; Lowell Parsons, C.; Frank, D.H. Subclinical Infection of the Silicone Breast Implant Surface as a Possible Cause of Capsular Contracture. *Aesthetic Plast. Surg.* **2020**, *44*, 1141–1147. [CrossRef]
56. Koves, B.; Magyar, A.; Tenke, P. Spectrum and antibiotic resistance of catheter-associated urinary tract infections. *GMS Infect. Dis.* **2017**, *5*, Doc06.
57. Cortese, Y.J.; Wagner, V.E.; Tierney, M.; Devine, D.; Fogarty, A. Review of Catheter-Associated Urinary Tract Infections and In Vitro Urinary Tract Models. *J. Healthc. Eng.* **2018**, *14*, 2986742. [CrossRef]
58. Gominet, M.; Compain, F.; Beloin, C.; Lebeaux, D. Central venous catheters and biofilms: Where do we stand in 2017? *APMIS* **2017**, *125*, 365–375. [CrossRef] [PubMed]
59. Rodrigues, M.E.; Lopes, S.P.; Pereira, C.R.; Azevedo, N.F.; Lourenco, A.; Henriques, M.; Pereira, M.O. Polymicrobial Ventilator Associated Pneumonia: Fighting In Vitro *Candida albicans-Pseudomonas aeruginosa* Biofilms with Antifungal-Antibacterial Combination Therapy. *PLoS ONE* **2017**, *12*, e0170433. [CrossRef]
60. Parker, L.A.; Magalhaes, M.; Desorcy-Scherer, K.; Torrez Lamberti, M.; Lorca, G.L.; Neu, J. Neonatal Feeding Tube Colonization and the Potential Effect on Infant Health: A Review. *Front. Nutr.* **2022**, *9*, 775014. [CrossRef] [PubMed]
61. Willcox, M.D. Microbial adhesion to silicone hydrogel lenses: A review. *Eye Contact Lenses* **2013**, *39*, 61–66. [CrossRef]
62. Kuwatani, M.; Kawakubo, K.; Sakamoto, N. Possible reasons for the regrettable results of patency of an inside stent in endoscopic transpapillary biliary stenting. *Dig. Endosc.* **2022**, *34*, 334–344. [CrossRef]

63. Yamabe, A.; Irisawa, A.; Wada, I.; Shibukawa, G.; Fujisawa, M.; Sato, A.; Igarashi, R.; Maki, T.; Hoshi, K. Application of a silver coating on plastic biliary stents to prevent biofilm formation: An experimental study using electron microscopy. *Endosc. Int. Open.* **2016**, *4*, 1090–1095. [CrossRef]
64. Wu, T.; Yang, Y.; Su, H.; Gu, Y.; Ma, Q.; Zhang, Y. Recent developments in antibacterial or antibiofilm compound coating for biliary stents. *Colloids Surf. B Biointerfaces* **2022**, *219*, 112837. [CrossRef] [PubMed]
65. Jirapinyo, P.; AlSamman, M.A.; Thompson, C.C. Impact of infected stent removal on recurrent cholangitis with time-to-event analysis. *Surg. Endosc.* **2019**, *33*, 4109–4115. [CrossRef]
66. Ciofu, O.; Moser, C.; Jensen, P.Ø.; Høiby, N. Tolerance and resistance of microbial biofilms. *Nat. Rev. Microbiol.* **2022**, *20*, 621–635. [CrossRef]
67. Lübbert, C.; Wendt, K.; Feisthammel, J.; Moter, A.; Lippmann, N.; Busch, T.; Mössner, J.; Hoffmeister, A.; Rodloff, A.C. Epidemiology and Resistance Patterns of Bacterial and Fungal Colonization of Biliary Plastic Stents: A Prospective Cohort Study. *PLoS ONE* **2016**, *11*, e0155479. [CrossRef]
68. Blanco-Míguez, A.; Carloni, S.; Cardenas, C.; Dioguardi, C.C.; Lambroia, L.; Capretti, G.; Nappo, G.; Fugazza, A.; Capogreco, A.; Armanini, F.; et al. Microbial composition associated with biliary stents in patients undergoing pancreatic resection for cancer. *NPJ Biofilms Microb.* **2024**, *10*, 35. [CrossRef]
69. Anderloni, A.; Fugazza, A.; Maroni, L.; Ormando, V.; Maselli, R.; Carrara, S.; Cappello, A.; Mangiavillano, B.; Omodei, P.; Preatoni, P.; et al. New biliary and pancreatic biodegradable stent placement: A single-center, prospective, pilot study (with video). *Gastrointest. Endosc.* **2020**, *92*, 405–411. [CrossRef] [PubMed]
70. Nakamura, K.; Sho, M.; Akahori, T.; Nagai, M.; Nishiwada, S.; Nakagawa, K.; Tanaka, T.; Kichikawa, K.; Tamamoto, T.; Hasegawa, M.; et al. A comparison betweenplasticandmetallicbiliary stent placement in patients receiving preoperative neoadjuvant chemoradiotherapy for resectable pancreatic cancer. *World J. Surg.* **2019**, *43*, 642–648. [CrossRef] [PubMed]
71. Cacaci, M.; De Maio, F.; Matteo, M.V.; Posteraro, B.; Di Vito, M.; Menchinelli, G.; Tringali, A.; Monzo, F.R.; Torelli, R.; Costamagna, G.; et al. Pilot study on cultural and metagenomic analysis of bile and biliary stentslead to unveiling the key players in stent occlusion. *Sci. Rep.* **2024**, *14*, 3344. [CrossRef]
72. Paluch, E.; Rewak-Soroczyńska, J.; Jędrusik, I.; Mazurkiewicz, E.; Jermakow, K. Prevention of biofilm formation by quorum quenching. *Appl. Microbiol. Biotechnol.* **2020**, *104*, 1871–1881. [CrossRef] [PubMed]
73. Ploux, L.; Beckendorff, S.; Nardin, M.; Neunlist, S. Quantitative and morphological analysis of biofilm formation on self-assembled monolayers. *Colloids Surf. B Biointerfaces* **2007**, *57*, 174–181. [CrossRef]
74. Casalini, S.; Bortolotti, C.A.; Leonardi, F.; Biscarini, F. Self-assembled monolayers in organic electronics. *Chem. Soc. Rev.* **2017**, *46*, 40–71. [CrossRef]
75. Nikawa, H.; Ishida, K.; Hamada, T.; Satoda, T.; Murayama, T.; Takemoto, T.; Tamamoto, M.; Tajima, H.; Shimoe, S.; Fujimoto, H.; et al. Immobilization of Octadecyl Ammonium Chloride on the Surface of Titanium and Its Effect on Microbial Colonization In Vitro. *Dent. Mater. J.* **2005**, *24*, 570–582. [CrossRef]
76. Celesti, C.; Gervasi, T.; Cicero, N.; Giofrè, S.V.; Espro, C.; Piperopoulos, E.; Gabriele, B.; Mancuso, R.; Lo Vecchio, G.; Iannazzo, D. Titanium Surface Modification for Implantable Medical Devices with Anti-Bacterial Adhesion Properties. *Materials* **2022**, *15*, 3283. [CrossRef]
77. Antoci, V.; Adams, C.S.; Parvizi, J.; Davidson, H.M.; Composto, R.J.; Freeman, T.A.; Wickstrom, E.; Ducheyne, P.; Jungkind, D.; Shapiro, I.M.; et al. The inhibition of Staphylococcus epidermidis biofilm formation by vancomycin-modified titanium alloy and implications for the treatment of periprosthetic infection. *Biomaterials* **2008**, *29*, 4684–4690. [CrossRef]
78. Wang, S.; Yang, Y.; Li, W.; Wu, Z.; Li, J.; Xu, K.; Zhang, W.; Zheng, X.; Chen, J. Study of the Relationship Between Chlorhexidine Grafted Amount and Biological Performances of Micro/Nanoporous Titanium Surfaces. *ACS Omega* **2019**, *4*, 18370–18380. [CrossRef] [PubMed]
79. Sorzabal-Bellido, I.; Diaz-Fernandez, Y.A.; Susarrey-Arce, A.; Skelton, A.A.; McBride, F.; Beckett, A.J.; Prior, I.A.; Raval, R. Exploiting Covalent, H-Bonding, and–Interactions to Design Antibacterial PDMS Interfaces That Load and Release Salicylic Acid. *ACS Appl. Bio Mater.* **2019**, *2*, 4801–4811. [CrossRef] [PubMed]
80. Slavin, Y.N.; Asnis, J.; Hafeli, U.O.; Bach, H. Metal nanoparticles: Understanding the mechanisms behind antibacterial activity. *J. Nanobiotechnol.* **2017**, *15*, 65. [CrossRef] [PubMed]
81. Crisan, C.M.; Mocan, T.; Manolea, M.; Lasca, L.I.; Tăbăran, F.-A.; Mocan, L. Review on Silver Nanoparticles as a Novel Class of Antibacterial Solutions. *Appl. Sci.* **2021**, *11*, 1120. [CrossRef]
82. Tang, S.; Zheng, J. Antibacterial Activity of Silver Nanoparticles: Structural Effects. *Adv. Healthc. Mater.* **2018**, *7*, e1701503. [CrossRef]
83. Tîlmaciu, C.-M.; Mathieu, M.; Lavigne, J.-P.; Toupet, K.; Guerrero, G.; Ponche, A.; Amalric, J.; Noël, D.; Mutin, P.H. In vitro and in vivo characterization of antibacterial activity and biocompatibility: A study on silver-containing phosphonate monolayers on titanium. *Acta Biomater.* **2015**, *15*, 266–277. [CrossRef]

84. Gargioni, C.; Borzenkov, M.; D'Alfonso, L.; Sperandeo, P.; Polissi, A.; Cucca, L.; Dacarro, G.; Grisoli, P.; Pallavicini, P.; D'Agostino, A.; et al. Self-Assembled Monolayers of Copper Sulfide Nanoparticles on Glass as Antibacterial Coatings. *Nanomaterials* **2020**, *10*, 352. [CrossRef]
85. Sharma, P.; Garg, N.; Sharma, A.; Capalash, N.; Singh, R. Nucleases of bacterial pathogens as virulence factors, therapeutic targets and diagnostic markers. *Int. J. Med. Microbiol.* **2019**, *309*, 151354. [CrossRef] [PubMed]
86. Sarkar, S. Release mechanisms and molecular interactions of *Pseudomonas aeruginosa* extracellular DNA. *Appl. Microbiol. Biotechnol.* **2020**, *104*, 6549–6564. [CrossRef]
87. Cegelski, L.; Pinkner, J.S.; Hammer, N.D.; Cusumano, C.K.; Hung, C.S.; Chorell, E.; Aberg, V.; Walker, J.N.; Seed, P.C.; Almqvist, F.; et al. Small-molecule inhibitors target *Escherichia coli* amyloid biogenesis and biofilm formation. *Nat. Chem. Biol.* **2009**, *5*, 913–919. [CrossRef]
88. Feng, R.; Zhang, T.; Kayani, M.U.R.; Wang, Z.; Shen, Y.; Su, K.L.; Bielike, K.; Chen, L. Patients with primary and secondary bile duct stones harbor distinct biliary microbial composition and metabolic potential. *Front. Cell. Infect. Microbiol.* **2022**, *12*, 881489. [CrossRef] [PubMed]
89. Deva, A.K.; Adams, W.P., Jr.; Vickery, K. The role of bacterial biofilms in device-associated infection. *Plast. Reconstr. Surg.* **2013**, *132*, 1319–1328. [CrossRef] [PubMed]
90. Srinivasan, R.; Santhakumari, S.; Poonguzhali, P.; Geetha, M.; Dyavaiah, M.; Xiangmin, L. Bacterial Biofilm Inhibition: A Focused Review on Recent Therapeutic Strategies for Combating the Biofilm Mediated Infections. *Front. Microbiol.* **2021**, *12*, 676458. [CrossRef]
91. Sun, M.; Chan, K.F.; Zhang, Z.; Wang, L.; Wang, Q.; Yang, S.; Chan, S.M.; Chiu, P.W.Y.; Sung, J.J.Y.; Zhang, L. Magnetic Microswarm and Fluoroscopy-Guided Platform for Biofilm Eradication in Biliary Stents. *Adv. Mater.* **2022**, *34*, e2201888. [CrossRef] [PubMed]
92. Sung, J.Y.; Leung, J.W.C.; Shaffer, E.A.; Lam, K.; Costerton, J.W. Bacterial biofilm, brown pigment stone and blockage of biliary stents. *J. Gastroenterol. Hepatol.* **1993**, *8*, 28–34. [CrossRef]

medicina

MDPI

Review

The Management of Cardiometabolic Risk in MAFLD: Therapeutic Strategies to Modulate Deranged Metabolism and Cholesterol Levels

Annalisa Pezzoli [1], Ludovico Abenavoli [2], Marialaura Scarcella [3], Carlo Rasetti [1], Gianluca Svegliati Baroni [4], Jan Tack [5] and Emidio Scarpellini [1,5,*]

[1] Internal Medicine Unit, " Madonna del Soccorso" General Hospital, 63074 San Benedetto del Tronto, Italy; annalisa.pezzoli@sanita.marche.it (A.P.); carlo.rasetti@sanita.marche.it (C.R.)
[2] Department of Health Sciences, University "Magna Graecia", 88100 Catanzaro, Italy; l.abenavoli@unicz.it
[3] Anesthesia, Intensive Care and Nutritional Science, Azienda Ospedaliera "Santa Maria", Via Tristano di Joannuccio, 05100 Terni, Italy; m.scarcella@aospterni.it
[4] "Hepatic Damage and Transplantation " Unit, Polytechnics University of Marche, 60121 Ancona, Italy; gsvegliati@staff.univpm.it
[5] Translational Research in Gastroeintestinal Disorders, Gasthuisberg University Hospital, KULeuven, Herestraat 49, 3000 Leuven, Belgium; jan.tack@med.kuleuven.be
* Correspondence: emidio.scarpellini@med.kuleuven.be

Abstract: *Background and Objectives:* Fatty Liver Disease is a major health problem worldwide. We can distinguish liver steatosis as non-associated or associated with chronic/acute alcohol consumption. These two entities share similar stages ranging from hepatic fat storage (namely, steatosis) to inflammation, necrosis, and fibrosis until hepatocellular carcinoma (HCC). Over time, "Metabolic Associated Fatty Liver Disease" (MAFLD) has replaced nonalcoholic fatty liver disease (NAFLD) nomenclature and has included cardiometabolic criteria in these patients definition. Thus, obesity, type 2 diabetes mellitus (T2DM), hypertension, and dyslipidemia are MAFLD features and are of the metabolic syndrome. Importantly, there is not a specific treatment for MAFLD, but there are therapeutic strategies that act on metabolic dysfunction related to MAFLD. They can reduce the progression of liver fibrosis and its complications. *Materials and Methods:* For all these reasons, we conducted a narrative review of the literature, and we focused on metabolic dysfunction related to MAFLD, with a special regard for cholesterol metabolism. *Results:* MAFLD is a recently redefined condition that better describes the metabolism derangement responsible for fatty liver disease. This distinguishes MAFLD from NAFLD. In fact, the diagnostic criteria for MAFLD require the presence of liver steatosis together with at least one of the following: obesity, T2DM, or evidence of metabolic disorder such as hypertriglyceridemia, low high-density lipoprotein cholesterol, or hypertension. As a result, MAFLD is closely linked to an increased cardiometabolic risk. Current therapeutic approaches can be used to reduce this risk, focusing on lifestyle interventions and pharmacological strategies. Several treatments in patients diagnosed with MAFLD are mainly cholesterol-lowering remedies. Among these, Pro-protein Convertase Subtilisin/Kexin type 9 inhibitors (PCSK9i) show the most promising efficacy profile but data on liver fibrosis are lacking. Agonists of GLP-1 receptor, Sodium-glucose cotransporter-2 inhibitors (SGLT2i) and Dipeptidyl Peptidase-4 inhibitors (DPP-4i) have a " multi-hit " action allowing their use also in diabetic patients with MAFLD. *Conclusions:* Lifestyle modifications, some nutraceuticals, statins, incretins, and PCSK9i have changed the natural course and significantly improved the cardiometabolic outcomes of MAFLD. Emerging cholesterol-lowering drugs, such as Bempedoic acid, can overcome low compliance to statins' use and their controversial effect on liver fibrosis. Finally, medications targeting insulin resistance allow for strategic interventions of the convoluted pathophysiology of MAFLD in multiple steps, with the potential to reduce liver steatosis, inflammation, and necrosis and, sometimes even to reverse liver fibrosis.

Keywords: fatty liver; NAFLD; MAFLD; dyslipidemia; diet; statin; PCSK9; nutraceuticals; bempedoic acid

1. Introduction

Fatty liver disease is a rising health problem. It was firstly classified as nonalcoholic fatty liver disease (NAFLD) or alcohol-related fatty liver disease (AFLD) depending on the absence/presence of alcohol use/abuse. In the last two decades, NAFLD prevalence (ranging from 12 to 22% of the general population) has been associated with various metabolism alterations (e.g., obesity, insulin resistance (IR), type 2 diabetes mellitus (T2DM), hypertension, hyperlipidemia and, more comprehensively, metabolic syndrome). Consequently, NALFD has been renamed as metabolic-associated fatty liver disease, MAFLD [1]. Obesity, diabetes, and Metabolic Syndrome (MetS) are key drivers of liver fat deposition in 'Metabolic Associated Fatty Liver Disease'. Therefore, MAFLD diagnosis is reached according to liver fat deposition detection (via radiology imaging, liver biopsy, or, more recently, blood biomarkers) and one of three major dysmetabolic conditions: obesity or being overweight, type-2 diabetes mellitus, and the presence of two or more metabolic abnormalities [2,3]. Thus, MAFLD prevalence has risen over that of NAFLD, overcoming 30% of the general population [2,3]. Robust evidence points out a strong association between increased cardiovascular risk and MAFLD. Atherosclerotic carotid plaques and a fatty liver are a common finding at baseline patients' evaluation [4] [Figure 1].

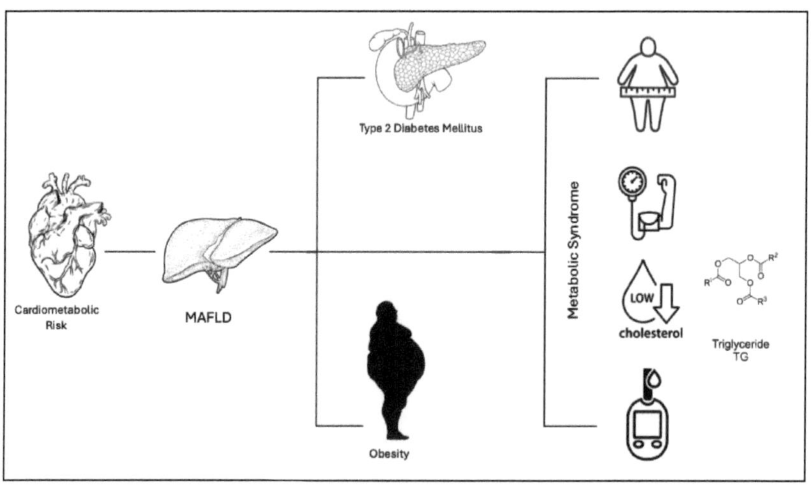

Figure 1. The Figure shows the link between cardiometabolic risk and MAFLD. According to the definition of "Metabolic Associated Fatty Liver Disease", MAFLD, the disease can present with various metabolic conditions including obesity, insulin resistance (IR), type 2 diabetes mellitus (T2DM), hypertension, hyper-lipidemia, and metabolic syndrome.

Indeed, the MAFLD definition has drawbacks: patients with fatty liver are not diagnosed according to the amount and frequency of alcohol use/abuse. Further, a vast group of etiologies are included in the definition. For these reasons, the new "metabolic associated steatotic liver disease", MASLD, requires at least one out of five cardiometabolic risk factors and distinguishes between alcohol use or abuse. Thus, the term "metabolic and alcohol-related/associated liver disease" (MetALD) has been introduced to describe patients with MASLD who consume greater amounts of alcohol per week (140–350 g/week and 210–420 g/week for females and males, respectively) [5]. In addition, the MASLD definition overcomes the stigma of the term "fatty" and, more importantly, includes the pathophysiologic dysmetabolic milieu of the disease [6,7].

MASLD prevalence reaches almost 25% of the global population with a growing perspective [6]. This prevalence follows that of obesity and is going to be the main contributor of increased prevalence of chronic hepatic diseases and hepatocellular carcinoma worldwide [7].

The age range for MAFLD patients is between 40 and 60 years. Indeed, the disease can be diagnosed also in children older than 10 years. MAFLD seems to significantly affect more males than females, perhaps at a younger age. This sex-based distinction is the opposite for patients older than 65 years [8]. This condition often progresses from simple fatty liver to more serious stages like steatohepatitis (NASH) and cirrhosis, with or without hepatocellular carcinoma. However, the transition from liver steatosis is not linear and sometimes steatohepatitis patients can develop hepatocellular carcinoma (HCC) without cirrhotic degeneration [9]. Given the increasing number of cases of MAFLD, several therapeutic strategies have been studied to act on metabolic dysfunctions affecting liver disease. The latter are centered on lifestyle modifications. Indeed, there are pharmacological interventions that address the shared underlying metabolic and hepatic pathologic pathways. The aim of these strategies is to focus on the most common metabolic disorders linked to MAFLD, such as: insulin resistance (often treated with metformin, thiazolidinediones), oxidative stress (counteracted with vitamin E, pentoxifylline), and inflammatory cytokines (modulated via anti-TNF-a, TGF-β, IL-11) [10]. Indeed, lowering cholesterol levels (with statins, ezetimibe or their combination, bempedoic acid and, due to statin-intolerance, the novel proprotein convertase subtilisin/kexin 9 inhibitors) is a prominent target.

Thus, we performed a narrative review of the literature's evidence on the definition of MAFLD, its diagnosis, the role of lipogenesis/hepatic lipid deposition and insulin resistance in its pathophysiology, and the relevance of remedies and drugs positively affecting deranged metabolism (e.g., lowering cholesterol and affecting insulin resistance). We have chosen the MAFLD and NAFLD nomenclature as it accounts for most of the reviewed studies.

2. Materials and Methods

We made a search on PubMed, Medline for the literature data (namely, original articles, reviews, meta-analyses, and case series) using the following keywords, their acronyms, and their associations (e.g., "and"): "Fatty liver", "NAFLD", "MAFLD", "Dyslipidemia", "Diet", "PCSK9", "Statin", "Nutraceutical Therapies", "Bempedoic acid". Importantly, we chose the MAFLD definition because most of the literature studies retrieved were performed in patients with this nomenclature of disease. We considered articles in the timeframe of 2000–2024 years. In these years, the terms NAFLD and MAFLD were the most frequently used. We selected articles published in English and involving human and animal models of NAFLD and MAFLD. The MASLD definition and physiopathologic data were included in the Introduction section to complete the NAFLD and MAFLD pathophysiology description.

3. Results

3.1. Metabolic Dysfunction-Associated Fatty Liver Disease (MAFLD)

3.1.1. Diagnostic Criteria for MAFLD

According to the most recent guidelines, NAFLD could be considered, according to liver fat accumulation detected at abdominal magnetic resonance imaging (MRI) and/or biopsy in the absence of other hepatic injury causes (e.g., alcohol, hepatotoxic drugs, toxins, viral infections, primary liver disease) [11,12]. Hepatic steatosis can be first diagnosed by ultrasound when the hepatic parenchyma shows supranormal brightness [12]. Indeed, liver steatosis can be underdiagnosed or missed when fat deposition spares more than 67% of hepatic parenchyma [13]. In addition, the term NAFLD has been replaced by MAFLD because it only describes liver fat accumulation, does not mention alcohol use/abuse, and does not rely on metabolism alterations. Thus, experts in the field have performed accurate consensus conferences to modify the definition. In particular, the MAFLD term was proposed in 2020 and connected the diagnosis of fatty liver disease and one of the following: Type 2 Diabetes Mellitus (T2DM), obesity, and metabolic dysfunction. Therefore, the stigma of "non-alcoholic "was removed [13].

To provide further detail and practically, liver steatosis must be accompanied by at least two dysmetabolic findings [13]: waist circumference $\geq$ 102/88 cm in Caucasian men/women or $\geq$ 90/80 cm in Asian men/women; blood pressure $\geq$ 130/85 mmHg or antihypertensive drugs; plasma triglycerides (TG) $\geq$ 150 mg/dL or TG lowering drugs; plasma high-density lipoprotein cholesterol (HDL-C) < 40 mg/dL for men and <50 mg/dL for women or lipid-lowering drugs; fasting plasma glucose (levels 100 and 125 mg/dL or 2 h post-load); glucose levels (140–199 mg/dL) or glycosylated hemoglobin (HbA1c) between 5.7 and 6.4%, Homa index score $\geq$ 2.5; and high- sensitivity C-reactive protein levels > 2 mg/L [Table 1].

Table 1. Criteria for MAFLD diagnosis, based on the metabolic alterations provided [13].

Metabolic Alteration	Criteria
Waist Circumference	≥102 cm in Caucasian men, ≥88 cm in Caucasian women; ≥90 cm in Asian men, ≥80 cm in Asian women
Blood Pressure	≥130/85 mmHg or use of antihypertensive drugs
Plasma Triglycerides (TG)	≥150 mg/dL or use of TG-lowering drugs
Plasma High-Density Lipoprotein Cholesterol (HDL-C)	<40 mg/dL for men and <50 mg/dL for women, or use of lipid-lowering drugs
Fasting Plasma Glucose	Between 100 and 125 mg/dL or 2 h post-load glucose levels
Glucose Levels	Between 140 and 199 mg/dLor HbA1c between 5.7 and 6.4%
HOMA Index	Insulin resistance score ≥2.5
High-Sensitivity C-Reactive Protein (hs-CRP)	Levels > 2 mg/L

Data on the relationship between sex, race, and other socioeconomic factors and liver steatosis are available mainly for NALFD. Indeed, there are substantial disparities in the development of NAFLD according to race and ethnicity because of genetics and environmental and social factors. In detail, the prevalence of NAFLD in the US population appears to be higher among Hispanics, followed by non-Hispanic Whites and Asians, and lastly, African Americans [14]. Similarly, the same disparity has also been observed in hospitalized patients [15]. Interestingly, Japanese Americans have a greater risk for NAFLD development because of high visceral adiposity prevalence [16]. Finally, Hispanics appear to have a higher NAFLD prevalence, an earlier onset, and a worse metabolic profile vs. other ethnicities. Unfortunately, data reported come almost exclusively from studies run within the US population [17].

The impact of socioeconomic status on NAFLD prevalence is varied around the world. Initially, NAFLD prevalence seemed to be higher among individuals with lower socioeconomic status in Western countries [18]. Similarly, South Korea data found people with a low socioeconomic status having a significantly higher risk of developing NAFLD (OR 1.7) [19]. Oppositely, a Chinese study found people with a higher median income having a two-fold higher risk of developing NAFLD vs. the low-income subjects [20]. Such disparities are linked to food security and food composition. Almost 30% of US adults with low-income have NAFLD and live in food-insecure households [18]. Conversely, the prevalence of food insecurity is much higher among the Iranian NAFLD population (56.8%) vs. non-NAFLD subjects (26.1%) [21].

3.1.2. Clinical and Laboratory Indexes for Steatosis Monitoring

Both NAFLD and MAFLD need an accurate grading and staging of steatosis activity and fibrosis. The grading and staging describe disease progression and allow physicians to predict patient outcomes and select the best therapeutic option. Thus, laboratory indexes and non-invasive scoring systems have emerged as valuable tools in this context. The disease is considered a continuum of stages from liver steatosis to steatohepatitis (namely, NASH in the case of NAFLD or Metabolic Dysfunction-Associated Steatohepatitis (MASH), in the case of MAFLD). In fact, disease activity and fibrosis can be evaluated through non-invasive methods or liver biopsy (e.g., NAFLD Activity Score and degree of fibrosis, respectively). In one regard, laboratory tests support the MAFLD diagnosis and allow the evaluation of dysmetabolic conditions associated with hepatic steatosis evolution. However, biomarkers used for liver fibrosis detection typically indicate matrix turnover, but not the extent of extracellular matrix deposition. Moreover, no biomarker is specific for liver fibrosis detection. In fact, extra-hepatic inflammatory and oxidative states can contribute to fibrosis development in MAFLD. There are two types of markers that determine fibrosis level in the liver: indirect and direct ones [Table 2].

Table 2. Available indirect and direct indexes for monitoring disease activity and fibrosis in MAFLD.

Indirect Markers [11]	**Direct Markers [22]**	
	Collagen Synthesis/ Degradation [23]	**Pro-Inflammatory Molecules [24]**
Aspartate amino Transferase (AST) [25]	PIIINP	TGF-Beta1
Alanine amino Transferase (AST) [25]	TIMP-1	GF-1
Platelet Count (PLT) [26]	TNF	CRP
Gamma Glutamil Transferase (GGT) [27]	MMP	Fibrinogen
Total Bilirubin [27]		Factor VIII
Alpha 2-macroglobulin and/or alpha 2 globulin [28]		PAI-1

The indirect markers describe deranged hepatic metabolism: aspartate amino Transferase (AST), Alanine amino Transferase (ALT), platelet count, Gamma-Glutamyl Transferase (GGT), total bilirubin, alpha 2-macroglobulin, or alpha 2-globulin (mainly haptoglobin). However, they do not help to detect the presence of fibrosis. Therefore, scores built from a combination of multiple biomarkers can have a higher diagnostic accuracy [28,29].

On the other hand, direct markers can help predict the presence of liver fibrosis. In fact, they are biomarkers of collagen synthesis/degradation, extracellular matrix glycoproteins, proteoglycans, and glycosaminoglycans (PIIINP: amino-terminal Propeptide of type III Procollagen; TIMP-1: Tissue Inhibitor of Metalloproteinase; TNF: Tumor Necrosis Factor; MMP: Matrix Metallo Proteinase). Furthermore, other biomarkers are pro-inflammatory molecules, such as Transforming Growth Factor beta-1 (TGF-β1), Insulin-Like Growth Factor (IGF-1), and endothelin-1 and inflammatory mediators such as C-Reactive Protein (CRP), Interleukin (IL)-6, and pro-coagulant factors such as fibrinogen, factor VIII, and plasminogen activator inhibitor-1 [29–31].

Finally, there are also scores that can assist physicians in evaluating the progression and the degree of hepatic steatosis, such as the Hepatic Steatosis Index (HIS) [32], Fatty Liver Index (FLI) [33], the FIB-4 index [34], ELF test [Lee J], and APRI [34,35].

3.1.3. Non-Invasive Imaging Techniques for MAFLD Diagnosis and Histological Findings

Several imaging techniques are currently employed to diagnose and assess the severity of MAFLD. Ultrasound (US) is the most used imaging available in current clinical practice due to its low cost and widespread availability. However, it has low sensitivity for liver steatosis detection and is not able to discriminate between liver steatosis and fibrosis [31]. Indeed, liver steatosis can be also frequently detected by computed tomography (CT) or MRI [36].

Alternatively, the controlled attenuation parameter (CAP) measured during elastography is a more sensitive radiologic tool, and the proton magnetic resonance spectroscopy (1H-MRS) is also an acceptable quantitative marker of steatosis. These measurements can be combined with laboratory biochemical testing in high-risk populations [37]. Indeed, the main European and American liver disease scientific societies (namely, EASL and AASLD) recommend the use of abdominal ultrasound and liver enzymes testing for all patients with documented metabolic risk factors [38]. CT is a more accurate diagnostic tool than ultrasound. However, its use is limited to mild steatosis patients because it requires radiation exposure [39]. While CT is more sensitive than ultrasound for evaluating hepatic fat content, substances like iron can interfere and impact the diagnosis [40]. The CAP technique, an ultrasound-based approach, measures steatosis (greater than 10%) but has been shown to be somewhat unreliable, even though it is still recommended in Asia-Pacific guidelines as a useful tool for NAFLD/MAFLD patients. MRI and magnetic resonance spectroscopy can detect liver fat and fibrosis, but their application in clinical practice is, as of yet, limited by the high costs. In detail, MRS has a complex protocol for routine clinical settings, despite its ability to detect 5.56% of liver fat content. Indeed, it can be considered the gold standard for diagnosing steatosis [41]. In recent years, transient elastography (Fibro Scan), an ultrasound-derived technique, has become more popular because it offers fast and convenient measurements of liver stiffness, which correlates closely with liver fibrosis stages [42]. It can be used either alone or in combination with a CAP measurement, giving a consensual liver stiffness and steatosis assessment [42].

Histological diagnosis remains the gold standard for confirming MAFLD and evaluating its severity, particularly in advanced stages. Liver biopsy is an invasive procedure with 0.05% risk of mortality linked to procedures' complications. It allows the assessment of key histopathological features, including steatosis, inflammation, the ballooning of hepatocytes, and fibrosis [43]. Although biopsy has a high diagnostic accuracy, it is an invasive and costly protocol with limited routine application [31,44,45].

More interestingly, liver biopsy shows greater issues when applied to bigger populations because of the sampling error (namely, the steatosis/fibrosis-spared liver segment receiving the biopsy). The latter produces false-negative cases with misread prognosis.

Histologically, we distinguish macrovesicular and microvesicular steatosis.

In microvacuolar (or macrovesicular) steatosis, triglycerides commonly accumulate as a single large lipid vacuole relocating the nucleus at the periphery of the hepatocyte [46]. This histological feature is the most common finding and is typically retrieved in obese subjects [47], alcoholic liver disease, Wilson's disease, and familial hypobetalipoproteinemia [47]. Usually, macrovacuolar steatosis is associated with hepatomegaly. In about 20% of cases, it may progress to steatohepatitis, with necroinflammation, ballooning degeneration of hepatocytes, and fibrosis [47]. It is important to acknowledge that fibrosis progresses to liver cirrhosis over weeks/months in drug-induced steatohepatitis. Indeed, obesity and other metabolic conditions favor cirrhosis development over decades [46].

The most common cause of microvesicular steatosis are drugs. Its clinical characteristics include liver failure, encephalopathy, multiorgan failure, and coma [47]. The rarer microvesicular steatosis recognizes the presence in the cytoplasm of several lipid droplets, which leave the nucleus at the center of the hepatocyte [46]. There is also a certain very rare association with macrovacuolar steatosis [46]. Small lipid droplets reflect severe mitochondrial dysfunction in the injured hepatocytes [48]. From a pathogenetic point of view, ATP deficiency has been linked to severe mitochondrial dysfunction and, altogether, these can lead to the growth of lipid droplets through the reduction of lipid synthesis or altered deranged expression of proteins and enzymes storing lipids [48]. In another hypothesis, triglyceride can be hydrolyzed in the largest lipid droplets to mobilize oxidable fatty acids [49]. Finally, lysosomal function derangement can favor small lipid droplets storage [49].

Other diseases with microvesicular steatosis are acute fatty liver of pregnancy, some inborn errors of mitochondrial fatty acids' oxidation, and several mitochondrial cytopathies (i.e., genetic disorders of the OXPHOS system) [4,7,49]. Typically, Reye's syndrome, triggered by an acute viral illness (e.g., influenza and varicella) and drugs (namely, the nonsteroidal anti-inflammatory drug (NSAID) aspirin (or herbal tea containing salicylate) and valproic acid) is associated with microvesicular steatosis [4,49].

In conclusion, non-invasive methods can be considered the first step for NAFLD/MAFLD diagnosis. Unfortunately, abdominal ultrasound has a significant inter-observer variability. Other techniques can result in over- or underestimation of the hepatic fibrosis stage [49]. When NASH/MASH and liver fibrosis are not diagnosed/detected by non-invasive methods, liver biopsy can be performed in individuals in whom the etiology of the liver disease needs to be clarified [44]. Non-invasive biomarkers and newer imaging techniques are becoming more and more popular for early patients' evaluation.

3.2. Metabolic Dysfunction in MAFLD

3.2.1. Lipid Metabolism and Insulin Resistance in MAFLD

MAFLD physiopathology encompasses a complex interplay of lipid metabolism pathways. The most advanced form of liver fibrosis, MASH and, more specifically MASH-cirrhosis, is initiated by lipotoxicity. The latter is characterized by cellular inflammation, oxidative stress, and hepatocellular ballooning. MASH can irreversibly progress to liver cirrhosis and/or to the development of HCC. MASH progresses to cirrhosis and/or HCC due to several pathologic processes: cellular senescence, oxidative stress, autophagy, and ferroptosis [50,51]. In detail, hepatic fat accumulation and steatosis are a result of excessive lipid uptake, de novo lipogenesis, impaired oxidation of fatty acids, and dysfunctional lipid export [51,52] [Figure 2].

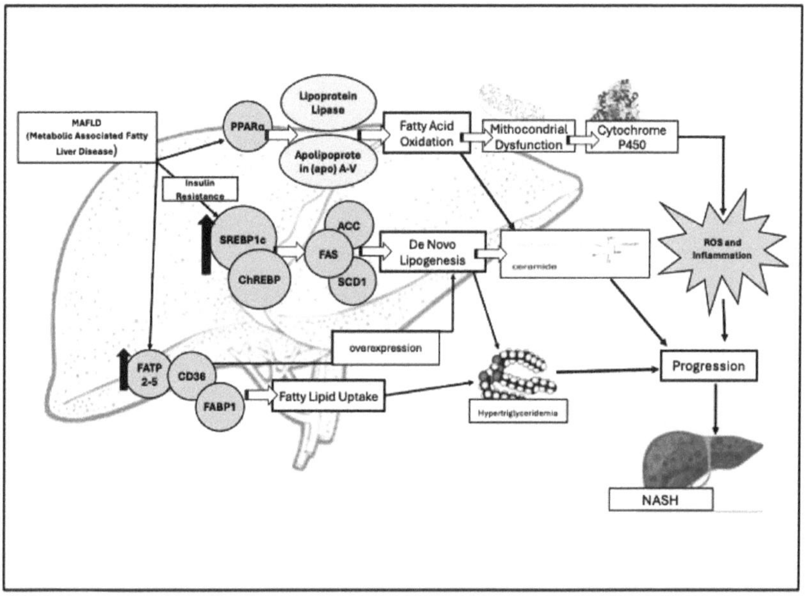

Figure 2. Main pathophysiologic key factors involved in deranged lipid metabolism and insulin resistance in MAFLD. Sterol regulatory element-binding protein 1c (SREBP1c) and fatty acid transport proteins (FATs) are upregulated (↑) in MAFLD patients and are responsible for deranged lipogenesis and lipid uptake. Lipogenesis is regulated by three enzymes: acetyl-CoA carboxylase (ACC), fatty acid synthase (FAS), and stearoyl-CoA desaturase-1 (SCD1). The result is the formation of ceramides (namely, palmitate, oleate, and palmitoleate), responsible for passage from liver steatosis to liver steatohepatitis (NASH/MASH). Their increased levels lead to hypertriglyceridemia. SREBP1c is primarily activated by insulin, especially in insulin resistance conditions. The latter impairs the suppression of lipolysis in adipose tissue, leading to an increased flux of free fatty acids (FFAs) up to the liver. Consensually, it has been determined that there is an increased expression of FA binding protein (FABP), FA transport protein (FATP), and CD36 genes responsible for fatty acid uptake and overexpression of de novo lipogenesis. FABP1 is specifically expressed in the liver where it transports fatty acids between organelles, binding cytotoxic free fatty acids, and aiding their oxidation or incorporation into triglycerides. The peroxisome proliferator-activated receptor-alfa (PPARα) regulates fatty acid oxidation across mitochondria, peroxisomes, and cytochrome pathways. PPARα activation induces the production of lipoprotein lipase (LPL) and Apolipoprotein (apo) A-V.

We must note that lipid metabolism plays a pivotal role in several crucial activities in humans. They include energy storage and release, cell membrane formation, hormone synthesis and transport, liposolubility of nutrients, and the regulation of inflammatory response [53]. MAFLD patients show their derangement.

Key molecular players, such as sterol regulatory element-binding protein 1c (SREBP1c) and fatty acid transport proteins (FATs), are upregulated in MAFLD, driving both lipogenesis and lipid uptake [50,52]. The liver's de novo lipogenesis is regulated by acetyl-CoA carboxylase (ACC), fatty acid synthase (FAS), and stearoyl-CoA desaturase-1 (SCD1). Their machinery results in the formation of palmitate, oleate, and palmitoleate whose storage ends up in hypertriglyceridemia and liver steatosis [54]. The two transcription factors that regulate the enzymes (precisely, FAS, and SCD1), are SREBP1c and the carbohydrate regulatory element-binding protein (ChREBP). As a master regulator of the de novo lipogenesis pathway, SREBP1c is primarily activated by insulin and shows a significant increase in MAFLD patients vs. healthy individuals [44]. Moreover, the overexpression of SREBP-1c is linked to the upregulation of key enzymes of de novo lipogenesis and results in hepatic lipid accumulation [55]. Indeed, mice with MAFLD undergoing single-cell RNA sequencing (scRNA-seq) and computational network analyses to assess lipid signatures showed that high SREBP1 expression is not predictive of liver lipids' accumulation. Further, the constitutive androstane receptor (CAR) is a key regulator of functional modules associated with cholesterol homeostasis, bile acid metabolism, fatty acid metabolism, and estrogens' response. Thus, there is a significant correlation of its activation with steatohepatitis development in humans [56]. In addition, among the other enzymes regulated by SREBP1c, the isoforms of ACC have a significantly increased expression in MAFLD patients vs. controls [57,58]. Interestingly, MAFLD patients also exhibited altered expressions the of FA binding protein (FABP), FA transport protein (FATP) [57,58], and cluster of differentiation 36 (CD36) genes. The latter are responsible for fatty acid uptake and the overexpression of de novo lipogenesis [56]. Fatty Acid Binding Protein 1 (FABP1) is specifically expressed within the liver and transports fatty acids between organelles, binding cytotoxic free fatty acids. The protein is also involved in their oxidation/incorporation into triglycerides [58]. In fact, knockout FABP1 mice showed a reduced response to fasting-induced increases in hepatic triglyceride uptake and oxidation [54]. In addition, hepatic lipid uptake is regulated by FATPs and CD36, and FATP isoforms 2 and 5 are the most abundant in the liver [59]. Increased FATP5 expression in humans significantly correlates with higher hepatic steatosis in male MAFLD patients [60,61]. Moreover, hepatic CD36 protein levels' expression rise under high-fat diet [62]. These findings suggest a connection between FATP5, CD36, and hepatic lipotoxicity [Figure 2].

De novo lipogenesis overexpression can lead to fat storage in MAFLD and, also to storage of toxic lipid species (e.g., ceramides), critical for liver fibrosis development.

The impaired oxidation of fatty acids has been detected within mitochondria in NAFLD research models. Very long-chain fatty acids are initially oxidized in peroxisomes before being further processed. Under conditions of lipid overload (e.g., high-fat diet), an alternative omega-oxidation pathway mediated by cytochrome P450 enzymes becomes active. This fuels fatty acid oxidation. Unfortunately, this pathway produces significant amount of reactive oxygen species (ROS) that trigger the inflammatory response and favor progression to NASH. The peroxisome proliferator-activated receptor-alfa (PPARα) plays a central role in regulating fatty acid oxidation across mitochondria, peroxisomes, and cytochrome pathways. PPARα modulates lipid and lipoprotein metabolism. It regulates the transcription of genes involved in the metabolism of triglycerides (TG)-rich lipoproteins and HDL [63]. Thus, PPARα activation induces lipoprotein lipase (LPL) and Apolipoprotein (apo) A-V synthesis expression. Conversely, its decrease is followed by apo C-III activation. This results in LPL activity inhibition and enhanced beta (β)-oxidation genes expressions [64] [Figure 2].

Current data from the literature demonstrate that also microRNAs (miRNAs) might be involved in NAFLD and MAFLD development. In detail, NAFLD animal models show miRNAs linked to deranged cholesterol metabolism and NAFLD [32,45,49]. Further, a comprehensive review of the literature including 19 articles demonstrated that 13 different miRNAs are related with the altered lipid metabolism typical of MAFLD. The most studied is miR122, one of the most abundant in the liver [57].

Insulin resistance (IR) is a central driver in the pathogenesis of MAFLD. It links hepatic steatosis to systemic metabolic dysregulation. IR impairs lipolysis inhibition within the adipose tissue, leading to an increased flux of FFAs up to the liver. The flux promotes hepatic triglyceride accumulation, exacerbating steatosis and triggering lipotoxicity. IR reduces hepatic glycogen synthesis and, on the other hand, enhances de novo lipogenesis via upregulation of transcription factors such as SREBP1c [65]. This imbalance is reinforced by oxidative stress, mitochondrial dysfunction, and chronic low-grade inflammation. Altogether, these processes create a vicious cycle that accelerates MAFLD progression toward MASH. Consensually, it has been determined that oxidative stress from mitochondrial dysfunction and reactive oxygen species reinforces inflammation and insulin resistance. IR downregulates lipases functioning and this results in the altered flow of fatty acids and of the intestinal production of chylomicrons (CM) and of hepatic very low-density lipoproteins (VLDL). Moreover, hyperinsulinemia increases fatty acid esterification and inhibits beta-oxidation that regulates triglycerides formation in liver [66]. In addition, dysmetabolic patients show increased oxidative stress and increased blood levels of glucose and lipoproteins, ending in foam cell formation and atherosclerotic disease [31].

Finally, the physiopathologic "multi-hit" hypothesis that integrates these pathways emphasizes the roles of insulin resistance, dyslipidemia, and inflammation in MAFLD pathogenesis, making lipid metabolism a crucial target for therapeutic strategies [67].

3.2.2. Cardiovascular Risk in MAFLD: The Link Among Physiopathology and Clinical Features

MAFLD has been significantly associated with cardiovascular disease-related mortality. In fact, it is a major risk factor for cardiovascular diseases (CVD) like myocardial infarction, stroke, and heart failure [51]. This is supported by meta-analysis and systematic review data [68]. Moreover, MAFLD contributes to an accelerated progression of coronary atherosclerosis, heart failure, and arrhythmia [68,69]. Similarly, MASLD shows a similar risk profile for CVD [70]. In a large study involving over 8.8 million South Korean adults, MAFLD was significantly associated with a higher incidence of cardiovascular events [69,71].

3.3. Current Therapeutic Strategies Targeting Metabolism in MAFLD

3.3.1. Diet and Lifestyle in the Treatment of MAFLD

To date, lifestyle interventions, particularly diet and physical activity, are the cornerstone for managing MAFLD. We aim to counteract the lifestyle changes derived from the rapid economic growth of the last 40 years. Specifically, a Westernized world has led to more meat and egg consumption. Conversely, ingestion of fruits, vegetables, and whole grains has decreased [72]. This alimentary shift is high in cholesterol [73]. Thus, a comprehensive approach addressing weight reduction, dietary changes, and increased physical activity is essential for improving liver health, metabolic dysfunction, and cardiovascular outcomes in MAFLD patients [74]. For example, overweight and obese MAFLD patients obtaining a weight loss of 7–10% show a decrease of hepatic steatosis grading and, of vascular and metabolic complications [75]. This phenotypic change correlates with reduced hepatic enzyme activity, improved histological liver steatosis and inflammation. Unfortunately, there is less certainty regarding the impact on fibrosis [75]. High-intensity interval exercise is able to improve plasmatic levels of triglyceride-rich VLDL1 particles and LDL cholesterol and insulin resistance and other CVD risk factors. Thus, it is strongly recommended together with a dietary approach (hypolipemic diet). In this regard, the Mediterranean diet and similar dietary approaches are gaining more and more attention from the scientific community [76].

3.3.2. The Use of Nutraceuticals in MAFLD

The use of nutraceuticals in managing MAFLD should not be underestimated. They can be effective either alone or in combination with dietary and lifestyle changes [77]. In particular, the nutraceuticals reviewed are those with a proven significant improvement in hepatic steatosis. However, their usage has some issues: they can have rather low bioavailability and limited effectiveness. In fact, individual genetics can affect nutraceuticals' absorption, storage, and excretion [31]. Nutraceuticals mainly target inflammation, glycemia and insulinemia, LDL-C, and blood pressure [78]. In MAFLD patients, they effectively address liver inflammation, steatosis, and insulin resistance [Table 3].

Table 3. Nutraceuticals used in the treatment of MAFLD.

Nutraceutical	Key Properties	Benefits in MAFLD	References
Silymarin	Antioxidant, anti-inflammatory, antifibrotic	Improves liver enzymes and reduces hepatic steatosis	[31,79]
Omega-3 Fatty Acids	Reduces triglycerides, anti-inflammatory action	Lowers triglycerides, improves hepatic steatosis and insulin resistance	[80,81]
Berberine	Lipid-lowering, insulin-sensitizing	Enhances metabolic profile, reduces hepatic fat accumulation	[31,82]
Curcumin	Anti-inflammatory, insulin-sensitizing	Reduces liver inflammation, improves insulin sensitivity and hepatic steatosis	[31,83]
Coenzyme Q10	Anti-inflammatory, antioxidant action	Regulates adipokine levels support metabolic balance, reduces oxidative stress	[31,84]
Nigella Sativa	Antioxidant, anti-inflammatory (contains Thymoquinone) action	Improves liver enzyme, reduces inflammation and lowers cardiovascular risk markers	[85]
Brown Algae (*Ascophyllum nodosum* and *Fucus vesiculosus)*	antioxidant, anti-inflammatory, and anti-cancer properties	lowers insulin levels, Homeostatic Model Assessment for Insulin Resistance (HOMA-IR), blood glucose, and waist circumference	[85]
Vitamin E	Antioxidant action	improves liver function, particularly in pediatric NASH patients; can reduce liver inflammation	[86–89]

Silymarin is known for its antioxidant, anti-inflammatory, and antifibrotic properties. It comprises seven flavonolignans (silybin A, silybin B, isosilybin A, isosilybin B, silychristin, iso-silychristin, and silydianin) and one flavonoid (taxifolin). It has shown benefits in improving liver enzymes and reducing hepatic steatosis. In particular, several clinical studies on NAFLD have shown that silymarin can delay the progression of liver disease, alleviate symptoms, and enhance the quality of life of patients [31]. Silymarin acts as a scavenger of free radicals. For this reason, it prevents lipid peroxidation and protects enzyme systems associated with hepatic cellular damage. Thus, it reduces oxidative stress and cytotoxicity [79]. The accepted and tolerated dosage of silymarin in NAFLD and MAFLD studies is 140 mg three times a day. The dosage can reduce deranged liver enzyme levels [79]. For example, Torre et al. showed that 4 months of silymarin administration significantly reduced transaminases and gamma-glutamyl transferase levels [90]. Further, Lee et al. found similar significant results after only 1 month of silymarin treatment and, importantly, the reduction was maintained for more than 4 years [91].

Omega-3 Fatty acids reduce triglyceride levels and improve hepatic steatosis and insulin resistance [31]. Further, several trials have investigated the role of omega-3 in the treatment of NASH in men. However, the variability of results among the studies can be explained by differences of product administration and experimental design (e.g., formulation of omega-3, the duration and dosage of the supplements, the endpoints, and the measured outputs such as exercise, dietary changes, and the genetic or epigenetic background of the participants) [80]. Although no study has demonstrated significant improvements in key histological prognostic features (namely, fibrosis), most trials have reported a reduction of steatosis grade. One study using biopsies found no change in steatosis after 12 months of treatment with a synthetic Eicosapentaenoic acid (EPA) supplement (up to 2700 mg/day vs. placebo) [81].

Berberine lowers circulating lipids' levels and enhances insulin sensitivity, contributing to improved metabolic profiles and a reduction in hepatic fat synthesis [31].

Curcumin is extracted from Curcuma Longa, has insulin-sensitizing effects and reduces liver inflammation and steatosis [31].

Coenzyme Q10 regulates adipokine levels and supports metabolic rebalancing in MAFLD [31].

Nigella Sativa with its active component, thymoquinone, possesses antioxidant and anti-inflammatory properties [31]. Specifically, it improves levels of transaminases, fasting glycemia, the lipid profile, the high-sensitivity C-reactive protein, and the degree of liver steatosis [84].

The combination of *Ascophyllum nodosum* and *Fucus vesiculosus* has been studied for its antioxidant, anti-inflammatory, and anti-cancer properties. These two types of brown algae enhance intestinal viscosity, slowing the absorption of cholesterol and inhibiting alfa (α)-amylase and α-glucosidase activity. The latter reduces sugar absorption. This combination significantly lowers insulin levels, blood glucose, the calculated HOMA-IR, and waist circumference. After six months of algae use, plasma HDL-C levels significantly rose in NAFLD subjects [85].

Vitamin E is a complex of tocopherols and tocotrienols extensively studied for the treatment of NASH due to its well-known antioxidant properties [86]. Most studies have focused on alfa (α)-tocopherol, obtaining inconsistent findings. Vitamin E (800 IU/day) and ursodeoxycholic acid (12–15 mg/kg) administered for two years, alone or in combination vs. a single/double placebo, demonstrated histological improvement in the combination group only [87]. Positive outcomes were also observed under E and C vitamin combined administration. More interestingly, in pediatric NASH patients, pioglitazone (belonging to thiazolidinediones, an insulin sensitizer used in type 2 diabetic patients) demonstrated significantly greater efficacy than vitamin E in reversing liver fibrosis. Indeed, fibrosis reduction was of 47% for pioglitazone vs. 36% for vitamin E vs. 21% for the placebo. Therefore, the data suggest that a high vitamin E dosage (precisely, 800 IU/day) can beneficially affect mild pediatric NASH patients with only a limited effect in adults [88,89].

3.3.3. Pharmacological Treatment of Cardiometabolic Profile in MAFLD: The Crucial Role of Lowering Cholesterol Remedies

Pharmacologic MAFLD treatment should target liver steatosis, the metabolic disturbances associated with the condition and prevention of liver fibrosis development.

The pharmacological treatment of MAFLD has evolved significantly in the last twenty years, with a wide range of agents now being considered for managing the disease and its associated metabolic complications.

Statins are commonly used in patients with MAFLD to manage hyperlipidemia and reduce cardiovascular risk. However, it is important to mention that their direct impact on liver histology is uncertain. In fact, statins can reduce liver fat content and improve liver enzymes but their effect on fibrosis is inconclusive. Epidemiological studies first supported the potential benefits of statins on liver function. These found statins' use to be associated with a decreased risk of NAFLD/NASH and MAFLD/MASH diagnosis according to ultrasonography or histology usage [92,93]. Furthermore, statins' use in diabetic patients was associated with a reduced risk for steatohepatitis and, even advanced liver fibrosis development [94]. In detail, atorvastatin reduces the expression of perilipin 5 in hepatocytes, contributes to increased lipolysis, and reduces triglyceride accumulation through protein kinase A phosphorylation [95]. Indeed, discontinuation of statin therapy remains a global issue in the frame of MAFLD treatment, mainly because of myalgia occurrence. Therefore, bempedoic acid (BA) usage has become more and more frequent. BA is an ATP citrate lyase inhibitor able to decrease the hepatic synthesis of cholesterol. In detail, it upregulates LDL receptor expression in the liver and clears circulating LDL-cholesterol from the bloodstream. Several randomized clinical trials showed a significant LDL level reduction (e.g., 17–28%) in statin-intolerant patients. Regarding the cholesterol biosynthesis cascade, ATP-citrate lyase (ACL) is an enzyme working two levels above HMG-CoA reductase. Indeed, its mechanism of action is similar to, but less efficient than, that of statins. Thus, BA mainly decreases the hepatic generation of cholesterol, upregulates the LDL receptor expression within the liver, and clears the circulating LDL-C from the systemic circulation [96] [Figure 3].

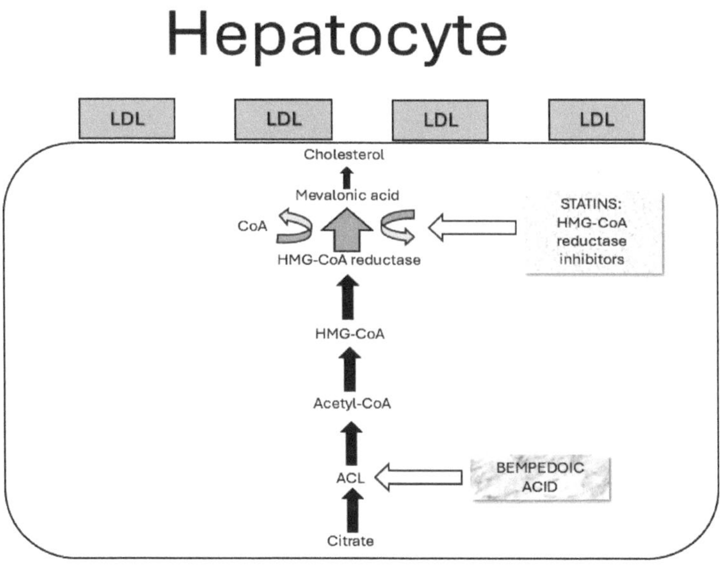

Figure 3. Mechanism of action of bempedoic acid. Bempedoic acid downregulates cholesterol biosynthesis by inhibiting ACL, a cytosolic enzyme that acts in the cholesterol synthesis chain on a phase prior to that of HMG-CoA reductase, the therapeutic target of statins.

A promising class of drugs includes Pro-protein Convertase Subtilisin/Kexin type 9 inhibitors (PCSK9i) that can offer additional benefits in patients with MAFLD. Although clinical studies using PCSK9i are, at the time of writing, few, their results indicate high efficacy and safety. PCSK9 inhibitors may reduce liver steatosis, inflammation, and fibrosis [97]. Although PCSK9i should be prescribed in patients without a compromised liver function, Shafiq et al. showed that this treatment can lower hepatic transaminases' levels [97]. Thus, PCSK9 inhibitors appear to have beneficial effects on patients with MAFLD. Notwithstanding, further research is necessary to gain more evidence for MAFLD treatment.

Although statins, the intestinal cholesterol transporter inhibitor (namely, ezetimibe) and PCSK9 inhibitors reduce serum levels of LDL-C, they do not act on Hypertriglyceridemia and HDL levels. The last two targets can be treated with fibrates, nicotinic acids, and n-3 polyunsaturated fatty acids. Fibrates have been shown to be PPARα agonists and can significantly lower triglycerides levels. However, they present some adverse effects, including an increase in creatinine levels and liver enzymes [65,97,98]. Intriguingly, several clinical trials demonstrate that Pemafibrate, the first Selective Peroxisome Proliferator-Activated Receptor Alpha Modulator (SPPARMα), can significantly lower liver enzymes and total bilirubin levels. The efficacy seems to be greater in more compromised patients [99]. These findings pave the road for its possible use in NAFLD/NASH patients that show more consistent evidence.

Antifibrotic therapies are gaining attention as they are showing promising preclinical studies. Agents that target TGF-beta and other fibrotic pathways could potentially reduce liver scarring and improve long-term outcomes in MAFLD patients [31]. To date, Resmetirom is the first Food and Drug Administration (FDA)-approved medication for the treatment of NASH/MASH. It is administered at either 80 mg or 100 mg per day. Resmetirom reduces liver fat accumulation by acting as an agonist of the thyroid hormone receptor (THR-β) [100]. In detail, the drug can provide NASH resolution (assessed by the NAFLD activity score) in 24.2% and 25.9% of patients treated with 80 and 100 mg, respectively, vs. 14.2% of those treated with a placebo ($p < 0.001$). Moreover, Resmetirom improves liver fibrosis (25.9% and 29.9%, after 80 mg and 100 mg, respectively, vs. 9.7% under placebo) ($p < 0.001$) in F2-F3 NASH patients [101].

Obeticholic acid (OCA), a semisynthetic derivative of the natural bile acid chenodeoxycholic acid, can improve insulin sensitivity and reduce biomarkers of liver fibrosis in NAFLD patients with type 2 diabetes mellitus [102]. However, we must bear in mind serious adverse events registered upon its use in non-cirrhotic patients with primary biliary cholangiopathy [103].

Antidiabetic medications are the most widely used pharmacological agents in the management of insulin resistance linked to MAFLD development. They can reduce hepatic glucose production and improve insulin resistance. Insulin sensitizers (e.g., pioglitazone and metformin) are recommended. Pioglitazone is recommended for patients diagnosed with MASH, while metformin has been found to markedly modify body composition and liver function in individuals with non-diabetic MAFLD [104]. Metformin reduces fat deposition and inhibits hepatic inflammation. This depends on enhanced phosphorylation of hepatic 5′ adenosine monophosphate-activated protein kinase (AMPK) and ACC and reduced lipogenic enzymes and proinflammatory cytokines [31].

Sodium-glucose cotransporter-2 (SGLT2) inhibitors improve liver enzymes and liver steatosis. For example, taking canagliflozin for 20 weeks delayed the onset of NASH and reduced liver enzymes, together with body weight [105]. Empagliflozin can improve both liver steatosis and fibrosis. It decreases transaminases in MAFLD patients with or without type 2 diabetes mellitus (T2DM) [106]. Furthermore, it decreases the hepatic expression of inflammatory cytokines (e.g., TNF-α, interleukin-6, and Monocyte Chemoattractant Protein-1 (MCP-1)) in NASH patients. When used in combination with linagliptin, a Dipeptidyl peptidase-4 (https://en.wikipedia.org/wiki/Dipeptidyl_peptidase-4 accessed on 1 May

2024) inhibitor (DPP-4i), it reduces mRNA expression of genes for fatty acid synthesis, collagen deposition, and the expression of Alpha Smooth Muscle Actin (αSMA). The latter is a biomarker of liver fibrosis [105,106]. Finally, empagliflozin attenuates inflammasome proteins' expression and the triglyceride NOD-like receptor (NLR) family pyrin domain containing NLRP-3 activation in the liver [106].

Glucagon-Like Peptide-1 (GLP-1) Receptor Agonists slow down the progression of MAFLD through several mechanisms: reducing inflammation, improving insulin sensitivity, and mitigating oxidative stress. Moreover, they inhibit enzymes involved in hepatic lipogenesis, activate the autophagy/mitophagy pathway, and enhance the activity of enzymes responsible for beta (β)-oxidation. MAFLD patients seem to be their favorite target because of the peculiar downregulation of GLP-1 receptors [31]. In fact, almost 40% of patients treated with liraglutide showed steatohepatitis reversal at liver biopsy and improved glycemic profile, liver enzymes, and increased HDL concentration. This was consensual with weight loss of approximately 5 kg [107]. In line with these results, s.c. semaglutide once weekly shows liver enzymes' normalization and reduced radiologic liver steatosis features in MASLD patients [108].

DPP-4i act via the suppression of the activity of DPP-4, leading to elevated incretin levels and reduced glucagon secretion. These increase insulin exocytosis, promote fatty acid oxidation in the liver, slow gastric emptying, and diminish hepatic glucose output [109]. Sitagliptin, a gliptin-based drug, can lower liver enzymes, reduce body weight, and reduce hepatocyte swelling in patients with concomitant diabetes and NASH [110].

4. Conclusions and Future Perspectives

In conclusion, the management of MAFLD requires a multifaceted approach, addressing all the metabolic components of the disease pathophysiology.

Solid evidence confirms that lifestyle modification, including correct nutrition combined with regular physical activity, has efficacy in reducing liver fat deposition and improving insulin sensitivity. Among the antioxidant substances available, high-dose Vitamin E has the most solid evidence, especially in pediatric NASH patients.

The use of pharmacological treatments such as statins, bempedoic acid, or PCSK9 inhibitors show promising results in MAFLD management and deserve further studies to confirm their capability to also revers liver fibrosis. In detail, statins can normalize liver function indexes but seem to not affect the liver fibrosis process. Indeed, they are a cornerstone in preventing fibrosis process in NAFLD/MAFLD subjects.

Newly released antifibrotic therapies show a promising impact on NASH/MASH patients for fibrosis reversal. More data are needed to confirm the first results.

Insulin sensitizers, such as Pioglitazone and metformin, can be used in diabetic and non-diabetic MASH patients, respectively, to reverse the disease course.

SGLT2 inhibitors and fibrates offer promising evidence for their use in treating dyslipidemia, insulin resistance, and inflammation in MAFLD.

GLP-1 receptor agonists have a very promising profile of action for liver fibrosis treatment in MAFLD and MASH patients, with special attention on body composition modification.

Dipeptidyl Peptidase-4 (DPP-4) inhibitors have a promising effect profile in MAFLD and NASH patients. Their potential usage can be considered in diabetic patients.

Finally, liver fat deposition, cholesterol synthesis, and transport until triglycerides' extracellular storage in the liver are targets for the reviewed treatments. However, a personalized therapeutic approach cannot be overstated due to the multifaceted physiopathology of MAFLD. Future artificial-intelligence-powered therapeutic flow-charts are warranted to fit a personalized approach to every single patient with an MAFLD diagnosis, according to the disease staging.

Author Contributions: Conceptualization, E.S. and A.P.; methodology, E.S., J.T. and L.A.; software, C.R.; validation, E.S., J.T. and G.S.B.; formal analysis, E.S.; investigation, A.P. and E.S.; resources, A.P., G.S.B. and E.S.; data curation, E.S., M.S. and J.T.; writing—original draft preparation, A.P. and E.S.; writing—review and editing, E.S. and J.T.; visualization, A.P. and M.S.; supervision, E.S. and J.T.; project administration, E.S. All authors have read and agreed to the published version of the manuscript.

Funding: This research received no external funding.

Institutional Review Board Statement: Not applicable.

Informed Consent Statement: Not applicable.

Data Availability Statement: All the data reviewed in the manuscript can be retrieved on the main medical database (e.g., PubMed, Medline) and on the website of the most important gastroenterology and hepatology international meetings (e.g., UEGW, DDW, AASLD, EASL).

Acknowledgments: We thank the unstoppable and passionate effort of researchers unraveling the pathophysiology of MAFLD and, lately, MASLD for the generations to come.

Conflicts of Interest: The authors declare no conflicts of interest.

References

1. Singh, S.; Osna, N.A.; Kharbanda, K.K. Treatment options for alcoholic and non-alcoholic fatty liver disease: A review. *World J. Gastroenterol.* **2017**, *23*, 6549–6570. [CrossRef] [PubMed]
2. Ceni, E.; Mello, T.; Galli, A. Pathogenesis of alcoholic liver disease: Role of oxidative metabolism. *World J. Gastroenterol.* **2014**, *20*, 17756–17772. [CrossRef] [PubMed]
3. Adhikari, R.; Mitra, R.; Bennett, R.G.; McVicker, B.L.; Tuma, P.L. Alcohol-induced tubulin post-translational modifications directly alter hepatic protein trafficking. *Hepatol. Commun.* **2023**, *7*, e0103. [CrossRef]
4. Salete-Granado, D.; Carbonell, C.; Puertas-Miranda, D.; Vega-Rodríguez, V.-J.; García-Macia, M.; Herrero, A.B.; Marcos, M. Autophagy, Oxidative Stress, and Alcoholic Liver Disease: A Systematic Review and Potential Clinical Applications. *Antioxidants* **2023**, *12*, 1425. [CrossRef]
5. European Association for the Study of the Liver (EASL); European Association for the Study of Diabetes (EASD); European Association for the Study of Obesity (EASO). EASL-EASD-EASO Clinical Practice Guidelines on the management of metabolic dysfunction-associated steatotic liver disease (MASLD). *J. Hepatol.* **2024**, *81*, 492–542. [CrossRef] [PubMed]
6. Radosavljevic, T.; Brankovic, M.; Samardzic, J.; Djuretić, J.; Vukicevic, D.; Vucevic, D.; Jakovljevic, V. Altered Mitochondrial Function in MASLD: Key Features and Promising Therapeutic Approaches. *Antioxidants* **2024**, *13*, 906. [CrossRef]
7. Buzzetti, E.; Pinzani, M.; Tsochatzis, E.A. The multiple-hit pathogenesis of non-alcoholic fatty liver disease (NAFLD). *Metabolism.* **2016**, *65*, 1038–1048. [CrossRef]
8. Lu, W.; Li, S.; Li, Y.; Zhou, J.; Wang, K.; Chen, N.; Li, Z. Associations of sex-related and thyroid-related hormones with risk of metabolic dysfunction-associated fatty liver disease in T2DM patients. *BMC Endocr. Disord.* **2024**, *24*, 84. [CrossRef]
9. Anastasopoulos, N.T.; Lianos, G.D.; Tatsi, V.; Karampa, A.; Goussia, A.; Glantzounis, G.K. Clinical heterogeneity in patients with non-alcoholic fatty liver disease-associated hepatocellular carcinoma. *Expert Rev. Gastroenterol. Hepatol.* **2020**, *14*, 1025–1033. [CrossRef]
10. Di Ciaula, A.; Passarella, S.; Shanmugam, H.; Noviello, M.; Bonfrate, L.; Wang, D.Q.; Portincasa, P. Nonalcoholic Fatty Liver Disease (NAFLD). Mitochondria as Players and Targets of Therapies? *Int. J. Mol. Sci.* **2021**, *22*, 5375. [CrossRef]
11. Kim, B.J.; Ryu, S.W.; Song, B.J. JNK- and p38 kinase-mediated phosphorylation of Bax leads to its activation and mitochondrial translocation and to apoptosis of human hepatoma HepG2 cells. *J. Biol. Chem.* **2006**, *281*, 21256–21265. [CrossRef] [PubMed]
12. National Workshop on Fatty Liver and Alcoholic Liver Disease; Chinese Society of Hepatology; Chinese Medical Association; Fatty Liver Expert Committee and Chinese Medical Doctor Association. Guidelines of prevention and treatment for nonalcoholic fatty liver disease: A 2018 update. *Zhonghua Gan Zang Bing Za Zhi* **2018**, *26*, 195–203.
13. Lim, S.; Kim, J.W.; Targher, G. Links between metabolic syndrome and metabolic dysfunction-associated fatty liver disease. *Trends Endocrinol. Metab.* **2021**, *32*, 500–514. [CrossRef] [PubMed]
14. Schneider, A.L.C.; Lazo, M.; Selvin, E.; Clark, J.M. Racial differences in nonalcoholic fatty liver disease in the U.S. population. *Obesity* **2014**, *22*, 292–299. [CrossRef] [PubMed]
15. Adejumo, A.C.; Samuel, G.O.; Adegbala, O.M.; Adejumo, K.L.; Ojelabi, O.; Akanbi, O.; Ogundipe, O.A.; Pani, L. Prevalence, trends, outcomes, and disparities in hospitalizations for nonalcoholic fatty liver disease in the United States. *Ann. Gastroenterol.* **2019**, *32*, 504–513. [CrossRef]

16. Lim, U.; Monroe, K.R.; Buchthal, S.; Fan, B.; Cheng, I.; Kristal, B.S.; Lampe, J.W.; Hullar, M.A.; Franke, A.A.; Stram, D.O.; et al. Propensity for Intra-abdominal and Hepatic Adiposity Varies Among Ethnic Groups. *Gastroenterology* **2019**, *156*, 966–975. [CrossRef] [PubMed]
17. Talens, M.; Tumas, N.; Lazarus, J.V.; Benach, J.; Pericàs, J.M. What Do We Know about Inequalities in NAFLD Distribution and Outcomes? A Scoping Review. *J. Clin. Med.* **2021**, *10*, 5019. [CrossRef]
18. Golovaty, I.; Tien, P.C.; Price, J.C.; Sheira, L.; Seligman, H.; Weiser, S.D. Food Insecurity May Be an Independent Risk Factor Associated with Nonalcoholic Fatty Liver Disease among Low-Income Adults in the United States. *J. Nutr.* **2020**, *150*, 91–98. [CrossRef]
19. Cho, J.; Lee, I.; Park, D.H.; Kwak, H.B.; Min, K. Relationships between socioeconomic status, handgrip strength, and non-alcoholic fatty liver disease in middle-aged adults. *Int. J. Environ. Res. Public Health* **2021**, *18*, 1892. [CrossRef] [PubMed]
20. Hu, W.; Liu, Z.; Hao, H.R.; Yu, W.N.; Wang, X.Q.; Shao, X.J.; Wu, X.J.; Wen, S.R.; Fan, Y.Q.; Ni, Y.J. Correlation between income and non-alcoholic fatty liver disease in a Chinese population. *Ann. Endocrinol.* **2020**, *81*, 561–566. [CrossRef] [PubMed]
21. Tutunchi, H.; Saghafi-Asl, M.; Ebrahimi-Mameghani, M.; Ostadrahimi, A. Food insecurity and lipid profile abnormalities are associated with an increased risk of non-alcoholic fatty liver disease (NAFLD): A case-control study in northwest of Iran. *Ecol. Food Nutr.* **2021**, *11*, 1–17.
22. Santos, R.D.; Valenti, L.; Romeo, S. Does nonalcoholic fatty liver disease cause cardiovascular disease? *Curr. Knowl. Gaps Atheroscler.* **2019**, *282*, 110–120. [CrossRef]
23. Vesković, M.; Pejović, M.; Šutulović, N.; Hrnčić, D.; Rašić-Marković, A.; Stanojlović, O.; Mladenović, D. Exploring Fibrosis Patho-physiology in Lean and Obese Metabolic-Associated Fatty Liver Disease: An In-Depth Comparison. *Int. J. Mol. Sci.* **2024**, *25*, 7405. [CrossRef]
24. Shaker, M.E. The contribution of sterile inflammation to the fatty liver disease and the potential therapies. *Biomed. Pharmacother.* **2022**, *148*, 112789. [CrossRef]
25. Shabbir, A.; Abbas, Z.; Khatoon, A.; Mirza, T. Role of Alanine Transaminase and Transient Elastography in Categorising Non-alcoholic Fatty Liver Disease Subgroups. *J. Coll. Physicians Surg. Pak.* **2024**, *34*, 22–26.
26. Zheng, M.H.; Lonardo, A. Red cell distribution width/platelet ratio predicts decompensation of metabolic dysfunction-associated steatotic liver disease-related compensated advanced chronic liver disease. *World J. Gastroenterol.* **2025**, *31*, 100393. [CrossRef]
27. Li, X.M.; Liu, S.L.; He, Y.J.; Shu, J.C. Using new indices to predict metabolism dysfunction-associated fatty liver disease (MAFLD): Analysis of the national health and nutrition examination survey database. *BMC Gastroenterol.* **2024**, *24*, 109. [CrossRef]
28. Meex, R.C.R.; Watt, M.J. Hepatokines: Linking nonalcoholic fatty liver disease and insulin resistance. *Nat. Rev. Endocrinol.* **2017**, *13*, 509–520. [CrossRef]
29. Sanyal, A.J.; Campbell–Sargent, C.; Mirshahi, F.; Rizzo, W.B.; Contos, M.J.; Sterling, R.K.; Luketic, V.A.; Shiffman, M.L.; Clore, J.N. Nonalcoholic steatohepatitis: Association of insulin resistance and mito-chondrial abnormalities. *Gastroenterology* **2001**, *120*, 1183–1192. [CrossRef] [PubMed]
30. Donnelly, K.L.; Smith, C.I.; Schwarzenberg, S.J.; Jessurun, J.; Boldt, M.D.; Parks, E.J. Sources of fatty acids stored in liver and secreted via lipoproteins in patients with nonalcoholic fatty liver disease. *J. Clin. Investig.* **2005**, *115*, 1343–1351. [CrossRef] [PubMed]
31. Al Hashmi, K.; Giglio, R.V.; Pantea Stoian, A.; Patti, A.M.; Al Waili, K.; Al Rasadi, K.; Ciaccio, M.; Rizzo, M. Metabolic dys-function-associated fatty liver disease: Current therapeutic strategies. *Front. Nutr.* **2024**, *11*, 1355732. [CrossRef] [PubMed]
32. Reinshagen, M.; Kabisch, S.; Pfeiffer, A.F.H.; Spranger, J. Liver Fat Scores for Noninvasive Diagnosis and Monitoring of Nonalcoholic Fatty Liver Disease in Epidemiological and Clinical Studies. *J. Clin. Transl. Hepatol.* **2023**, *11*, 1212–1227. [PubMed]
33. Biciusca, T.; Stan, S.I.; Balteanu, M.A.; Cioboata, R.; Ghenea, A.E.; Danoiu, S.; Bumbea, A.M.; Biciusca, V. The Role of the Fatty Liver Index (FLI) in the Management of Non-Alcoholic Fatty Liver Disease: A Systematic Review. *Diagnostics* **2023**, *13*, 3316. [CrossRef] [PubMed]
34. Lee, J.; Vali, Y.; Boursier, J.; Spijker, R.; Anstee, Q.M.; Bossuyt, P.M.; Zafarmand, M.H. Prognostic accuracy of FIB-4, NAFLD fibrosis score and APRI for NAFLD-related events: A systematic review. *Liver Int.* **2021**, *41*, 261–270. [CrossRef] [PubMed]
35. Boeckmans, J.; Natale, A.; Buyl, K.; Rogiers, V.; De Kock, J.; Vanhaecke, T.; Rodrigues, R.M. Human-based systems: Mechanistic NASH modelling just around the corner? *Pharmacol. Res.* **2018**, *134*, 257–267. [CrossRef] [PubMed]
36. Esterson, Y.B.; Grimaldi, G.M. Radiologic Imaging in Nonalcoholic Fatty Liver Disease and Nonalcoholic Steatohepatitis. *Clin. Liver Dis.* **2018**, *22*, 93–108. [CrossRef]
37. Leoni, S.; Tovoli, F.; Napoli, L.; Serio, I.; Ferri, S.; Bolondi, L. Current guidelines for the management of non-alcoholic fatty liver disease: A systematic review with comparative analysis. *World J. Gastroenterol.* **2018**, *24*, 3361–3373. [CrossRef] [PubMed]
38. Ofosu, A.; Ramai, D.; Reddy, M. Non-alcoholic fatty liver disease: Controlling an emerging epidemic, challenges, and future directions. *Ann. Gastroenterol.* **2018**, *31*, 288–295. [CrossRef]
39. Lee, D.H. Imaging evaluation of non-alcoholic fatty liver disease: Focused on quantification. *Clin. Mol. Hepatol.* **2017**, *23*, 290–301. [CrossRef] [PubMed]

40. Newsome, P.N.; Sasso, M.; Deeks, J.J.; Paredes, A.; Boursier, J.; Chan, W.K.; Yilmaz, Y.; Czernichow, S.; Zheng, M.H.; Wong, V.W.; et al. FibroScan-AST (FAST) score for the non-invasive identification of patients with non-alcoholic steatohepatitis with significant activity and fibrosis: A prospective derivation and global validation study. *Lancet Gastroenterol. Hepatol.* **2020**, *5*, 362–373. [CrossRef] [PubMed]
41. Noureddin, M.; Lam, J.; Peterson, M.R.; Middleton, M.; Hamilton, G.; Le, T.A.; Bettencourt, R.; Changchien, C.; Brenner, D.A.; Sirlin, C.; et al. Utility of magnetic resonance imaging versus histology for quantifying changes in liver fat in nonalcoholic fatty liver disease trials. *Hepatology* **2013**, *58*, 1930–1940. [CrossRef] [PubMed]
42. Mikolasevic, I.; Milic, S.; Orlic, L.; Stimac, D.; Franjic, N.; Targher, G. Factors associated with significant liver steatosis and fibrosis as assessed by transient elastography in patients with one or more components of the metabolic syndrome. *J. Diabetes Its Complicat.* **2016**, *30*, 1347–1353. [CrossRef]
43. Zhang, J.Z.; Cai, J.J.; Yu, Y.; She, Z.G.; Li, H. Nonalcoholic Fatty Liver Disease: An Update on the Diagnosis. *Gene Expr.* **2019**, *19*, 187–198. [CrossRef]
44. Boyd, A.; Cain, O.; Chauhan, A.; Webb, G.J. Medical liver biopsy: Background, indications, procedure and histopathology. *Frontline Gastroenterol.* **2020**, *11*, 40–47. [CrossRef]
45. Castera, L. Diagnosis of non-alcoholic fatty liver disease/non-alcoholic steatohepatitis: Non-invasive tests are enough. *Liver Int.* **2018**, *38* (Suppl. S1), 67–70. [CrossRef]
46. Ramachandran, R.; Kakar, S. Histological patterns in drug-induced liver disease. *J. Clin. Pathol.* **2009**, *62*, 481–492. [CrossRef] [PubMed]
47. Borgne-Sanchez, A.; Fromenty, B. Mitochondrial dysfunction in drug-induced hepatic steatosis: Recent findings and current concept. *Clin. Res. Hepatol. Gastroenterol.* **2025**, *49*, 102529. [CrossRef] [PubMed]
48. Fromenty, B. Inhibition of mitochondrial fatty acid oxidation in drug-induced hepatic steatosis. *Liver Res.* **2019**, *3*, 157–169. [CrossRef]
49. Schott, M.B.; Weller, S.G.; Schulze, R.J.; Krueger, E.W.; Drizyte-Miller, K.; Casey, C.A.; McNiven, M.A. Lipid droplet size directs lipolysis and lipophagy catabolism in hepatocytes. *J. Cell Biol.* **2019**, *218*, 3320–3335. [CrossRef]
50. Syed-Abdul, M.M. Lipid Metabolism in Metabolic-Associated Steatotic Liver Disease (MASLD). *Metabolites* **2024**, *14*, 12. [CrossRef]
51. Konings, M.C.J.M.; Baumgartner, S.; Mensink, R.P.; Plat, J. Investigating microRNAs to Explain the Link between Cholesterol Metabolism and NAFLD in Humans: A Systematic Review. *Nutrients* **2022**, *14*, 4946. [CrossRef] [PubMed]
52. Habibullah, M.; Jemmieh, K.; Ouda, A.; Haider, M.Z.; Malki, M.I.; Elzouki, A.-N. Metabolic-associated fatty liver disease: A selective review of pathogenesis, diagnostic approaches, and therapeutic strategies. *Front. Med.* **2024**, *11*, 1291501. [CrossRef]
53. Gurr, M.I. Lipid metabolism in man. *Proc. Nutr. Soc.* **1988**, *47*, 277–285. [CrossRef]
54. Hong, S.; Gordon, D.; Stec, D.E.; Hinds, T.D. Bilirubin: A ligand of the PPARα nuclear receptor. In *Nuclear Receptors: The Art and Science of Modulator Design and Discovery*; Badr, M.Z., Ed.; Springer International Publishing: Cham, Switzerland, 2021; pp. 463–482.
55. Knebel, B.; Haas, J.; Hartwig, S.; Jacob, S.; Köllmer, C.; Nitzgen, U.; Muller–Wieland, D.; Kotzka, J. Liver-specific expression of transcriptionally active SREBP-1c is associated with fatty liver and increased visceral fat mass. *PLoS ONE* **2012**, *7*, e31812. [CrossRef]
56. Coassolo, L.; Liu, T.; Jung, Y.; Taylor, N.P.; Zhao, M.; Charville, G.W.; Nissen, S.B.; Yki-Jarvinen, H.; Altman, R.B.; Svensson, K.J. Mapping transcriptional heterogeneity and metabolic networks in fatty livers at single-cell resolution. *iScience* **2023**, *26*, 105802. [CrossRef]
57. Zhu, L.; Baker, S.S.; Liu, W.; Tao, M.H.; Patel, R.; Nowak, N.J.; Baker, R.D. Lipid in the livers of adolescents with nonalcoholic steatohepatitis: Combined effects of pathways on steatosis. *Metabolism* **2011**, *60*, 1001–1011. [CrossRef] [PubMed]
58. Mitsuyoshi, H.; Yasui, K.; Harano, Y.; Endo, M.; Tsuji, K.; Minami, M.; Itoh, Y.; Okanoue, T.; Yoshikawa, T. Analysis of hepatic genes involved in the metabolism of fatty acids and iron in nonalcoholic fatty liver disease. *Hepatol. Res.* **2009**, *39*, 366–373. [CrossRef] [PubMed]
59. Ipsen, D.H.; Lykkesfeldt, J.; Tveden-Nyborg, P. Molecular mechanisms of hepatic lipid accumulation in non-alcoholic fatty liver disease. *Cell Mol. Life Sci.* **2018**, *75*, 3313–3327. [CrossRef]
60. Newberry, E.P.; Xie, Y.; Kennedy, S.; Han, X.; Buhman, K.K.; Luo, J.; Gross, R.W.; Davidson, N.O. Decreased hepatic triglyceride accumulation and altered fatty acid uptake in mice with deletion of the liver fatty acid-binding protein gene. *J. Biol. Chem.* **2003**, *278*, 51664–51672. [CrossRef]
61. Auinger, A.; Valenti, L.; Pfeuffer, M.; Helwig, U.; Herrmann, J.; Fracanzani, A.L.; Dongiovanni, P.; Fargion, S.; Schrezenmeir, J.; Rubin, D. A promoter polymorphism in the liver-specific fatty acid transport protein 5 is associated with features of the metabolic syndrome and steatosis. *Horm. Metab. Res.* **2010**, *42*, 854–859. [CrossRef] [PubMed]
62. Koonen, D.P.; Jacobs, R.L.; Febbraio, M.; Young, M.E.; Soltys, C.-L.M.; Ong, H.; Vance, D.E.; Dyck, J.R. Increased hepatic CD36 expression contributes to dyslipidemia associated with diet-induced obesity. *Diabetes* **2007**, *56*, 2863–2871. [CrossRef] [PubMed]

63. Abdel-Hamed, A.R.; Mesbah, N.M.; Ghattas, M.H.; Abo-elmatty, D.M.; Saleh, S.M. Serum miRNA-122 expression in non-alcoholic fatty liver disease among Egyptian patients and its correlation with interleukin-1A gene polymorphism. *Meta Gene* **2017**, *14*, 19–23. [CrossRef]
64. Yamashita, S.; Rizzo, M.; Su, T.-C.; Masuda, D. Novel Selective PPARα Modulator Pemafibrate for Dyslipidemia, Nonalcoholic Fatty Liver Disease (NAFLD), and Atherosclerosis. *Metabolites* **2023**, *13*, 626. [CrossRef]
65. Baffy, G. MicroRNAs in Nonalcoholic Fatty Liver Disease. *J. Clin. Med.* **2015**, *4*, 1977–1988. [CrossRef]
66. Dongiovanni, P.; Meroni, M.; Longo, M.; Fargion, S.; Fracanzani, A.L. miRNA Signature in NAFLD: A Turning Point for a Non-Invasive Diagnosis. *Int. J. Mol. Sci.* **2018**, *19*, 3966. [CrossRef]
67. Cai, J.; Zhang, X.J.; Ji, Y.X.; Zhang, P.; She, Z.G.; Li, H. Nonalcoholic fatty liver disease pandemic fuels the upsurge in cardiovascular diseases. *Circ. Res.* **2020**, *126*, 679–704. [CrossRef]
68. Quek, J.; Ng, C.H.; Tang, A.S.P.; Chew, N.; Chan, M.; Khoo, C.M.; Wei, C.P.; Chin, Y.H.; Tay, P.; Lim, G.; et al. Metabolic associated fatty liver disease increases the risk of systemic complications and mortality. A Meta-analysis and systematic review of 12,620,736 individuals. *Endocr. Pract.* **2022**, *28*, 667–672. [CrossRef]
69. Targher, G.; Byrne, C.D.; Tilg, H. MASLD: A systemic metabolic disorder with cardiovascular and malignant complications. *Gut* **2024**, *73*, 691–702. [CrossRef]
70. Anstee, Q.M.; Mantovani, A.; Tilg, H.; Targher, G. Risk of cardiomyopathy and cardiac arrhythmias in patients with Nonalcoholic fatty liver disease. *Nat. Rev. Gastroenterol. Hepatol.* **2018**, *15*, 425–439. [CrossRef]
71. Li, W.; Zhang, S.; Zhang, T.; Shen, Y.; Han, L.; Peng, Z.; Xie, Z.; Zhong, C.; Jia, S. Exenatide once-weekly improves metabolic parameters, endothelial dysfunction and carotid intima-media thickness in patients with type-2 diabetes: An 8-month prospective study. *Diabetes Res. Clin. Pract.* **2019**, *149*, 163–169. [CrossRef]
72. Spence, J.D. Diet for stroke prevention. *Stroke Vasc. Neurol.* **2018**, *3*, 44–50. [CrossRef] [PubMed]
73. Carson, J.A.S.; Lichtenstein, A.H.; Anderson, C.A.; Appel, L.J.; Kris-Etherton, P.M.; Meyer, K.A.; Petersen, K.; Polonsky, T.; Van Horn, L. Dietary cholesterol and cardiovascular risk: A science advisory from the American Heart Association. *Circulation.* **2020**, *141*, e39–e53. [CrossRef] [PubMed]
74. Eslam, M.; Sarin, S.K.; Wong, V.W.; Fan, J.G.; Kawaguchi, T.; Ahn, S.H.; Zheng, M.H.; Shiha, G.; Yilmaz, Y.; Gani, R.; et al. The Asian Pacific Association for the Study of the Liver clinical practice guidelines for the diagnosis and management of metabolic associated fatty liver disease. *Hepatol. Int.* **2020**, *14*, 889–919. [CrossRef]
75. Koutoukidis, D.A.; Astbury, N.M.; Tudor, K.E.; Morris, E.; Henry, J.A.; Noreik, M.; Jebb, S.A.; Aveyard, P. Association of Weight Loss Interventions with Changes in biomarkers of nonalcoholic fatty liver disease: A systematic review and Meta-analysis. *JAMA Intern. Med.* **2019**, *179*, 1262–1271. [CrossRef] [PubMed]
76. Seebacher, F.; Zeigerer, A.; Kory, N.; Krahmer, N. Hepatic lipid droplet homeostasis and fatty liver disease. *Semin. Cell Dev. Biol.* **2020**, *108*, 72–81. [CrossRef] [PubMed]
77. Rizzo, M.; Colletti, A.; Penson, P.E.; Katsiki, N.; Mikhailidis, D.P.; Toth, P.P.; Gouni-Berthold, I.; Mancini, J.; Marais, D.; Ruscica, M.; et al. Nutraceutical approaches to non-alcoholic fatty liver disease (NAFLD): A position paper from the international lipid expert panel (ILEP). *Pharmacol. Res.* **2023**, *189*, 106679. [CrossRef] [PubMed]
78. Giglio, R.V.; Stoian, A.P.; Al-Rasadi, K.; Banach, M.; Patti, A.M.; Ciaccio, M.; Rizvi, A.A.; Rizzo, M. Novel therapeutical approaches to managing atherosclerotic risk. *Int. J. Mol. Sci.* **2021**, *22*, 4633. [CrossRef]
79. Del Ben, M.; Baratta, F.; Polimeni, L.; Pastori, D.; Loffredo, L.; Averna, M.; Violi, F.; Angelico, F. Under-prescription of statins in patients with non-alcoholic fatty liver disease. *Nutr. Metab. Cardiovasc. Dis.* **2017**, *27*, 161–167. [CrossRef]
80. Browning, L.M.; Walker, C.G.; Mander, A.P.; West, A.L.; Gambell, J.; Madden, J.; Calder, P.C.; Jebb, S.A. Compared with daily, weekly n-3 PUFA intake affects the in-corporation of eicosapentaenoic acid and docosahexaenoic acid into platelets and mononuclear cells in humans. *J. Nutr.* **2014**, *144*, 667–672. [CrossRef]
81. Sanyal, A.J.; Abdelmalek, M.F.; Suzuki, A.; Cummings, O.W.; Chojkier, M. No significant effects of ethyl-eicosapentanoic acid on histologic features of nonalcoholic steatohepatitis in a phase 2 trial. *Gastroenterology* **2014**, *147*, 377–384. [CrossRef]
82. Koperska, A.; Moszak, M.; Seraszek-Jaros, A.; Bogdanski, P.; Szulinska, M. Does berberine impact anthropometric, hepatic, and metabolic parameters in patients with metabolic dysfunction-associated fatty liver disease? Randomized, double-blind placebo-controlled trial. *J. Physiol. Pharmacol.* **2024**, *75*.
83. Różański, G.; Tabisz, H.; Zalewska, M.; Niemiro, W.; Kujawski, S.; Newton, J.; Zalewski, P.; Słomko, J. Meta-Analysis of Exploring the Effect of Curcumin Supplementation with or without Other Advice on Biochemical and Anthropometric Parameters in Patients with Metabolic-Associated Fatty Liver Disease (MAFLD). *Int. J. Environ. Res. Public Health* **2023**, *20*, 4266. [CrossRef]
84. Hallajzadeh, J.; Milajerdi, A.; Mobini, M.; Amirani, E.; Azizi, S.; Nikkhah, E.; Bahadori, B.; Sheikhsoleimani, R.; Mirhashemi, S.M. Effects of *Nigella sativa* on glycemic control, lipid profiles, and biomarkers of inflammatory and oxidative stress: A systematic review and meta-analysis of randomized controlled clinical trials. *Phytother. Res.* **2020**, *34*, 2586–2608. [CrossRef] [PubMed]

85. Farsi, F.; Mohammadshahi, M.; Alavinejad, P.; Rezazadeh, A.; Zarei, M.; Engali, K.A. Functions of co-enzyme Q10 supplementation on liver enzymes, markers of systemic inflammation, and Adipokines in patients affected by nonalcoholic fatty liver disease: A double-blind, placebo-controlled, randomized clinical trial. *J. Am. Coll. Nutr.* **2016**, *35*, 346–353. [CrossRef]
86. Brigelius-Flohé, R.; Traber, M.G. Vitamin E: Function and metabolism. *FASEB J.* **1999**, *13*, 1145–1155. [CrossRef] [PubMed]
87. Dufour, J.-F.; Oneta, C.M.; Gonvers, J.-J.; Bihl, F.; Cerny, A.; Cereda, J.-M.; Zala, J.-F.; Helbling, B.; Steuerwald, M.; Zimmermann, A.; et al. Randomized placebo-controlled trial of ursodeoxycholic acid with vitamin e in nonalcoholic steatohepatitis. *Clin. Gastroenterol. Hepatol.* **2006**, *4*, 1537–1543. [CrossRef]
88. Miller, E.R.; Pastor-Barriuso, R.; Dalal, D.; Riemersma, R.A.; Appel, L.J.; Gual-lar, E. Meta-Analysis: High-dosage Vitamin E supplementation may increase all-cause mortality. *Ann. Intern. Med.* **2005**, *142*, 37–46. [CrossRef]
89. Brown, B.G.; Crowley, J. Is there any hope for vitamin E? *JAMA* **2005**, *293*, 1387–1390. [CrossRef] [PubMed]
90. Torre, A. Silymarin in the management of liver enzyme activity in steatohepatitis: A case report. *Drugs Context* **2023**, *12*, 2023-1-5. [CrossRef] [PubMed]
91. Lee, Y.Y.; Tee, V. Management of non-alcoholic fatty liver disease incidentally detected during other medical assessments. *Drugs Context* **2023**, *12*, 2023-1-3. [CrossRef]
92. Dongiovanni, P.; Petta, S.; Mannisto, V.; Mancina, R.M.; Pipitone, R.; Karja, V.; Maggioni, M.; Kakela, P.; Wiklund, O.; Mozzi, E.; et al. Statin use and non-alcoholic steatohepatitis in at risk individuals. *J. Hepatol.* **2015**, *63*, 705–712. [CrossRef] [PubMed]
93. Bril, F.; Sanchez, P.P.; Lomonaco, R.; Orsak, B.; Hecht, J.; Tio, F.; Cusi, K. Liver safety of statins in prediabetes or T2DM and nonalcoholic steatohepatitis: Post hoc analysis of a randomized trial. *J. Clin. Endocrinol. Metab.* **2017**, *102*, 2950–2961. [CrossRef]
94. Nascimbeni, F.; Aron-Wisnewsky, J.; Pais, R.; Tordjman, J.; Poitou, C.; Charlotte, F.; Bedossa, P.; Poynard, T.; Clément, K.; Ratziu, V.; et al. Statins, antidiabetic medications and liver histology in patients with diabetes with non-alcoholic fatty liver disease. *BMJ Open Gastroenterol.* **2016**, *3*, e000075. [CrossRef]
95. Gao, X.; Nan, Y.; Zhao, Y.; Yuan, Y.; Ren, B.; Sun, C.; Cao, K.; Yu, M.; Feng, X.; Ye, J. Atorvastatin reduces lipid accumulation in the liver by activating protein kinase A-mediated phosphorylation of perilipin 5. *Biochim. Biophys. Acta* **2017**, *1862*, 1512–1519. [CrossRef] [PubMed]
96. Pinkosky, S.L.; Newton, R.S.; Day, E.A.; Ford, R.J.; Lhotak, S.; Austin, R.C.; Birch, C.M.; Smith, B.K.; Filippov, S.; Groot, P.H.; et al. Liver-specific ATP-citrate lyase inhibition by bempedoic acid decreases LDL-C and attenuates ath-erosclerosis. *Nat. Commun.* **2016**, *7*, 13457. [CrossRef] [PubMed]
97. Zhang, X.L.; Zhu, Q.Q.; Zhu, L.; Chen, J.Z.; Chen, Q.H.; Li, G.N.; Xie, J.; Kang, L.N.; Xu, B. Safety and efficacy of anti-PC-SK9 antibodies: A meta-analysis of 25 randomized, controlled trials. *BMC Med.* **2015**, *13*, 123. [CrossRef]
98. Jun, M.; Foote, C.; Lv, J.; Neal, B.; Patel, A.; Nicholls, S.J.; Grobbee, D.E.; Cass, A.; Chalmers, J.; Perkovic, V. Effects of fibrates on cardiovascular outcomes: A systematic review and meta-analysis. *Lancet* **2010**, *375*, 1875–1884. [CrossRef]
99. Yokote, K.; Yamashita, S.; Arai, H.; Araki, E.; Matsushita, M.; Nojima, T.; Suganami, H.; Ishibashi, S. Effects of pemafibrate on glucose metabolism markers and liver function tests in patients with hypertriglyceridemia: A pooled analysis of six phase 2 and phase 3 randomized double-blind placebo-controlled clinical trials. *Cardiovasc. Diabetol.* **2021**, *20*, 96. [CrossRef]
100. Madrigal Pharmaceuticals Announces FDA Approval of RezdiffraTM (Resmetirom) for the Treatment of Patients with Noncirrhotic Nonalcoholic Steatohepatitis (NASH) with Moderate to Advanced Liver Fibrosis. Madrigal Pharmaceuticals. Available online: https://www.rezdiffra.com/about-rezdiffra (accessed on 19 March 2024).
101. Guirguis, E.; Dougherty, J.; Thornby, K.; Grace, Y.; Mack, K. Resmetirom: The First Food and Drug Administration-Approved Medication for Nonalcoholic Steatohepatitis (NASH). *Ann. Pharmacother.* **2025**, *59*, 162–173. [CrossRef] [PubMed]
102. Mudaliar, S.; Henry, R.R.; Sanyal, A.J.; Morrow, L.; Marschall, H.; Kipnes, M.; Adorini, L.; Sciacca, C.I.; Clopton, P.; Castelloe, E.; et al. Efficacy and safety of the farnesoid X receptor agonist obeticholic acid in patients with type 2 diabetes and nonalcoholic fatty liver disease. *Gastroenterology* **2013**, *145*, 574–582.e1. [CrossRef] [PubMed]
103. Available online: https://www.fda.gov/drugs/drug-safety-and-availability/serious-liver-injury-being-observed-patients-without-cirrhosis-taking-ocaliva-obeticholic-acid-treat#:~:text=This%20information%20is%20an%20update,I%20sued%20on%20May%2026,%202021 (accessed on 1 May 2024).
104. Lee, C.H.; Lui, D.T.; Mak, L.Y.; Fong, C.H.; Chan, K.S.; Mak, J.H.; Cheung, C.Y.; Chow, W.S.; Woo, Y.C.; Yuen, M.F.; et al. Benefits of combining SGLT2 inhibitors and pioglitazone on risk of MASH in type 2 diabetes-A real-world study. *Diabetes Obes. Metab.* **2025**, *27*, 574–582. [CrossRef] [PubMed]
105. Shiba, K.; Tsuchiya, K.; Komiya, C.; Miyachi, Y.; Mori, K.; Shimazu, N.; Yamaguchi, S.; Ogasawara, N.; Katoh, M.; Itoh, M.; et al. Canagliflozin, an SGLT2 inhibitor, attenuates the development of hepatocellular carcinoma in a mouse model of human NASH. *Sci. Rep.* **2018**, *8*, 2362. [CrossRef] [PubMed]
106. Liu, W.; You, D.; Lin, J.; Zou, H.; Zhang, L.; Luo, S.; Yuan, Y.; Wang, Z.; Qi, J.; Wang, W.; et al. SGLT2 inhibitor promotes ketogenesis to improve MASH by suppressing CD8+ T cell activation. *Cell Metab.* **2024**, *36*, 2245–2261.e6. [CrossRef] [PubMed]

107. Armstrong, M.J.; Gaunt, P.; Aithal, G.P.; Barton, D.; Hull, D.; Parker, R.; Hazlehurst, J.M.; Guo, K.; Abouda, G.; Aldersley, M.A.; et al. Liraglutide safety and efficacy in patients with non-alcoholic steatohepatitis (LEAN): A multicentre, double-blind, randomised, placebo-controlled phase 2 study. *Lancet* **2016**, *387*, 679–690. [CrossRef] [PubMed]
108. Dutta, D.; Kumar, M.; Shivaprasad, K.S.; Kumar, A.; Sharma, M. Impact of semaglutide on biochemical and radiologic measures of metabolic-dysfunction associated fatty liver disease across the spectrum of glycaemia: A meta-analysis. *Diabetes Metab. Syndr.* **2022**, *16*, 102539. [CrossRef]
109. Sayari, S.; Neishaboori, H.; Jameshorani, M. Combined effects of synbiotic and sitagliptin versus sitagliptin alone in patients with nonalcoholic fatty liver disease. *Clin. Mol. Hepatol.* **2018**, *24*, 331–338. [CrossRef] [PubMed]
110. Yilmaz, Y.; Yonal, O.; Deyneli, O.; Celikel, C.A.; Kalayci, C.; Duman, D.G. Effects of sitagliptin in diabetic patients with nonalcoholic steatohepatitis. *Acta Gastroenterol. Belg.* **2012**, *75*, 240–244. [PubMed]

medicina

MDPI

Review

Helicobacter pylori, Atherosclerosis, and Coronary Artery Disease: A Narrative Review

Angela Saviano [1], Maria Rita Morabito Loprete [1], Giulia Pignataro [1], Andrea Piccioni [1], Antonio Gasbarrini [2], Francesco Franceschi [1] and Marcello Candelli [1,*]

[1] Emergency, Anesthesiological and Reanimation Sciences Department, Fondazione Policlinico Universitario A. Gemelli—IRCCS of Rome, 00168 Rome, Italy; angela.saviano@policlinicogemelli.it (A.S.); maria.morabitoloprete01@icatt.it (M.R.M.L.); giulia.pignataro@policlinicogemelli.it (G.P.); andrea.piccioni@policlinicogemelli.it (A.P.); francesco.franceschi@policlinicogemelli.it (F.F.)

[2] Medical and Surgical Science Department, Fondazione Policlinico Universitario A. Gemelli—IRCCS of Rome, 00168 Rome, Italy; antonio.gasbarrini@policlinicogemelli.it

* Correspondence: marcello.candelli@policlinicogemelli.it

Abstract: Coronary artery disease (CAD) is one of the leading causes of death worldwide, significantly contributing to mortality in both developed and developing nations. CAD arises from a combination of risk factors, including atherosclerosis, dyslipidemia, hypertension, diabetes, and smoking. In recent years, growing evidence has suggested a potential link between infectious agents and cardiovascular diseases. Among these, *Helicobacter pylori* (*H. pylori*) infection has been hypothesized for over a decade to play a role in the pathogenesis of CAD. This hypothesis is based on the bacterium's ability to trigger host inflammatory or autoimmune responses, potentially contributing to the progression of atherosclerotic plaques and coronary events. The association between *H. pylori* infection and CAD is of considerable interest as it opens new avenues for prevention and management strategies in cardiovascular health. Understanding this relationship could lead to innovative approaches to reducing the burden of CAD, particularly in populations with a high prevalence of *H. pylori*. In this review, we aim to provide a comprehensive overview of the most recent evidence on the involvement of *H. pylori* in the development and prognosis of CAD. By analyzing and synthesizing current findings, we seek to shed light on unresolved questions and clarify the ambiguous aspects of this potential connection. Our goal is to contribute to a deeper understanding of how *H. pylori*, may influence cardiovascular disease and to inspire further research in this critical area.

Keywords: *Helicobacter pylori*; CAD; infection; stroke; CagA

1. Introduction

Helicobacter pylori (*H. pylori*) is a gram-negative, acidophilic, spiral-shaped bacterium that primarily colonizes the stomach and duodenum. It is a major cause of acute gastritis, one of the most prevalent infections globally, affecting a significant portion of the population. In addition to acute gastritis, it is implicated in chronic gastritis, peptic ulcer disease, gastric adenocarcinoma, and gastric mucosa-associated lymphoid tissue (MALT) lymphoma [1]. It is estimated that in developing countries, between 70% and 90% of the population is infected with *H. pylori* [2]. The bacterium can be transmitted through both oral–oral and fecal–oral routes, either directly from person to person or indirectly through contaminated surroundings [3]. *H. pylori* is capable of colonizing and persisting in the gastric lumen. The expression of urease activity and flagellar motility are essential for its survival and function, allowing *H. pylori* to penetrate the mucus layer of the stomach [4]. Moreover, flagella and adhesins such as SabA, BabA, and HopQ facilitate colonization and promote the formation of biofilms, which are aggregates of microorganisms within a hydrated matrix of extracellular substances that protect *H. pylori* against antibiotics and harsh environments [5]. Several risk factors for infection include dietary habits, smoking, water contamination [6], and gut microbiota [7]. *H. pylori* infection can be diagnosed using both non-invasive and invasive methods. Non-invasive methods include the detection of *H. pylori* antigens in stool samples, the urea breath test (UBT), or the detection of antibodies in serum, urine, and oral samples [8]. Invasive tests include histopathology, biopsy cultures, rapid urease tests, and fluorescent in situ hybridization [8]. Treatment of *H. pylori* infection typically involves a combination of antibiotics and proton pump inhibitors. Treatment regimens are tailored based on factors such as the patient's age, symptoms, concomitant medications, local antibiotic resistance patterns, treatment availability, and associated costs [9]. As previously mentioned, *H. pylori* has been associated with gastrointestinal conditions (such as acute and chronic gastritis, peptic ulcers, and cancer) but it is involved in the pathogenesis of various extra-gastric conditions. These include idiopathic iron deficiency anemia, vitamin B12 deficiency, immune thrombocytopenic purpura (ITP), neurodegenerative diseases such as Alzheimer's and Parkinson's disease, an increased risk of preeclampsia in infected women, and cardiovascular diseases [10–12]. Several studies have reported an association between *H. pylori* and coronary artery disease (CAD), though the relationship remains a topic of debate. The prevalence of CAD varies significantly across different geographical regions, ethnicities, and genders, but it remains one of the leading diseases affecting the global population. Risk factors for the development of CAD include lifestyle choices, environmental influences, and genetic predispositions. The widespread prevalence of these risk factors in otherwise healthy individuals highlights the potential for an increased incidence of CAD shortly [13]. Numerous projects have been conducted to assess the incidence of cardiovascular diseases across various populations. The World Health Organization (WHO) has coordinated the "MONICA (Monitoring Trends and Determinants in Cardiovascular Diseases) Project" which consists of a multicenter international collaborative study to measure the risk factors (cigarette smoking, blood pressure, and serum lipids and cholesterol) and determinants of cardiovascular diseases (coronary heart attacks and strokes), over 10 years. The population included women and men aged 25–64 years. About 39 collaborating centers from 26 countries in North America, Europe, and the Western Pacific are collaborating in this project, using a standardized protocol and covering a population of about 10 million to identify trends in mortality and morbidity for cardiovascular diseases in defined communities in different countries and to measure how these trends are related to changes in both risk factor and/or medical care [14]. Whereas the INTERHEART study provided valuable insights into the prevalence of CAD across diverse populations. The INTERHEART study was a case-control study conducted in 52 countries, (15,152 cases, 14,820 controls) to evaluate the effect of potentially modifiable risk factors associated with myocardial infarction. Both in women and in men, and for all ages in all regions, the authors reported an association with risk of coronary heart disease and hypertension, diabetes, abdominal obesity, smoking, abnormal cholesterol, alcohol use, consumption of fruits, and vegetables, and regular physical activity. The authors measured the odds ratio (OR) and population-attributable risks (PAR). They found for hypertension (OR 1.91, PAR 17.9%), diabetes (OR 2.37, PAR 9.9%), smoking (OR 2.87 for current vs. never, PAR 35.7% for current vs. never), alcohol consumption (OR 0.91, PAR 6.7%), daily consumption of fruits and vegetables (OR 0.70, PAR 13.7% for lack of daily consumption), regular physical activity (OR 0.86, PAR 12.2%) that prevention can be based on similar principles almost worldwide [15]. In 2016, the American Heart Association published an updated report on heart disease and stroke statistics, revealing that CAD affects 15.5 million individuals over the age of 20 in the United States. This prevalence was found to be increasing with age in both men and women [16]. Intensive epidemiological research has linked CAD to risk factors such as smoking, diabetes, hyperlipidemia, and hypertension [17–20]. Interestingly, a growing number of studies have demonstrated a link between CAD and various infectious agents, including *H. pylori*, *Chlamydia pneumoniae*, and cytomegalovirus. For example, Eeskandarian et al. [21], showed in a prospective study, the effects of *H. pylori* on the incidence of cardiovascular events in 433 patients presenting with acute coronary syndrome (ACS). The study's key finding was a positive association between *H. pylori* seropositivity and the incidence of short-term adverse cardiovascular events within the first month after an ACS episode. This review aims to summarize the most important and recent research linking *H. pylori* infection and CAD.

2. Pathophysiology

Several studies have explored the mechanisms by which *H. pylori* may contribute to the development of atherothrombosis, including chronic inflammation and direct injury to the vessel wall. These processes can promote the progression or rupture of atherosclerotic plaques, as well as trigger systemic inflammation, both of which are linked to *H. pylori* colonization of the stomach [21]. These inflammatory processes can induce prothrombotic changes in the blood, affecting both plasma (e.g., hyperfibrinogenemia and altered coagulation) and platelets (e.g., increased platelet count, activation, and aggregation), thereby contributing to the development of acute coronary syndrome (ACS) [21,22]. It has been demonstrated that *H. pylori* can stimulate inflammatory cells and trigger the excessive production of cytokines within atherosclerotic plaques, leading to local endothelial and vascular dysfunction. *H. pylori* infection also induces inflammatory mediators such as interleukin-1 (IL-1), interleukin-6 (IL-6), C-reactive protein (CRP), and tumor necrosis factor-alpha (TNF-α), all of which contribute to plaque instability [23]. In addition, *H. pylori* can enter endothelial cells via exosomes containing the cytotoxin-associated gene A (CagA), which causes endothelial damage. The bacterium also secretes another virulence factor, vacuolating cytotoxin A (VacA), which reduces nitric oxide (NO) levels, thereby impairing endothelial function [24,25]. The expression of P-selectin increases following *H. pylori* infection, and the interaction between von Willebrand factor (vWF), released by platelets, and P-selectin promotes platelet aggregation, elevating the risk of thrombosis and, consequently, increasing the risk of CAD [26]. It should also be mentioned that *H. pylori* infection can exacerbate conditions such as hypertension, dyslipidemia, hyper-homocysteinemia, diabetes, and impaired glucose tolerance, all of which are established risk factors for cardiovascular disease [27]. A recent meta-analysis confirmed that *H. pylori* infection was significantly associated with arterial hypertension [28]. Additionally, it has been shown that the eradication of *H. pylori* in patients with essential hypertension can reduce blood pressure values, particularly diastolic pressure [29]. Izhari et al. [30] reported that patients infected with *H. pylori* had a higher risk of elevated serum levels of total cholesterol, triglycerides, and low-density lipoprotein (LDL) cholesterol, as well as reduced levels of high-density lipoprotein (HDL) cholesterol. This effect on lipids could be attributed to a reduction in the activity of paraoxonase and arylesterase, along with an increase in lipid hydroperoxide and total thiol (SH) levels [30]. The interaction between *H. pylori* infection and diabetes has been extensively studied. Evidence suggests that diabetic patients exhibit poorer glycemic control, which in turn serves as a risk factor for the development of CAD [31]. Finally, *H. pylori* infection has been associated with low serum levels of vitamin B12 and folic acid, which consequently lead to hyper-homocysteinemia. Elevated homocysteine levels are a significant factor in the development of atherosclerosis due to their harmful effects on endothelial cells, promoting the formation of atherosclerotic plaques and contributing to vascular diseases [32–34].

2.1. Cytotoxin-Associated Gene Antigen and Atherosclerosis

The CagA protein is a virulence factor produced by *H. pylori*. It is encoded by the CagA gene, which is part of the cag pathogenicity island (PAI), a region in the bacterial genome associated with increased virulence. When *H. pylori* infects gastric epithelial cells, CagA is delivered into the host cells via a type IV secretion system. Once inside, CagA undergoes tyrosine phosphorylation and interacts with multiple signaling pathways, causing a range of effects. It disrupts cytoskeletal organization and tight junctions, compromising epithelial barrier integrity. In addition, CagA induces the release of pro-inflammatory cytokines such as IL-1α, IL-8, and IL-18, contributing to chronic inflammation. CagA has also been implicated in the development of gastric cancer due to its ability to interfere with cellular signaling pathways [35]. It inhibits autophagy in host cells, promotes uncontrolled cell proliferation, and suppresses apoptosis. The presence of the CagA gene is associated with more severe clinical outcomes, including peptic ulcers, gastric adenocarcinoma, and a heightened inflammatory response compared to *H. pylori* strains lacking this virulence factor [35]. CagA also promotes the activation of c-Met, which triggers the PI3K/AKT/mTOR signaling pathway. This induces a reduction in autophagy within the host cell and leads to the accumulation of the sequestosome-1 (SQSTM1) protein, further enhancing the production of NF-κB-dependent cytokines [36]. As reported by Xia et al. [37], CagA-positive *H. pylori* strains promote atherosclerosis through exosome-mediated reactive oxygen species (ROS) formation. A study conducted by Rozankovic et al. [38] suggests the existence of autoimmune mechanisms that contribute not only to the pathogenesis of atherosclerotic plaques but also to their destabilization. Other research indicates that CagA antibodies cross-react with antigens present in both normal and atherosclerotic blood vessels. This suggests that the presence of CagA-positive *H. pylori* may influence the progression of atherosclerosis in patients infected with CagA-positive strains [39]. Specifically, cross-reactivity may occur between antibodies targeting lipopolysaccharide-binding protein (LBP) and those directed against *H. pylori* heat shock protein 60 (HSP60), as well as antigens present on endothelial cells and arterial smooth muscle [40]. It has been shown that CagA-positive *H. pylori* strains have an increased ability to stimulate IL-6 production, a cytokine associated with the aging of both vascular and myeloid cells [40–42] and induces macrophage cell formation by downregulating the expression of the transcription factors peroxisome proliferator-activated receptor (PPAR)γ and liver-X receptor (LXR)α [43]. Gastric epithelial cells injected with CagA release exosomes containing protein to the systemic circulation, which facilitates the transport of CagA into endothelial cells [44]. In a study conducted on transgenic mice expressing CagA in their endothelial cells, exposure to a high-fat diet induced the development of proatherogenic lesions in the aorta, which were absent in non-transgenic mice exposed to the same diet. These lesions were characterized by an increase in the thickness of the tunica media and a reduction in its elasticity. When the high-fat diet was administered for a longer period, mice with CagA-expressing endothelial cells exhibited greater macrophage infiltration and the development of atherosclerotic plaques [45]. It has also been demonstrated that CagA-positive *H. pylori* strains are associated with increased expression of endothelial adhesion molecules such as ICAM-1 and VCAM-1. These molecules facilitate the binding of circulating monocytes, promoting macrophage infiltration into the endothelium. The upregulation of adhesion molecules is driven by the activation of the NLRP3/Caspase-1/IL-1β pathway, which leads to elevated IL-6 production. This, in turn, promotes local inflammation and, together with macrophage infiltration, contributes to the progression of atherosclerosis [46]. Moreover, it has been demonstrated that the inhibition of exosome secretion with GW4869 effectively prevented excessive aortic ROS production, endothelial dysfunction, and atherosclerosis in mice with CagA-positive *H. pylori* infection [37]. In the meta-analysis published by Shi et al. [47] it has been demonstrated that *H. pylori* can promote atherosclerosis in people under the age of 60 without other cardiovascular risk factors. Recent studies have demonstrated that outer membrane vesicles (OMVs) derived from *H. pylori*-infected gastric epithelial cells encapsulating the CagA are present in the blood of both patients and animal models. This suggests that these OMVs can facilitate the systemic dissemination of CagA into the bloodstream. Building on this evidence, some researchers have hypothesized that *H. pylori* may contribute to the development of atherosclerosis (AS) through OMV-mediated mechanisms [48]. The support this hypothesis it has been shown that the administration of OMVs from CagA-positive *H. pylori* accelerated atherosclerosis plaque formation in ApoE $-/-$ mice [49]. Although these pathophysiological mechanisms have been elucidated and hypothesized, clinical evidence supporting the idea that *H. pylori* eradication improves survival is still lacking. On one hand, this absence of evidence may be attributed to the lack of sufficiently long randomized controlled trials. On the other hand, some authors suggest that the lack of clinical benefit may be due to the dysregulation effects induced by the eradication of antibiotic therapy on the gut microbiota [50]. Figure 1 summarizes the diseases associated with *H. pylori* infection, as well as the effects induced by *H. pylori* that may contribute to the development of coronary artery disease (CAD).

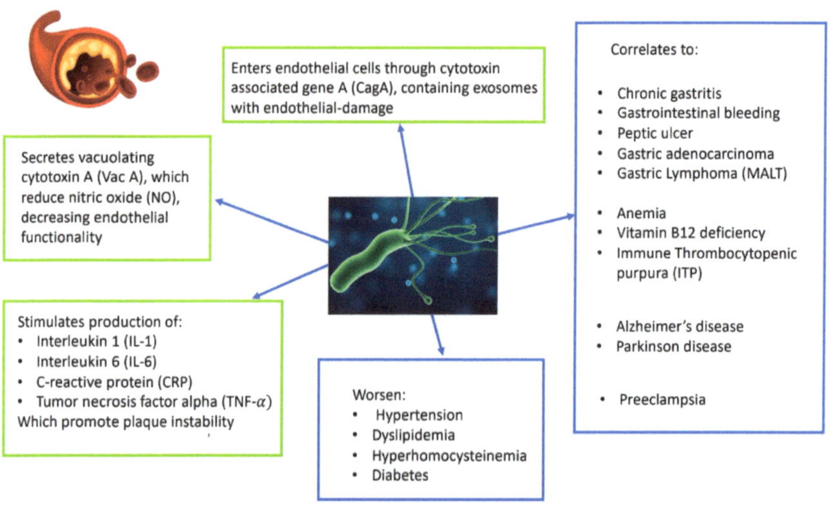

Figure 1. *H. pylori*-related pathophysiological mechanisms associated with coronary artery disease.

Several clinical studies have highlighted an association between CagA-positive *H. pylori* infection and extra-gastric diseases that share atherosclerotic pathogenesis and risk factors with ischemic heart disease. For example, *H. pylori* DNA has been found in carotid plaques, and CagA-positive strains have shown a higher prevalence in patients with non-cardioembolic stroke, reinforcing this connection [51,52]. A 2019 cross-sectional study and subsequent meta-analyses identified *H. pylori* infection, particularly with CagA-positive strains, as independent risk factors for non-cardioembolic ischemic stroke [53,54]. Collectively, these findings underscore the critical role of *H. pylori* virulence factors, especially CagA, in the development of atherosclerosis and related diseases.

2.2. *H. pylori and Autoimmunity*

Several studies have been led about the link between *H. pylori* infection and autoimmune diseases. Higher serological prevalence rates of *H. pylori* infection have been reported in patients with type 1 diabetes (T1DM) and autoimmune thyroiditis (AT) [55,56]. In particular, Choi YM. et al. [57] found a higher prevalence of *H. pylori* infection in patients with autoimmune thyroid disease and controls. Another study led by El-Eshmawy et al. [58] proved a connection between both type 1 diabetes and autoimmune thyroids, supporting the idea of a connection between *H.pylori* infection and the occurrence of anti- TPO, anti-Tg autoantibodies, and AT in young patients with T1DM. Many autoimmune mechanisms, some of which were described in the previous paragraph, have been considered a potential link to coronary artery disease (CAD). For example, it has been observed that anti-glycan antibodies produced after immunizing animals with heat-killed *H. pylori* strains cross-reacted with histological preparations of infarcted myocardial tissue. Additionally, a cross-reaction between *H. pylori* antibodies and blood group Lewis's antigens has been documented. These studies suggest that autoimmune phenomena, such as cross-mimicry with *H. pylori*, may play a role in the pathogenesis of CAD by directly promoting thrombotic occlusion through endothelial damage, as well as through local procoagulant phenomena [59].

3. *H. pylori* and Predisposition to CAD

Chronic inflammation associated with persistent infections is believed to contribute to the progression of atherosclerotic disease, which may eventually manifest as coronary artery disease (CAD). *H. pylori* presence was detected by polymerase chain reaction (PCR) in 29.5% of 105 patients who underwent coronary artery bypass grafting (CABG) [60]. Additionally, serological evidence of infection was found in 53.3% of these patients. These findings suggest that *H. pylori* infection could play a role in plaque rupture and subsequent ischemic heart disease. Notably, high levels of anti-CagA antibody titers were observed in CAD patients compared to healthy individuals and those with anti-CagA positivity exhibited more severe CAD lesions [61]. Another study evaluated the effects of *H. pylori* eradication, revealing an improvement in CAD. Interestingly, a greater loss of coronary lumen was noted in patients with serological evidence of *H. pylori* infection. However, *H. pylori* eradication attenuated the reduction in coronary artery lumen compared to the placebo group [62]. Further studies are required to explore whether early *H. pylori* eradication may help reduce CAD morbidity.

4. Myocardial Infarction

Myocardial infarction (MI) is an acute and severe cardiovascular disease, brought on by ischemia of the heart muscle and blockage of the coronary arteries, poses a significant threat to patients' lives, and is a serious public health problem [63]. There is a variety of risk factors for MI, such as lifestyle, diet, genetics, and environmental factors [64,65]. Risk factors are divided into modifiable, such as age and ethnicity, and modifiable, such as diet, smoking, and exercise. Reducing these last ones can improve MI prevention and control [66–68]. It has been proved that inflammation has a key role in atherosclerosis progression [66]. Inflammation is involved in restenosis or vessel narrowing, after initially successful balloon angioplasty or coronary stenting [69,70]. *H. pylori* gastric infection is one of the common chronic infections and can induce a pro-inflammatory role, which leads to the development of atherosclerosis and the progression of coronary heart disease (CHD) [71–73]. A previous study, led by Franceschi et al. demonstrated that anti-CagA IgG can recognize vascular wall self-antigens of 160 and 180 kDa [39], meaning the autoimmune system may also be involved. It is a phenomenon known as "epitope spreading" [74]. In case of an *H. pylori* infection, T lymphocytes activated against CagA positive strains of the bacteria could, therefore, recognize fragments of these self-antigens presented on type I major histocompatibility complex by antigen-presenting cells and stimulate an atherogenic inflammatory response within vascular walls. Moreover, pathogen-induced tissue inflammation may result from local activation of antigen-presenting cells and enhanced processing/presentation of self-antigens that cause T-cell priming, followed by T-cell activation and expansion of additional specificities [75]. A study led by Wang et al. analyzes whether there is a causality of anti-*H. pylori* IgG levels on MI and HDL cholesterol levels. Increased anti-*H. Pylori* IgG levels are significantly associated with an increased risk of MI and decreases in HDL cholesterol levels [76]. A meta-analysis led by Liu J. et al. estimated an approximately 70–100% MI risk increase for *H. pylori* infection [77]. Other studies, not included in this meta-analysis, led in two different areas of Northern Italy and in USA, proved an increased MI risk in patients having a *H. pylori* infection [78–81]. A study led by Niccoli G. et al. [82] enrolled 181 consecutive patients (155 men, mean age 64 ± 13 years) presenting with STEMI. In all patients, serum levels of IgG anti-CagA were assessed. The results showed that anti-CagA IgG seropositive patients presented more frequently a previous history of acute coronary syndrome compared with seronegative patients. Also, the major adverse cardiovascular event rate was higher in anti-CagA IgG seropositive compared with seronegative patients. These data suggest that CagA-positive strains of *H. pylori* seem to be involved in the pathogenesis of recurring ACS, and CagA seropositivity predicts the outcome of STEMI patients undergoing primary PCI. Table 1 shows the main findings of studies that analyzed the relationship between *H. pylori*, atherosclerosis, and coronary artery disease.

Table 1. Studies about the relationship among *H. pylori*, atherosclerosis, coronary artery disease and autoimmune disturbances.

Evidence	Studies	Main Findings
H. pylori infection and atherosclerosis	*Shi H. Helicobacter. 2022* [47]	*H. pylori* promotes atherosclerosis in people under the age of 60 without cardiovascular risk factors
	N. Wang. Front. Cell Dev. Biol. 2021 [49]	Outer membrane vesicles secreted by *H. pylori* exert an effect on distant organ and tissue
	M. Candelli. Int. J. Mol. 2023 [50]	*H. pylori* has a role in the pathogenesis and progress of atherosclerosis
H. pylori infection and myocardial infection	*Q. Wang. Front. Microbiol., 2023* [76]	Anti-*H. pylori* IgG levels are associated with an increased risk of MI
	J. Liu. Helicobacter, 2015 [77]	70–100% increased MI risk for *H. pylori* infected patients
	R. Pellicano. Int. J. Clin. Lab. 1999 [80]	Increased MI risk in patients with *H. pylori* infection
	R. Pellicano. Panminerva 1999 [81]	Increased MI risk in patients with *H. pylori* infection
H. pylori infection and CAD	*G. Niccoli. Eur. Heart J. ACC, 2017* [82]	In patients with STEMI, Anti-Cag-Ab is associated with history of ACS
	G. Niccoli. Coron. Artery Dis, 2010 [61]	*H. pylori* is associated with plaque rupture
	M Kowalski. J. Physiol. Pharmacol. 2001 [62]	*H. pylori* infection was associated with higher loss of coronary lumen
	NP. Tobin. Am. J. Physiol., 2008 [24]	Eradication of *H. pylori* attenuated the reduction in the coronary artery lumen

Table 1. *Cont.*

Evidence	Studies	Main Findings
H. pylori and autoimmune pathologies	*N. Figura et al. Antibiotics 2019* [56]	High prevalence of *H. pylori* infection in patients with Hashimoto thyroiditis
	M.M. El-Eshmawy. Diabet Met Syn. 2011 [58]	*H. pylori* infection is the link between type1 diabetes and thyroiditis

Legend: MI: myocardial infarction, STEMI: ST elevation myocardial infarction, CAD: Coronary artery disease, ACS: acute coronary syndrome.

5. Methods

This review included papers published from 2000 to 2025 about the relationship between *H. pylori* and CAD. We searched literature reviews, observational studies (case-control, cross-sectional), retrospective and prospective studies, and clinical trials. We selected studies containing data on the association between *H. pylori* infection and CAD, ranging from pathophysiological mechanisms (autoimmune, inflammatory) to clinical aspects. Studies were chosen based on the research period, title, abstract, study type, and English language. We searched on Up-to-Date®, PubMed®, Web of Science®, and Cochrane®. This review does not need ethical approval. We included as principal words of research: Helicobacter pylori AND coronary artery disease OR myocardial infarction OR atherosclerosis OR heart; infectious diseases AND coronary artery disease OR atherosclerosis; Helicobacter pylori AND coronary plaques. The authors first selected relevant studies by analyzing the titles, followed by a review of the abstracts. A final selection was made based on a full-text assessment of the remaining papers.

6. Discussion and Future Research

The available data seemed to suggest that patients with *H. pylori* infection have an increased risk factor for CAD. Meanwhile, studies are reporting that this association is casual. The relationship between *H. pylori* infection and CAD has been controversial for years. Some studies suggest that the association between *H. pylori* infection and CAD is casual. Studies led on population, have found a higher prevalence of *H. pylori* infection in men than in women, as for CAD. However, it has been shown that *H. pylori* infection is more common in non-Hispanic blacks and Hispanic ethnics and it is linked to poor hygienic conditions. CAD was demonstrated to be prevalent in the Asiatic population and to be linked to psychosocial conditions and factors such as smoking, abdominal obesity, and a raised ApoB/ApoA1 ratio. It has been proven that *H. pylori* infection links to atherothrombosis and brings the organism to a status of chronic inflammation, by stimulating inflammatory mediators, such as in-terleukin-1 (IL-1), interleukin-6 (IL-6), C-reactive protein (CRP), and tumor necrosis factor alpha (TNF-α). This leads to atherosclerotic plaque instability, which is one of the major risk factors for CAD. It has also been proven that *H. pylori* infection worsens hypertension, dyslipidemia, hyper-homocysteinemia, diabetes, and impaired glucose tolerance. These are all factors that lead to a higher risk of CAD. It has been widely proven that there is a relationship between *H. pylori* infection and CAD risk factors, and indirectly to CAD. However, more studies are needed to show whether *H. pylori* infection may be considered a direct risk factor or not and how strongly can be *H. pylori* infection be related to a higher risk of CAD. Whether there could be a strong relationship between *H. pylori* infection and CAD, it would be possible to start screening for *H. pylori* infection as prevention for CAD. This could lead to the use of different treatments, such as antibiotics for atherosclerosis. Such an early screening could also lead to a decrease in CAD risk factors, such as hypertension, dyslipidemia, hyper-homocysteinemia, and diabetes; which are worsened by *H. pylori* infection. This means improved general health for a big slice of the worldwide population. More comprehensive and large studies are required to investigate better this association and to clarify the role of this microorganism in such pathogeneses. This may be helpful to screen for *H. pylori* infection, consider the use of different treatments as antibiotics for atherosclerosis, and improve the lives of many patients with CAD with a strong positive impact all over the world. Future research should include (1) prospective population-based studies in which the incidence or the recurrence of CAD has to be evaluated in correlation with *H. pylori* infection, (2) intervention trials, focusing separately on the chronic and acute phases of CAD; and (3) studies of physiopathology (both in the animal model and humans) to understand the potential biological plausibility.

7. Conclusions

Despite the numerous pathophysiological mechanisms underlying the association between *Helicobacter pylori* (*H. pylori*) infection and coronary artery disease (CAD) that have been studied, clinical evidence demonstrating that eradication of the infection provides a tangible clinical benefit is still lacking. Research in this area should be encouraged through prospective studies, as identifying a risk factor as easily addressable as a bacterial infection could represent a crucial step in the prevention of the world's leading cause of mortality.

Author Contributions: Conceptualization, M.C. and G.P.; writing—original draft preparation, A.S. and M.R.M.L.; methodology, A.P.; validation, F.F. and G.P.; writing—review and editing, M.C. and A.P.; visualization, F.F. and A.G. All authors have read and agreed to the published version of the manuscript.

Funding: This research received no external funding.

Conflicts of Interest: The authors declare no conflicts of interest.

References

1. Fischbach, W.; Malfertheiner, P. *Helicobacter Pylori* Infection. *Dtsch. Ärztebl. Int.* **2018**, *115*, 429. [CrossRef] [PubMed]
2. Kuo, Y.C.; Ko, H.J.; Yu, L.Y.; Shih, S.C.; Wang, H.Y.; Lin, Y.C.; Hu, K.C. Kill Two Birds with One Stone? The Effect of *Helicobacter pylori* Eradication in Decreased Prevalence of Gastric Cancer and Colorectal Cancer. *Cancers* **2024**, *16*, 3881. [CrossRef] [PubMed]
3. Duan, M.; Li, Y.; Liu, J.; Zhang, W.; Dong, Y.; Han, Z.; Wan, M.; Lin, M.; Lin, B.; Kong, Q.; et al. Transmission routes and patterns of *helicobacter pylori*. *Helicobacter* **2023**, *28*, e12945. [CrossRef] [PubMed]
4. Dunne, C.; Dolan, B.; Clyne, M. Factors that mediate colonization of the human stomach by *Helicobacter pylori*. *World J. Gastroenterol.* **2014**, *20*, 5610–5624. [CrossRef] [PubMed]
5. Elshenawi, Y.; Hu, S.; Hathroubi, S. Biofilm of *Helicobacter pylori*: Life Cycle, Features, and Treatment Options. *Antibiotics* **2023**, *12*, 1260. [CrossRef] [PubMed]
6. Sun, Q.; Yuan, C.; Zhou, S.; Lu, J.; Zeng, M.; Cai, X.; Song, H. *Helicobacter pylori* infection: A dynamic process from diagnosis to treatment. *Front. Cell Infect. Microbiol.* **2023**, *13*, 1257817. [CrossRef]
7. Leja, M.; Grinberga-Derica, I.; Bilgilier, C.; Steininger, C. Review: Epidemiology of *Helicobacter pylori* infection. *Helicobacter* **2019**, *24* (Suppl. S1), e12635. [CrossRef] [PubMed]
8. Elbehiry, A.; Marzouk, E.; Aldubaib, M.; Abalkhail, A.; Anagreyyah, S.; Anajirih, N.; Almuzaini, A.M.; Rawway, M.; Alfadhel, A.; Draz, A.; et al. *Helicobacter pylori* Infection: Current Status and Future Prospects on Diagnostic, Therapeutic and Control Challenges. *Antibiotics* **2023**, *12*, 191. [CrossRef]
9. Nista, E.C.; Pellegrino, A.; Giuli, L.; Candelli, M.; Schepis, T.; De Lucia, S.S.; Ojetti, V.; Franceschi, F.; Gasbarrini, A. Clinical Implications of *Helicobacter pylori* Antibiotic Resistance in Italy: A Review of the Literature. *Antibiotics* **2022**, *11*, 1452. [CrossRef]
10. Campuzano-Maya, G. Hematologic manifestations of Helicobacter pylori infection. *World J. Gastroenterol.* **2014**, *20*, 12818. [CrossRef]
11. Suzuki, H.; Franceschi, F.; Nishizawa, T.; Gasbarrini, A. Extragastric Manifestations of *Helicobacter pylori* Infection: Extragastric Manifestations of *H. pylori*. *Helicobacter* **2011**, *16*, 65–69. [CrossRef] [PubMed]
12. Tan, H.-J.; Goh, K.-L. Extragastrointestinal manifestations of *Helicobacter pylori* infection: Facts or myth? A critical review: Extra GI features of *H. pylori* infection. *J. Dig. Dis.* **2012**, *13*, 342–349. [CrossRef]
13. Malakar, A.K.; Choudhury, D.; Halder, B.; Paul, P.; Uddin, A.; Chakraborty, S. A review on coronary artery disease, its risk factors, and therapeutics. *J. Cell. Physiol.* **2019**, *234*, 16812–16823. [CrossRef] [PubMed]

14. WHO MONICA Project Principal Investigators. The world health organization monica project (monitoring trends and determinants in cardiovascular disease): A major international collaboration. *J. Clin. Epidemiol.* **1988**, *41*, 105–114. [CrossRef]
15. Yusuf, S.; Hawken, S.; Ounpuu, S.; Dans, T.; Avezum, A.; Lanas, F.; McQueen, M.; Budaj, A.; Pais, P.; Varigos, J.; et al. Effect of potentially modifiable risk factors associated with myocardial infarction in 52 countries (the INTERHEART study): Case-control study. *Lancet* **2004**, *364*, 937–952. [CrossRef] [PubMed]
16. Mozaffarian, D.; Benjamin, E.J.; Go, A.S.; Arnett, D.K.; Blaha, M.J.; Cushman, M.; Das, S.R.; de Ferranti, S.; Després, J.P.; Fullerton, H.J.; et al. Executive Summary: Heart Disease and Stroke Statistics—2016 Update: A Report from the American Heart Association. *Circulation* **2016**, *133*, 447–454. [CrossRef] [PubMed]
17. Sheng, X.; Yang, G.; Zhang, Q.; Zhou, Y.; Pu, J. Impact of risk factors on intervened and non-intervened coronary lesions. *Am. J. Cardiovasc. Dis.* **2024**, *14*, 255–266. [CrossRef] [PubMed]
18. Roth, G.A.; Mensah, G.A.; Johnson, C.O.; Addolorato, G.; Ammirati, E.; Baddour, L.M.; Barengo, N.C.; Beaton, A.Z.; Benjamin, E.J.; Benziger, C.P.; et al. Global Burden of Cardiovascular Diseases and Risk Factors, 1990–2019: Update From the GBD 2019 Study. *J. Am. Coll. Cardiol.* **2020**, *76*, 2982–3021. [CrossRef]
19. Shah, M.U.; Roebuck, A.; Srinivasan, B.; Ward, J.K.; Squires, P.E.; Hills, C.E.; Lee, K. Diagnosis and management of type 2 diabetes mellitus in patients with ischaemic heart disease and acute coronary syndromes—A review of evidence and recommendations. *Front. Endocrinol.* **2025**, *15*, 1499681. [CrossRef]
20. Wang, Q.; Cao, Y.; Jia, L. Lipidomics-based investigation of its impact on the pathogenesis of coronary atherosclerosis: A Mendelian randomization study. *Hereditas* **2025**, *162*, 13. [CrossRef]
21. Budzyński, J.; Koziński, M.; Kłopocka, M.; Kubica, J.M.; Kubica, J. Clinical significance of *Helicobacter pylori* infection in patients with acute coronary syndromes: An overview of current evidence. *Clin. Res. Cardiol.* **2014**, *103*, 855–886. [CrossRef] [PubMed]
22. Hofmann, R.; Bäck, M. Time for Routine Helicobacter pylori Screening in Coronary Artery Disease? *Circulation* **2023**, *147*, 1731–1733. [CrossRef]
23. Aggarwal, K.; Singh, S.; Singla, A.; Kanagala, S.G.; Anamika, F.; Singh, B.; Aggarwal, P.; Jain, R. Unveiling the Silent Intruder: *H. pylori*'s Hidden Link to Ischemic Heart Disease. *Cardiol. Rev.* **2024**. [CrossRef] [PubMed]
24. Tobin, N.P.; Henehan, G.T.; Murphy, R.P.; Atherton, J.C.; Guinan, A.F.; Kerrigan, S.W.; Cox, D.; Cahill, P.A.; Cummins, P.M. *Helicobacter pylori*-induced inhibition of vascular endothelial cell functions: A role for VacA-dependent nitric oxide reduction. *Am. J. Physiol. Heart Circ. Physiol.* **2008**, *295*, H1403–H1413. [CrossRef]
25. Figura, N.; Palazzuoli, A.; Vaira, D.; Campagna, M.; Moretti, E.; Iacoponi, F.; Giordano, N.; Clemente, S.; Nuti, R.; Ponzetto, A. Cross-sectional study: CagA-positive *Helicobacter pylori* infection, acute coronary artery disease and systemic levels of B-type natriuretic peptide. *J. Clin. Pathol.* **2014**, *67*, 251–257. [CrossRef] [PubMed]
26. Yeh, J.-J.; Tsai, S.; Wu, D.-C.; Wu, J.-Y.; Liu, T.-C.; Chen, A. P-selectin–dependent platelet aggregation and apoptosis may explain the decrease in platelet count during *Helicobacter pylori* infection. *Blood* **2010**, *115*, 4247–4253. [CrossRef] [PubMed]
27. Tong, L.; Wang, B.-B.; Li, F.-H.; Lv, S.-P.; Pan, F.-F.; Dong, X.-J. An Updated Meta-Analysis of the Relationship Between *Helicobacter pylori* Infection and the Risk of Coronary Heart Disease. *Front. Cardiovasc. Med.* **2022**, *9*, 794445. [CrossRef] [PubMed]
28. Huang, M.; Zhu, L.; Jin, Y.; Fang, Z.; Chen, Y.; Yao, Y. Associação entre Infecção por Heicobacter Pylori e Hipertensão Arterial Sistêmica: Metanálise. *Arq. Bras. Cardiol.* **2021**, *117*, 626–636. [CrossRef]
29. Migneco, A.; Ojetti, V.; Specchia, L.; Franceschi, F.; Candelli, M.; Mettimano, M.; Montebelli, R.; Savi, L.; Gasbarrini, G. Eradication of *Helicobacter pylori* infection improves blood pressure values in patients affected by hypertension. *Helicobacter* **2003**, *8*, 585–589. [CrossRef] [PubMed]
30. Izhari, M.A.; Al Mutawa, O.A.; Mahzari, A.; Alotaibi, E.A.; Almashary, M.A.; Alshahrani, J.A.; Gosady, A.R.A.; Almutairi, A.M.; Dardari, D.M.M.; AlGarni, A.K.A. *Helicobacter pylori* (*H. pylori*) Infection-Associated Dyslipidemia in the Asir Region of Saudi Arabia. *Life* **2023**, *13*, 2206. [CrossRef] [PubMed]
31. Song, X.; Cai, C.; Jin, Q.; Chen, X.; Yu, C. The efficacy of Helicobacter pylori eradication in diabetics and its effect on glycemic control: A systematic review and meta-analysis. *Helicobacter* **2021**, *26*, e12781. [CrossRef] [PubMed]
32. Chen, Y.; Xu, C.; Xu, H.; Chen, W.; Wang, H.; Wang, Z.; Zhang, J. Persistent *Helicobacter pylori* infection for more than 3 years leads to elevated serum homocysteine concentration: A retrospective cohort study based on a healthy Chinese population. *J. Gastroenterol. Hepatol.* **2021**, *36*, 3077–3083. [CrossRef] [PubMed]
33. Habib, S.S.; Al-khlaiwi, T.; Almushawah, A.; Alsomali, A.; Habib, S.A. Homocysteine as a predictor and prognostic marker of atherosclerotic cardiovascular disease: A systematic review and meta-analysis. *Eur. Rev. Med. Pharmacol. Sci.* **2023**, *27*, 8598–8608. [CrossRef]
34. Santarelli, L.; Gabrielli, M.; Cremonini, F.; Santoliquido, A.; Candelli, M.; Nista, E.C.; Pola, P.; Gasbarrini, G.; Gasbarrini, A. Atrophic gastritis as a cause of hyperhomocysteinaemia. *Aliment. Pharmacol. Ther.* **2004**, *19*, 107–111. [CrossRef]
35. Hatakeyama, M. Structure and function of Helicobacter pylori CagA, the first-identified bacterial protein involved in human cancer. *Proc. Jpn. Acad. Ser. B* **2017**, *93*, 196–219. [CrossRef] [PubMed]

36. Li, N.; Tang, B.; Jia, Y.P.; Zhu, P.; Zhuang, Y.; Fang, Y.; Li, Q.; Wang, K.; Zhang, W.J.; Guo, G.; et al. *Helicobacter pylori* CagA Protein Negatively Regulates Autophagy and Promotes Inflammatory Response via c-Met-PI3K/Akt-mTOR Signaling Pathway. *Front. Cell. Infect. Microbiol.* **2017**, *7*, 417. [CrossRef] [PubMed]
37. Xia, X.; Zhang, L.; Wu, H.; Chen, F.; Liu, X.; Xu, H.; Cui, Y.; Zhu, Q.; Wang, M.; Hao, H.; et al. CagA+*Helicobacter pylori*, Not CagA-*Helicobacter pylori*, Infection Impairs Endothelial Function Through Exosomes-Mediated ROS Formation. *Front. Cardiovasc. Med.* **2022**, *9*, 881372. [CrossRef] [PubMed]
38. Rožanković, P.B.; Huzjan, A.L.; Cupić, H.; Benčić, I.J.; Bašić, S.; Demarin, V. Influence of CagA-positive *Helicobacter pylori* strains on atherosclerotic carotid disease. *J. Neurol.* **2011**, *258*, 753–761. [CrossRef] [PubMed]
39. Franceschi, F.; Sepulveda, A.R.; Gasbarrini, A.; Pola, P.; Silveri, N.G.; Gasbarrini, G.; Graham, D.Y.; Genta, R.M. Cross-reactivity of anti-CagA antibodies with vascular wall antigens: Possible pathogenic link between *Helicobacter pylori* infection and atherosclerosis. *Circulation* **2002**, *106*, 430–434. [CrossRef]
40. Chmiela, M.; Gonciarz, W. Molecular mimicry in Helicobacter pylori infections. *World J. Gastroenterol.* **2017**, *23*, 3964. [CrossRef] [PubMed]
41. Tyrrell, D.J.; Goldstein, D.R. Ageing and atherosclerosis: Vascular intrinsic and extrinsic factors and potential role of IL-6. *Nat. Rev. Cardiol.* **2021**, *18*, 58–68. [CrossRef] [PubMed]
42. Amedei, A.; Munari, F.; Bella, C.D.; Niccolai, E.; Benagiano, M.; Bencini, L.; Cianchi, F.; Farsi, M.; Emmi, G.; Zanotti, G.; et al. *Helicobacter pylori* secreted peptidyl prolyl cis, trans-isomerase drives Th17 inflammation in gastric adenocarcinoma. *Intern. Emerg. Med.* **2014**, *9*, 303–309. [CrossRef] [PubMed]
43. Pandolfi, F.; Franza, L.; Carusi, V.; Altamura, S.; Andriollo, G.; Nucera, E. Interleukin-6 in Rheumatoid Arthritis. *Int. J. Mol. Sci.* **2020**, *21*, 5238. [CrossRef]
44. Yang, S.; Xia, Y.P.; Luo, X.Y.; Chen, S.L.; Li, B.W.; Ye, Z.M.; Chen, S.C.; Mao, L.; Jin, H.J.; Li, Y.N.; et al. Exosomal CagA derived from *Helicobacter pylori*-infected gastric epithelial cells induces macrophage foam cell formation and promotes atherosclerosis. *J. Mol. Cell. Cardiol.* **2019**, *135*, 40–51. [CrossRef] [PubMed]
45. Tahmina, K.; Hikawa, N.; Takahashi-Kanemitsu, A.; Knight, C.T.; Sato, K.; Itoh, F.; Hatakeyama, M. Transgenically expressed *Helicobacter pylori* CagA in vascular endothelial cells accelerates arteriosclerosis in mice. *Biochem. Biophys. Res. Commun.* **2022**, *618*, 79–85. [CrossRef] [PubMed]
46. Li, B.W.; Liu, Y.; Zhang, L.; Guo, X.Q.; Wen, C.; Zhang, F.; Luo, X.Y.; Xia, Y.P. Cytotoxin-associated gene A (CagA) promotes aortic endothelial inflammation and accelerates atherosclerosis through the NLRP3/caspase-1/IL-1β axis. *FASEB J.* **2021**, *35*, e21942. [CrossRef] [PubMed]
47. Shi, H.; Li, Y.; Dong, C.; Si, G.; Xu, Y.; Peng, M.; Li, Y. *Helicobacter pylori* infection and the progression of atherosclerosis: A systematic review and meta-analysis. *Helicobacter* **2022**, *27*, e12865. [CrossRef] [PubMed]
48. Qiang, L.; Hu, J.; Tian, M.; Li, Y.; Ren, C.; Deng, Y.; Jiang, Y. Extracellular vesicles from *Helicobacter pylori*-infected cells and *Helicobacter pylori* outer membrane vesicles in atherosclerosis. *Helicobacter* **2022**, *27*, e12877. [CrossRef] [PubMed]
49. Wang, N.; Zhou, F.; Chen, C.; Luo, H.; Guo, J.; Wang, W.; Yang, J.; Li, L. Role of Outer Membrane Vesicles From *Helicobacter pylori* in Atherosclerosis. *Front. Cell. Dev. Biol.* **2021**, *9*, 673993. [CrossRef] [PubMed]
50. Candelli, M.; Franza, L.; Cianci, R.; Pignataro, G.; Merra, G.; Piccioni, A.; Ojetti, V.; Gasbarrini, A.; Franceschi, F. The Interplay between *Helicobacter pylori* and Gut Microbiota in Non-Gastrointestinal Disorders: A Special Focus on Atherosclerosis. *Int. J. Mol. Sci.* **2023**, *24*, 17520. [CrossRef] [PubMed]
51. Ameriso, S.F.; Fridman, E.A.; Leiguarda, R.C.; Sevlever, G.E. Detection of *Helicobacter pylori* in Human Carotid Athero-sclerotic Plaques. *Stroke* **2001**, *32*, 385–391. [CrossRef]
52. De Bastiani, R.; Gabrielli, M.; Ubaldi, E.; Benedetto, E.; Sanna, G.; Cottone, C.; Candelli, M.; Zocco, M.A.; Saulnier, N.; Santoliquido, A.; et al. High prevalence of Cag-A positive H. pylori strains in ischemic stroke: A primary care multicenter study. *Helicobacter* **2008**, *13*, 274–277. [CrossRef] [PubMed]
53. Shindler-Itskovitch, T.; Chodick, G.; Shalev, V.; Muhsen, K. *Helicobacter pylori* infection and prevalence of stroke. *Helicobacter* **2019**, *24*, e12553. [CrossRef]
54. Doheim, M.F.; Altaweel, A.A.; Elgendy, M.G.; Elshanbary, A.A.; Dibas, M.; Ali, A.A.H.A.; Dahy, T.M.; Sharaf, A.K.; Hassan, A.E. Association between *Helicobacter pylori* infection and stroke: A meta-analysis of 273,135 patients. *J. Neurol.* **2021**, *268*, 3238–3248. [CrossRef] [PubMed]
55. Chua, W.K.; Hong, Y.K.; Hu, S.W.; Fan, H.C.; Ting, W.H. A Significant Association between Type 1 Diabetes and *Helicobacter pylori* Infection: A Meta-Analysis Study. *Medicina* **2024**, *60*, 119. [CrossRef]
56. Figura, N.; Di Cairano, G.; Moretti, E.; Iacoponi, F.; Santucci, A.; Bernardini, G.; Gonnelli, S.; Giordano, N.; Ponzetto, A. *Helicobacter pylori* Infection and Autoimmune Thyroid Diseases: The Role of Virulent Strains. *Antibiotics* **2019**, *9*, 12. [CrossRef] [PubMed]
57. Choi, Y.M.; Kim, T.Y.; Kim, E.Y.; Jang, E.K.; Jeon, M.J.; Kim, W.G.; Shong, Y.K.; Kim, W.B. Association between thyroid autoimmunity and *Helicobacter pylori* infection. *Korean J. Intern. Med.* **2017**, *32*, 309–313. [CrossRef] [PubMed]

58. El-Eshmawy, M.M.; El-Hawary, A.K.; Abdel Gawad, S.S.; El-Baiomy, A.A. *Helicobacter pylori* infection might be responsible for the interconnection between type 1 diabetes and autoimmune thyroiditis. *Diabetol. Metab. Syndr.* **2011**, *3*, 28. [CrossRef]
59. Negrini, R.; Villanacci, V.; Poiesi, C.; Savio, A. Anti-Glycan Autoantibodies Induced by *Helicobacter pylori* as a Potential Risk Factor for Myocardial Infarction. *Front. Immunol.* **2020**, *11*, 597. [CrossRef] [PubMed]
60. Izadi, M.; Fazel, M.; Sharubandi, S.H.; Saadat, S.H.; Farahani, M.M.; Nasseri, M.H.; Dabiri, H.; SafiAryan, R.; Esfahani, A.A.; Ahmadi, A.; et al. Helicobacter species in the atherosclerotic plaques of patients with coronary artery disease. *Cardiovasc. Pathol.* **2012**, *21*, 307–311. [CrossRef]
61. Niccoli, G.; Franceschi, F.; Cosentino, N.; Giupponi, B.; De Marco, G.; Merra, G.; Conte, M.; Montone, R.A.; Ferrante, G.; Bacà, M.; et al. Coronary atherosclerotic burden in patients with infection by CagA-positive strains of *Helicobacter pylori*. *Coron. Artery Dis.* **2010**, *21*, 217–221. [CrossRef]
62. Kowalski, M. *Helicobacter pylori* (*H. pylori*) infection in coronary artery disease: Influence of *H. pylori* eradication on coronary artery lumen after percutaneous transluminal coronary angioplasty. The detection of *H. pylori* specific DNA in human coronary atherosclerotic plaque. *J. Physiol. Pharmacol.* **2001**, *52* (Suppl. 1), 3–31. [PubMed]
63. Lu, L.; Liu, M.; Sun, R.; Zheng, Y.; Zhang, P. Myocardial Infarction: Symptoms and Treatments. *Cell Biochem. Biophys.* **2015**, *72*, 865–867. [CrossRef] [PubMed]
64. Ballantyne, C.M.; Nambi, V. Markers of inflammation and their clinical significance. *Atheroscler. Suppl.* **2005**, *6*, 21–29. [CrossRef]
65. Anand, S.S.; Islam, S.; Rosengren, A.; Franzosi, M.G.; Steyn, K.; Yusufali, A.H.; Keltai, M.; Diaz, R.; Rangarajan, S.; Yusuf, S.; et al. Risk factors for myocardial infarction in women and men: Insights from the INTERHEART study. *Eur. Heart J.* **2008**, *29*, 932–940. [CrossRef] [PubMed]
66. Smyth, A.; O'Donnell, M.; Lamelas, P.; Teo, K.; Rangarajan, S.; Yusuf, S. Physical Activity and Anger or Emotional Upset as Triggers of Acute Myocardial Infarction: The INTERHEART Study. *Circulation* **2016**, *134*, 1059–1067. [CrossRef] [PubMed]
67. Teo, K.K.; Liu, L.; Chow, C.K.; Wang, X.; Islam, S.; Jiang, L.; Sanderson, J.E.; Rangarajan, S.; Yusuf, S.; INTERHEART Investigators in China. Potentially modifiable risk factors associated with myocardial infarction in China: The INTERHEART China study. *Heart* **2009**, *95*, 1857–1864. [CrossRef] [PubMed]
68. Colombo, A.; Proietti, R.; Čulić, V.; Lipovetzky, N.; Viecca, M.; Danna, P. Triggers of acute myocardial infarction: A neglected piece of the puzzle. *J. Cardiovasc. Med.* **2014**, *15*, 1–7. [CrossRef] [PubMed]
69. Kubica, J.; Kozinski, M.; Krzewina-Kowalska, A.; Zbikowska-Gotz, M.; Dymek, G.; Sukiennik, A.; Piasecki, R.; Bogdan, M.; Grzesk, G.; Chojnicki, M.; et al. Combined periprocedural evaluation of CRP and TNF-alpha enhances the prediction of clinical restenosis and major adverse cardiac events in patients undergoing percutaneous coronary interventions. *Int. J. Mol. Med.* **2005**, *16*, 173–180. [PubMed]
70. Kozinski, M.; Krzewina-Kowalska, A.; Kubica, J.; Zbikowska-Gotz, M.; Dymek, G.; Piasecki, R.; Sukiennik, A.; Grzesk, G.; Bogdan, M.; Chojnicki, M.; et al. Percutaneous coronary intervention triggers a systemic inflammatory response in patients treated for in-stent restenosis—Comparison with stable and unstable angina. *Inflamm. Res.* **2005**, *54*, 187–193. [CrossRef] [PubMed]
71. Malfertheiner, P.; Megraud, F.; O'Morain, C.A.; Atherton, J.; Axon, A.T.; Bazzoli, F.; Gensini, G.F.; Gisbert, J.P.; Graham, D.Y.; Rokkas, T.; et al. Management of *Helicobacter pylori* infection—The Maastricht IV/Florence Consensus Report. *Gut* **2012**, *61*, 646–664. [CrossRef] [PubMed]
72. Epstein, S.E. The Multiple Mechanisms by Which Infection May Contribute to Atherosclerosis Development and Course. *Circ. Res.* **2002**, *90*, 2–4. [CrossRef] [PubMed]
73. Zhu, W.; Liu, S. The role of human cytomegalovirus in atherosclerosis: A systematic review. *Acta Biochim. Biophys. Sin.* **2020**, *52*, 339–353. [CrossRef]
74. Youssefi, M.; Tafaghodi, M.; Farsiani, H.; Ghazvini, K.; Keikha, M. *Helicobacter pylori* infection and autoimmune diseases; Is there an association with systemic lupus erythematosus, rheumatoid arthritis, autoimmune atrophy gastritis and autoimmune pancreatitis? A systematic review and meta-analysis study. *J. Microbiol. Immunol. Infect.* **2021**, *54*, 359–369. [CrossRef] [PubMed]
75. Amedei, A.; Bergman, M.P.; Appelmelk, B.J.; Azzurri, A.; Benagiano, M.; Tamburini, C.; van der Zee, R.; Telford, J.L.; Vandenbroucke-Grauls, C.M.; D'Elios, M.M.; et al. Molecular mimicry between *Helicobacter pylori* antigens and H^+, K^+—Adenosine triphosphatase in human gastric autoimmunity. *J. Exp. Med.* **2003**, *198*, 1147–1156. [CrossRef] [PubMed]
76. Wang, Q.; Liu, Y.; Xu, Z.; Wang, Z.; Xue, M.; Li, X.; Wang, Y. Causality of anti-Helicobacter pylori IgG levels on myocardial infarction and potential pathogenesis: A Mendelian randomization study. *Front. Microbiol.* **2023**, *14*, 1259579. [CrossRef] [PubMed]
77. Liu, J.; Wang, F.; Shi, S. Helicobacter pylori Infection Increase the Risk of Myocardial Infarction: A Meta-Analysis of 26 Studies Involving more than 20,000 Participants. *Helicobacter* **2015**, *20*, 176–183. [CrossRef]
78. Wärme, J.; Sundqvist, M.; Mars, K.; Aladellie, L.; Pawelzik, S.C.; Erlinge, D.; Jernberg, T.; James, S.; Hofmann, R.; Bäck, M. *Helicobacter pylori* screening in clinical routine during hospitalization for acute myocardial infarction. *Am. Heart J.* **2021**, *231*, 105–109. [CrossRef] [PubMed]

79. Tabata, N.; Sueta, D.; Akasaka, T.; Arima, Y.; Sakamoto, K.; Yamamoto, E.; Izumiya, Y.; Yamamuro, M.; Tsujita, K.; Kojima, S.; et al. *Helicobacter pylori* Seropositivity in Patients with Interleukin-1 Polymorphisms Is Significantly Associated with ST-Segment Elevation Myocardial Infarction. *PLoS ONE* **2016**, *11*, e0166240. [CrossRef]
80. Pellicano, R.; Mazzarello, M.G.; Morelloni, S.; Allegri, M.; Arena, V.; Ferrari, M.; Rizzetto, M.; Ponzetto, A. Acute myocardial infarction and *Helicobacter pylori* seropositivity. *Int. J. Clin. Lab. Res.* **1999**, *29*, 141–144. [CrossRef] [PubMed]
81. Pellicano, R.; Parravicini, P.P.; Bigi, R.; La Rovere, M.T.; Baduini, G.; Gandolfo, N.; Casaccia, M.; Reforzo, F.; Santoriello, L.; Aruta, E.; et al. Patients with acute myocardial infarction in northern Italy are often infected by *Helicobacter pylori*. *Panminerva Med.* **1999**, *41*, 279–282. [PubMed]
82. Niccoli, G.; Roberto, M.; D'Amario, D.; Scalone, G.; Fracassi, F.; Cosentino, N.; Candelli, M.; Franceschi, F.; Crea, F. Cytotoxin-associated gene antigen-positive strains of *Helicobacter pylori* and recurring acute coronary syndromes. *Eur. Heart J. Acute Cardiovasc. Care* **2017**, *6*, 535–544. [CrossRef]

medicina

MDPI

Review

Effects of Selected Food Additives on the Gut Microbiome and Metabolic Dysfunction-Associated Steatotic Liver Disease (MASLD)

Sara Jarmakiewicz-Czaja [1], Aneta Sokal-Dembowska [1] and Rafał Filip [2,3,*]

[1] Faculty of Health Sciences and Psychology, University of Rzeszow, 35-959 Rzeszow, Poland; sjczaja@ur.edu.pl (S.J.-C.); asokal@ur.edu.pl (A.S.-D.)
[2] Gastroenterology Clinic, Center for Comprehensive Treatment of Inflammatory, Bowel Disease Regional Hospital No. 2 in Rzeszow, 35-301 Rzeszow, Poland
[3] Department of Internal Medicine, Faculty of Medicine, University of Rzeszow, 35-959 Rzeszow, Poland
* Correspondence: r.s.filip@wp.pl

Abstract: The purpose of this article is to present selected food additives as disruptors of normal intestinal homeostasis with a potential impact on the development of metabolic dysfunction-associated steatotic liver disease (MASLD). A comprehensive literature search was conducted in three major electronic databases: PubMed, ScienceDirect, and Google Scholar. MASLD is a prevalent liver condition that is closely related to the global rise in obesity. Its pathogenesis is multifactorial, with genetic, environmental, and metabolic factors playing a key role. The "multiple-hit" hypothesis suggests that a Western-style diet, rich in ultra-processed foods, saturated fats, and food additives, combined with low physical activity, contributes to obesity, which promotes lipid accumulation in the liver. Recent studies underscore the role of impaired intestinal homeostasis in the development of MASLD. Food additives, including preservatives, emulsifiers, and sweeteners, affect gut health and liver function. Selected preservatives inhibit pathogenic microorganisms but disrupt the intestinal microbiota, leading to changes in intestinal permeability and liver dysfunction. Some emulsifiers and thickeners can cause inflammation and alter the gut microbiome, contributing to liver steatosis. Furthermore, the use of sweeteners such as sucralose and aspartame has been linked to changes in liver metabolism and intestinal microbial composition, which in turn promotes metabolic disorders.

Keywords: colors; emulsifiers; food additives; MASLD; microbiota; preservatives; sweeteners; taste enhancers

1. Introduction

Metabolic dysfunction-associated steatotic liver disease (MASLD) was previously referred to as nonalcoholic fatty liver disease (NAFLD). NAFLD was first described by Ludwig et al. in 1980 [1]. In 2020, the term steatohepatitis associated with metabolic dysfunction (MAFLD) was introduced, while in 2023, an international panel of experts proposed a new nomenclature—MASLD [2]. In 2018, the estimated prevalence of NAFLD worldwide was 24% [3]. Currently, is one of the most common liver diseases in the world, and its incidence is increasing every year not only in the adult population but also in children, in parallel with the obesity epidemic [4,5]. MASLD is associated with a greater predisposition to cardiovascular disease, type 2 diabetes, kidney disease, and other diseases [6,7]. The diagnosis of MASLD requires the presence of one or more metabolic risk factors for the disease, i.e., hyperglycemia, excessive body weight, abnormal lipid

 https://doi.org/10.3390/medicina61020192

metabolism, and high blood pressure [8]. It can also predispose people to steatohepatitis associated with metabolic dysfunction (MASH), organ fibrosis, cirrhosis, and hepatocellular carcinoma (HCC) [9]. There are several risk factors for MASLD, but in recent years, researchers have increasingly turned their attention to the disruption of the intestinal microbiota and the disruption of the integrity of the intestinal barrier. The purpose of this article is to present selected food additives as disruptors of normal intestinal homeostasis with a potential impact on the development of MASLD.

2. Methods

The data collection process took place from October to December 2024. A comprehensive literature search was conducted in three major electronic databases: PubMed, ScienceDirect, and Google Scholar. The search strategy used a combination of relevant keywords and phrases in two main categories: (i) intervention terms: food additives, colors, emulsifiers, preservatives, sweeteners, or taste enhancers; (ii) condition terms: MASLD and microbiota, MASLD, or microbiota. Articles published in peer-reviewed journals were included if they met the following criteria: (i) written in English, (ii) published between 2019 and 2024, and (iii) directly addressed the relationship of food additives to gut microbiota and MASLD. Publications published prior to 2019 were also reviewed if they were considered crucial to providing basic knowledge. The search for publications was carried out by selecting titles and abstracts, followed by a review of the full text to ensure appropriate methodological quality. The literature was also hand-searched to find other relevant articles. Randomized controlled trials, meta-analyses, systematic reviews, prospective cohort studies, and animal and in vitro studies were analyzed.

3. Pathogenesis

The pathogenesis of MASLD is multifactorial and complex. The most commonly indicated link is between genetic, environmental, immunological, and metabolic factors, so the hypothesis of "multiple hits" has been proposed [10,11].

One of the main modulators of pathogenesis is the adherence to a Western-type diet, containing ultra-processed foods, high in saturated and trans fatty acids and food additives, and with a lack or low levels of physical activity. This type of lifestyle predisposes people to insulin resistance, being overweight, or obesity. This, in turn, affects organokines (gut cytokines, osteokines, adipokines, and myokines) and, more specifically, the amount of their secretion. An example is the reduction in Nrg4 (neuregulin 4), which is secreted by brown adipose tissue, which can consequently cause disturbances in liver metabolic homeostasis. Another example is cell communication network factor 4, which is secreted by adipose tissue, altering the action of, among other things, insulin [12]. Excess body weight predisposes people to the hepatic accumulation of triglycerides from non-esterified fatty acids secreted from adipose tissue [13]. MASLD also has an increased rate of de novo hepatic lipogenesis, which accelerates lipid accumulation in the organ [14]. Other risk factors are the gene polymorphisms present, e.g., patatin-like phospholipase domain-containing protein 3, membrane-bound O-acyltransferase domain-containing protein 7, or transmembrane 6 superfamily 2. Some also show an increased predisposition to liver fibrosis and the appearance of HCC [15–18]. Furthermore, gut microbiota metabolites can affect liver lipogenesis. One of them is short-chain fatty acids (SCFAs), which are formed from dietary fiber; more specifically, propionic acid is formed, which can predispose people to gluconeogenesis and adipogenesis, which have been shown to have an effect on the development of MASLD [19]. Another metabolite is bile acids, whose primary role is the digestion of lipids, but they have also been shown to help maintain the normal homeostasis of the intestinal microbiota [20]. A different example is choline and its metabolites, mainly trimethylamine (TMA). TMA is metabolized by the gut microbiota, absorbed through the intestines, and transported to the liver. Therefore, diets deficient in choline affect the occurrence of intestinal barrier disorders, which predispose people to the accumulation of lipids in the liver and the onset of organ steatosis [21]. Additionally, the sheer disturbance in the quantity and quality of the gut microbial composition can influence the onset of MASLD. Proteobacteria are observed more frequently in the presence of hepatic steatosis [22]. Boursier et al., in their work, observed that *Bacteroides* can be associated with nonalcoholic steatohepatitis, while *Ruminococcus* can be associated with the progression of organ fibrosis [23]. In another paper, Zhang et al., after observing animal models fed a high-fat diet, found that there was an increase in *Mucispirillum and Desulfovibrio,* among others, while there was a decrease in *Bacteroides* and *Bifidobacterium,* which was associated with the development and progression of MASLD to the onset of HCC [24]. However, there is currently a lack of long-term studies identifying a clear link between the gut microbiota and the occurrence of MASLD.

The mechanism of the MASLD formation process is complex and multifactorial. An influx of released free fatty acids (FFAs) from adipose tissue into the liver, along with the body's hyperinsulinemia, causes an imbalance between hepatic lipid absorption and excretion. Donnelly et al. indicate that approximately 59% of hepatic lipids from MASLD come from FFA, about 26% from the de novo lipogenesis pathway, and only about 15% from the diet [14]. Due to the accumulation of lipotoxic fatty acid β-oxidation intermediates, mitochondria malfunction, which, in turn, reinforces the lack of fatty acid burning efficiency [25,26]. Mitochondria produce 90% of cellular reactive oxygen species (ROS); with uncontrolled mitochondrial oxidative stress, there is oxidative damage to hepatocytes [27]. Endoplasmic reticulum stress triggers inflammatory cascades by increasing the activity of nuclear factor κB, c-Jun N-terminal kinase, and others, which can regulate inflammatory macrophage activation [28]. Activated macrophages after organ damage secrete pro-inflammatory cytokines, such as interleukin-6 (IL6), human tumor necrosis factor-α (TNFα), and interleukin-1β (IL1β), and so inflammation increases [29]. Macrophages and Kupffer cells (KCs) are also involved in the inflammatory process. Activated KCs, through liver injury, like macrophages, promote inflammatory reactions occurring in the organ, which causes the activation of hepatic stellate cells (HSCs) [30,31]. These, in turn, are a key element for liver fibrosis, which in advanced cases can lead to organ cirrhosis [32,33]. The simplified connections between the gut and liver are shown in Figure 1.

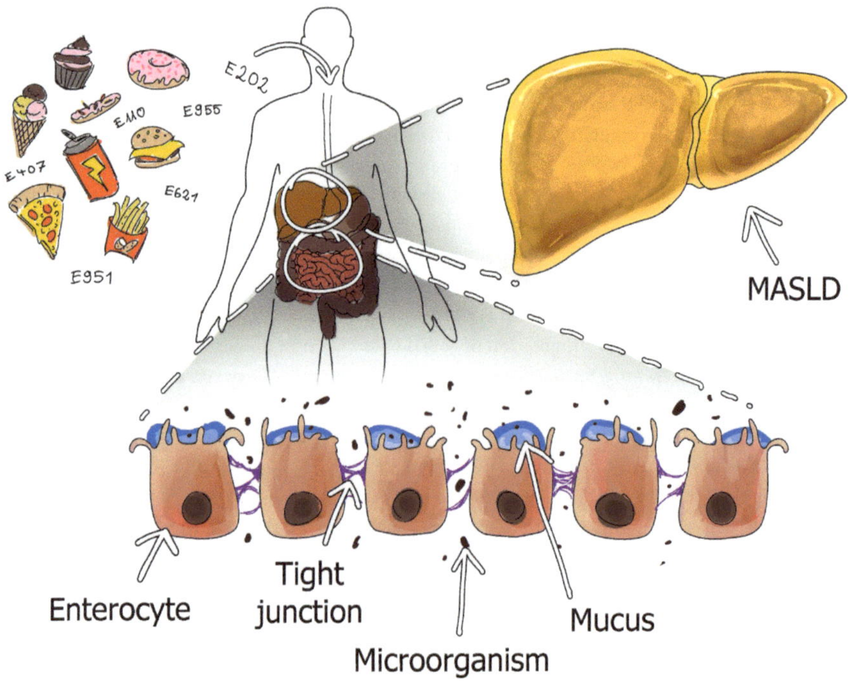

Figure 1. Potential influence of selected food additives on the association with intestinal homeostasis disorders and the occurrence of MASLD. Selected food additives can adversely affect the maintenance of intestinal homeostasis by reducing normal tight junction function, decreasing mucus, and altering the composition and quantity of certain intestinal microorganisms. And these changes affect the liver, leading to an increased risk of MASLD.

4. Nutritional Models Used in the Management of MASLD

As has been demonstrated, diet is one of the most significant modifiable factors in lifestyle that influence the diversity of the host microbiome [34]. There is clear evidence that the proper state of the microbiota plays a crucial role in the pathogenesis of numerous diseases, as well as in their prevention and management [35,36]. Dietary models based on a high supply of dietary fiber are an important part of the prevention of intestinal diseases [35] and diet therapy for chronic liver diseases, including MASLD [37]. An analysis of the association between dietary fiber intake and MASLD was evaluated in the NHANES study among 5935 participants, which confirmed that there is an inverse relationship between dietary fiber intake and changes in liver steatosis [38].

Dietary fiber modulates the gut microbiota, as it is one of the main substrates used by gut microbes. The fermentation of dietary fiber and resistant starch in the large intestine results in the production of SCFAs, including acetate, butyrate, and propionate [35]. These substances demonstrate anti-inflammatory, immunomodulatory effects and improve intestinal barrier function [39,40].

Fiber also acts as a prebiotic, promoting the growth of beneficial gut bacteria such as *Bifidobacterium*, *Lactobacillus*, *Faecalibacterium*, *Ruminococcus*, *Akkermania*, or *Roseburia*, which contribute to improved intestinal health and the overall health of the body [41,42]. Regular consumption of fiber can improve the composition of the microbiota, which can have a long-term impact on reducing the risk of chronic diseases such as obesity, type 2 diabetes, and heart disease, which are often associated with MASLD [43].

New therapeutic approaches to the nutritional management of MASLD are constantly being sought. Hansen et al. showed that a low-carbohydrate, high-fat diet (50–60% fat, less than 20% carbohydrate, and 25–30% protein) can result in significant improvements in glycemic and weight control compared to a high-carbohydrate, low-fat diet (50–60% carbohydrate, 20–30% fat, and 20–25% protein) [44]. Similarly, Chen et al. reported improvements in body composition parameters and alanine transaminase (ALT), aspartate aminotransferase (AST), uric acid, and insulin levels in patients with MASLD following a low-carbohydrate, high-fiber diet combined with nutritional education [45]. Cunha et al. reported better weight reduction accompanied by substantial decreases in visceral adipose tissue and liver fat fractions in comparison with the standard diet [46].

Nevertheless, the low-calorie Mediterranean diet (MED) remains the most frequently recommended dietary intervention for managing MASLD [47,48]. Montemayor et al. conducted a multicenter (Mallorca and Navarra, Spain) prospective randomized trial to test the effect of the MED diet in patients with MASLD and metabolic syndrome. A six-month follow-up showed that adherence to the diet led to lower levels of parameters, such as body mass index (BMI), body weight, waist circumference, and intrahepatic fat content, and lower levels of blood pressure (systolic blood pressure and diastolic blood pressure) [49]. However, preliminary results of a multicenter RCT (randomized controlled trial) conducted by Rosi et al. showed no difference between the use of an MED diet and a low-fat diet in counteracting obesity in children and adolescents [50]. However, it should be borne in mind that the use of dietary questionnaires to assess dietary habits may be subject to measurement error.

Cheng et al. showed that dietary intervention combined with physical activity positively affects the stability of the ecosystem interaction network, thereby improving host metabolism [51]. An interesting intervention was carried out by Chooi et al. The researchers included supplementation with pentadecanoic acid (C15:0), which naturally occurs in milk fat and ruminant meat, in the diets of women with MASLD. The use of a diet rich in dietary fiber and unsaturated fatty acids in combination with C15:0 led to a reduction in low-density lipoprotein (LDL) cholesterol levels and an increase in the abundance of *Bifidobacterium adolescentis* [52]. In the Gómez-Pérez study, an MED diet combined with physical activity for 12 months in patients with clinically suspected MASLD and MASH resulted in improved gut microbiota composition, which was associated with changes in MASLD/MASH biochemical indices (non-suspected fibrosis and indeterminate or suspected fibrosis). There was an increase in the genus *Coprococcus* and *Lachnospira*, as well as *Oscillospira*, and a decrease in *Proteobacteria* and its family *Enterobacteriaceae* [53].

An analysis of recent data by Bialczyk et al. showed that the introduction of probiotic therapy in patients with MASLD can favorably affect liver enzyme levels, improving insulin sensitivity, lipid profile parameters, and BMI. Probiotic supplementation can reduce inflammatory markers such as IL-6, c-reactive protein (CRP), and TNF-α, and, when combined with prebiotics, can also improve histological markers [54]. *Lactobacillus* may be beneficial in alleviating MASLD through their effects on various tissues and organs in the body, and their effectiveness may vary depending on the strain and the patient's current condition. *Lactobacillus* can restore intestinal homeostasis by modulating Mucin2 and intestinal tight junctions [55]. In patients with MASLD, the use of multi-strain probiotic therapy (six different species of *Lactobacillus* and *Bifidobacterium*) at a concentration of 30 billion CFU for 6 months did not significantly affect the degree of steatosis/fibrosis or improve laboratory parameters. However, due to the significant reduction in the expression of CD8+ T lymphocytes and zonula occludens-1 (ZO-1) in the placebo group, the authors suggest that probiotics may play a role in stabilizing the immune function of the mucosa and also prevent intestinal permeability in MASLD patients [56]. An interesting observation was made by Xue et al., who performed fecal microbiota transplantation (FMT) on patients with MASLD from healthy donors. Patients who received FMT showed a reduction in fat accumulation in the liver by improving intestinal microbiota dysbiosis, thereby reducing hepatic steatosis, especially in patients suffering from obesity. The control group who received probiotic therapy (*Bifidobacterium viable* and *Lactobacillus acidophilus* capsules) showed no differences in blood lipid levels, liver function, and fat suppression before treatment [57].

According to the available data, a low-processed diet, rich in dietary fiber, focusing on the intake of whole grain products, vegetables, fruits, and unsaturated fatty acids through the modulation of the intestinal microbiota, promotes intestinal health and may be key in the prevention of MASLD (Figure 2).

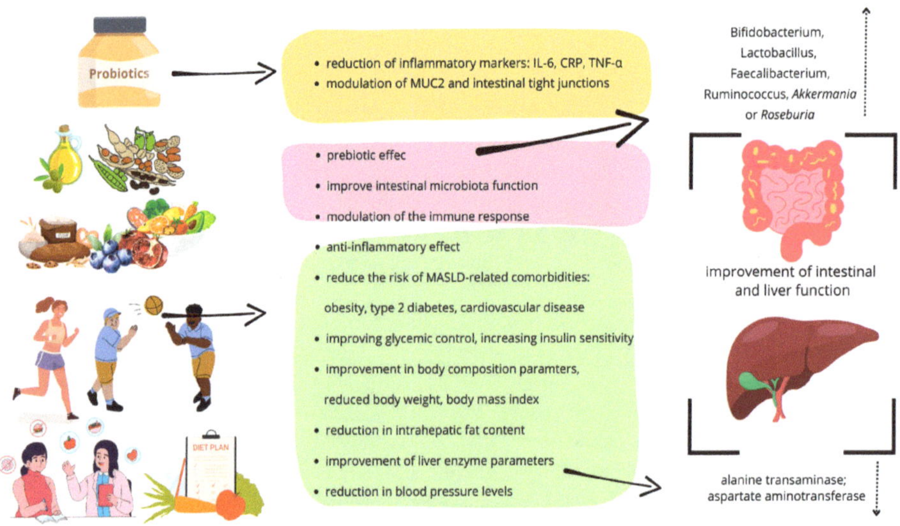

Figure 2. The benefits of a diet rich in dietary fiber and unsaturated fatty acids, combined with exercise in the context of improving intestinal and liver function and preventing metabolic disorders associated with MASLD. Il-6—interleukin 6; CRP—c-reactive protein; TNF-α—tumor necrosis factor alpha; MUC2—Mucin2.

5. Food Additives in Association with the Gut Microbiome and MASLD

5.1. Colors (E100–E199)

It is widely accepted that food colors are among the most toxic food additives used in the food industry, with those belonging to the 'azo' group being considered to be the most genotoxic. Tatrazine (TS) (E102) is an artificial dye that contains an azo group and is soluble in water [58,59]. It is a compound commonly used in the food industry. Due to its yellow color, it is often added to yellow cheeses, sauces, jellies, chewing gums, fish products, flavored wines, and other drinks, such as soft drinks and sports drinks. It is also an additive to vegetable and fruit products—both canned and bottled [58,60]. The acceptable daily intake (ADI) level according to the Food and Drug Administration was recognized at 5 mg/kg bw in 2011, while the European Food Safety Authority approved it at 7.5 mg/kg bw [60]. TS, to a small extent, can be reduced by various bacterial taxa [61]. Azoreduction of TS in the gut leads to the formation of sulfanilic acid and 4-amino-3-carboxy-5-hydroxy-1-(4-sulfophenyl)pyrazole (SCAP) [62]. TS metabolites are considered potentially hazardous to health, especially for children who consume significant amounts of colored foods and soft drinks [63]. Nitrogen dyes account for 10–22% of the maximum

ADI in beverages [64]. Subsequently, toxic concentrations of SCAP can occur in the gut when TS is consumed at the limit of the recommended daily dose [62].

A study by Wu et al. showed that the ingestion of TS (1.4, 5.5 and 10 mg/kg/day) could cause severe histopathological and cellular changes in the intestines and liver of goldfish. Some epithelial cells became vacuolized and the intestinal villi was ruptured. In addition, TS supply induced oxidative stress (proportional dose-dependent increase in malondialdehyde (MDA)) and led to changes in the intestinal microbiota, an increase in *Actinobacteriota* and *Proteobacteria*, and a significant decrease in *Planctomycetota* and *Fusobacteriota*, as well as the bacteria responsible for SCFA production (*Bacteroides* and *Clostridium_sensu_stricto_1*) [65]. In a subsequent study, Wu et al. reached similar conclusions. TS supply was associated with changes in the intestinal microbiota and the development of inflammation, which was associated with the up-regulation of pro-inflammatory cytokines (IL1 and IL6), lysozymes (lyz), β-defensin 3 (defb3), and complement component 3 (c3) [66].

In a study by El-Desoky et al., the administration of TS at a dose of 7.5 mg/kg bw for 50 days resulted in increased liver function enzymes (aspartate aminotransferase (AST), alanine aminotransferase (ALT), gamma-glutamyl transferase, and alkaline phosphatase), bilirubin levels, abnormal lipid profile, and serum glucose. There was a decrease in body weight with an increase in liver weight, indicating its toxic effects. An increase in protein kinase C (PKC) isoforms was indicative of ROS generation, and alpha-fetoprotein was indicative of liver failure [67].

As with TS, Allura Red (AR) (E129) can be metabolized by an intestinal bacterium through nitrogen reduction, including *O. splanchnicus* and *P. vulgatus*. This results in the formation of two compounds, creisin-4-sulfonic acid and 1-amino-2-naphthol-6-sulfonic acid [61,68]. Hofseth et al. believe that the association of AR with inflammation, DNA damage, and the concomitant disruption of the microbiome is notable [68]. Kwon et al. observed that chronic, long-term exposure to AR promotes experimental colitis via serotonin in the colon in a pathway, whether dependent on or independent of the gut microbiota, in mice [69]. He et al. noted that the risk of developing colitis was increased in mice expressing the increased expression of IL-23, leading to the increased generation of activated CD4 T cells that expressed interferon-γ. In turn, the induction of colitis was dependent on the commensal microbiota, promoting AR azo reduction and the production of the metabolite 1-amino-2-naphthol-6-sulfonate sodium [70].

The administration of AR to albino rats for 4 weeks at a dose of 7 mg/kg bw resulted in an increase in biochemical markers of liver function (ALT and AST) and MDA levels and a decrease in serum antioxidants. There were changes in the histological structure with a decrease in Bcl2 expression and an increase in cytochrome c oxidase subunit II expression [71].

Sunset yellow (SY) (E110), also commonly used in the food industry, is added to foods such as desserts, chips, cookies, ice cream, and soft drinks and also to pharmaceutical products and drugs and syrups for children [72]. Due to its potential mutagenic and carcinogenic effects, SY has been banned for use in Norway and Finland [73]. The ADI of SY is 4 mg/kg/bw [74].

According to Sensoy et al., SY can affect the intestinal epithelium, inducing changes in intestinal secretion. SY has also been shown to interfere with intestinal signaling interactions by exerting antagonistic effects on the glucagon-like peptide-1 (GLP-1) receptor, a peptide hormone [72]. In the study by Zahran et al., SY administration at a dose of 6.17 mg/kg (equivalent to human ADI of 1 mg/kg) for 12 weeks increased serum LPD and altered the intestinal microbiome, leading to the disruption of intestinal integrity by altering the jejunal E-cadherin/β-catenin adhesion junction complex and decreasing cloverleaf factor (TFF)-3. SY decreased the abundance of beneficial taxa, including Treponema 2 and *Anaerobiospirillum*, while increasing the abundance of potentially pathogenic microorganisms *Prevotella* and *Oribacterium* [75].

Abdelhamid et al. have shown that SY can induce a number of adverse structural and biochemical changes in the liver [76]. In the previously mentioned study, Khayytat et al. observed similar changes when SY was administered at a dose of 2.5 mg/kg body weight as when AR was used. However, an additional genotoxic effect was observed for SY, which was not detected for AR [71]. Huessein et al. reported that the long-term oral administration of SY above ADI is hepatotoxic and has negative effects on immunity. However, the authors noted an increase in ALT and AST and, even at low doses of SY, an increase in the mRNA expression of proapoptotic protein [77].

Data on the effects of food dyes on the gut microbiota and liver function come mainly from studies on animal models. However, the results of these studies indicate that these substances can cause negative health effects in humans as well. Dyes commonly used in the industry can change the composition of the intestinal microbiota, leading to structural changes in the intestines, with the liver disrupting their function. Thus, they lead to the development of inflammation and a decrease in antioxidants in the body.

5.2. Preservatives (E200–299)

This section concerns a group of food additives labeled E200 to E299. They have their use in cured meats, sauces, marinades, and processed foods [78]. Some preservatives have been shown to inhibit the growth of beneficial microorganisms, thereby disrupting intestinal microbial homeostasis and causing predisposition to inflammation [79]. Another example of substances used in the food industry with a preservative effect are sulfites. Their main function is to inhibit the growth of pathogenic microorganisms. However, they can exhibit a number of other undesirable functions. Irwin et al. point out that sulfites can potentially lead to cell damage reactions; additionally, due to their bacteriostatic and bactericidal effects, they can potentially alter the oral and intestinal microbiome [80,81]. In a study by Nagpal et al., in animal models, the authors observed a qualitative and quantitative change in the composition of the intestinal microbiota after treatment with potassium sorbate (E202), benzoic acid (E210), or sodium nitrate (E251). Changes related to intestinal epithelial permeability and the expression of markers of intestinal tight junctions were also demonstrated [82]. In addition to their indirect effects on the liver through the modulation of the gut microbiota, preservatives added for food preservation can also have direct effects. Hrncir et al. studied liver function after exposure to fructose and preservatives such as sodium benzoate, sodium nitrite, and potassium sorbate. They noted that preservatives can amplify the adverse effects of fructose on liver function and lipid metabolism. Such synergistic adverse effects have also been observed to increase intestinal permeability [83]. Crowe et al. examined the effects of sodium nitrite found in sausages on the colorectal cancer status in mice. They showed that the consumption of meat products with sodium nitrite added was associated with intestinal dysbiosis and higher lipid peroxidation [84]. On the contrary, Van Hecke et al. show that the consumption of peated beef compared to fresh beef alters the composition of the intestinal microbiota in animal models with a higher relative abundance of *Ruminococcaceae* [85]. Preservatives can have an indirect effect on the onset of homeostatic disorders leading to MASLD, but some studies indicate a direct effect by disrupting normal organ function.

5.3. Emulsifiers, Thickeners, Stabilizers (E400–499)

Food additives in the group of emulsifiers, thickeners, and stabilizers are designated as E400-E499. An example of a food additive with thickening and emulsifying properties is carrageenan (E407). It is extracted from the cell walls of red seaweed. It is often an ingredient in fat-reduced food products, for example dairy products, cured meats, dietary supplements, jams, jellies, powdered products, instant drinks, and sauces [86]. Borsani et al. in their paper presented that carrageenan can induce inflammation, predisposing people to the onset of exacerbation in ulcerative colitis (UC) and thus disrupting intestinal homeostasis and the integrity of the intestinal barrier, and can cause an increased risk of bacterial translocation, which would also have consequences for the liver [87]. Ariffin et al. studied the effects of carrageenan on intestinal and hepatic cell lines. They noted that the acid hydrolysis products of k-carrageenin could exhibit cytotoxic effects on both cell lines, while unintegrated carrageenin did not show such properties [88]. In a review article, Liu et al. conclude that the long-term consumption of carrageenan may be associated with inflammation in the gut and changes in the composition of the gut microbiota [89]. Furthermore, Naimi et al. indicate that carrageenan may have adverse effects on the intestinal epithelium and the composition of the intestinal microbiota and may show increased expression of pro-inflammatory molecules [90]. Through this type of interaction, it can potentially cause inflammation that is histopathologically similar to inflammatory bowel disease [91]. All of these disorders can trigger further consequences in the pathogenesis of liver disease [92,93].

Polysorbate 80 (P-80), labeled E433 on the list of food additives, is used as an emulsion stabilizer. It is added primarily during the production of sauces, ice cream, and confectionery products [94]. P-80 or carboxymethylcellulose (CMC, E-466) can cause a predisposition to type 1 diabetes, cardiovascular disease, intestinal disease, and metabolic syndrome [95–98]. In the case of the direct effects of emulsifiers on the liver, Vilas-Boas et al. describe that there was a predisposition to liver toxicity with formulations in which P-80 was added, which may be the reason for increased membrane permeability [99]. Lv et al., in their study, observed that P-80 can stimulate colitis synergistically with a high-fat diet, affect weight gain, and predispose people to changes in their bile acid profile [100]. In another study in animal models, Singh et al. indicate that P-80-fed mice showed an association with faster fat growth. They also observed elevated parameters indicative of metabolic syndrome and low-grade inflammation, which can lead to MASLD. The authors showed the appearance of abnormalities in the gut microbiota in mice fed P-80 [101]. On the other hand, a direct link to intestinal dysbiosis and the occurrence of MASLD may be caused by the incorrect detection of receptor containing protein 6 (NLRP6) and receptor containing protein 3 (NLRP3) inflammasomes [102].

Other Substances with a Thickening Effect Using Maltodextrin as an Example

Although not classified as an emulsifier, maltodextrin (MDX) exhibits thickening properties in starchy products and is widely used as a food additive. Nickerson et al., in their work, indicate that up to 60% of packaged food products may contain MDX or modified starch in their composition [103]. Arnold et al. describe that MDX consumption may predispose people to low-grade inflammation and may be a risk factor for IBD; it may also increase the risk of metabolic disease, which is also associated with MASLD [104]. In another study in animal models, Singh et al. showed that in mouse pups fed a predominantly MDX mixture, it could induce intestinal damage similar to necrotizing enterocolitis (NEC). The consequences included bacterial translocation and the altered functioning of tight junction (TJ) proteins. The factor leading to intestinal damage altered homeostasis between pro-inflammatory cytokines and anti-inflammatory cytokines [105]. In another study, Zangara et al. reached similar conclusions. In addition to increasing susceptibility to colitis in genetically susceptible individuals, it can also alter intestinal mucus production [106]. Almutairi et al. conducted a systematic review of randomized placebo trials in which MDX was used as a placebo. They concluded that because MDX can induce modifications in the gut microbiota and immune factors, it should not be used as a placebo in clinical trials [107]. All of these impacts on the gut microbiota and the disruption of gut barrier integrity could also potentially affect the liver and the appearance of MASLD. However, in many studies on liver effects, MDX is used as a placebo or with other prebiotic substances, and so further studies are needed to determine the actual effects of MDX on the occurrence of MASLD.

5.4. Taste Enhancers (E600–699)

The group of food additives known as flavor enhancers is classified in the list from E600 to E699. These are substances added to foods to increase the intensity of certain flavor characteristics and aromas. The flavor enhancers most commonly studied are monoammonium glutamate (MAG) (E624) and monosodium glutamate (MSG) (E621). They are found mainly in powdered soups and sauces, salty snacks, seasoning mixes, stock cubes, and fast food dishes [108,109]. Ahangari et al., in their review article, indicate that the long-term intake of MSG can cause changes in the gut microbiota and have effects on liver metabolism and hepatocyte damage [110]. Nahok et al., in a study using animal models, showed that the metabolic changes that occurred after MSG ingestion are related to gluconeogenesis and branched-chain amino acid metabolism with concomitant intestinal dysbiosis [111]. In another study, the authors indicate that the addition of MSG to the diet causes a reduction in *Akkermansia muciniphila*, which, among other things, produces mucin [112]. In a study by Coelho et al., the researchers observed that in mice with prevalent obesity with MSG in the diet, there is a premature induction of fat accumulation in the liver, predisposing them to the induction of MASLD and subsequent disease progression in the form of hepatitis [113]. MSG causes oxidative stress, which can also contribute to liver damage and changes in liver fat metabolism [114,115]. Olowofolahan et al., after orally administering MSG in various doses to rats, observed that MSG in low doses is tolerated by the animals, while in high doses, it causes cytotoxicity by opening the hepatic mitochondrial permeability transition pore; a similar trend occurred for lipid peroxidation, among other things [116]. Obesity-induced hepatic steatosis while consuming MSG may predispose to the infiltration of lymphocytes, macrophages, and eosinophils, increasing inflammation, which can then cause a predisposition to HCC [117]. Flavor enhancers can have a negative impact on the gut microbiota, and chronic and excessive consumption of these food additives can predispose people to oxidative stress and cause damage to hepatocytes, leading to altered organ function.

5.5. Sweeteners (E900–999)

Sweeteners are substances used to add a sweet taste to food products. In this capacity, they function as substitutes for sucrose and other naturally occurring saccharides. Sucralose (E955) is one of the most prevalent sweeteners in the industry. Sucralose is a non-nutritive sweetener, 600 times sweeter than sucrose. For years, it has been recommended to patients suffering from diabetes and obesity. Sucralose is poorly absorbed and enters the lower gastrointestinal tract practically unchanged, where it can potentially alter the composition of the microbiota [118,119]. In the 1990s, it was considered safe by the Food and Agriculture Organization of the United Nations and World Health Organization, and its ADI was set at 15 mg/kg body weight (bw) [118]. Sucralose is often added to foods and beverages, including products designed for patients with diabetes or people who want to reduce energy consumption in their diet [120]. It is also used in the production of alcoholic and non-alcoholic beverages, dairy drinks, chewing gums, ice cream, jams, and jellies [121]. However, in 2023, the WHO has published new guidelines on non-sugar sweeteners (NSS), which advise against the use of NSS for weight management or to reduce the risk of noncommunicable diseases [122]. Feng et al. highlighted the important issue of the influence of other factors, both external and internal, that can affect the microbiome with NSS exposure, such as environmental factors, diet, and stress [123].

According to Shiffman et al., the amount of sucralose 6-acetate in a single daily sucralose-sweetened beverage can far exceed the toxicological risk threshold for genotoxicity (TTC genotox) of 0.15 μg/person/day. Sucralose 6-acetate was shown to significantly increase the expression of genes related to inflammation, oxidative stress, and cancer, with the highest expression observed for the metallothionein 1 G gene (MT1G) [124].

Bian et al. demonstrated that sucralose has the ability to influence the composition of the gut mycobiome and modify its metabolic functions. A six-month supply of sucralose was administered to mice, resulting in alterations in hepatic gene expression (matrix metalloproteinase 2 and inducible nitric oxide synthase (iNOS)) [125]. Sucralose consumption significantly increased the abundance of the intestinal genera *Bacteroides* and *Clostridioides*, which are responsible for the production of deoxycholic acid, in a study of MISICG models, and its increase has been linked to the development of MASLD [126].

The role of gut bacteria in the biotransformation of bile acids is also important. By cleaving amino acid residues, gut bacteria can lead to the uncoupling of taurine/glycine-conjugated bile acids. In a study by Chi et al., sucralose was found to have a reducing effect on the abundance of bacteria associated with the metabolism of bile acids. Furthermore, sucralose administration was found to result in reduced levels of hepatic farnesoid X receptor activation, which was associated with an increase in intrahepatic cholesterol levels [127].

A randomized case–control study by Suez et al. showed that sucralose, at a dose below the ADI, impaired glycemic responses in healthy subjects over two weeks of administration. The increase in glycaemia was accompanied by an increase in plasma trichloroacetic acid and changes in the oral microbiome in the relative abundance of six Streptococcus species in the sucralose group [128]. Different results were obtained by Orku et al. They showed that the regular consumption of water sweetened with sweeteners, including saccharin, in doses similar to those consumed in everyday life had no significant effect on glycemic response, insulin sensitivity, GLP-1 release, and body weight in healthy subjects [129]. Moreover, of all the substances tested (sucralose, aspartame, saccharin, and stevia), dietary supplementation with sucralose had the greatest impact on the functional potential of the human microbiome in an NNS-specific manner [128]. However, the results of a randomized, double-blind, cross-over clinical trial showed that the oral consumption of sweetened beverages at a dose of 136 mg/day had no measurable effect on gut microbiota in healthy participants [130].

Aspartame (E951) is a sweetener composed of two naturally occurring amino acids, L-phenylalanine and L-aspartic acid. It is about 200 times sweeter than sucrose, and the ADI is 40 mg/kg bw [131]. Aspartame is used in the food and pharmaceutical industries and is added to chewing gum, ice cream, dairy products, cough drops, and chewable vitamins [132]. After absorption in the intestinal lumen, aspartame is hydrolyzed to phenylalanine (50%), aspartic acid (40%), and methanol (10%) [133].

Finamor et al. showed that aspartame can negatively affect liver health. Feeding aspartame to mice at a dose of 80 mg/kg for 12 weeks led to increased levels of liver enzymes ALT and AST and liver fibrosis. Increases in profibrotic markers were observed, including transforming growth factor β 1, collagen type I alpha 1, and alpha-smooth muscle actin. Aspartame decreased the activation of erythroid nuclear factor 2-related factor 2 (Nrf2) and increased lipid peroxidation, thereby affecting the activation of NLRP3 [134]. The initiation of the NLRP3 inflammasome has been linked to liver cancer, particularly HCC [135]. Aspartame also reduced levels of peroxisome proliferator-activated receptor gamma coactivator 1 alpha (PGC-1α), and its deficiency may be responsible for changes in the serum lipid profile, as well as lipid accumulation and impaired hepatic gluconeogenesis. Palmnäs et al. demonstrated that the administration of aspartame at a dose of 5–7 mg/kg/d, via drinking water over a period of 8 weeks, was associated with an increase in propionate, a substrate with a high gluconeogenic potential. The increase in propionate, in conjunction with the increase in fasting glucose levels, has the potential to impair insulin-stimulated glucose disposal on both standard and low-fat diets, irrespective of body composition. Furthermore, aspartame intake was found to be associated with an increase in the total number of *Enterobacteriaceae* and *Clostridium leptum* bacteria [136]. In another study, Finamor et al. showed that chronic aspartame administration can lead to hepatic glutathione depletion, which is associated with reduced glutamate cysteine ligase and cysteine levels. Moreover, aspartame induced a blockade of the trans-sulfuration pathway at two steps, namely methionine adenosyltransferase and cystathionine γ-lyase [137].

In the previously cited double-blind study conducted by Ahmad et al., an evaluation of the effects of the oral consumption of aspartame-sweetened beverages at a dose of 425 mg/day showed no effect on the gut microbiota in healthy participants [130]. A randomized, case–control study by Tey et al. demonstrated that the consumption of artificially and naturally sweetened non-calorie beverages, including aspartame, had minimal effects on postprandial glucose and insulin levels in comparison to sucrose-sweetened beverages [138]. Interestingly, changes in insulin sensitivity caused by aspartame supply have been linked to the development of carcinogenesis [135]. Orku et al. [129] and Suez et al. reported comparable outcomes on the impact on the glycemic response. However, the ingestion of sweeteners, including aspartame, has been shown to induce substantial alterations in the composition of the fecal and oral microbiome, as well as the plasma metabolome [128].

Saccharin (E954) is a commonly used sweetener in the food industry, and the ADI is 5 mg/kg bw [131]. It is added to a wide range of foods, from dairy drinks to unripened cheese, jams, confectionery, and even breakfast cereals [139]. Saccharin is not metabolized in the body, but can pass through the placenta and breast milk, so it is not recommended for pregnant or breastfeeding women [131,140]. There is evidence that saccharin can be a metabolic disruptor and can alter the composition of the gut microbiome in offspring [140].

The administration of saccharin in drinking water at a concentration of 0.3 mg/mL over a period of six months resulted in a substantial increase in iNOS and TNF-α levels in the liver of mice. This increase was concomitant with the onset of inflammation and the disruption of the microbiome. The composition of the gut microbiota and the metabolome exhibited alterations. The study noted an increase in bacteria associated with inflammation, including *Corynebacterium*, *Turicibacter*, and *Roseburia* [141]. Serrano et al. conducted a double-blind, placebo-controlled, parallel study with healthy men and women, as well as an animal model. The study found no effect of saccharin supply on glucose tolerance or gut microbiota composition in humans or mice. However, it should be noted that the study evaluated the effects of short-term saccharin consumption in the maximum permitted amounts [142]. Similarly, Orku et al. did not show an effect of saccharin on glycemic response or insulin sensitivity in healthy subjects [129]. Suez et al. obtained divergent results. They reported a reduced relative abundance of *Fusobacterium* in the oral mycobacterium after saccharin ingestion and an impaired glycemic response during a short-term (two week) supply of saccharin [128]. The recommendation to use sweeteners as substitutes for sugars continues to appear in recommendations, including for people suffering from obesity. Based on the available research results, it is worth noting that the consumption of sweeteners instead of sugar should not be a long-term change, but only a temporary one during a modification of eating habits.

Table 1 presents the results of studies on the effects of food additives on gut microbiota homeostasis and MASLD discussed in this review.

Table 1. The results of studies on the effects of food additives on gut microbiota homeostasis and MASLD discussed in this review.

Food Additive Example	Acceptable Daily Intake (ADI) [143]	Potential Impact on Gut Microbiota Disorders in Association with MASLD
Tatrazine (E102)	7.5 mg/kg body weight	• Histopathological and cellular changes in the gut and liver [86]; • Increase in oxidative stress [86]; • Changes in the composition of the intestinal microbiota [66,86]; • Liver dysfunction (increased liver enzymes) [67]; • Abnormal lipid profile and increased serum glucose levels [67]; • Up-regulation of pro-inflammatory cytokines [66]; • Increased ROS production [67]; • Increase in alpha-fetoprotein in serum [67].
Allura Red (E129)	7 mg/kg body weight	• Increased inflammation, disruption of the intestinal microbiota [68]; • Liver dysfunction (increase in liver function enzymes) [71]; • Increase in serum MDA and NO levels [71]; • Decrease in serum antioxidants [71]; • Changes in histological structure (disorganization of hepatic strands, necrotic and hydropic degeneration of hepatic cells) [71]; • Decreased Bcl2 expression and increased COX2 expression [71].
Sunset Yellow (E110)	4 mg/kg body weight	• Disruption of intestinal signaling interactions—antagonistic effect on the glucagon-like peptide-1 (GLP-1) receptor, a peptide hormone [72]; • Increase in serum LPS, increase in intestinal permeability, change in composition of the microbiota [75]; • Decrease in total serum antioxidants, increase in serum MDA and NO levels [71]; • Infiltration of leukocytes and increase in Kupffer cells [71]; • Changes in histological structure [71]; • Reduction in Bcl2 expression [71].
Potassium sorbate (E202)/ benzoic acid (E210)/ sodium nitrate (E251)	1 mg/kg body weight/ 5 mg/kg body weight/ 3.7 mg/kg body weight	• Qualitative and quantitative change in the composition of the intestinal microbiota changes associated with intestinal epithelial permeability and in the expression of intestinal tight junction markers [82].
Carrageenan (E407)	75 mg/kg body weight	• Adverse effects on the intestinal epithelium, composition of the intestinal microbiota and increased expression of pro-inflammatory molecules [90].
Polysorbate 80 (E433)	25 mg/kg body weight	• Faster increase in body fat, elevated parameters indicating metabolic syndrome and low-grade inflammation, the presence of abnormalities in the gut microbiota [101].
Maltodextrin (E1400)	-	• Increased susceptibility to colitis in genetically susceptible individuals [106]; • Modifications in gut microbiota and immune factors [107].
Monosodium glutamate (E621)	30 mg/kg body weight	• Infiltration of lymphocytes, macrophages, eosinophils, increasing inflammation which predisposes to hepatocellular carcinoma [117]; • Gut dysbiosis [111].

Table 1. *Cont.*

Food Additive Example	Acceptable Daily Intake (ADI) [143]	Potential Impact on Gut Microbiota Disorders in Association with MASLD
Sucralose (E955)	15 mg/kg body weight	• Increased expression of genes related to inflammation and oxidative stress [134]; • Reducing the abundance of bacterial communities associated with bile acid metabolism [127]; • A reduction in the level of hepatic FXR activation [127]; • Disturbance of the composition of the intestinal microbiota; increased production of deoxycholic acid [126]; • Impaired glycemic response [128].
Aspartame (E951)	40 mg/kg body weight	• Liver dysfunction—increase in liver enzymes (ALT and AST) [134]; • Impaired hepatic gluconeogenesis, decreased GSH and GCLc and cysteine levels [137]; • Blockage of the trans-sulfuration pathway, increased levels of oxidative stress due to increased MDA and decreased Nrf2 activation [134]; • Increased levels of lipid peroxidation [134]; • Disturbances in the composition of the microbiota [136].
Saccharin (E954)	9 mg/kg body weight	• Increased levels of iNOS and TNF-α in the liver [141]; • An increase in the development of inflammation and disturbances in the composition of the intestinal microbiome [141]; • Impaired glycemic response [129].

FXR—farnesoid X receptor; ALT—alanine transaminase; AST—aspartate aminotransferase; GSH—glutathione; GCLc—glutamate cysteine ligase; ROS—reactive oxygen species; MDA—malondialdehyde; NO—nitric oxide; LPS—lipopolysaccharide; Nrf2—nuclear factor erythroid 2-related factor 2; iNOS—inducible nitric oxide synthase; TNF-α—tumor necrosis factor alpha.

6. Limitations

A limitation is the lack of animal and human studies examining the effects of the food additives analyzed in the publication, their impact on the gut microbiota, and their association with the occurrence of MASLD. There is also a lack of long-term studies on these associations; moreover, some of the mechanisms of association are still poorly understood, and so, in the future, it would be worthwhile to expand research to include other food additives as well.

7. Conclusions

Foods rich in food additives, thanks to their properties, are particularly attractive to consumers, especially children and adolescents. Some food additives including emulsifiers, preservatives, flavor enhancers, dyes, or artificial sweeteners may predispose people to dysfunction in the integrity of the intestinal barrier and may affect the composition of the intestinal microbiota, leading to inflammation. These mechanisms can lead to oxidative stress that is not sufficiently compensated for, predisposing the individual to impaired lipid metabolism in the liver. However, further research is still needed to fully elucidate the extent of the health effects of certain food additives and to understand the mechanisms by which they affect the intestinal microbiota and the pathogenesis of MASLD.

Given the increasing incidence of obesity and MASLD, there is a need to effectively control the intake of these substances, particularly among young people, and to further restrict their use in food production. It seems that the recommendations for acceptable intake should be reviewed as well.

Author Contributions: Conceptualization, S.J.-C. and A.S.-D.; writing—original draft preparation, S.J.-C. and A.S.-D.; and writing—review and editing, A.S-D., S.J.-C. and R.F.; supervision, R.F. All authors have read and agreed to the published version of the manuscript.

Funding: This research received no external funding.

Institutional Review Board Statement: Not applicable.

Informed Consent Statement: Not applicable.

Data Availability Statement: Not applicable.

Acknowledgments: We would like to express our gratitude to Alicja Jarmakiewicz for her assistance with the figure entitled "Potential influence of selected food additives on the association with intestinal homeostasis disorders and the occurrence of MASLD".

Conflicts of Interest: The authors declare no conflicts of interest.

References

1. Ludwig, J.; Viggiano, T.R.; McGill, D.B.; Oh, B.J. Nonalcoholic steatohepatitis: Mayo Clinic experiences with a hitherto unnamed disease. *Mayo Clin. Proc.* **1980**, *55*, 434–438. [CrossRef] [PubMed]
2. Rinella, M.E.; Lazarus, J.V.; Ratziu, V.; Francque, S.M.; Sanyal, A.J.; Kanwal, F.; Romero, D.; Abdelmalek, M.F.; Anstee, Q.M.; NAFLD Nomenclature Consensus Group; et al. A multisociety Delphi consensus statement on new fatty liver disease nomenclature. *Hepatology* **2023**, *78*, 1966–1986. [CrossRef]
3. Younossi, Z.; Anstee, Q.M.; Marietti, M.; Hardy, T.; Henry, L.; Eslam, M.; George, J.; Bugianesi, E. Global burden of NAFLD and NASH: Trends, predictions, risk factors and prevention. *Nat. Rev. Gastroenterol. Hepatol.* **2018**, *15*, 11–20. [CrossRef] [PubMed]
4. Chan, W.K.; Chuah, K.H.; Rajaram, R.B.; Lim, L.L.; Ratnasingam, J.; Vethakkan, S.R. Metabolic Dysfunction-Associated Steatotic Liver Disease (MASLD): A State-of-the-Art Review. *J. Obes. Metab. Syndr.* **2023**, *32*, 197–213. [CrossRef]
5. Benedé-Ubieto, R.; Cubero, F.J.; Nevzorova, Y.A. Breaking the barriers: The role of gut homeostasis in Metabolic-Associated Steatotic Liver Disease (MASLD). *Gut Microbes* **2024**, *16*, 2331460. [CrossRef]
6. European Association for the Study of the Liver (EASL); European Association for the Study of Diabetes (EASD); European Association for the Study of Obesity (EASO). EASL-EASD-EASO Clinical Practice Guidelines on the management of metabolic dysfunction-associated steatotic liver disease (MASLD). *J. Hepatol.* **2024**, *81*, 492–542. [CrossRef]
7. Bilson, J.; Mantovani, A.; Byrne, C.D.; Targher, G. Steatotic liver disease, MASLD and risk of chronic kidney disease. *Diabetes Metab.* **2024**, *50*, 101506. [CrossRef]
8. Loomba, R.; Wong, V.W. Implications of the new nomenclature of steatotic liver disease and definition of metabolic dysfunction-associated steatotic liver disease. *Aliment. Pharmacol. Ther.* **2024**, *59*, 150–156. [CrossRef] [PubMed]
9. Provera, A.; Vecchio, C.; Sheferaw, A.N.; Stoppa, I.; Pantham, D.; Dianzani, U.; Sutti, S. From MASLD to HCC: What's in the middle? *Heliyon* **2024**, *10*, e35338. [CrossRef] [PubMed]
10. Tilg, H.; Moschen, A.R. Evolution of inflammation in nonalcoholic fatty liver disease: The multiple parallel hits hypothesis. *Hepatology* **2010**, *52*, 1836–1846. [CrossRef] [PubMed]
11. Buzzetti, E.; Pinzani, M.; Tsochatzis, E.A. The multiple-hit pathogenesis of non-alcoholic fatty liver disease (NAFLD). *Metabolism* **2016**, *65*, 1038–1048. [CrossRef] [PubMed]
12. Wang, Y.D.; Wu, L.L.; Qi, X.Y.; Wang, Y.Y.; Liao, Z.Z.; Liu, J.H.; Xiao, X.H. New insight of obesity-associated NAFLD: Dysregulated "crosstalk" between multi-organ and the liver? *Genes Dis.* **2022**, *10*, 799–812. [CrossRef] [PubMed]
13. Huttasch, M.; Roden, M.; Kahl, S. Obesity and MASLD: Is weight loss the (only) key to treat metabolic liver disease? *Metabolism* **2024**, *157*, 155937. [CrossRef]
14. Donnelly, K.L.; Smith, C.I.; Schwarzenberg, S.J.; Jessurun, J.; Boldt, M.D.; Parks, E.J. Sources of fatty acids stored in liver and secreted via lipoproteins in patients with nonalcoholic fatty liver disease. *J. Clin. Investig.* **2005**, *115*, 1343–1351. [CrossRef]
15. Takahashi, Y.; Dungubat, E.; Kusano, H.; Fukusato, T. Pathology and Pathogenesis of Metabolic Dysfunction-Associated Steatotic Liver Disease-Associated Hepatic Tumors. *Biomedicines* **2023**, *11*, 2761. [CrossRef]
16. Rashu, E.B.; Werge, M.P.; Hetland, L.E.; Thing, M.; Nabilou, P.; Kimer, N.; Junker, A.E.; Jensen, A.H.; Nordestgaard, B.G.; Stender, S.; et al. Use of PNPLA3, TM6SF2, and HSD17B13 for detection of fibrosis in MASLD in the general population. *Clin. Res. Hepatol. Gastroenterol.* **2024**, *48*, 102389. [CrossRef] [PubMed]
17. Castanho Martins, M.; Dixon, E.D.; Lupo, G.; Claudel, T.; Trauner, M.; Rombouts, K. Role of PNPLA3 in Hepatic Stellate Cells and Hepatic Cellular Crosstalk. *Liver Int.* **2024**, 1–11. [CrossRef]

18. Sherman, D.J.; Liu, L.; Mamrosh, J.L.; Xie, J.; Ferbas, J.; Lomenick, B.; Ladinsky, M.S.; Verma, R.; Rulifson, I.C.; Deshaies, R.J. The fatty liver disease-causing protein PNPLA3-I148M alters lipid droplet-Golgi dynamics. *Proc. Natl. Acad. Sci. USA* **2024**, *121*, e2318619121. [CrossRef]
19. Fang, J.; Yu, C.H.; Li, X.J.; Yao, J.M.; Fang, Z.Y.; Yoon, S.H.; Yu, W.Y. Gut dysbiosis in nonalcoholic fatty liver disease: Pathogenesis, diagnosis, and therapeutic implications. *Front. Cell. Infect. Microbiol.* **2022**, *12*, 997018. [CrossRef]
20. Schoeler, M.; Caesar, R. Dietary lipids, gut microbiota and lipid metabolism. *Rev. Endocr. Metab. Disord.* **2019**, *20*, 461–472. [CrossRef]
21. Arias, N.; Arboleya, S.; Allison, J.; Kaliszewska, A.; Higarza, S.G.; Gueimonde, M.; Arias, J.L. The Relationship between Choline Bioavailability from Diet, Intestinal Microbiota Composition, and Its Modulation of Human Diseases. *Nutrients* **2020**, *12*, 2340. [CrossRef]
22. Aron-Wisnewsky, J.; Vigliotti, C.; Witjes, J.; Le, P.; Holleboom, A.G.; Verheij, J.; Nieuwdorp, M.; Clément, K. Gut microbiota and human NAFLD: Disentangling microbial signatures from metabolic disorders. *Nat. Rev. Gastroenterol. Hepatol.* **2020**, *17*, 279–297. [CrossRef]
23. Boursier, J.; Mueller, O.; Barret, M.; Machado, M.; Fizanne, L.; Araujo-Perez, F.; Guy, C.D.; Seed, P.C.; Rawls, J.F.; David, L.A.; et al. The severity of nonalcoholic fatty liver disease is associated with gut dysbiosis and shift in the metabolic function of the gut microbiota. *Hepatology* **2016**, *63*, 764–775. [CrossRef]
24. Zhang, X.; Coker, O.O.; Chu, E.S.; Fu, K.; Lau, H.C.H.; Wang, Y.X.; Chan, A.W.H.; Wei, H.; Yang, X.; Sung, J.J.Y.; et al. Dietary cholesterol drives fatty liver-associated liver cancer by modulating gut microbiota and metabolites. *Gut* **2021**, *70*, 761–774. [CrossRef] [PubMed]
25. Caputo, V.; Tarantino, G.; Santini, S.J.; Fracassi, G.; Balsano, C. The Role of Epigenetic Control of Mitochondrial (Dys)Function in MASLD Onset and Progression. *Nutrients* **2023**, *15*, 4757. [CrossRef]
26. Vidal-Cevallos, P.; Sorroza-Martínez, A.P.; Chávez-Tapia, N.C.; Uribe, M.; Montalvo-Javé, E.E.; Nuño-Lámbarri, N. The Relationship between Pathogenesis and Possible Treatments for the MASLD-Cirrhosis Spectrum. *Int. J. Mol. Sci.* **2024**, *25*, 4397. [CrossRef]
27. Nassir, F. NAFLD: Mechanisms, Treatments, and Biomarkers. *Biomolecules* **2022**, *12*, 824. [CrossRef] [PubMed]
28. Horn, P.; Tacke, F. Metabolic reprogramming in liver fibrosis. *Cell Metab.* **2024**, *36*, 1439–1455. [CrossRef]
29. Park, S.J.; Garcia Diaz, J.; Um, E.; Hahn, Y.S. Major roles of kupffer cells and macrophages in NAFLD development. *Front. Endocrinol.* **2023**, *14*, 1150118. [CrossRef] [PubMed]
30. Li, W.; Yang, Y.; Yang, L.; Chang, N.; Li, L. Monocyte-derived Kupffer cells dominate in the Kupffer cell pool during liver injury. *Cell Rep.* **2023**, *42*, 113164. [CrossRef]
31. Xu, G.X.; Wei, S.; Yu, C.; Zhao, S.Q.; Yang, W.J.; Feng, Y.H.; Pan, C.; Yang, K.X.; Ma, Y. Activation of Kupffer cells in NAFLD and NASH: Mechanisms and therapeutic interventions. *Front. Cell Dev. Biol.* **2023**, *11*, 1199519. [CrossRef]
32. Higashi, T.; Friedman, S.L.; Hoshida, Y. Hepatic stellate cells as key target in liver fibrosis. *Adv. Drug Deliv. Rev.* **2017**, *121*, 27–42. [CrossRef]
33. Smith, A.; Baumgartner, K.; Bositis, C. Cirrhosis: Diagnosis and Management. *Am. Fam. Physician* **2019**, *100*, 759–770.
34. Ross, F.C.; Patangia, D.; Grimaud, G.; Lavelle, A.; Dempsey, E.M.; Ross, R.P.; Stanton, C. The interplay between diet and the gut microbiome: Implications for health and disease. *Nat. Rev. Microbiol.* **2024**, *22*, 671–686. [CrossRef]
35. Rinninella, E.; Tohumcu, E.; Raoul, P.; Fiorani, M.; Cintoni, M.; Mele, M.C.; Cammarota, G.; Gasbarrini, A.; Ianiro, G. The role of diet in shaping human gut microbiota. *Best. Pract. Res. Clin. Gastroenterol.* **2023**, *62–63*, 101828. [CrossRef] [PubMed]
36. Hou, K.; Wu, Z.X.; Chen, X.Y.; Wang, J.Q.; Zhang, D.; Xiao, C.; Zhu, D.; Koya, J.B.; Wei, L.; Li, J.; et al. Microbiota in health and diseases. *Signal Transduct. Target. Ther.* **2022**, *7*, 135. [CrossRef]
37. Zhu, Y.; Yang, H.; Zhang, Y.; Rao, S.; Mo, Y.; Zhang, H.; Liang, S.; Zhang, Z.; Yang, W. Dietary fiber intake and non-alcoholic fatty liver disease: The mediating role of obesity. *Front. Public Health* **2023**, *10*, 1038435. [CrossRef]
38. Chen, X.; Fu, L.; Zhu, Z.; Wang, Y. Exploring the association between dietary fiber intake and hepatic steatosis: Insights from NHANES. *BMC Gastroenterol.* **2024**, *24*, 160. [CrossRef] [PubMed]
39. Xiong, R.G.; Zhou, D.D.; Wu, S.X.; Huang, S.Y.; Saimaiti, A.; Yang, Z.J.; Shang, A.; Zhao, C.N.; Gan, R.Y.; Li, H.B. Health Benefits and Side Effects of Short-Chain Fatty Acids. *Foods* **2022**, *11*, 2863. [CrossRef]
40. Zhao, Y.; Jayachandran, M.; Xu, B. In vivo antioxidant and anti-inflammatory effects of soluble dietary fiber Konjac glucomannan in type-2 diabetic rats. *Int. J. Biol. Macromol.* **2020**, *159*, 1186–1196. [CrossRef]
41. Wang, H.; Huang, X.; Tan, H.; Chen, X.; Chen, C.; Nie, S. Interaction between dietary fiber and bifidobacteria in promoting intestinal health. *Food Chem.* **2022**, *393*, 133407. [CrossRef] [PubMed]
42. Fu, J.; Zheng, Y.; Gao, Y.; Xu, W. Dietary Fiber Intake and Gut Microbiota in Human Health. *Microorganisms* **2022**, *10*, 2507. [CrossRef] [PubMed]
43. Glass, L.M.; Hunt, C.M.; Fuchs, M.; Su, G.L. Comorbidities and Nonalcoholic Fatty Liver Disease: The Chicken, the Egg, or Both? *Fed. Pract.* **2019**, *36*, 64–71.

44. Hansen, C.D.; Gram-Kampmann, E.M.; Hansen, J.K.; Hugger, M.B.; Madsen, B.S.; Jensen, J.M.; Olesen, S.; Torp, N.; Rasmussen, D.N.; Kjærgaard, M.; et al. Effect of Calorie-Unrestricted Low-Carbohydrate, High-Fat Diet Versus High-Carbohydrate, Low-Fat Diet on Type 2 Diabetes and Nonalcoholic Fatty Liver Disease: A Randomized Controlled Trial. *Ann. Intern. Med.* **2023**, *176*, 10–21. [CrossRef]
45. Chen, J.; Huang, Y.; Xie, H.; Bai, H.; Lin, G.; Dong, Y.; Shi, D.; Wang, J.; Zhang, Q.; Zhang, Y.; et al. Impact of a low-carbohydrate and high-fiber diet on nonalcoholic fatty liver disease. *Asia Pac. J. Clin. Nutr.* **2020**, *29*, 483–490. [CrossRef]
46. Cunha, G.M.; Guzman, G.; Correa De Mello, L.L.; Trein, B.; Spina, L.; Bussade, I.; Marques Prata, J.; Sajoux, I.; Countinho, W. Efficacy of a 2-Month Very Low-Calorie Ketogenic Diet (VLCKD) Compared to a Standard Low-Calorie Diet in Reducing Visceral and Liver Fat Accumulation in Patients with Obesity. *Front. Endocrinol.* **2020**, *11*, 607. [CrossRef]
47. Hepburn, C.; von Roenn, N. Nutrition in Liver Disease—A Review. *Curr. Gastroenterol. Rep.* **2023**, *25*, 242–249. [CrossRef] [PubMed]
48. Montemayor, S.; García, S.; Monserrat-Mesquida, M.; Tur, J.A.; Bouzas, C. Dietary Patterns, Foods, and Nutrients to Ameliorate Non-Alcoholic Fatty Liver Disease: A Scoping Review. *Nutrients* **2023**, *15*, 3987. [CrossRef] [PubMed]
49. Montemayor, S.; Mascaró, C.M.; Ugarriza, L.; Casares, M.; Llompart, I.; Abete, I.; Zulet, M.Á.; Martínez, J.A.; Tur, J.A.; Bouzas, C. Adherence to Mediterranean Diet and NAFLD in Patients with Metabolic Syndrome: The FLIPAN Study. *Nutrients* **2022**, *14*, 3186. [CrossRef]
50. Rosi, A.; Teixo, R.; Batista, N.; Calderón-Pérez, L.; Caimari, A.; Scazzina, F. Multicenter Randomized Controlled Trial to Tackle Obesity through a Mediterranean Diet vs. A Low-Fat Diet in Children and Adolescents: Preliminary Results from the MED4YOUTH STUDY. *Proceedings* **2023**, *91*, 126. [CrossRef]
51. Cheng, R.; Wang, L.; Le, S.; Yang, Y.; Zhao, C.; Zhang, X.; Yang, X.; Xu, T.; Xu, L.; Wiklund, P.; et al. A randomized controlled trial for response of microbiome network to exercise and diet intervention in patients with nonalcoholic fatty liver disease. *Nat. Commun.* **2022**, *13*, 2555. [CrossRef] [PubMed]
52. Chooi, Y.C.; Zhang, Q.A.; Magkos, F.; Ng, M.; Michael, N.; Wu, X.; Volchanskaya, V.S.B.; Lai, X.; Wanjaya, E.R.; Elejalde, U.; et al. Effect of an Asian-adapted Mediterranean diet and pentadecanoic acid on fatty liver disease: The TANGO randomized controlled trial. *Am. J. Clin. Nutr.* **2024**, *119*, 788–799. [CrossRef]
53. Gómez-Pérez, A.M.; Ruiz-Limón, P.; Salas-Salvadó, J.; Vioque, J.; Corella, D.; Fitó, M.; Vidal, J.; Atzeni, A.; Torres-Collado, L.; Álvarez-Sala, A.; et al. Gut microbiota in nonalcoholic fatty liver disease: A PREDIMED-Plus trial sub analysis. *Gut Microbes* **2023**, *15*, 2223339. [CrossRef]
54. Bialczyk, A.; Rajewska, A.; Junik, R.; Suwała, S. The Role of Probiotics in Managing Metabolic-Associated Fatty Liver Disease: An Updated Review. *Curr. Res. Nutr. Food Sci.* **2024**, *12*, 490–501. [CrossRef]
55. Santos, A.A.; Duarte, R.; Arella, F.; Margues, V.; Roos, S.; Rodrigues, C.M.P. Impact of Lactobacillaceae supplementation on the multi-organ axis during MASLD. *Life Sci.* **2024**, *354*, 122948. [CrossRef]
56. Mohamad Nor, M.H.; Ayob, N.; Mokhtar, N.M.; Raja Ali, R.A.; Tan, G.C.; Wong, Z.; Shafiee, N.H.; Wong, Y.P.; Mustangin, M.; Nawawi, K.N.M. The Effect of Probiotics (MCP® BCMC® Strains) on Hepatic Steatosis, Small Intestinal Mucosal Immune Function, and Intestinal Barrier in Patients with Non-Alcoholic Fatty Liver Disease. *Nutrients* **2021**, *13*, 3192. [CrossRef] [PubMed]
57. Xue, L.; Deng, Z.; Luo, W.; He, X.; Chen, Y. Effect of Fecal Microbiota Transplantation on Non-Alcoholic Fatty Liver Disease: A Randomized Clinical Trial. *Front. Cell. Infect. Microbiol.* **2022**, *12*, 759306. [CrossRef] [PubMed]
58. Dos Santos, J.R.; de Sousa Soares, L.; Soares, B.M.; de Gomes Farias, M.; de Oliveira, V.A.; de Sousa, N.A.B.; Negreiros, H.A.; da Silva, F.C.C.; Peron, A.P.; Pacheco, A.C.L.; et al. Cytotoxic and mutagenic effects of the food additive tartrazine on eukaryotic cells. *BMC Pharmacol. Toxicol.* **2022**, *23*, 95. [CrossRef]
59. Vander Leek, T. Food additives and reactions: Antioxidants, benzoates, parabens, colorings, flavorings, natural protein-based additives. In *Encyclopedia of Food Allergy*, 1st ed.; Elsevier: Amsterdam, The Netherlands, 2024; pp. 862–881, ISBN 9780323960199. [CrossRef]
60. Amchova, P.; Siska, F.; Ruda-Kucerova, J. Safety of tartrazine in the food industry and potential protective factors. *Heliyon* **2024**, *10*, e38111. [CrossRef]
61. Elder, R.; Vancuren, S.J.; Botschner, A.J.; Josephy, P.D.; Allen-Vercoe, E. Metabolism of azo food dyes by bacterial members of the human gut microbiome. *Anaerobe* **2023**, *83*, 102783. [CrossRef] [PubMed]
62. Pay, R.; Sharrock, A.V.; Elder, R.; Maré, A.; Bracegirdle, J.; Torres, D.; Malone, N.; Vorster, J.; Kelly, L.; Ryan, A.; et al. Preparation, analysis and toxicity characterisation of the redox metabolites of the azo food dye tartrazine. *Food Chem. Toxicol.* **2023**, *182*, 114193. [CrossRef]
63. Lehmkuhler, A.L.; Miller, M.D.; Bradman, A.; Castorina, R.; Chen, M.A.; Xie, T.; Mitchell, A.E. Dataset of FD&C Certified Food Dyes in Foods Commonly Consumed by Children. *Data Brief.* **2022**, *46*, 108806. [CrossRef] [PubMed]
64. de Oliveira, Z.B.; Silva da Costa, D.V.; da Silva Dos Santos, A.C.; da Silva Júnior, A.Q.; de Lima Silva, A.; de Santana, R.C.F.; Costa, I.C.G.; de Sous Ramos, S.F.; Padilla, G.; da Silva, S.K.R. Synthetic Colors in Food: A Warning for Children's Health. *Int. J. Environ. Res. Public Health* **2024**, *21*, 682. [CrossRef]

65. Wu, L.; Xu, Y.; Lv, X.; Chang, X.; Ma, X.; Tian, X.; Shi, X.; Li, X.; Kong, X. Impacts of an azo food dye tartrazine uptake on intestinal barrier, oxidative stress, inflammatory response, and intestinal microbiome in crucian carp (*Carassius auratus*). *Ecotoxicol. Environ. Saf.* **2021**, *223*, 112551. [CrossRef] [PubMed]
66. Wu, L.; Lv, X.; Zhang, Y.; Xin, Q.; Zou, Y.; Li, X. Tartrazine exposure results in histological damage, oxidative stress, immune disorders, and gut microbiota dysbiosis in juvenile crucian carp (*Carassius carassius*). *Aquat. Toxicol.* **2021**, *241*, 105998. [CrossRef]
67. El-Desoky, G.E.; Wabaidur, S.M.; AlOthman, Z.A.; Habila, M.A. Evaluation of Nano-curcumin effects against Tartrazine-induced abnormalities in liver and kidney histology and other biochemical parameters. *Food Sci. Nutr.* **2022**, *10*, 1344–1356. [CrossRef] [PubMed]
68. Hofseth, L.J.; Hebert, J.R.; Murphy, E.A.; Trauner, E.; Vikas, A.; Harris, Q.; Chumanevich, A.A. Allura Red AC is a xenobiotic. Is it also a carcinogen? *Carcinogenesis* **2024**, *45*, 711–720. [CrossRef]
69. Kwon, Y.H.; Banskota, S.; Wang, H.; Rossi, L.; Grondin, J.A.; Syed, S.A.; Yousefi, Y.; Schertzer, J.D.; Morrison, K.M.; Wade, M.G.; et al. Author Correction: Chronic exposure to synthetic food colorant Allura Red AC promotes susceptibility to experimental colitis via intestinal serotonin in mice. *Nat. Commun.* **2023**, *14*, 3125. [CrossRef]
70. He, Z.; Chen, L.; Catalan-Dibene, J.; Bongers, G.; Faith, J.J.; Suebsuwong, C.; DeVita, R.J.; Shen, Z.; Fox, J.G.; Lafaille, J.J.; et al. Food colorants metabolized by commensal bacteria promote colitis in mice with dysregulated expression of interleukin-23. *Cell Metab.* **2021**, *33*, 1358–1371.e5. [CrossRef]
71. Khayyat, L.I.; Essawy, A.E.; Sorour, J.M.; Soffar, A. Sunset Yellow and Allura Red modulate Bcl2 and COX2 expression levels and confer oxidative stress-mediated renal and hepatic toxicity in male rats. *PeerJ* **2018**, *6*, e5689. [CrossRef] [PubMed]
72. Şensoy, E. Comparison of the effect of Sunset Yellow on the stomach and small intestine of developmental period of mice. *Heliyon* **2024**, *10*, e31998. [CrossRef]
73. Wu, J.; Lee, H. Determination of Sunset Yellow and Tartrazine in drinks using screen-printed carbon electrodes modified with reduced graphene oxide and NiBTC frameworks. *Microchem. J.* **2020**, *158*, 105133. [CrossRef]
74. SUNSET YELLOW FCF. Available online: https://apps.who.int/food-additives-contaminants-jecfa-database/Home/Chemical/2703 (accessed on 17 December 2024).
75. Zahran, S.A.; Mansour, S.M.; Ali, A.E.; Kamal, S.M.; Römling, U.; El-Abhar, H.S.; Ali-Tammam, M. Sunset Yellow dye effects on gut microbiota, intestinal integrity, and the induction of inflammasomopathy with pyroptotic signaling in male Wistar rats. *Food Chem. Toxicol.* **2024**, *187*, 114585. [CrossRef] [PubMed]
76. Abdelhamid, W.; Abdel Wahab, M.; Moussa, M.; Elkhateb, L.; Sadek, D. A Comparative Study of the Toxic Effects of Monosodium Glutamate and Sunset Yellow on the Structure and Function of the Liver, Kidney, and Testis and the Possible Protective Role of Curcumin in Rats. *Egypt. J. Histol.* **2023**, *46*, 2094–2114. [CrossRef]
77. Hussein, M.M.A.; Arisha, A.H.; Tayel, E.M.; Abdo, S.A. Effect of long-term oral exposure to carmoisine or Sunset Yellow on different hematological parameters and hepatic apoptotic pathways in mice. *J. Anim. Health Prod.* **2021**, *9*, 80–86. [CrossRef]
78. Dey, N.B.; Nagababu, B.H. Applications of food color and bio-preservatives in the food and its effect on the human health. *Food Chem. Adv.* **2022**, *1*, 100019. [CrossRef]
79. Wang, H.; Bai, J.; Miao, P.; Wei, Y.; Chen, X.; Lan, H.; Qing, Y.; Zhao, M.; Li, Y.; Tang, R.; et al. The key to intestinal health: A review and perspective on food additives. *Front. Nutr.* **2024**, *11*, 1420358. [CrossRef]
80. Irwin, S.V.; Fisher, P.; Graham, E.; Malek, A.; Robidoux, A. Sulfites inhibit the growth of four species of beneficial gut bacteria at concentrations regarded as safe for food. *PLoS ONE* **2017**, *12*, e0186629. [CrossRef] [PubMed]
81. Irwin, S.V.; Deardorff, L.M.; Deng, Y.; Fisher, P.; Gould, M.; June, J.; Kent, R.S.; Qin, Y.; Yadao, F. Sulfite preservatives effects on the mouth microbiome: Changes in viability, diversity and composition of microbiota. *PLoS ONE* **2022**, *17*, e0265249. [CrossRef]
82. Nagpal, R.; Indugu, N.; Singh, P. Distinct Gut Microbiota Signatures in Mice Treated with Commonly Used Food Preservatives. *Microorganisms* **2021**, *9*, 2311. [CrossRef]
83. Hrncir, T.; Trckova, E.; Hrncirova, L. Synergistic Effects of Fructose and Food Preservatives on Metabolic Dysfunction-Associated Steatotic Liver Disease (MASLD): From Gut Microbiome Alterations to Hepatic Gene Expression. *Nutrients* **2024**, *16*, 3722. [CrossRef]
84. Crowe, W.; Pan, X.; Mackle, J.; Harris, A.; Hardiman, G.; Elliott, C.T.; Green, B.D. Dietary inclusion of nitrite-containing frankfurter exacerbates colorectal cancer pathology and alters metabolism in APCmin mice. *npj Sci. Food* **2022**, *6*, 60. [CrossRef] [PubMed]
85. Van Hecke, T.; Vossen, E.; Goethals, S.; Boon, N.; De Vrieze, J.; De Smet, S. In vitro and in vivo digestion of red cured cooked meat: Oxidation, intestinal microbiota and fecal metabolites. *Food Res. Int.* **2021**, *142*, 110203. [CrossRef]
86. Wu, L.; Zhang, C.; Long, Y.; Chen, Q.; Zhang, W.; Liu, G. Food additives: From functions to analytical methods. *Crit. Rev. Food Sci. Nutr.* **2022**, *62*, 8497–8517. [CrossRef]
87. Borsani, B.; De Santis, R.; Perico, V.; Penagini, F.; Pendezza, E.; Dilillo, D.; Bosetti, A.; Zuccotti, G.V.; D'Auria, E. The Role of Carrageenan in Inflammatory Bowel Diseases and Allergic Reactions: Where Do We Stand? *Nutrients* **2021**, *13*, 3402. [CrossRef]
88. Ariffin, S.H.; Yeen, W.W.; Abidin, I.Z.; Abdul Wahab, R.M.; Ariffin, Z.Z.; Senafi, S. Cytotoxicity effect of degraded and undegraded kappa and iota carrageenan in human intestine and liver cell lines. *BMC Complement. Altern. Med.* **2014**, *14*, 508. [CrossRef]

89. Liu, F.; Hou, P.; Zhang, H.; Tang, Q.; Xue, C.; Li, R.W. Food-grade carrageenans and their implications in health and disease. *Compr. Rev. Food Sci. Food Saf.* **2021**, *20*, 3918–3936. [CrossRef]
90. Naimi, S.; Viennois, E.; Gewirtz, A.T.; Chassaing, B. Direct impact of commonly used dietary emulsifiers on human gut microbiota. *Microbiome* **2021**, *9*, 66. [CrossRef]
91. Martino, J.V.; Van Limbergen, J.; Cahill, L.E. The Role of Carrageenan and Carboxymethylcellulose in the Development of Intestinal Inflammation. *Front. Pediatr.* **2017**, *5*, 96. [CrossRef]
92. Wang, R.; Tang, R.; Li, B.; Ma, X.; Schnabl, B.; Tilg, H. Gut microbiome, liver immunology, and liver diseases. *Cell. Mol. Immunol.* **2021**, *18*, 4–17. [CrossRef]
93. Tilg, H.; Adolph, T.E.; Trauner, M. Gut-liver axis: Pathophysiological concepts and clinical implications. *Cell Metab.* **2022**, *34*, 1700–1718. [CrossRef]
94. Ogulur, I.; Yazici, D.; Pat, Y.; Bingöl, E.N.; Babayev, H.; Ardicli, S.; Heider, A.; Rückert, B.; Sampath, V.; Dhir, R.; et al. Mechanisms of gut epithelial barrier impairment caused by food emulsifiers polysorbate 20 and polysorbate 80. *Allergy* **2023**, *78*, 2441–2455. [CrossRef] [PubMed]
95. Delaroque, C.; Chassaing, B. Dietary emulsifier consumption accelerates type 1 diabetes development in NOD mice. *npj Biofilms Microbiomes* **2024**, *10*, 1. [CrossRef]
96. Sellem, L.; Srour, B.; Javaux, G.; Chazelas, E.; Chassaing, B.; Viennois, E.; Debras, C.; Salamé, C.; Druesne-Pecollo, N.; Esseddik, Y.; et al. Food additive emulsifiers and risk of cardiovascular disease in the NutriNet-Santé cohort: Prospective cohort study. *BMJ* **2023**, *382*, e076058. [CrossRef]
97. Bancil, A.S.; Sandall, A.M.; Rossi, M.; Chassaing, B.; Lindsay, J.O.; Whelan, K. Food Additive Emulsifiers and Their Impact on Gut Microbiome, Permeability, and Inflammation: Mechanistic Insights in Inflammatory Bowel Disease. *J. Crohns Colitis* **2021**, *15*, 1068–1079. [CrossRef]
98. Warner, J.O. Artificial food additives: Hazardous to long-term health? *Arch. Dis. Child.* **2024**, *109*, 882–885. [CrossRef] [PubMed]
99. Vilas-Boas, V.; Gijbels, E.; Jonckheer, J.; De Waele, E.; Vinken, M. Cholestatic liver injury induced by food additives, dietary supplements and parenteral nutrition. *Environ. Int.* **2020**, *136*, 105422. [CrossRef]
100. Lv, W.; Song, J.; Nowshin, R.R.; Sun, J.; Shi, G.; Wu, H.; Xiao, J.; Xu, D. Effects of food emulsifiers on high fat-diet-induced obesity, intestinal inflammation, changes in bile acid profile, and liver dysfunction. *Food Res. Int.* **2023**, *173*, 113302. [CrossRef] [PubMed]
101. Singh, R.K.; Wheildon, N.; Ishikawa, S. Food Additive P-80 Impacts Mouse Gut Microbiota Promoting Intestinal Inflammation, Obesity and Liver Dysfunction. *SOJ Microbiol. Infect. Dis.* **2016**, *4*, 1–10. [CrossRef]
102. Henao-Mejia, J.; Elinav, E.; Jin, C.; Hao, L.; Mehal, W.Z.; Strowig, T.; Thaiss, C.A.; Kau, A.L.; Eisenbarth, S.C.; Jurczak, M.J.; et al. Inflammasome-mediated dysbiosis regulates progression of NAFLD and obesity. *Nature* **2012**, *482*, 179–185. [CrossRef]
103. Nickerson, K.P.; Chanin, R.; McDonald, C. Deregulation of intestinal anti-microbial defense by the dietary additive, maltodextrin. *Gut Microbes* **2015**, *6*, 78–83. [CrossRef] [PubMed]
104. Arnold, A.R.; Chassaing, B. Maltodextrin, Modern Stressor of the Intestinal Environment. *Cell. Mol. Gastroenterol. Hepatol.* **2019**, *7*, 475–476. [CrossRef] [PubMed]
105. Singh, P.; Sanchez-Fernandez, L.L.; Ramiro-Cortijo, D.; Ochoa-Allemant, P.; Perides, G.; Liu, Y.; Medina-Morales, E.; Yakah, W.; Freedman, S.D.; Martin, C.R. Maltodextrin-induced intestinal injury in a neonatal mouse model. *Dis. Model. Mech.* **2020**, *13*, dmm044776. [CrossRef]
106. Zangara, M.T.; Ponti, A.K.; Miller, N.D.; Engelhart, M.J.; Ahern, P.P.; Sangwan, N.; McDonald, C. Maltodextrin Consumption Impairs the Intestinal Mucus Barrier and Accelerates Colitis Through Direct Actions on the Epithelium. *Front. Immunol.* **2022**, *13*, 841188. [CrossRef]
107. Almutairi, R.; Basson, A.R.; Wearsch, P.A.; Cominelli, F.; Rodriguez-Palacios, A. Correction to: Validity of food additive maltodextrin as placebo and effects on human gut physiology: Systematic review of placebo-controlled clinical trials. *Eur. J. Nutr.* **2023**, *62*, 2345. [CrossRef]
108. Vasilaki, A.; Panagiotopoulou, E.; Koupantsis, T.; Katsanidis, E.; Mourtzinos, I. Recent insights in flavor-enhancers: Definition, mechanism of action, taste-enhancing ingredients, analytical techniques and the potential of utilization. *Crit. Rev. Food Sci. Nutr.* **2022**, *62*, 9036–9052. [CrossRef]
109. Rocha, R.A.R.; Ribeiro, M.N.; Silva, G.A.; Rocha, L.C.R.; Pinheiro, A.C.M.; Nunes, C.A.; Carneiro, J.D.S. Temporal profile of flavor enhancers MAG, MSG, GMP, and IMP, and their ability to enhance salty taste, in different reductions of sodium chloride. *J. Food Sci.* **2020**, *85*, 1565–1575. [CrossRef]
110. Ahangari, H.; Bahramian, B.; Khezerlou, A.; Tavassoli, M.; Kiani-Salmi, N.; Tarhriz, V.; Ehsani, A. Association between monosodium glutamate consumption with changes in gut microbiota and related metabolic dysbiosis—A systematic review. *Food Sci. Nutr.* **2024**, *12*, 5285–5295. [CrossRef] [PubMed]
111. Nahok, K.; Phetcharaburanin, J.; Li, J.V.; Silsirivanit, A.; Thanan, R.; Boonnate, P.; Joonhuathon, J.; Sharma, A.; Anutrakulchai, S.; Selmi, C.; et al. Monosodium Glutamate Induces Changes in Hepatic and Renal Metabolic Profiles and Gut Microbiome of Wistar Rats. *Nutrients* **2021**, *13*, 1865. [CrossRef] [PubMed]

112. Kyaw, T.S.; Sukmak, M.; Nahok, K.; Sharma, A.; Silsirivanit, A.; Lert-Itthiporn, W.; Sansurin, N.; Senthong, V.; Anutrakulchai, S.; Sangkhamanon, S.; et al. Monosodium glutamate consumption reduces the renal excretion of trimethylamine N-oxide and the abundance of *Akkermansia muciniphila* in the gut. *Biochem. Biophys. Res. Commun.* **2022**, *630*, 158–166. [CrossRef]
113. Coelho, C.F.F.; França, L.M.; Nascimento, J.R.; Dos Santos, A.M.; Azevedo-Santos, A.P.S.; Nascimento, F.R.F.; Paes, A.M.A. Early onset and progression of non-alcoholic fatty liver disease in young monosodium l-glutamate-induced obese mice. *J. Dev. Orig. Health Dis.* **2019**, *10*, 188–195. [CrossRef]
114. Ugur Calis, I.; Turgut Cosan, D.; Saydam, F.; Kerem Kolac, U.; Soyocak, A.; Kurt, H.; Veysi Gunes, H.; Sahinturk, V.; Sahin Mutlu, F.; Ozdemir Koroglu, Z.; et al. The Effects of Monosodium Glutamate and Tannic Acid on Adult Rats. *Iran. Red Crescent Med. J.* **2016**, *18*, e37912. [CrossRef]
115. Henry-Unaeze, H.N. Update on food safety of monosodium l-glutamate (MSG). *Pathophysiology* **2017**, *24*, 243–249. [CrossRef]
116. Olowofolahan, A.O.; Adeosun, O.A.; Olorunsogo, O.O. Monosodium Glutamate Induces Cytotoxicity in Rat Liver via Mitochondrial Permeability Transition Pore Opening. *Cell Biochem. Biophys.* **2020**, *78*, 429–437. [CrossRef]
117. Banerjee, A.; Mukherjee, S.; Maji, B.K. Worldwide flavor enhancer monosodium glutamate combined with high lipid diet provokes metabolic alterations and systemic anomalies: An overview. *Toxicol. Rep.* **2021**, *8*, 938–961. [CrossRef]
118. Aguayo-Guerrero, J.A.; Méndez-García, L.A.; Solleiro-Villavicencio, H.; Viurcos-Sanabria, R.; Escobedo, G. Sucralose: From Sweet Success to Metabolic Controversies-Unraveling the Global Health Implications of a Pervasive Non-Caloric Artificial Sweetener. *Life* **2024**, *14*, 323. [CrossRef]
119. Del Pozo, S.; Gómez-Martínez, S.; Díaz, L.E.; Nova, E.; Urrialde, R.; Marcos, A. Potential Effects of Sucralose and Saccharin on Gut Microbiota: A Review. *Nutrients* **2022**, *14*, 1682. [CrossRef]
120. Magnuson, B.A.; Roberts, A.; Nestmann, E.R. Critical review of the current literature on the safety of sucralose. *Food Chem. Toxicol.* **2017**, *106 Pt A*, 324–355. [CrossRef]
121. Sucralose. Available online: https://www.fao.org/4/y0474s/y0474s6j.htm (accessed on 27 December 2024).
122. Al-Domi, H.; Cummings, J.H.; Elmadfa, I.; Hooper, L.; Kumanyika, S.; L'Abbé, M.; Laneroll, eP.; Li, D.; Mann, J.; Meerpohl, J. *Use of Non-Sugar Sweeteners: WHO Guideline*; World Health Organization: Geneva, Switzerland, 2023; Recommendation and Supporting Information. Available online: https://www.ncbi.nlm.nih.gov/books/NBK592246 (accessed on 23 December 2024).
123. Feng, J.; Peng, J.; Hsiao, Y.C.; Liu, C.W.; Yang, Y.; Zhao, H.; Teitelbaum, T.; Wang, X.; Lu, K. Non/Low-Caloric Artificial Sweeteners and Gut Microbiome: From Perturbed Species to Mechanisms. *Metabolites* **2024**, *14*, 544. [CrossRef]
124. Schiffman, S.S.; Scholl, E.H.; Furey, T.S.; Nagle, H.T. Toxicological and pharmacokinetic properties of sucralose-6-acetate and its parent sucralose: In vitro screening assays. *J. Toxicol. Environ. Health B Crit. Rev.* **2023**, *26*, 307–341. [CrossRef]
125. Bian, X.; Chi, L.; Gao, B.; Tu, P.; Ru, H.; Lu, K. Gut Microbiome Response to Sucralose and Its Potential Role in Inducing Liver Inflammation in Mice. *Front. Physiol.* **2017**, *8*, 487. [CrossRef] [PubMed]
126. Shi, Z.; Chen, G.; Cao, Z.; Wu, F.; Lei, H.; Chen, C.; Song, Y.; Liu, C.; Li, J.; Zhou, J.; et al. Gut Microbiota and Its Metabolite Deoxycholic Acid Contribute to Sucralose Consumption-Induced Nonalcoholic Fatty Liver Disease. *J. Agric. Food Chem.* **2021**, *69*, 3982–3991. [CrossRef] [PubMed]
127. Chi, L.; Yang, Y.; Bian, X.; Gao, B.; Tu, P.; Ru, H.; Lu, K. Chronic sucralose consumption inhibits farnesoid X receptor signaling and perturbs lipid and cholesterol homeostasis in the mouse livers, potentially by altering gut microbiota functions. *Sci. Total Environ.* **2024**, *919*, 169603. [CrossRef]
128. Suez, J.; Cohen, Y.; Valdés-Mas, R.; Mor, U.; Dori-Bachash, M.; Federici, S.; Zmora, N.; Leshem, A.; Heinemann, M.; Linevsky, R.; et al. Personalized microbiome-driven effects of non-nutritive sweeteners on human glucose tolerance. *Cell* **2022**, *185*, 3307–3328.e19. [CrossRef] [PubMed]
129. Orku, S.E.; Suyen, G.; Bas, M. The effect of regular consumption of four low- or no-calorie sweeteners on glycemic response in healthy women: A randomized controlled trial. *Nutrition* **2023**, *106*, 111885. [CrossRef]
130. Ahmad, S.Y.; Friel, J.; Mackay, D. The Effects of Non-Nutritive Artificial Sweeteners, Aspartame and Sucralose, on the Gut Microbiome in Healthy Adults: Secondary Outcomes of a Randomized Double-Blinded Crossover Clinical Trial. *Nutrients* **2020**, *12*, 3408. [CrossRef] [PubMed]
131. Ruiz-Ojeda, F.J.; Plaza-Díaz, J.; Sáez-Lara, M.J.; Gil, A. Effects of Sweeteners on the Gut Microbiota: A Review of Experimental Studies and Clinical Trials. *Adv. Nutr.* **2019**, *10* (Suppl. S1), S31–S48. [CrossRef]
132. Aspartame Hazard and Risk Assessment Results Released. Available online: https://www.who.int/news/item/14-07-2023-aspartame-hazard-and-risk-assessment-results-released (accessed on 17 December 2024).
133. Czarnecka, K.; Pilarz, A.; Rogut, A.; Maj, P.; Szymańska, J.; Olejnik, Ł.; Szymański, P. Aspartame-True or False? Narrative Review of Safety Analysis of General Use in Products. *Nutrients* **2021**, *13*, 1957. [CrossRef]
134. Finamor, I.A.; Bressan, C.A.; Torres-Cuevas, I.; Rius-Pérez, S.; da Veiga, M.; Rocha, M.I.; Pavanato, M.A.; Pérez, S. Long-Term Aspartame Administration Leads to Fibrosis, Inflammasome Activation, and Gluconeogenesis Impairment in the Liver of Mice. *Biology* **2021**, *10*, 82. [CrossRef]
135. Sergi, C.M. MASLD and aspartame: Are new studies in the horizon? *Front. Med.* **2023**, *10*, 1266918. [CrossRef]

136. Palmnäs, M.S.; Cowan, T.E.; Bomhof, M.R.; Su, J.; Reimer, R.A.; Vogel, H.J.; Hittel, D.S.; Shearer, J. Low-dose aspartame consumption differentially affects gut microbiota-host metabolic interactions in the diet-induced obese rat. *PLoS ONE* **2014**, *9*, e109841. [CrossRef]
137. Finamor, I.; Pérez, S.; Bressan, C.A.; Brenner, C.E.; Rius-Pérez, S.; Brittes, P.C.; Cheiran, G.; Rocha, M.I.; da Veiga, M.; Sastre, J.; et al. Chronic aspartame intake causes changes in the trans-sulphuration pathway, glutathione depletion and liver damage in mice. *Redox Biol.* **2017**, *11*, 701–707. [CrossRef]
138. Tey, S.L.; Salleh, N.B.; Henry, J.; Forde, C.G. Effects of aspartame-, monk fruit-, stevia- and sucrose-sweetened beverages on postprandial glucose, insulin and energy intake. *Int. J. Obes.* **2017**, *41*, 450–457. [CrossRef] [PubMed]
139. Saccharin. Available online: https://www.fao.org/4/y0474s/y0474s5t.htm#TopOfPage (accessed on 17 December 2024).
140. Yin, X.; Shi, Y.; Sheng, T.; Ji, C. Early-Life Gut Microbiota: A Possible Link Between Maternal Exposure to Non-Nutritive Sweeteners and Metabolic Syndrome in Offspring. *Nutr. Rev.* **2024**, *30*, nuae140. [CrossRef]
141. Bian, X.; Tu, P.; Chi, L.; Gao, B.; Ru, H.; Lu, K. Saccharin induced liver inflammation in mice by altering the gut microbiota and its metabolic functions. *Food Chem. Toxicol.* **2017**, *107 Pt B*, 530–539. [CrossRef]
142. Serrano, J.; Smith, K.R.; Crouch, A.L.; Sharma, V.; Yi, F.; Vargova, V.; LaMoia, T.E.; Dupont, L.M.; Serna, V.; Tang, F.; et al. High-dose saccharin supplementation does not induce gut microbiota changes or glucose intolerance in healthy humans and mice. *Microbiome* **2021**, *9*, 11. [CrossRef] [PubMed]
143. EFSA. Available online: https://www.efsa.europa.eu/en (accessed on 16 January 2025).

medicina

MDPI

Review

Lynch Syndrome—Impact of the Type of Deficient Mismatch Repair Gene Mutation on Diagnosis, Clinical Presentation, Surveillance and Therapeutic Approaches

Tudor Razvan Grigorie [1,2], Gheorghe Potlog [3] and Sorin Tiberiu Alexandrescu [1,2,*]

1 Department of Surgery, Faculty of Medicine, Carol Davila University of Medicine and Pharmacy, 050474 Bucharest, Romania; mihai.grig@gmail.com
2 Department of Hepato-Bilio-Pancreatic Surgery, Emergency University Hospital Bucharest, Splaiul Independentei 169, Sector 5, 050098 Bucharest, Romania
3 Center for Digestive Diseases and Liver Transplantation, Fundeni Clinical Institute, 022328 Bucharest, Romania; g.potlog@yahoo.com
* Correspondence: sorin.alexandrescu@umfcd.ro

Abstract: In today's world, with its continuing advancements in genetics, the identification of Lynch syndrome (LS) increasingly relies on sophisticated genetic testing techniques. Most guidelines recommend a tailored surveillance program, as well as personalized prophylactic and therapeutic approaches, according to the type of dMMR gene mutation. Carriers of path_MLH1 and path_MSH2 genes have a higher risk of developing colorectal cancer (CRC), despite intensive colonoscopic surveillance. Conversely, carriers of path_MSH6 and path_PMS2 genes have a lower risk of developing CRC, which may be due to their lower penetrance and later age of onset. Thus, carriers of path_MLH1 or path_MSH2 would theoretically derive greater benefits from total colectomy, compared to low-risk carriers (path_MSH6 and path_PMS2), in which colonoscopic surveillance might achieve an efficient prophylaxis. Furthermore, regarding the risk of endometrial/ovarian cancer development, there is a global agreement to offer both hysterectomy and bilateral salpingo-oophorectomy to path_MLH1, path_MSH2 and path_MSH6 carriers after the age of 40. In patients with CRC, preoperative knowledge of the diagnosis of LS is of tremendous importance, due to the high risk of metachronous CRC. However, this risk depends on the type of dMMR gene mutation. For carriers of the high-risk variants (MLH1, MSH2 and EPCAM) who have already developed colon cancer, it is strongly recommended a subtotal or total colectomy is performed, while partial colectomy followed by endoscopic surveillance is an appropriate management approach to treat colon cancer in carriers of the low-risk variants (MSH6 and PMS2). On the other hand, extended surgery for index rectal cancer (such as total proctocolectomy) is less effective than extended surgery for index colon cancer from the point of view of metachronous CRC risk reduction, and is associated with a decreased quality of life.

Keywords: Lynch syndrome; deficient mismatch repair gene; extended colectomy; surveillance; prophylactic colectomy; prophylactic total hysterectomy; bilateral salpingo-oophorectomy

1. Introduction

The tremendous developments in understanding the molecular basis of cancers over the last decade allow for a refined prognostic estimation and personalized therapeutic approach in most oncologic patients [1–4]. Lynch syndrome (LS), also known as hereditary non-polyposis colorectal cancer (HNPCC), is an inherited genetic disorder that significantly increases an individual's risk of developing various types of cancer, particularly colorectal cancer (CRC) and endometrial cancer (EC) [5]. It is caused by inherited mutations in one of the mismatch repair (MMR) genes (MLH1, MSH2, MSH6, or PMS2) or in the epithelial cell adhesion molecule (EPCAM), which leads to the epigenetic silencing of MSH2. The condition follows an autosomal dominant pattern of inheritance. First-degree relatives (parents, siblings and children) have a 50% chance of being affected by LS.

Individuals with LS have an increased lifetime risk of developing CRC (up to 80%) and EC (up to 60%), as well as other cancers, including ovarian (up to 15%), gastric (up to 18%), urinary tract (up to 20%), pancreatic (4%), small intestine cancers, glioblastoma (Turcot syndrome) and sebaceous neoplasms (Muir–Torre syndrome) [6–12].

LS is the most common hereditary form of CRC. It accounts for 2–4% of all CRC diagnoses [7,13]. In patients with LS, CRCs have an adenoma-carcinoma progression ratio of almost 1:1, with an estimated adenoma-to-cancer transformation time of 1–3 years. This contrasts with sporadic cases, which have a ratio of 30:1 and an estimated transformation time of 8–17 years. If left untreated, the majority of polyps will become malignant, with about 70% of patients developing cancer by age 70 and 80% by age 85. Additionally, there is a higher incidence of metachronous and synchronous colon cancers, with a second primary CRC occurring in up to 30% of patients within 10 years and 50% within 15 years [14].

LS carriers/patients have distinct clinic, evolutive and prognostic features, depending on the type of deficient MMR (dMMR) gene. In this narrative review, we present the clinical impact of each of these genes' mutations, as well as the personalized therapeutic approach according to the type of genetic mutation that led to the development of LS. We also present the available data on the usefulness of screening and surveillance programs for patients with LS. Finally, we discuss prophylactic approaches that should be employed in case of each gene's mutations.

2. The Clinical Impact of Distinct Genetic Mutations

LS is caused by germline mutations in DNA mismatch repair genes, including MLH1, MSH2, MSH6 and PMS2 [15,16]. Mutations in these genes lead to microsatellite instability, a hallmark of Lynch syndrome-associated tumors [17].

MLH1 and MSH2 are the most frequently mutated genes in patients with LS, accounting for approximately 70% of the identified mutations (32% in MLH1 and 38% in MSH2) [13,18]. The carriers of pathogenic variants in MLH1 and MSH2 genes have a significantly higher risk of developing CRC at a younger age compared to carriers of pathogenic variants in MSH6 or PMS2 genes [19]. Individuals with mutations in the MSH2 gene have a higher likelihood of developing extracolonic cancers and a lower frequency of CRC compared to those with mutations in the MLH1 gene [20,21]. MSH6 mutations are more commonly associated with gastrointestinal and endometrial cancers that typically occur at a later age [22,23].

Some studies have shown that constitutional 3′ deletions of EPCAM can lead to LS by causing epigenetic silencing of MSH2 in EPCAM-expressing tissues, which results in a tissue-specific deficiency of MSH2. Kempers et al. conducted a cohort study comparing 194 patients with an EPCAM deletion to 473 patients with mutations in MLH1, MSH2, MSH6, or a combined EPCAM-MSH2 deletion. The study found that carriers of an EPCAM deletion had a 75% cumulative risk of developing CRC before the age of 70, similar to that of carriers of combined EPCAM-MSH2 deletions or MSH2 mutations, but higher than that observed in MSH6 mutation carriers. However, only those with deletions extending near the MSH2 promoter showed an increased risk of endometrial cancer [24].

A change in any of the above-mentioned MMR genes can lead to the accumulation of numerous errors in the DNA repetitive sequences known as microsatellites, which occur throughout the genome. This process is known as microsatellite instability (MSI) and is present in LS, but not exclusive to it. Therefore, not all the patients with MSI have LS. To enhance the detection of individuals with LS, "universal tumor screening" is recommended. In this approach, all individuals newly diagnosed with CRC undergo either tumor-based dMMR genetic testing or immunohistochemistry (IHC) testing to check for the absence of DNA mismatch repair (MMR) proteins (MLH1, MSH2, MSH6 or PMS2). The latter method achieves a sensitivity of 100% (95% CI, 99.3–100%) and a specificity of 93.0% (95% CI, 92.0–93.7%) for identifying individuals with Lynch syndrome [25–28].

Traditionally, the diagnosis of LS typically began with clinical suspicion, particularly in individuals with a family history of CRC or other Lynch syndrome-associated cancers. The Amsterdam II criteria and the revised Bethesda guidelines were commonly used to identify individuals who may benefit from further genetic evaluation [29–31]. The shortcomings of using this strategy were, on one hand, that 50% of patients who met these criteria do not actually have LS and, on the other hand, that these criteria were missing in 50% of LS patients. For these reasons, testing for the MMR status of the tumors is nowadays recommended. If IHC staining for MLH1 (alone or with PMS2) is abnormal, testing for the BRAF mutation or MLH1-promoter methylation should be performed to detect tumors lacking DNA-MMR.

Somatic BRAF mutations occur in a small fraction of CRCs overall [32], but are present in 69% to 78% of CRCs with MLH1 promoter methylation. These mutations are virtually never seen in Lynch syndrome-associated cancers, making the presence of a BRAF mutation highly predictive of a sporadic origin and a high negative predictive value for LS [33,34]. If the test is negative, germline mutation testing for LS should be conducted. A multigene panel test is available, particularly for individuals diagnosed before the age of 50 [35,36].

In the absence of available tumor data or known mutations, online tools such as PREMM5 and MMR Predict help in estimating an individual's risk of carrying an MMR mutation [6,37–39]. Given the complexities involved in selecting and interpreting the tests, as well as the potential implications of the results for the family, genetic counseling should precede and also succeed germline mutation testing.

3. Screening and Surveillance in Lynch Syndrome Patients

In LS patients, screening for CRC by colonoscopic surveillance has been generally accepted as a method for providing greater life expectancy. But the benefits offered by screening methods and surveillance are debatable for other cancers that put LS patients at risk.

Screening for CRC by colonoscopy is recommended for people at risk of (first-degree relatives who have not had genetic testing of known MMR gene mutation carriers) or with LS. Colonoscopies should be performed every 1 to 2 years, starting at age 25 or 5 years before the youngest case in family. Recent European guidelines from the EHTG and ESCP, based on PLSD studies (the prospective LS database), recommend a tailored surveillance program according to the type of dMMR gene mutation. Thus, for MLH1, MSH2 and MSH6, colonoscopic surveillance is recommended every 2 to 3 years, and for PMS2, surveillance every 5 years may be considered [5]. The fecal immunochemical test (FIT) is extensively utilized as a screening tool for CRC in the general population; however, its role in LS surveillance remains under investigation and is not yet well established. Recent studies have demonstrated that the FIT has low sensitivity (23%) for detecting adenomas. Although sensitivity for advanced adenomas reached 66.7%, the overall detection rates for adenomas are insufficient to replace colonoscopy as the primary surveillance method [40,41]. Nevertheless, the FIT may hold potential as an augmentative tool to complement colonoscopy in specific scenarios, warranting further investigation into its supplementary role and integration into LS surveillance strategies.

Although endometrial and ovarian cancer screening does not have proven benefits in women with LS according to some studies [38], more recent data suggest that yearly gynecological examination, pelvic ultrasound, CA125 and endometrial biopsy from age 30 to 35 may be useful [39,42–46]. In regions with a high incidence of gastric cancer and in families with a history of gastric neoplasms, upper endoscopy surveillance may be recommended every 2–4 years, with gastric biopsying of the antrum starting at the age of 30–40 years [47–49]. For the urinary tract, no consensus currently exists regarding the proper screening protocol, and great variability still exists regarding the methods and the starting age of screening, ranging from 25 to 50 years [50]. Annual magnetic resonance imaging (MRI) and/or endoscopic ultrasound (EUS) surveillance may be considered for individuals with LS who have one first-degree relative affected by pancreatic cancer, although additional supporting evidence is needed to back up this recommendation [51]. Routine screening for prostate and breast cancer is not recommended beyond what is advised for the general population. A skin exam every 1 to 2 years with a healthcare provider experienced in recognizing Lynch syndrome-associated skin manifestations is recommended. The optimal age to begin surveillance is uncertain and can be individualized based on personal and family history [39].

Recent observations suggest that future knowledge about the changes of gut microbiota in LS may be a useful tool for the surveillance of these patients. Research over the last few years suggests that the gut microbiota may have a different pattern in LS and non-LS patients, probably due to the underlying differences in epithelial biology and immunology [52–54]. For example, Rifkin et al. showed that Veillonella was enriched and Faecalibacterium and Romboutsia were depleted in LS [52], whereas Mori et al. suggested that microbiota pattern associated with LS is characterized by an over-representation of Faecalibacterium prausnitzii, Parabacteroides distasonis, Ruminococcus bromii, Bacteroides plebeius, Bacteroides fragilis and Bacteroides uniformis species [53]. The interaction between the specific fecal microbiota pattern and the altered immune surveillance of LS patients may play a critical role in CRC development [55]. Thus, Yan et al. found that a subset of Clostridiaceae was depleted in stool biopsies, corresponding with baseline adenomas, while Desulfovibrio was enriched both in stool and in mucosal biopsies [54]. Their observations suggest that although prospective monitoring of microbiome has limited benefit in the early detection of adenoma, these early changes in microbiota may play a causal role in colonic neoplasia [4]. Moreover, Mori et al. suggested that despite the possible existence of a fecal microbiota pattern associated with a LS genetic background, there were no differences between microbial communities of patients with LS and CRC, and those observed in patients with LS and gynecologic malignancies [53]. However, future studies are needed to better understand the relationship between microbiota and cancer development in LS patients, and how the changes in microbiota can be used in the early detection of Lynch syndrome-related malignancies. Furthermore, the adequate manipulation of microbiota could represent a future therapeutic option to avoid the development of some malignancies related with LS.

4. Preventive Measures for Lynch Syndrome

Aspirin may be considered as a preventive measure against cancer in individuals with LS, although the optimal dosage remains unclear. The Colorectal Adenoma/Carcinoma Prevention Programme 2 (CAPP2) demonstrated a 60% reduction in the incidence of CRC and other Lynch syndrome-associated tumors in individuals who took 600 mg of aspirin daily for at least two years, compared to those who received a placebo [56]. The ongoing CAPP3 study aims to determine the most effective dose by comparing daily aspirin intake at 600, 300 and 100 mg. The European guidelines from the European Hereditary Tumour Group (EHTG) and the European Society of Coloproctology (ESCP) recommend the acetylsalicylic acid dose should be a minimum of 75–100 mg daily and this dose should be increased for people with above-average body mass [5]. However, the American College of Gastroenterology does not recommend the routine use of aspirin for the chemoprevention in LS [57].

In certain cases, prophylactic surgical interventions such as total colectomy or risk-reducing salpingo-oophorectomy may be considered for individuals with LS who are at particularly high risk of developing certain cancers.

4.1. Prophylactic Surgery for CRC in Patients with LS

Prophylactic surgery aims to remove organs before cancer develops, reducing the potential risk. The decision regarding which operation is preferable should be made on the basis of individual patient's factors and preferences, with special emphasis on the risk of metachronous CRC, the functional consequences of surgery, the patient's age and the commitment of the patient to continue colonoscopic surveillance [58]. The term of prophylactic total colectomy is defined either by total colectomy with ileo-rectal anastomosis, or proctocolectomy ended with ileal-anal pouch anastomosis (IPAA) or with ileostomy. In contrast to FAP, in which proctolectomy is the procedure of choice, Syngal et al. [59] showed that minimal benefit is derived from performing proctocolectomy rather than subtotal colectomy on patients with LS. By contrary, Henegan et al. [60] consider that a true prophylactic surgery is total proctocolectomy with IPAA or with end ileostomy because we can no longer talk about prophylaxis in rectal sparing surgery as it leaves the rectum in the LS patient, who then has a 1% per year risk of developing metachronous rectal cancer for the first 12 years [61]. However, patients with the rectum left in place could be regularly surveilled via a rectoscopy, which is more easily performed and accepted by the patient than a full colonoscopy.

The timing for surgery should be evaluated on an individual basis, taking into consideration gender, familial pattern of cancer and the age when cancers occur in relatives. Prophylactic surgery needs to be performed at an earlier age than the age of cancer occurrence in the youngest relative [61]. Prophylactic colectomy requires the careful evaluation of its implications, as it can significantly impact quality of life, lead to considerable morbidity and carries mortality risks. Individuals with LS have a lifetime colorectal cancer (CRC) risk of about 70%, indicating that nearly 30% out of these surgeries may be unnecessary for patients that would never develop CRC. Moreover, some of patients could eventually develop types of cancer other than CRC, and the prophylactic colectomy would not only be futile, but could also worsen their quality of life [62].

Llach et al. considered that prophylactic colectomy or proctocolectomy in healthy LS patients is not indicated due to the efficacy of colonoscopy on CRC mortality reduction, but they argue that there may be a role for prophylactic colorectal surgery in the secondary prevention of CRC [57,63–65].

Prophylactic proctocolectomy is recommended for patients with a pathologic germline mutation in the APC gene leading to familial adenomatous polyposis (FAP) [66]. Similarly, risk stratification by affected MMR gene may help identify the LS patients more prone to developing CRC. Thus, carriers of path_MLH1 or path_MSH2, who have a higher risk of developing CRC, would theoretically derive a greater benefit from total colectomy compared to carriers of the low-risk variants (path_MSH6 and path_PMS2), in which colonoscopic surveillance might achieve an efficient prophylaxis (Table 1).

Table 1. Impact of genetic mutations on the risk of colon cancer development and prophylactic surgery (+++ high risk/strong recommendation, + low risk/weak recommendation).

Genes	Risk of Colon Cancer	Prophylactic Colectomy	Prophylactic THBSO
MLH 1	+++	+++	+++
MSH 2	+++	+++	+++
MSH 6	+	+	+++
PMS 2	+	+	+

Prophylactic colorectal surgery might be considered in some particular situations, e.g., for mutation carriers who are unable to undergo surveillance, for patients who are non-compliant with surveillance examinations or have endoscopically unresectable adenomas with severe dysplasia, or for patients with severe distress regarding the development CRC who prefer surgery to surveillance [63,66].

Further prospective studies are necessary in order to elaborate clear guidelines concerning the role of prophylactic CRC surgery for patients with LS before developing CRC.

4.2. Prophylactic Surgery for Gynecologic Cancers in Lynch Syndrome

Women with LS have a 40 to 60% lifetime risk of EC and a 10 to 12% lifetime risk of ovarian cancer [62].

Therefore, for women with LS, prophylactic hysterectomy and bilateral salpingo-oophorectomy (BSO) can, at least theoretically, significantly reduce the risk of endometrial and ovarian cancers. Such surgical interventions may provide a substantial risk reduction for gynecological cancers, which are common in LS, and may be an important consideration for women with this condition, particularly those who have completed childbearing (>40 years) [8,64,67,68].

Women who carry an MSH2, MLH1 or MSH6 germline mutation and who present with CRC in the absence of distant metastases will present an extraordinarily high lifetime risk for carcinoma of the endometrium and/or ovary, therefore prophylactic hysterectomy and oophorectomy may also be considered for female patients with LS [61,69]. These procedures aim to reduce the risk of gynecologic cancers in this high-risk population.

Several surgical techniques have been proposed and implemented to achieve this goal. The most common surgical techniques include prophylactic total hysterectomy and bilateral salpingo-oophorectomy (THBSO). The opportunity for combining THBSO with colectomy should be discussed with the patient, taking into account the patient's age, comorbid conditions, plans for fertility and specific family history of cancer.

Although studies conducted before 2006 showed uncertain benefits in reducing gynecologic cancer risk after prophylactic surgery, the benefit of prophylactic THBSO was clearly demonstrated in a case-matched study reported by Schmeler et al. [67].

The timing of surgery should be individualized based on comorbidities, family history, LS gene and whether childbearing is complete [70].

In 2021, the ESGO-ESTRO-ESP consensus recommended surveillance for endometrial cancer in patients with LS starting at the age of 35 by annual transvaginal ultrasound (TVUS) and annual or biennial endometrial biopsy. Prophylactic THBSO should be considered at the end of childbearing and preferably before 40 [71], or even sooner if the patient does not wish to preserve fertility. The lack of consensus and evidence for the effectiveness of surveillance may also be used to enhance the argument for prophylactic THBSO when the opportunity arises, either at the time of prophylactic or curative colectomy in women with LS or as a separate operation once childbearing is complete for patients wanting to preserve fertility [62,72].

In a 2020 survey study that involved 18 countries, there was global agreement (>90%) in favor of offering both hysterectomy and BSO to carriers of path_MLH1, path_MSH2 and path_MSH6 genes after the age of 40 [73] (Table 1).

Because there is a wide variation in how, when and to whom risk-reducing gynecological surgery is offered, there is a clear need for further research in the field of care for the management of gynecological cancer risk.

5. Curative Surgery for Colorectal Cancer in Lynch Syndrome

The primary focus for individuals with LS is to prevent and/or detect cancer at an early stage. This involves using pre-symptomatic screening methods and opting for surgical removal when feasible.

Colorectal resection surgery on patients with Lynch syndrome who have been diagnosed with colorectal cancer should adhere to at least the same oncological standards used for those with non-hereditary colorectal cancer. It is crucial to ensure that these patients receive comprehensive and personalized care, considering the unique aspects of LS.

Moreover, a multidisciplinary approach involving a team of experts, including colorectal surgeons, gastroenterologists, oncologists and genetic counselors is fundamental for optimizing the surgical outcomes and long-term prognosis of these patients.

Additionally, post-operative surveillance is of utmost importance to facilitate early detection of any possible recurrence or the development of secondary malignancies.

With advances in medical research, targeted therapies and immunotherapies are emerging as potential treatment options for patients with CRC associated with Lynch syndrome, further emphasizing the need for individualized treatment plans.

5.1. Surgical Management of Primary Colon Cancer in Lynch Syndrome

Individuals with LS face a significant risk of developing life-threatening colorectal and endometrial cancers, with incidences reaching 40–80% by the age of 75 [74]. Despite surveillance efforts, the effectiveness of early detection remains limited, mainly because accelerated adenoma-to-carcinoma progression has been reported in patients with LS, with estimated polyp-to-cancer dwell times of 35 months compared with 10 to 15 years in sporadic cancer [75,76]. The limited effectiveness of colonoscopy can be explained by missed lesions on exploration, fast progression of newly formed adenomas, the fact that not all CRC in LS follow an adenoma-carcinoma pathway and the occurrence of induced lesions by multiple colonoscopies in MMR carriers [77]. Carriers of path_MLH1 and path_MSH2 genes have a higher risk of developing colorectal cancer, despite intensive surveillance colonoscopy [8,77–79]. Conversely, carriers of path_MSH6 [8,79] and path_PMS2 [80] genes have a lower risk of developing CRC, which may be due to their lower penetrance and later age of onset, and can be further reduced by regular colonoscopic surveillance or even become near to zero in carriers of PMS2 [78,81,82]. Characteristically, in LS, CRC develops at an early age, with right-sided tumor predominance (60–65%), along with extracolonic tumors of the endometrium, ovary, stomach, renal pelvis, ureter and other organs [62,74].

The surgical principles required when considering a case with CRC in the setting of LS should respect the following desideration: (1) the appropriate treatment of the primary tumor according to oncological principles applied in sporadic cases; (2) consideration of further risk reduction with prophylactic removal of larger parts of the non-neoplastic colon; (3) decrease morbidity and increase quality of life after colectomy [1]; (4) patient gender, age and general status; and (5) patient choice. In order to respect these principles, the range of surgical removal extends from limited resection/segmental colectomy, towards total colectomy with ileo-rectal anastomosis and finally to proctocolectomy completed with an IPAA or with end ileostomy. The extent of colorectal resection should be thoroughly discussed with the patient and the decision on this issue should consider patient's gender, age, general status, willingness to adhere to the program of colonoscopic surveillance and the degree of distress regarding the development of metachronous CRC.

To avoid the misleading interpretation of the terminology used for surgical approaches, we will further define the terms "segmental colectomy" and "extended colectomy". Segmental colectomy includes right or left hemicolectomy (for right or left colon cancer, respectively), segmental colectomy (for transverse colic cancers), and anterior resection or abdominoperineal resection (for rectal cancer). Extended colectomy includes extended right hemicolectomy, subtotal colectomy or total colectomy with ileo-rectal anastomosis (for colon cancer), and total proctocolectomy ended with IPAA or with end ileostomy (for rectal cancer).

When feasible, a minimally invasive approach (MIS) should be favored for patients with LS. Overall, the implementation of a MIS for patients with LS is highly recommended, as it optimizes surgical outcomes while prioritizing patient safety and well-being [83]. In the context of advancements in CRC surgery for patients with LS, the utilization of laparoscopic and robotic techniques has shown promising results in terms of reducing postoperative complications and improving recovery times [84].

5.1.1. Segmental Colectomy for Index Colonic Cancer

The selection of the suitable surgical procedure should be made after carefully considering the patient's unique factors and preferences. It is crucial that several aspects are taken into account, such as the risk of developing metachronous CRC, the age of the patient, the pathologic gene that determined LS and readiness to undergo surveillance colonoscopy. By thoroughly analyzing these factors, one can determine the most appropriate surgical procedure that will ensure optimal outcomes for the patient.

However, the vast majority of primary CRCs in LS patients are managed with segmental colectomy, simply because of the lack of preoperative recognition of the syndrome [62]. Some patients are susceptible to LS based on family history, but this can often be incomplete. Moreover, a majority of the patients with an unknown family history will be diagnosed by genetic testing of the colorectal specimen only after surgical removal.

Therefore, at present, it is strongly recommended that an immunohistochemical (IHC) evaluation of MMR genes expression is performed on the specimen attained by colonoscopic biopsy. If IHC staining revealed MSI-high status, genetic testing for germline mutations of MMR genes should be performed, in order to have a precise diagnosis before a surgical intervention.

Preoperative knowledge of the diagnosis of LS is of tremendous importance, due to the high risk of metachronous CRC in these patients. Thus, Parry et al. reveal that the risk of metachronous CRC is significantly reduced by 31% for every 10 cm of bowel removed, and Kim et al. report that a bowel resection of 25 cm or longer decreases the risk, as compared to less extensive resections [85,86]. Therefore, extended colectomy or even total/near total colectomy should be advocated in these patients.

Furthermore, every genetic variation in the MMR genes linked to LS (MLH1, MSH2, MSH6, PMS2 and EPCAM) carries a distinct risk of developing metachronous cancer. As a result, current guidelines distinguish between these genes when making recommendations for extension of colonic resection and surveillance programs. MLH1, MSH2 and EPCAM are classified as high-risk variants, and MSH6 and PMS2 as low-risk variants [5,20,74].

The latest NCCN version on Genetic/Familial High-Risk Assessment (version 1.2024—9 Spetember 2024 https://www.nccn.org/guidelines/guidelines-detail?category=2&id=15

44) [39] predicted an up to 43% cumulative lifetime risk of metachronous CRC for MLH1 and MSH2 carriers who have segmental resection, and a lower risk for MSH6 carriers. For this reason, it is strongly recommended that a subtotal or total colectomy is performed for carriers of the high-risk variants who have developed CRC. There are limited data on PMS2, but no marked increase in risk for metachronous CRC has been reported in the available literature. Eikenboom et al. show that the risk of metachronous colorectal cancer did not differ between carriers of low-risk variants who had segmental colectomy and those of high-risk variants who had extensive colectomy, and they conclude that a partial colectomy followed by endoscopic surveillance is an appropriate management approach to treat colorectal cancer in carriers of low-risk Lynch syndrome variants [74] (Table 2).

Table 2. Extent of colorectal resection according to the index colorectal tumor location and genetic mutations (APR = abdominoperineal resection).

Index Tumor	MLH1	MSH2	MSH6	PMS2
Primary Colon Cancer	Total/subtotal colectomy	Total/subtotal colectomy	Segmental colectomy	Segmental colectomy
Primary Rectal Cancer	Anterior resection/APR	Anterior resection/APR	Anterior resection/APR	Anterior resection/APR

Therefore, the decision to perform segmental versus total/near total colectomy should balance the risks of metachronous cancer according to the pathologic gene, the functional consequences of surgery and the patient's age and wishes.

Compared to young and fit patients, elderly and frail patients are more susceptible to experience adverse outcomes following surgery. Such outcomes include postoperative complications, functional decline and worse quality of life after surgery. Advanced age in LS typically refers to patients over 60–65 years of age [8,87]. Although the Mallorca Group Surgery [88] recommends total abdominal colectomy with ileo-rectal anastomosis regardless of patient age, older patients have a relatively short life expectancy in comparison to younger patients, and quality of life means more than longevity for many of these patients [89].

Quality of life encompasses various aspects such as physical function, psychological well-being and social interactions. Surgery, especially in older, frail patients, can disrupt these areas and lead to a decrease in overall quality of life. In light of these challenges, it is important for healthcare professionals to carefully assess and manage the risks associated with surgery in older CRC patients. This includes implementing tailored approaches to optimize outcomes and minimize potential complications. By taking a comprehensive and individualized approach, healthcare teams can strive to improve the overall prognosis and postoperative experience for older LS patients with CRC [90,91]. For these frequently frail patients, it seems reasonable to perform a segmental colectomy instead of an extended colectomy, in order to minimize the postoperative morbidity and offer a better quality of life.

5.1.2. Extended Colectomy for Index Colonic Cancer

Theoretically, extended colectomy reduces the amount of future colorectal tissue exposed to carcinogenesis, thus reducing the recurrence of CRC. The more extended the CRC resection, the lower the risk of metachronous CRC development. Subtotal colectomy with ileo-sigmoidostomy or total colectomy with ileo-rectal anastomosis significantly lowers the risk of developing future CRC, but does not eliminate it completely. However, surveillance is easier in these patients, as there is only a small portion of rectum (and sigmoid) that has to be monitored regularly by recto-sigmoidoscopy, and this investigation is better tolerated by patients. Although the recurrence rate is higher in patients treated by limited resection versus extended colonic resection (the rates of recurrent CRC after 10 years were 16% vs. 4%, respectively), the overall survival benefit of extended resection was not demonstrated by Natarjan et al. [92], probably due to the relatively small number of available subjects. However, de Vos tot Nederveen Cappel and colleagues show that life expectancy increased up to 2.3 years for patients who underwent extended colectomy at a younger age (under 47 years) compared to their counterparts treated with segmental resection [87].

Furthermore, Natarajan et al. recommend extended colectomy due to the increased incidence of metachronous CRC and the frequent necessity for a second abdominal surgery on patients who undergo limited resection [92].

The advantage of extended colectomy may be influenced by the age at first CRC. A decision analysis model pointed out that subtotal colectomy performed at 25 years of age in LS patients with CRC led to the greatest life expectancy [8].

Extended colectomy is also indicated in recurrent CRC because it is cost effective, and is favored by patients because it spares them from repeated colonoscopies and laparotomies [62,93]. As already mentioned, the EHTG and ESCP guidelines recommend extended colectomy (either total or subtotal colectomy ended with ileo-rectal anastomosis or with ileo-sigmoidostomy) for high-risk patients (path_MLH1 and path_MSH2 carriers) [5] (Table 2).

Thus, the extent of colonic resection in individuals with LS remains a complex and nuanced topic. While the fundamental principles of oncologic colorectal surgery apply, the unique considerations of this high-risk population must be carefully weighed. A tailored surgical approach based on tumor characteristics, gene mutations and patient's risk factors and desires, as well as the potential for neoadjuvant therapy and organ-preserving strategies for patients with rectal cancer may optimize both oncologic and functional outcomes for individuals with LS [94,95].

Despite active surveillance in LS, more frequent colonoscopic surveillance did not reduce the incidence of metachronous CRC or stage at detection [96]. This is another reason for opting for extended CRC surgery in LS, even at the time of metachronous CRC. Nevertheless, many guidelines recommend colonoscopic surveillance every 1–2 years in LS patients [5,25,83].

As described above, the choice between segmental and extended colectomy for patients with colon cancer in LS involves weighing the benefits of reduced metachronous CRC risk against the potential for worse bowel function and lower quality of life [97].

5.2. *Surgical Management of Primary Rectal Cancer in Lynch Syndrome*

Although 60% of CRCs in Lynch syndrome occur on the right colon, about 10–15% of LS patients present with rectal cancer as an index tumor [98,99]. It is associated mostly with mutations in the MSH2 and MSH6 genes that are also present in extracolonic malignancies [100].

Some authors suggest that rectal cancer should be managed in the usual way, based on standard oncologic principles for sporadic rectal cancer, without a requirement for LS-specific approaches [63]. Thus, different guidelines recommend segmental resection (either anterior resection or abdominoperineal resection) in LS patients presented with index rectal cancer [5,101]. On the other hand, You et al. [102] consider that in dMMR genes carriers with rectal cancer, the surgical strategy should be tailored by addressing not only the rectal cancer (loco-regional control, distant metastases control and functional outcome) but also the issues associated with LS—the risk of metachronous CRC and the risk of cancer occurrence in other organs (especially ovarian or endometrial).

Thus, the surgeon should decide whether to perform a standard rectal resection according to the location of the primary tumor in the rectum or extend the resection to the remaining colon in order to reduce the risk of metachronous CRC (performing a total proctocolectomy ended with IPAA or an end ileostomy). The choice of localized vs. extended resection should be discussed with the patient and explained thoroughly, given the issues of regular colonoscopic surveillance, the risk of missed lesions at colonoscopy (in case of a limited resection) and the decreased quality of life and higher morbidity rates (observed after extended resections). Moreover, surgeons should discuss the possibility of performing prophylactic THBSO at the same time as the CRC surgery, given the increased risk of uterine and ovarian cancer in women with LS.

A recent study that compared the risk of metachronous CRC after surgical resection in two groups [colonic cancer (CC) and rectal cancer (RC) index group] found that the incidence of metachronous CRC was lower in the rectal group [103]. Another finding was that cause of death was associated with extracolonic LS tumors (mainly gynecologic in women) without any deaths due to CRC in the RC group, whereas in the CC group, 28.6% of deaths were associated with metachronous CRC. Therefore, considering the above-mentioned results, extended surgery for index rectal cancer (such as total proctocolectomy) is less effective than extended surgery in index colon cancer group, from the point of view of metachronous CRC risk reduction, and is associated with a decreased quality of life (Table 2). Also, prophylactic THBSO at the time of index surgery seem to be lifesaving in both groups. Although Kalady et al. and Win et al. propose total proctocolectomy with IPAA as a treatment of index rectal cancer, Chikatani et al. consider that this extensive surgery cannot be recommended as a standard treatment [103–105].

Since the published results are contradictory, further prospective studies are needed, with larger cohorts, in order to achieve definitive conclusions.

Other issues concerning the treatment of rectal cancer in patients with LS are the multimodal treatment and alternative approaches to TME.

The standard treatment of locally advanced (stage II and III) rectal cancer is preoperative (chemo)radiotherapy followed by surgery and adjuvant systemic chemotherapy [106,107]. The question of whether pelvic radiation can be skipped for certain patients is really important for LS patients with rectal cancer, mainly in sphincter-saving procedures in which radiation therapy may led to bowel dysfunction. Another aspect in LS patients is that they are generally young and the patient should be informed about the risks related to long term consequences of pelvic radiation: sexual dysfunction, hip fracture and fibrosis [108,109].

Regarding the surgical approach, radical total mesorectal excision (TME) is the mainstay treatment for patients with rectal cancer. Although laparoscopic, robotic and transanal-TME (TaTME) approaches improved the surgical armamentarium in TME, there is significant debate regarding the approach that achieves the best oncologic results [110].

Even though TME is the procedure of choice for patients with resectable rectal cancer without metastases and local excision (LE) techniques have been associated with inferior oncologic outcomes, in select cases, LE may be a recommended surgical alternative, due to the lower morbidity rates and better quality of life for select patients [110–113]. This could be particularly useful when treating elderly LS patients with multiple prior surgical interventions, or those who refuse stoma formation or an extended resection that can lead to bowel and functional disturbances [102].

Only 4% [114] of LS patients present with stage IV rectal cancer. In patients with unresectable metastases, the role of surgery is minimal and patients could benefit from systemic therapy, although up to 40% of these patients could be rendered to resectability [115,116].

Nevertheless, after curative intent treatment, such patients could achieve 5-year overall survival and disease-free survival rates up to 60% and up to 40%, respectively [94,117,118].

6. Conclusions

As the field is ever-evolving, it is crucial for clinicians to stay informed about the latest guidelines and recommendations regarding the management of LS.

Presently, the diagnosis of LS is genetic. The pathologic MMR gene has a huge impact on clinical presentation, on the risk of developing different types of cancers and, consequently, on the surveillance programs, prophylactic approaches and extent of colorectal resection. Pathologic variants of MLH1 or MSH2 genes are associated with a significantly higher risk of developing CRC and metachronous CRC after the resection of the index colon cancer. For these reasons, carriers of these high-risk variants would derive a greater benefit from total colectomy compared to carriers of the low-risk genes. Total or subtotal colectomies are recommended for treating index colon cancer in such patients. By contrary, extended surgery for index rectal cancer seems to be less effective than extended surgery in the index colon cancer group, from the point of view of metachronous CRC risk reduction. Furthermore, the performance of a total proctocolectomy (ended with either IPAA or an end ileostomy) is associated with a decreased quality of life and, for these reasons, this extensive surgery cannot be currently recommended as standard treatment. Prophylactic THBSO should be considered at the end of the childbearing age and preferably before 40 years of age or sooner if the patients do not wish to preserve fertility, especially in path_MLH1, path_MSH2 and path_MSH6 carriers.

Future studies should focus on refining the criteria for surveillance and intervention, and ensuring that patients receive individualized care based on their unique genetic profiles and personal circumstances.

Author Contributions: Conceptualization: S.T.A.; data curation: T.R.G. and G.P.; methodology: S.T.A.; formal analysis and investigation: T.R.G. and G.P.; writing—original draft preparation: T.R.G. and G.P.; writing—review and editing: S.T.A.; supervision: S.T.A. All authors have read and agreed to the published version of the manuscript.

Funding: This research received no funding.

Institutional Review Board Statement: Not applicable.

Informed Consent Statement: Not applicable.

Data Availability Statement: No new data were created or analyzed in this study. Data sharing is not applicable to this article.

Conflicts of Interest: The authors declare no conflicts of interest.

References

1. Negura, I.; Ianole, V.; Danciu, M.; Preda, C.; Iosep, D.G.; Dănilă, R.; Grigorovici, A.; Ciobanu Apostol, D.G. Thyroid Collision Tumors: The Presence of the Medullary Thyroid Carcinoma Component Negatively Influences the Prognosis. *Diagnostics* **2023**, *13*, 285. [CrossRef]
2. McCabe, A.; Quinn, G.; Jain, S.; Dálaigh, M.; Dean, K.; Murphy, R.; Mcdade, S. ClassifieR 2.0: Expanding interactive gene expression-based stratification to prostate and high-grade serous ovarian cancer. *BMC Bioinform.* **2024**, *25*, 362. [CrossRef] [PubMed]
3. Liu, X.-Y.; Yu, T.-J.; Shao, Z.-M. Precision medicine for breast cancer: Advances and challenges. *Transl. Breast Cancer Res.* **2024**, *5*, 35. [CrossRef]
4. Alexandrescu, S.T.; Dinu, I.M.; Diaconescu, A.S.; Micu, A.; Pasare, E.; Durdu, C.; Dorobantu, B.M.; Popescu, I. Embryologic Origin of the Primary Tumor and RAS Status Predict Survival after Resection of Colorectal Liver Metastases. *Medicina* **2022**, *58*, 1100. [CrossRef]

5. Seppälä, T.T.; Latchford, A.; Negoi, I.; Sampaio Soares, A.; Jimenez-Rodriguez, R.; Sánchez-Guillén, L.; Evans, D.G.; Ryan, N.; Crosbie, E.J.; Dominguez-Valentin, M.; et al. European guidelines from the EHTG and ESCP for Lynch syndrome: An updated third edition of the Mallorca guidelines based on gene and gender. *Br. J. Surg.* **2021**, *108*, 484–498. [CrossRef]
6. Stjepanovic, N.; Moreira, L.; Carneiro, F.; Balaguer, F.; Cervantes, A.; Balmaña, J.; Martinelli, E. Hereditary gastrointestinal cancers: ESMO Clinical Practice Guidelines for diagnosis, treatment and follow-up. *Ann. Oncol.* **2019**, *30*, 1558–1571. [CrossRef]
7. Lynch, H.T.; de la Chapelle, A. Hereditary colorectal cancer. *N. Engl. J. Med.* **2003**, *348*, 919–932. [CrossRef] [PubMed]
8. Giardiello, F.M.; Allen, J.I.; Axilbund, J.E.; Boland, C.R.; Burke, C.A.; Burt, R.W.; Church, J.M.; Dominitz, J.A.; Johnson, D.A.; Kaltenbach, T.; et al. Guidelines on genetic evaluation and management of Lynch syndrome: A consensus statement by the US Multi-Society Task Force on colorectal cancer. *Gastroenterology* **2014**, *147*, 502–526. [CrossRef] [PubMed]
9. Engel, C.; Loeffler, M.; Steinke, V.; Rahner, N.; Holinski-Feder, E.; Dietmaier, W.; Schackert, H.K.; Goergens, H.; von Knebel Doeberitz, M.; Goecke, T.O.; et al. Risks of less common cancers in proven mutation carriers with lynch syndrome. *J. Clin. Oncol.* **2012**, *30*, 4409–4415. [CrossRef]
10. Win, A.K.; Young, J.P.; Lindor, N.M.; Tucker, K.M.; Ahnen, D.J.; Young, G.P.; Buchanan, D.D.; Clendenning, M.; Giles, G.G.; Winship, I.; et al. Colorectal and other cancer risks for carriers and noncarriers from families with a DNA mismatch repair gene mutation: A prospective cohort study. *J. Clin. Oncol.* **2012**, *30*, 958–964. [CrossRef] [PubMed]
11. Samadder, N.J.; Smith, K.R.; Wong, J.; Thomas, A.; Hanson, H.; Boucher, K.; Kopituch, C.; Cannon-Albright, L.A.; Burt, R.W.; Curtin, K. Cancer Risk in Families Fulfilling the Amsterdam Criteria for Lynch Syndrome. *JAMA Oncol.* **2017**, *3*, 1697–1701. [CrossRef] [PubMed]
12. Lupuşoru, I.; Ciobanu, D.; Ursaru, M.; Bălan, G.G.; Grigorovici, A. Difficulties in Treating a Patient with Multiple Cancers in the COVID-19 Pandemic. *Chirurgia* **2020**, *115*, 670–676. [CrossRef] [PubMed]
13. Hampel, H.; Frankel, W.L.; Martin, E.; Arnold, M.; Khanduja, K.; Kuebler, P.; Nakagawa, H.; Sotamaa, K.; Prior, T.W.; Westman, J.; et al. Screening for the Lynch syndrome (hereditary nonpolyposis colorectal cancer). *N. Engl. J. Med.* **2005**, *352*, 1851–1860. [CrossRef] [PubMed]
14. De Jong, A.E.; Morreau, H.; Van Puijenbroek, M.; Eilers, P.H.c.; Wijnen, J.; Nagengast, F.M.; Griffioen, G.; Cats, A.; Menko, F.H.; Kleibeuker, J.H.; et al. The role of mismatch repair gene defects in the development of adenomas in patients with HNPCC. *Gastroenterology* **2004**, *126*, 42–48. [CrossRef] [PubMed]
15. Grosse, S.D. When is Genomic Testing Cost-Effective? Testing for Lynch Syndrome in Patients with Newly-Diagnosed Colorectal Cancer and Their Relatives. *Healthcare* **2015**, *3*, 860–878. [CrossRef] [PubMed]
16. Valdez, J.M.; Nichols, K.E.; Kesserwan, C. Li-Fraumeni syndrome: A paradigm for the understanding of hereditary cancer predisposition. *Br. J. Haematol.* **2017**, *176*, 539–552. [CrossRef] [PubMed]
17. Markowitz, S.D.; Bertagnolli, M.M. Molecular origins of cancer: Molecular basis of colorectal cancer. *N. Engl. J. Med.* **2009**, *361*, 2449–2460. [CrossRef] [PubMed]
18. Barnetson, R.A.; Tenesa, A.; Farrington, S.M.; Nicholl, I.D.; Cetnarskyj, R.; Porteous, M.E.; Campbell, H.; Dunlop, M.G. Identification and survival of carriers of mutations in DNA mismatch-repair genes in colon cancer. *N. Engl. J. Med.* **2006**, *354*, 2751–2763. [CrossRef]
19. Møller, P.; Seppälä, T.; Bernstein, I.; Holinski-Feder, E.; Sala, P.; Evans, D.G.; Lindblom, A.; Macrae, F.; Blanco, I.; Sijmons, R.; et al. Cancer incidence and survival in Lynch syndrome patients receiving colonoscopic and gynaecological surveillance: First report from the prospective Lynch syndrome database. *Gut* **2017**, *66*, 464–472. [CrossRef] [PubMed]
20. Vasen, H.F.; Stormorken, A.; Menko, F.H.; Nagengast, F.M.; Kleibeuker, J.H.; Griffioen, G.; Taal, B.G.; Moller, P.; Wijnen, J.T. MSH2 mutation carriers are at higher risk of cancer than MLH1 mutation carriers: A study of hereditary nonpolyposis colorectal cancer families. *J. Clin. Oncol.* **2001**, *19*, 4074–4080. [CrossRef]
21. Lin, K.M.; Shashidharan, M.; Ternent, C.A.; Thorson, A.G.; Blatchford, G.J.; Christensen, M.A.; Lanspa, S.J.; Lemon, S.J.; Watson, P.; Lynch, H.T. Colorectal and extracolonic cancer variations in MLH1/MSH2 hereditary nonpolyposis colorectal cancer kindreds and the general population. *Dis. Colon Rectum* **1998**, *41*, 428–433. [CrossRef] [PubMed]
22. Hendriks, Y.M.C.; Wagner, A.; Morreau, H.; Menko, F.; Stormorken, A.; Quehenberger, F.; Sandkuijl, L.; Møller, P.; Genuardi, M.; Van Houwelingen, H.; et al. Cancer risk in hereditary nonpolyposis colorectal cancer due to MSH6 mutations: Impact on counseling and surveillance. *Gastroenterology* **2004**, *127*, 17–25. [CrossRef] [PubMed]
23. Buchanan, D.D.; Tan, Y.Y.; Walsh, M.D.; Clendenning, M.; Metcalf, A.M.; Ferguson, K.; Arnold, S.T.; Thompson, B.A.; Lose, F.A.; Parsons, M.T.; et al. Tumor mismatch repair immunohistochemistry and DNA MLH1 methylation testing of patients with endometrial cancer diagnosed at age younger than 60 years optimizes triage for population-level germline mismatch repair gene mutation testing. *J. Clin. Oncol.* **2014**, *32*, 90–100. [CrossRef] [PubMed]
24. Kempers, M.J.E.; Kuiper, R.P.; Ockeloen, C.W.; Chappuis, P.O.; Hutter, P.; Rahner, N.; Schackert, H.K.; Steinke, V.; Holinski-Feder, E.; Morak, M.; et al. Risk of colorectal and endometrial cancers in EPCAM deletion-positive Lynch syndrome: A cohort study. *Lancet Oncol.* **2011**, *12*, 49–55. [CrossRef] [PubMed]

25. Gupta, S.; Provenzale, D.; Llor, X.; Halverson, A.L.; Grady, W.; Chung, D.C.; Haraldsdottir, S.; Markowitz, A.J.; Slavin, T.P.; Hampel, H.; et al. NCCN Guidelines Insights: Genetic/Familial High-Risk Assessment: Colorectal, Version 2.2019. *J. Natl. Compr. Cancer Netw.* **2019**, *17*, 1032–1041. [CrossRef]
26. Moreira, L.; Balaguer, F.; Lindor, N.; de la Chapelle, A.; Hampel, H.; Aaltonen, L.A.; Hopper, J.L.; Le Marchand, L.; Gallinger, S.; Newcomb, P.A.; et al. Identification of Lynch syndrome among patients with colorectal cancer. *JAMA* **2012**, *308*, 1555–1565. [CrossRef] [PubMed]
27. Mvundura, M.; Grosse, S.; Hampel, H.; Palomaki, G. The cost-effectiveness of genetic testing strategies for Lynch syndrome among newly diagnosed patients with colorectal cancer. *Genet. Med.* **2010**, *12*, 93–104. [CrossRef]
28. Pérez-Carbonell, L.; Ruiz-Ponte, C.; Guarinos, C.; Alenda, C.; Payá, A.; Brea, A.; Egoavil, C.M.; Castillejo, A.; Barberá, V.M.; Bessa, X.; et al. Comparison between universal molecular screening for Lynch syndrome and revised Bethesda guidelines in a large population-based cohort of patients with colorectal cancer. *Gut* **2012**, *61*, 865–872. [CrossRef] [PubMed]
29. Vasen, H.F.; Watson, P.; Mecklin, J.P.; Lynch, H.T. New clinical criteria for hereditary nonpolyposis colorectal cancer (HNPCC, Lynch syndrome) proposed by the International Collaborative group on HNPCC. *Gastroenterology* **1999**, *116*, 1453–1456. [CrossRef] [PubMed]
30. Menon, G.; Carr, S.; Kasi, A. Familial Adenomatous Polyposis. In *StatPearls*; StatPearls Publishing: Treasure Island, FL, USA, 2024. Available online: http://www.ncbi.nlm.nih.gov/books/NBK538233/ (accessed on 17 September 2024).
31. Umar, A.; Boland, C.R.; Terdiman, J.P.; Syngal, S.; de la Chapelle, A.; Rüschoff, J.; Fishel, R.; Lindor, N.M.; Burgart, L.J.; Hamelin, R.; et al. Revised Bethesda Guidelines for hereditary nonpolyposis colorectal cancer (Lynch syndrome) and microsatellite instability. *J. Natl. Cancer Inst.* **2004**, *96*, 261–268. [CrossRef] [PubMed]
32. Alexandrescu, S.; Anastase, D.; Grigorie, R.; Zlate, A.-C.; Andrei, S.; Costea, R.; Gramaticu, I.; Croitoru, A.; Popescu, I. Influence of the Primary Tumor Location on the Pattern of Synchronous Metastatic Spread in Patients with Stage IV Colorectal Carcinoma, According to the 8 th Edition of the AJCC Staging System. *J. Gastrointest. Liver Dis.* **2020**, *29*, 561–568. [CrossRef] [PubMed]
33. Adar, T.; Rodgers, L.H.; Shannon, K.M.; Yoshida, M.; Ma, T.; Mattia, A.; Lauwers, G.Y.; Iafrate, A.J.; Chung, D.C. A tailored approach to BRAF and MLH1 methylation testing in a universal screening program for Lynch syndrome. *Mod. Pathol.* **2017**, *30*, 440–447. [CrossRef] [PubMed]
34. Palomaki, G.E.; McClain, M.R.; Melillo, S.; Hampel, H.L.; Thibodeau, S.N. EGAPP supplementary evidence review: DNA testing strategies aimed at reducing morbidity and mortality from Lynch syndrome. *Genet. Med.* **2009**, *11*, 42–65. [CrossRef] [PubMed]
35. Pearlman, R.; Frankel, W.L.; Swanson, B.; Zhao, W.; Yilmaz, A.; Miller, K.; Bacher, J.; Bigley, C.; Nelsen, L.; Goodfellow, P.J.; et al. Prevalence and Spectrum of Germline Cancer Susceptibility Gene Mutations Among Patients With Early-Onset Colorectal Cancer. *JAMA Oncol.* **2017**, *3*, 464–471. [CrossRef]
36. Yurgelun, M.B.; Kulke, M.H.; Fuchs, C.S.; Allen, B.A.; Uno, H.; Hornick, J.L.; Ukaegbu, C.I.; Brais, L.K.; McNamara, P.G.; Mayer, R.J.; et al. Cancer Susceptibility Gene Mutations in Individuals With Colorectal Cancer. *J. Clin. Oncol.* **2017**, *35*, 1086–1095. [CrossRef] [PubMed]
37. Kastrinos, F.; Steyerberg, E.W.; Mercado, R.; Balmaña, J.; Holter, S.; Gallinger, S.; Siegmund, K.D.; Church, J.M.; Jenkins, M.A.; Lindor, N.M.; et al. The PREMM1,2,6 Model Predicts Risk of MLH1, MSH2, and MSH6 Germline Mutations Based on Cancer History. *Gastroenterology* **2011**, *140*, 73–81. [CrossRef]
38. Kastrinos, F.; Uno, H.; Ukaegbu, C.; Alvero, C.; McFarland, A.; Yurgelun, M.B.; Kulke, M.H.; Schrag, D.; Meyerhardt, J.A.; Fuchs, C.S.; et al. Development and Validation of the PREMM5 Model for Comprehensive Risk Assessment of Lynch Syndrome. *J. Clin. Oncol.* **2017**, *35*, 2165–2172. [CrossRef] [PubMed]
39. NCCN Guidelines: Genetic/Familial High-Risk Assessment: Colorectal, Endometrial and Gastric. Available online: https://www.nccn.org/professionals/physician_gls/pdf/genetics_ceg.pdf (accessed on 9 September 2024).
40. Gerrard, A.D.; Maeda, Y.; Strachan, J.; Speake, D.; Dunlop, M.G.; Din, F.V.N. Diagnostic Performance of Faecal Immunochemical Testing (FIT) in Patients with Lynch Syndrome Scheduled for Colonoscopic Surveillance. *Diagnostics* **2024**, *14*, 2431. [CrossRef]
41. Lincoln, A.G.; Benton, S.C.; Piggott, C.; Sheikh, S.R.; Beggs, A.D.; Buckley, L.; DeSouza, B.; East, J.E.; Sanders, P.; Lim, M.; et al. Risk-stratified faecal immunochemical testing (FIT) for urgent colonoscopy in Lynch syndrome during the COVID-19 pandemic. *BJS Open* **2023**, *7*, zrad079; Erratum in *BJS Open* **2023**, *7*, zrad153. [CrossRef]
42. Auranen, A.; Joutsiniemi, T. A systematic review of gynecological cancer surveillance in women belonging to hereditary nonpolyposis colorectal cancer (Lynch syndrome) families. *Acta Obs. Gynecol. Scand.* **2011**, *90*, 437–444. [CrossRef]
43. Chen, L.; Yang, K.Y.; Little, S.E.; Cheung, M.K.; Caughey, A.B. Gynecologic cancer prevention in Lynch syndrome/hereditary nonpolyposis colorectal cancer families. *Obs. Gynecol.* **2007**, *110*, 18–25. [CrossRef] [PubMed]
44. Järvinen, H.J.; Renkonen-Sinisalo, L.; Aktán-Collán, K.; Peltomäki, P.; Aaltonen, L.A.; Mecklin, J.-P. Ten years after mutation testing for Lynch syndrome: Cancer incidence and outcome in mutation-positive and mutation-negative family members. *J. Clin. Oncol.* **2009**, *27*, 4793–4797. [CrossRef]
45. Renkonen-Sinisalo, L.; Bützow, R.; Leminen, A.; Lehtovirta, P.; Mecklin, J.-P.; Järvinen, H.J. Surveillance for endometrial cancer in hereditary nonpolyposis colorectal cancer syndrome. *Int. J. Cancer* **2007**, *120*, 821–824. [CrossRef] [PubMed]

46. Rijcken, F.E.M.; Mourits, M.J.E.; Kleibeuker, J.H.; Hollema, H.; van der Zee, A.G.J. Gynecologic screening in hereditary nonpolyposis colorectal cancer. *Gynecol. Oncol.* **2003**, *91*, 74–80. [CrossRef]
47. Ladigan-Badura, S.; Vangala, D.B.; Engel, C.; Bucksch, K.; Hueneburg, R.; Perne, C.; Nattermann, J.; Steinke-Lange, V.; Rahner, N.; Schackert, H.K.; et al. Value of upper gastrointestinal endoscopy for gastric cancer surveillance in patients with Lynch syndrome. *Int. J. Cancer* **2021**, *148*, 106–114. [CrossRef]
48. Farha, N.; Hrabe, J.; Sleiman, J.; Beard, J.; Lyu, R.; Bhatt, A.; Church, J.; Heald, B.; Liska, D.; Mankaney, G.; et al. Clinically actionable findings on surveillance EGD in asymptomatic patients with Lynch syndrome. *Gastrointest. Endosc.* **2022**, *95*, 105–114. [CrossRef] [PubMed]
49. Kumar, S.; Dudzik, C.M.; Reed, M.; Long, J.M.; Wangensteen, K.J.; Katona, B.W. Upper Endoscopic Surveillance in Lynch Syndrome Detects Gastric and Duodenal Adenocarcinomas. *Cancer Prev. Res.* **2020**, *13*, 1047–1054. [CrossRef] [PubMed]
50. Lonati, C.; Simeone, C.; Suardi, N.; Spiess, P.E.; Necchi, A.; Moschini, M. Genitourinary manifestations of Lynch syndrome in the urological practice. *Asian J. Urol.* **2022**, *9*, 443–450. [CrossRef]
51. Canto, M.I.; Harinck, F.; Hruban, R.H.; Offerhaus, G.J.; Poley, J.-W.; Kamel, I.; Nio, Y.; Schulick, R.S.; Bassi, C.; Kluijt, I.; et al. International Cancer of the Pancreas Screening (CAPS) Consortium summit on the management of patients with increased risk for familial pancreatic cancer. *Gut* **2013**, *62*, 339–347. [CrossRef] [PubMed]
52. Rifkin, S.B.; Sze, M.A.; Tuck, K.; Koeppe, E.; Stoffel, E.M.; Schloss, P.D. Gut Microbiome Composition in Lynch Syndrome With and Without History of Colorectal Neoplasia and Non-Lynch Controls. *J. Gastrointest. Cancer* **2024**, *55*, 207–218. [CrossRef]
53. Mori, G.; Orena, B.S.; Cultrera, I.; Barbieri, G.; Albertini, A.M.; Ranzani, G.N.; Carnevali, I.; Tibiletti, M.G.; Pasca, M.R. Gut Microbiota Analysis in Postoperative Lynch Syndrome Patients. *Front. Microbiol.* **2019**, *10*, 1746. [CrossRef] [PubMed]
54. Yan, Y.; Drew, D.A.; Markowitz, A.; Lloyd-Price, J.; Abu-Ali, G.; Nguyen, L.H.; Tran, C.; Chung, D.C.; Gilpin, K.K.; Meixell, D.; et al. Structure of the Mucosal and Stool Microbiome in Lynch Syndrome. *Cell Host Microbe* **2020**, *27*, 585–600.e4. [CrossRef] [PubMed]
55. Hanus, M.; Parada-Venegas, D.; Landskron, G.; Wielandt, A.M.; Hurtado, C.; Alvarez, K.; Hermoso, M.A.; López-Köstner, F.; De la Fuente, M. Immune System, Microbiota, and Microbial Metabolites: The Unresolved Triad in Colorectal Cancer Microenvironment. *Front. Immunol.* **2021**, *12*, 612826. [CrossRef] [PubMed]
56. Burn, J.; Gerdes, A.-M.; Macrae, F.; Mecklin, J.-P.; Moeslein, G.; Olschwang, S.; Eccles, D.; Evans, D.G.; Maher, E.R.; Bertario, L.; et al. Long-term effect of aspirin on cancer risk in carriers of hereditary colorectal cancer: An analysis from the CAPP2 randomised controlled trial. *Lancet* **2011**, *378*, 2081–2087. [CrossRef] [PubMed]
57. Syngal, S.; Brand, R.E.; Church, J.M.; Giardiello, F.M.; Hampel, H.L.; Burt, R.W.; American College of Gastroenterology. ACG clinical guideline: Genetic testing and management of hereditary gastrointestinal cancer syndromes. *Am. J. Gastroenterol.* **2015**, *110*, 223–262; quiz 263. [CrossRef]
58. Monahan, K.J.; Bradshaw, N.; Dolwani, S.; Desouza, B.; Dunlop, M.G.; East, J.E.; Ilyas, M.; Kaur, A.; Lalloo, F.; Latchford, A.; et al. Guidelines for the management of hereditary colorectal cancer from the British Society of Gastroenterology (BSG)/Association of Coloproctology of Great Britain and Ireland (ACPGBI)/United Kingdom Cancer Genetics Group (UKCGG). *Gut* **2020**, *69*, 411–444. [CrossRef]
59. Syngal, S. Benefits of Colonoscopic Surveillance and Prophylactic Colectomy in Patients with Hereditary Nonpolyposis Colorectal Cancer Mutations. *Ann. Intern. Med.* **1998**, *129*, 787. [CrossRef] [PubMed]
60. Heneghan, H.M.; Martin, S.T.; Winter, D.C. Segmental vs extended colectomy in the management of hereditary nonpolyposis colorectal cancer: A systematic review and meta-analysis. *Color. Dis.* **2015**, *17*, 382–389. [CrossRef] [PubMed]
61. Lynch, H.T.; Lynch, J.F.; Fitzgibbons, R. Role of Prophylactic Colectomy in Lynch Syndrome. *Clin. Color. Cancer* **2003**, *3*, 99–101. [CrossRef]
62. Cirocco, W.C.; Hampel, H. Lynch Syndrome: Management of the Colon, What Operation? In *Management of Hereditary Colorectal Cancer: A Multidisciplinary Approach*; Guillem, J.G., Friedman, G., Eds.; Springer International Publishing: Cham, Switzerland, 2020; pp. 149–174. [CrossRef]
63. Llach, J.; Pellisé, M.; Monahan, K. Lynch syndrome; towards more personalized management? *Best. Pract. Res. Clin. Gastroenterol.* **2022**, *58–59*, 101790. [CrossRef] [PubMed]
64. Stoffel, E.M.; Mangu, P.B.; Gruber, S.B.; Hamilton, S.R.; Kalady, M.F.; Lau, M.W.Y.; Lu, K.H.; Roach, N.; Limburg, P.J. Hereditary Colorectal Cancer Syndromes: American Society of Clinical Oncology Clinical Practice Guideline Endorsement of the Familial Risk–Colorectal Cancer: European Society for Medical Oncology Clinical Practice Guidelines. *JCO* **2015**, *33*, 209–217. [CrossRef]
65. Yurgelun, M.B.; Hampel, H. Recent Advances in Lynch Syndrome: Diagnosis, Treatment, and Cancer Prevention. *Am. Soc. Clin. Oncol. Educ. Book* **2018**, *38*, 101–109. [CrossRef] [PubMed]
66. Doerner, J. Risk of Metachronous Colorectal Cancer in Lynch Syndrome: Who Needs an Extended Resection? *Surgeries* **2022**, *3*, 185–191. [CrossRef]

67. Schmeler, K.M.; Lynch, H.T.; Chen, L.; Munsell, M.F.; Soliman, P.T.; Clark, M.B.; Daniels, M.S.; White, K.G.; Boyd-Rogers, S.G.; Conrad, P.G.; et al. Prophylactic surgery to reduce the risk of gynecologic cancers in the Lynch syndrome. *N. Engl. J. Med.* **2006**, *354*, 261–269. [CrossRef]
68. Crosbie, E.J.; Ryan, N.A.J.; Arends, M.J.; Bosse, T.; Burn, J.; Cornes, J.M.; Crawford, R.; Eccles, D.; Frayling, I.M.; Ghaem-Maghami, S.; et al. The Manchester International Consensus Group recommendations for the management of gynecological cancers in Lynch syndrome. *Genet. Med.* **2019**, *21*, 2390–2400. [CrossRef] [PubMed]
69. Jass, J.R.; Stewart, S.M. Evolution of hereditary non-polyposis colorectal cancer. *Gut* **1992**, *33*, 783–786. [CrossRef] [PubMed]
70. Capasso, I.; Santoro, A.; Lucci Cordisco, E.; Perrone, E.; Tronconi, F.; Catena, U.; Zannoni, G.F.; Scambia, G.; Fanfani, F.; Lorusso, D.; et al. Lynch Syndrome and Gynecologic Tumors: Incidence, Prophylaxis, and Management of Patients with Cancer. *Cancers* **2023**, *15*, 1400. [CrossRef]
71. Concin, N.; Matias-Guiu, X.; Vergote, I.; Cibula, D.; Mirza, M.R.; Marnitz, S.; Ledermann, J.; Bosse, T.; Chargari, C.; Fagotti, A.; et al. ESGO/ESTRO/ESP guidelines for the management of patients with endometrial carcinoma. *Int. J. Gynecol. Cancer* **2021**, *31*, 12–39. [CrossRef]
72. Church, J.; Simmang, C.; Standards Task Force; American Society of Colon and Rectal Surgeons; Collaborative Group of the Americas on Inherited Colorectal Cancer and the Standards Committee of The American Society of Colon and Rectal Surgeons. Practice parameters for the treatment of patients with dominantly inherited colorectal cancer (familial adenomatous polyposis and hereditary nonpolyposis colorectal cancer). *Dis. Colon. Rectum* **2003**, *46*, 1001–1012. [CrossRef] [PubMed]
73. Dominguez-Valentin, M.; Seppälä, T.T.; Engel, C.; Aretz, S.; Macrae, F.; Winship, I.; Capella, G.; Thomas, H.; Hovig, E.; Nielsen, M.; et al. Risk-Reducing Gynecological Surgery in Lynch Syndrome: Results of an International Survey from the Prospective Lynch Syndrome Database. *J. Clin. Med.* **2020**, *9*, 2290. [CrossRef] [PubMed]
74. Eikenboom, E.L.; Moen, S.; Van Leerdam, M.E.; Papageorgiou, G.; Doukas, M.; Tanis, P.J.; Dekker, E.; Wagner, A.; Spaander, M.C.W. Metachronous colorectal cancer risk according to Lynch syndrome pathogenic variant after extensive versus partial colectomy in the Netherlands: A retrospective cohort study. *Lancet Gastroenterol. Hepatol.* **2023**, *8*, 1106–1117. [CrossRef] [PubMed]
75. Lepore Signorile, M.; Disciglio, V.; Di Carlo, G.; Pisani, A.; Simone, C.; Ingravallo, G. From Genetics to Histomolecular Characterization: An Insight into Colorectal Carcinogenesis in Lynch Syndrome. *Int. J. Mol. Sci.* **2021**, *22*, 6767. [CrossRef]
76. Edelstein, D.L.; Axilbund, J.; Baxter, M.; Hylind, L.M.; Romans, K.; Griffin, C.A.; Cruz-Correa, M.; Giardiello, F.M. Rapid Development of Colorectal Neoplasia in Patients with Lynch Syndrome. *Clin. Gastroenterol. Hepatol.* **2011**, *9*, 340–343. [CrossRef]
77. Ahadova, A.; Seppälä, T.T.; Engel, C.; Gallon, R.; Burn, J.; Holinski-Feder, E.; Steinke-Lange, V.; Möslein, G.; Nielsen, M.; ten Broeke, S.W.; et al. The "unnatural" history of colorectal cancer in Lynch syndrome: Lessons from colonoscopy surveillance. *Int. J. Cancer* **2021**, *148*, 800–811. [CrossRef] [PubMed]
78. Dowty, J.G.; Win, A.K.; Buchanan, D.D.; Lindor, N.M.; Macrae, F.A.; Clendenning, M.; Antill, Y.C.; Thibodeau, S.N.; Casey, G.; Gallinger, S.; et al. Cancer risks for MLH1 and MSH2 mutation carriers. *Hum. Mutat.* **2013**, *34*, 490–497. [CrossRef] [PubMed]
79. Bonadona, V.; Bonaïti, B.; Olschwang, S.; Grandjouan, S.; Huiart, L.; Longy, M.; Guimbaud, R.; Buecher, B.; Bignon, Y.-J.; Caron, O.; et al. Cancer risks associated with germline mutations in MLH1, MSH2, and MSH6 genes in Lynch syndrome. *JAMA* **2011**, *305*, 2304–2310. [CrossRef]
80. Ten Broeke, S.W.; van der Klift, H.M.; Tops, C.M.J.; Aretz, S.; Bernstein, I.; Buchanan, D.D.; de la Chapelle, A.; Capella, G.; Clendenning, M.; Engel, C.; et al. Cancer Risks for PMS2-Associated Lynch Syndrome. *J. Clin. Oncol.* **2018**, *36*, 2961–2968. [CrossRef]
81. Dominguez-Valentin, M.; Sampson, J.R.; Seppälä, T.T.; Ten Broeke, S.W.; Plazzer, J.-P.; Nakken, S.; Engel, C.; Aretz, S.; Jenkins, M.A.; Sunde, L.; et al. Cancer risks by gene, age, and gender in 6350 carriers of pathogenic mismatch repair variants: Findings from the Prospective Lynch Syndrome Database. *Genet. Med.* **2020**, *22*, 15–25. [CrossRef]
82. Goverde, A.; Eikenboom, E.L.; Viskil, E.L.; Bruno, M.J.; Doukas, M.; Dinjens, W.N.M.; Dubbink, E.J.; van den Ouweland, A.M.W.; Hofstra, R.M.W.; Wagner, A.; et al. Yield of Lynch Syndrome Surveillance for Patients With Pathogenic Variants in DNA Mismatch Repair Genes. *Clin. Gastroenterol. Hepatol.* **2020**, *18*, 1112–1120.e1. [CrossRef]
83. Kudchadkar, S.; Ahmed, S.; Mukherjee, T.; Sagar, J. Current guidelines in the surgical management of hereditary colorectal cancers. *WJGO* **2022**, *14*, 833–841. [CrossRef]
84. Creavin, B.; Kelly, M.E.; Ryan, E.; Winter, D.C. Meta-analysis of the impact of surgical approach on the grade of mesorectal excision in rectal cancer. *Br. J. Surg.* **2017**, *104*, 1609–1619. [CrossRef] [PubMed]
85. Parry, S.; Win, A.K.; Parry, B.; Macrae, F.A.; Gurrin, L.C.; Church, J.M.; Baron, J.A.; Giles, G.G.; Leggett, B.A.; Winship, I.; et al. Metachronous colorectal cancer risk for mismatch repair gene mutation carriers: The advantage of more extensive colon surgery. *Gut* **2011**, *60*, 950–957. [CrossRef] [PubMed]
86. Kim, T.J.; Kim, E.R.; Hong, S.N.; Kim, Y.-H.; Huh, J.W.; Park, Y.A.; Cho, Y.B.; Yun, S.H.; Kim, H.C.; Lee, W.Y.; et al. Survival Outcome and Risk of Metachronous Colorectal Cancer After Surgery in Lynch Syndrome. *Ann. Surg. Oncol.* **2017**, *24*, 1085–1092. [CrossRef]

87. de Vos tot Nederveen Cappel, W.H.; Buskens, E.; van Duijvendijk, P.; Cats, A.; Menko, F.H.; Griffioen, G.; Slors, J.F.; Nagengast, F.M.; Kleibeuker, J.H.; Vasen, H.F.A. Decision analysis in the surgical treatment of colorectal cancer due to a mismatch repair gene defect. *Gut* **2003**, *52*, 1752–1755. [CrossRef]
88. Vasen, H.F.A.; Blanco, I.; Aktan-Collan, K.; Gopie, J.P.; Alonso, A.; Aretz, S.; Bernstein, I.; Bertario, L.; Burn, J.; Capella, G.; et al. Revised guidelines for the clinical management of Lynch syndrome (HNPCC): Recommendations by a group of European experts. *Gut* **2013**, *62*, 812–823. [CrossRef] [PubMed]
89. Mohile, S.G.; Hurria, A.; Cohen, H.J.; Rowland, J.H.; Leach, C.R.; Arora, N.K.; Canin, B.; Muss, H.B.; Magnuson, A.; Flannery, M.; et al. Improving the quality of survivorship for older adults with cancer. *Cancer* **2016**, *122*, 2459–2568. [CrossRef] [PubMed]
90. Van Der Vlies, E.; Vernooij, L.M.; Van Erning, F.N.; Vink, G.R.; Bos, W.J.W.; Portielje, J.E.A.; Noordzij, P.G.; Los, M. Survival of surgical and non-surgical older patients with non-metastatic colorectal cancer: A population-based study in the Netherlands. *Eur. J. Surg. Oncol.* **2021**, *47*, 3144–3150. [CrossRef]
91. Kristjansson, S.R.; Nesbakken, A.; Jordhøy, M.S.; Skovlund, E.; Audisio, R.A.; Johannessen, H.-O.; Bakka, A.; Wyller, T.B. Comprehensive geriatric assessment can predict complications in elderly patients after elective surgery for colorectal cancer: A prospective observational cohort study. *Crit. Rev. Oncol./Hematol.* **2010**, *76*, 208–217. [CrossRef]
92. Natarajan, N.; Watson, P.; Silva-Lopez, E.; Lynch, H.T. Comparison of Extended Colectomy and Limited Resection in Patients With Lynch Syndrome. *Dis. Colon. Rectum* **2010**, *53*, 77–82. [CrossRef]
93. Mecklin, J.P.; Järvinen, H. Treatment and follow-up strategies in hereditary nonpolyposis colorectal carcinoma. *Dis. Colon. Rectum* **1993**, *36*, 927–929. [CrossRef]
94. Ruo, L.; Guillem, J.G. Surgical management of primary colorectal cancer. *Surg. Oncol.* **1998**, *7*, 153–163. [CrossRef] [PubMed]
95. Comparison of the Pathological Response to 2 or 4 Cycles of Neoadjuvant CAPOX in II/III Rectal Cancer Patients with low/Intermediate Risks: Study Protocol for a Prospective, Non-Inferior, Randomized Control Trial (COPEC Trial) | Trials | Full Text. Available online: https://trialsjournal.biomedcentral.com/articles/10.1186/s13063-023-07405-x (accessed on 10 September 2024).
96. Quezada-Diaz, F.F.; Hameed, I.; von Mueffling, A.; Salo-Mullen, E.E.; Catalano, J.D.; Smith, J.J.; Weiser, M.R.; Garcia-Aguilar, J.; Stadler, Z.K.; Guillem, J.G. Risk of Metachronous Colorectal Neoplasm After a Segmental Colectomy in Lynch Syndrome Patients According to Mismatch Repair Gene Status. *J. Am. Coll. Surg.* **2020**, *230*, 669–675. [CrossRef] [PubMed]
97. Urso, E.D.L.; Celotto, F.; Giandomenico, F.; Gavaruzzi, T.; Del Bianco, P.; Lotto, L.; Spolverato, G.; Pucciarelli, S.; Bao, Q.R. Analysis of morbidity and mortality, quality of life and bowel function after total colectomy with ileorectal anastomosis versus right and left hemicolectomy: A study to optimise the treatment of lynch syndrome and attenuated polyposis coli. *Eur. J. Surg. Oncol.* **2020**, *46*, 1613–1619. [CrossRef]
98. Kalady, M.F.; McGannon, E.; Vogel, J.D.; Manilich, E.; Fazio, V.W.; Church, J.M. Risk of colorectal adenoma and carcinoma after colectomy for colorectal cancer in patients meeting Amsterdam criteria. *Ann. Surg.* **2010**, *252*, 507–511; discussion 511–513. [CrossRef]
99. Lee, J.S.; Petrelli, N.J.; Rodriguez-Bigas, M.A. Rectal cancer in hereditary nonpolyposis colorectal cancer. *Am. J. Surg.* **2001**, *181*, 207–210. [CrossRef]
100. de Rosa, N.; Rodriguez-Bigas, M.A.; Chang, G.J.; Veerapong, J.; Borras, E.; Krishnan, S.; Bednarski, B.; Messick, C.A.; Skibber, J.M.; Feig, B.W.; et al. DNA Mismatch Repair Deficiency in Rectal Cancer: Benchmarking Its Impact on Prognosis, Neoadjuvant Response Prediction, and Clinical Cancer Genetics. *J. Clin. Oncol.* **2016**, *34*, 3039–3046. [CrossRef]
101. Møller, P.; Seppälä, T.T.; Bernstein, I.; Holinski-Feder, E.; Sala, P.; Gareth Evans, D.; Lindblom, A.; Macrae, F.; Blanco, I.; Sijmons, R.H.; et al. Cancer risk and survival in path_MMR carriers by gene and gender up to 75 years of age: A report from the Prospective Lynch Syndrome Database. *Gut* **2018**, *67*, 1306–1316. [CrossRef]
102. You, Y.N.; Marcante, M.; George, T.J. Lynch Syndrome: Management of Rectum, What Operation? In *Management of Hereditary Colorectal Cancer: A Multidisciplinary Approach*; Guillem, J.G., Friedman, G., Eds.; Springer International Publishing: Cham, Switzerland, 2020; pp. 175–200. [CrossRef]
103. Chikatani, K.; Ishida, H.; Mori, Y.; Nakajima, T.; Ueki, A.; Akagi, K.; Takao, A.; Yamada, M.; Taniguchi, F.; Komori, K.; et al. Risk of metachronous colorectal cancer after surgical resection of index rectal cancer in Lynch syndrome: A multicenter retrospective study in Japan. *Surg. Today* **2024**, *54*, 1075–1083. [CrossRef]
104. Win, A.K.; Parry, S.; Parry, B.; Kalady, M.F.; Macrae, F.A.; Ahnen, D.J.; Young, G.P.; Lipton, L.; Winship, I.; Boussioutas, A.; et al. Risk of metachronous colon cancer following surgery for rectal cancer in mismatch repair gene mutation carriers. *Ann. Surg. Oncol.* **2013**, *20*, 1829–1836. [CrossRef]
105. Kalady, M.F.; Lipman, J.; McGannon, E.; Church, J.M. Risk of Colonic Neoplasia After Proctectomy for Rectal Cancer in Hereditary Nonpolyposis Colorectal Cancer. *Ann. Surg.* **2012**, *255*, 1121–1125. [CrossRef]
106. Wang, X.; Yu, Y.; Meng, W.; Jiang, D.; Deng, X.; Wu, B.; Zhuang, H.; Wang, C.; Shen, Y.; Yang, L.; et al. Total neoadjuvant treatment (CAPOX plus radiotherapy) for patients with locally advanced rectal cancer with high risk factors: A phase 2 trial. *Radiother. Oncol.* **2018**, *129*, 300–305. [CrossRef]

107. Altomare, N.J.; Mulcahy, M.F. Evolution of therapy for locally advanced rectal cancer. *J. Surg. Oncol.* **2024**, *129*, 78–84. [CrossRef]
108. Bailey, C.E.; Tran Cao, H.S.; Hu, C.-Y.; Chang, G.J.; Feig, B.W.; Rodriguez-Bigas, M.A.; Nguyen, S.T.; Skibber, J.M.; You, Y.N. Functional deficits and symptoms of long-term survivors of colorectal cancer treated by multimodality therapy differ by age at diagnosis. *J. Gastrointest. Surg.* **2015**, *19*, 180–188. [CrossRef] [PubMed]
109. Survivorship, Version 2.2018, NCCN Clinical Practice Guidelines in Oncology—PubMed. Available online: https://pubmed.ncbi.nlm.nih.gov/30323092/ (accessed on 13 September 2024).
110. Stitzenberg, K.B.; Barnes, E. Advances in Rectal Cancer Surgery. *Clin. Color. Cancer* **2022**, *21*, 55–62. [CrossRef]
111. Young, D.O.; Kumar, A.S. Local Excision of Rectal Cancer. *Surg. Clin. N. Am.* **2017**, *97*, 573–585. [CrossRef]
112. Jones, H.J.S.; Hompes, R.; Mortensen, N.; Cunningham, C. Modern management of T1 rectal cancer by transanal endoscopic microsurgery: A 10-year single-centre experience. *Color. Dis.* **2018**, *20*, 586–592. [CrossRef] [PubMed]
113. Alexandrescu, S.T.; Dumitru, A.V.; Babiuc, R.D.; Costea, R.V. Assessment of clinical and pathological complete response after neoadjuvant chemoradiotherapy in rectal adenocarcinoma and its therapeutic implications. *Rom. J. Morphol. Embryol.* **2021**, *62*, 411–425. [CrossRef] [PubMed]
114. Goldstein, J.; Tran, B.; Ensor, J.; Gibbs, P.; Wong, H.L.; Wong, S.F.; Vilar, E.; Tie, J.; Broaddus, R.; Kopetz, S.; et al. Multicenter retrospective analysis of metastatic colorectal cancer (CRC) with high-level microsatellite instability (MSI-H). *Ann. Oncol.* **2014**, *25*, 1032–1038. [CrossRef]
115. Popescu, I.; Alexandrescu, S. Metastazele hepatice colorectale—Posibilităti terapeutice actuale [Hepatic metastasis of colorectal cancer—Current therapeutic possibilities]. *Chirurgia* **2010**, *105*, 155–169.
116. Popescu, I.; Alexandrescu, S.; Croitoru, A.; Boros, M. Strategies to convert to resectability the initially unresectable colorectal liver metastases. *Hepatogastroenterology* **2009**, *56*, 739–744.
117. Alexandrescu, S.; Diaconescu, A.; Ionel, Z.; Zlate, A.-C.; Grigorie, R.; Hrehoret, D.; Brasoveanu, V.; Dima, S.; Botea, F.; Ionescu, M.; et al. Comparative Analysis between Simultaneous Resection and Staged Resection for Synchronous Colorectal Liver Metastases—A Single Center Experience on 300 Consecutive Patients. *Chirurgia* **2017**, *112*, 278–288. [CrossRef] [PubMed]
118. Alexandrescu, S.; Hrehoreţ, D.; Ionel, Z.; Croitoru, A.; Anghel, R.; Popescu, I. Simultaneous resection of the primary colorectal tumor and liver metastases—A safe and effective operation. *Chirurgia* **2012**, *107*, 298–307.

medicina

MDPI

Review

Impact of Microbiota on Irritable Bowel Syndrome Pathogenesis and Management: A Narrative Review

Mhd Bashir Almonajjed [1], Mahdi Wardeh [1], Abdallah Atlagh [1], Abdulrahman Ismaiel [2,*], Stefan-Lucian Popa [2], Flaviu Rusu [2] and Dan L. Dumitrascu [2]

1 Faculty of Medicine, "Iuliu Hatieganu" University of Medicine and Pharmacy, 400006 Cluj-Napoca, Romania; m.basheralmounajed@gmail.com (M.B.A.); mahdi.wardeh2003@gmail.com (M.W.); atlagha1@yahoo.fr (A.A.)

2 2nd Department of Internal Medicine, "Iuliu Hatieganu" University of Medicine and Pharmacy, 400006 Cluj-Napoca, Romania; popa.stefan@umfcluj.ro (S.-L.P.); flaviurusu@yahoo.com (F.R.); ddumitrascu@umfcluj.ro (D.L.D.)

* Correspondence: abdulrahman.ismaiel@yahoo.com

Abstract: Irritable bowel syndrome (IBS) is a prevalent gastrointestinal disorder, affecting 3–5% of the global population and significantly impacting patients' quality of life and healthcare resources. Alongside physical symptoms such as abdominal pain and altered bowel habits, many individuals experience psychological comorbidities, including anxiety and depression. Recent research has highlighted the critical role of the gut microbiota in IBS, with dysbiosis, characterized by an imbalance in microbial diversity, frequently observed in patients. The gut–brain axis, a bidirectional communication network between the gut and central nervous system, plays a central role in the development of IBS symptoms. Although interventions such as probiotics, prebiotics, synbiotics, and fecal microbiota transplantation (FMT) have demonstrated potential in modulating the gut microbiota and alleviating symptoms, their efficacy remains an area of ongoing investigation. This review examines the interactions between the gut microbiota, immune system, and brain, emphasizing the need for personalized therapeutic strategies. Future research should aim to identify reliable microbiota-based biomarkers for IBS and refine microbiome-targeted therapies to enhance patient outcomes.

Keywords: irritable bowel syndrome; microbiota; gut–brain axis; disorders of gut brain interaction (DGBI)

1. Introduction

Irritable bowel syndrome (IBS) affects 3–5% of the world's population and is diagnosed using the Rome IV criteria, which are symptom-based [1]. While the impact on mortality is unknown, IBS has a major influence on quality of life, especially through linked psychiatric illnesses such as anxiety and depression, resulting in higher healthcare usage and decreased productivity.

Research emphasizes the significance of gut microbiota in IBS, with dysbiosis, characterized by reduced microbial diversity, reported in many patients. Specific bacterial families, such as *Firmicutes* and *Proteobacteria*, are implicated, and a decrease in butyrate-producing bacteria may contribute to symptom onset by compromising intestinal barrier integrity [2]. Probiotics have shown promise in treating IBS symptoms, highlighting the microbiome's role in the condition [2].

The brain–gut connection, which has a long history, has been scientifically validated using modern imaging techniques. According to research, gut stimuli can trigger brain regions involved in emotion regulation, and gastrointestinal dysfunction frequently precedes neurological diseases such as Parkinson's disease [3]. This emphasizes the significance of gut health in general neurological and emotional well-being, underlining the need to better understand gut–brain connections in IBS patients.

The global prevalence of IBS varies due to factors such as food, ethnicity, and healthcare systems. IBS-D (diarrhea-predominant) and IBS-C (constipation-predominant) account for around 30% of cases, with women having a greater incidence [4]. The disorder has a substantial impact on daily living, limiting productivity and social participation. Some patients are willing to give up years of life for symptom alleviation. Recent research has connected altered gut microbiota to IBS, implying that bile acids, psychosocial variables, and genetic predispositions all contribute to IBS pathogenesis [4]. However, the particular microbiome signature associated with IBS severity and treatment response is still being investigated.

The pathogenesis of IBS is complicated, involving elements such as visceral hypersensitivity and gut microbiota changes, with the gut–brain axis playing a key role in symptom development. While studies on probiotics such as *Lactobacillus* and *Bifidobacterium* spp. demonstrated promise, their clinical importance is unknown [5]. IBS patients typically have altered gut microbiota, including decreased *Bifidobacterium* and increased *Bacteroides*, but the cause and stability of these alterations are still being explored [5]. The lack of specific biomarkers interferes with the diagnosis and the management, while genetic predisposition and psychological variables also contribute to the disease progression. The Rome IV criteria also indicate a continuum in gastrointestinal diseases, with symptom overlap common, confounding diagnosis and comprehension of disorders of gut–brain interaction (DGBI) [6].

While the significance of probiotics in IBS treatment is debated, synbiotics have shown promise in relieving symptoms, particularly in IBS-D patients [6]. Furthermore, fecal microbiota transplantation (FMT) is emerging as a potential therapy option [6]; however, its efficacy and safety must be further investigated.

This review explores the function of gut microbiota in the development and treatment of IBS. It investigates differences in gut microbiota composition among IBS subtypes, the gut–brain axis, and their roles in symptom development. Furthermore, it emphasizes the roles of biofilms and small intestinal bacterial overgrowth (SIBO) in IBS pathogenesis. Current microbiome-targeted therapeutics, including probiotics, prebiotics, synbiotics, and fecal microbiota transplantation (FMT), are evaluated alongside dietary interventions to determine their impact on gut microbiota and symptom alleviation. The emphasis is on personalized therapy techniques, with a focus on identifying research gaps and providing future approaches for improving microbiome-based diagnostics and therapeutics in IBS.

2. The Gut Microbiome in Health and Disease

The gut microbiota is critical to human health since it ferments dietary fibers, produces short-chain fatty acids (SCFAs), and regulates the immune system [6]. A healthy and diversified microbiome is essential for intestinal health and general well-being. Healthy individuals often have a diverse microbial makeup, characterized by beneficial bacteria such as *Lactobacilli* and *Bifidobacteria* [7]. The gut microbiota is predominantly composed of four phyla: *Firmicutes*, *Bacteroidetes*, *Actinobacteria*, and *Proteobacteria*, all of which play important roles in metabolic processes and immunological function [8].

The gut microbiota develops early in childhood and is impacted by a variety of factors, including nutrition and antibiotic exposure, which can alter microbial populations [8]. Individual microbial makeup varies significantly, influenced by eating habits, age, and lifestyle choices. According to research, microbiome alterations might cause imbalances, which contribute to gastrointestinal problems [6]. In healthy individuals, the microbiota makeup is dominated by beneficial species that aid in gastrointestinal homeostasis [9].

Perinatal variables, such as method of delivery and maternal education, have a major impact on the development of IBS, with a strong link to cesarean delivery. These factors have an impact on early life microbial profiles, which may increase IBS risk later in life [10].

3. Alterations in Gut Microbiota in IBS Patients

Alterations in gut microbiota have also been associated to immunological activation and intestinal barrier dysfunction, particularly in post-infectious IBS (PI-IBS), where previous infections can cause chronic symptoms [11]. Evidence indicates that molecular mimicry between microbial antigens and host proteins may contribute to chronic inflammation and nerve damage [11]. A systematic review found a relationship between gastroenteritis and IBS, and the overall prevalence of PI-IBS was 14.5% [12]. Compared to bacterial and viral enteritis, protozoal infections pose a greater risk for the development of PI-IBS [11]. This is linked to the stimulation of inflammatory processes, exposure to exogenous substances, and increased intestinal permeability [11]. Furthermore, gut dysbiosis, which includes reductions in good bacteria like *Bifidobacterium* and *Lactobacillus* and increases in dangerous species like *Enterobacteriaceae*, has been linked to IBS [11]. However, the variability of available studies makes it difficult to identify a consistent microbial signature for IBS, underlining the importance of subtype-specific research and microbiota-targeted therapy.

Functional gastrointestinal disorders (FGIDs) are common but poorly understood due to the absence of obvious organic abnormalities, complicating diagnosis and therapy. The Rome IV criteria redefined FGIDs as DGBI, emphasizing the importance of psychological comorbidities including anxiety and depression [13]. This review also emphasizes the gut–brain axis and the microbiome's critical roles in DGBI, which includes IBS and functional dyspepsia [13]. New research suggests that the gut microbiota regulates gut motility, visceral sensitivity, and even brain activity, with neurological, immunological, and metabolic pathways supporting bidirectional communication [13]. Microbial metabolites that affect the gut–brain axis, such as serotonin, tryptophan, tryptamine, and SCFAs, modify these effects. Stress-induced dysbiosis and comorbid illnesses are among the psychiatric diseases linked to gut microbiota composition [11]. Given that each person's microbiome is distinct, individualized treatment approaches, particularly diet-based therapies, are recommended to address individual variability.

Numerous phyla that make up the gut microbiota are essential to preserving gut homeostasis. *Enterococcus, Ruminococcus, Clostridium, Lactobacillus, Faecalibacterium, Roseburia,* and *Eubacterium* are among the *Firmicutes* that are involved in the metabolism of amino acids, carbohydrates, and lipids as well as the transformation of bile acids and the creation of cholesterol [2,8]. Along with aiding in the synthesis of vitamins K2, B1, B2, B6, B7, B9, and B12, they also support the integrity of the intestinal epithelial barrier and immunological response, which guards against enteric infections [2,8]. *Bacteroidetes*, which include taxa like *Bacteroides* and *Prevotella,* have related roles in immunological modulation, metabolic pathways, and appetite regulation [8,11]. Vitamin production (K2, B1, B2, B6, B7, B9, and B12), bile acid metabolism, and protection against enteric infections are all aided by *Actinobacteria,* which are represented by *Bifidobacterium* and *Corynebacterium* [2,8]. Finally, *Proteobacteria,* which include *Shigella, Escherichia,* and *Desulfovibrio,* are important in the metabolism of amino acids and can affect gut disease when their populations become dysregulated [8,11].

In contrast to healthy microbiota, IBS patients have dysbiosis, which is characterized by diminished microbial diversity and microbial population imbalance. (Figure 1 demonstrates the alterations that can happen to the microbiome in IBS.) Studies show that IBS patients have a different gut microbiota makeup, with decreased levels of beneficial bacteria and an increase in pro-inflammatory species [4,14]. Dysbiosis in IBS patients causes changes in microbial metabolites that impact the mucosal and systemic levels [11]. Visceral hypersensitivity and increased inflammatory cytokine production are frequent in PI-IBS [11]. Distinct microbial patterns have also been noted across IBS subtypes. In particular, one study discovered that fecal *Lactobacillus* and *Bifidobacterium* were correlated with IL-10 in IBS-C patients, while Gram-positive and Gram-negative bacteria are correlated with C-X-C motif chemokine ligand 11 in IBS-D patients [11]. The altered microbiota of some IBS patients may be associated with clinical severity and psychosocial factors; changes in brain regions related to emotional responses are correlated with changes in the microbiota, such as the prevalence of *Prevotella* over *Bacteroides* [11]. This microbiota is frequently unstable, influenced by environmental factors such as nutrition and antibiotic use, confounding comprehension of its function in the illness [5,14].

According to research, certain bacterial families, such as *Bifidobacteria* and *Faecalibacterium*, represented lower numbers than in healthy subjects, whereas *Lactobacilli* and *Bacteroides* were found to be increased [15]. IBS symptoms include decreased microbial diversity, increased pro-inflammatory species such as *Bacteroides*, and a decline in the number of anti-inflammatory species such as the butyrate-producing bacteria *Faecalibacterium prausnitzii* [16]. Subtype-specific differences in transcriptomics and metabolomics show various microbiota-related processes that underpin IBS symptoms. Additionally, one study showed an increase in *Clostridium* [17] and higher levels of *Streptococcus* and *Gardnerella vaginalis* [18], all of which are linked to IBS symptoms. One study indicated decreasing levels of *Lactobacilli* in IBS-D patients [19], while another mentioned a trend of decreased beneficial species such as *Lactobacilli* and *Bifidobacteria* in its findings, highlighting discrepancy [9]. Notably, IBS frequently results in a reduction in butyrate-producing bacteria and an increase in pro-inflammatory *Enterobacteriaceae* [8].

Through the fermentation of polysaccharides, *Methanobrevibacter* species like *Methanobrevibacter smithii* and *M. stadtmanae* play important roles in the gut microbiota by creating hydrogen (H_2) and methane (CH_4), which can affect gut permeability and bowel motility [2]. Furthermore, metabolite-sensing G-protein-coupled receptors (GPR43, GPR41, and GPR109A) interact with butyrate and other SCFAs generated by colonic bacteria, such as strains of *Bifidobacterium* and *Lactobacillus*, to control inflammatory responses and support gut homeostasis [2]. Because IBS is characterized by inflammation and immune system activation, these pathways are especially pertinent in this condition [2,11].

Methane-producing bacteria are more abundant in IBS-C than in IBS-D, with an overall decrease in butyrate-producing bacteria [9]. IBS patients had a higher *Firmicutes*-to-*Bacteroidetes* ratio and higher levels of certain *Streptococci* and *Ruminococcus* species than healthy individuals [9]. *Ruminococcaceae* levels have been found significantly reduced in IBS-D patients, as has bile acid metabolism, which is linked to symptoms such as diarrhea and visceral hypersensitivity [20].

Microbes and their constituents can enter the mucosa due to increased permeability caused by the intestinal epithelial barrier being disrupted in IBS [2,8,11]. The immune system is triggered by this exposure, which results in aberrant inflammatory reactions that exacerbate IBS symptoms [3,8,11]. This problem has been made worse by the discovery that IBS patients have abnormalities in tight junction proteins, which are essential for preserving barrier function [8,11]. IBS is characterized by immune activation in the intestinal mucosa, wherein pro-inflammatory cytokines are produced in greater quantities in response to either direct microbial stimulation or indirect activation via microbial antigens [8,11]. Visceral hypersensitivity and bowel pain are linked to this elevated immune response, which is biased toward pro-inflammatory Th1 and Th17 pathways [2,11]. Moreover, cross-reactive immune responses may be triggered by molecular mimicry between host proteins and pathogen antigens. For instance, host proteins like vinculin may be mistakenly targeted by antibodies produced against bacterial toxins, impairing intestinal neuronal activity [11]. Furthermore, by blocking histone deacetylases, microbial metabolites like butyrate can cause epigenetic modifications, modifying gene expression and adding to the molecular alterations observed in the gut and brain systems of individuals with IBS [11].

The gut–brain axis plays a part in the pathophysiology of IBS, as evidenced by elevated levels of pro-inflammatory cytokines such as IL-6, TNF-α, and IL-1β, which are associated with anxiety and depression [2]. Furthermore, intestinal permeability and somatic hypersensitivity are made worse by genetic and epigenetic variables, such as dysregulated microRNA production and changes in the serotonin receptor gene, which result in symptoms of the disease and a lower quality of life for IBS patients [2].

Despite proven abnormalities in gut microbiota in IBS patients, a particular microbial signature that distinguishes these individuals has yet to be found, with no clear signature identified for IBS subgroups [15,21]. Small intestine bacterial overgrowth (SIBO) is significantly more common in IBS patients [22], particularly in IBS-D [23]. However, its role is debatable due to diagnostic limitations [21].

The interplay of gut microbiota and bile acids (BAs) is important in IBS pathogenesis. Certain BAs, particularly CDCA and DCA, can cause cellular damage and compromise tight junction integrity, resulting in increased intestinal permeability that contributes to IBS symptoms [17]. According to research, impaired intestinal barrier integrity and increased immune activation may contribute to IBS symptoms, especially in instances triggered by past infections [11]. PI-IBS is a significant risk factor that can develop following a variety of gastrointestinal infections, with meta-analyses revealing a fourfold increase in IBS risk after infection [21]. PI-IBS can cause long-term symptoms due to dysbiosis and inflammation [23].

The gut microbiota has a major impact on intestinal barrier integrity and immune system modulation, two important aspects of IBS pathogenesis. In IBS patients, mast cells are more prevalent close to enteric nerve fibers [8,10]. These cells release mediators including serotonin and histamine, which cause cytokine imbalance and lymphocyte activation, changing pain thresholds and escalating visceral hypersensitivity [8,10]. By releasing tryptase, mast cell degranulation and eosinophil activation further weaken tight junction proteins, which increases intestinal permeability [8,10]. Certain microorganisms help to regulate these processes: tryptophan is converted by *Lactobacilli* species into indole-3-aldehyde, which activates the aryl hydrocarbon receptor (AHR), which controls intraepithelial lymphocyte populations and stimulates the production of the anti-inflammatory IL-22 [8].

Mucus layer composition and thickness are influenced by *Ruminococcus* species, *Bacteroides thetaiotaomicron*, and *Faecalibacterium prausnitzii* [8,18]. One of the most prevalent bacterial species in the gut is *Faecalibacterium prausnitzii.* Through the activation of regulatory T cells, the promotion of IL-10 secretion, and the inhibition of IL-8 synthesis, it demonstrates anti-inflammatory properties [8]. *Firmicutes* create SCFAs, which improve epithelial integrity by upregulating the expression of tight junction proteins such as occludin and claudins [8,17]. E-cadherin production is stimulated by polyamines produced by genera such as *Lactobacillus* and *Clostridium,* which strengthen barrier function [5,8]. *Lactobacillus rhamnosus, Bifidobacterium breve,* and other probiotic strains control pro- and anti-inflammatory cytokines, preserving the integrity of the intestinal barrier and reducing the symptoms of IBS [5,8].

Intestinal barrier dysfunction is common in IBS, particularly in IBS-D, and correlates with increased permeability, which contributes to low-grade inflammation and symptom exacerbation [24]. Tight junction protein expression changes and enhanced mast cell activation have been seen inside the intestinal mucosa [24]. Notably, SCFAs play an important function in increasing tight junction protein expression [8]. Dysbiosis can lead to the generation of proteases that weaken the intestinal barrier, emphasizing the intricate relationship between microbiota composition and gut integrity [24]. Furthermore, a reduction in the number of butyrate-producing bacteria may compromise intestinal barrier function [2].

Recent research found that higher serum levels of D-lactate and diamine oxidase are related to increased IBS severity [25], implying a possible relationship between microbiota composition and symptom intensity. Certain bacterial families, both beneficial and harmful, have been linked to IBS severity, supporting the concept of a microbiota-based biomarker for the illness. However, significant studies have found that IBS patients have an excess of harmful bacteria, such as *Enterobacteriaceae*, and fewer good bacteria, such as *Bifidobacterium* and *Lactobacillus* [25]. Dietary therapies, such as gluten-free or low fermentable oligosaccharides, disaccharides, monosaccharides, and polyols (FODMAP) diets, have not consistently improved dysbiosis indices in IBS patients [25], highlighting the complexities of the microbiome's participation in the condition.

Research has also shown that psychological problems such as anxiety and depression are common among IBS patients, with a substantial relationship between symptom severity and psychological distress [2,13,26].

Food hypersensitivity adds to dysbiosis, complicating treatment options and highlighting the importance of personalized dietary approaches to properly control symptoms [27]. The gluten-free diet (GFD) has become popular among IBS patients, with some research suggesting symptom relief. However, this relief may be due to a reduction in fructans, which are FODMAPs rather than gluten itself, confounding dietary assessments [28].

In simple terms, changes in gut microbiota composition and functionality are critical to the pathophysiology of IBS, demanding additional research to understand these complex interactions. The link between gut microbiota composition and IBS symptoms is variable across the literature, underscoring the need for additional study to determine causality [15].

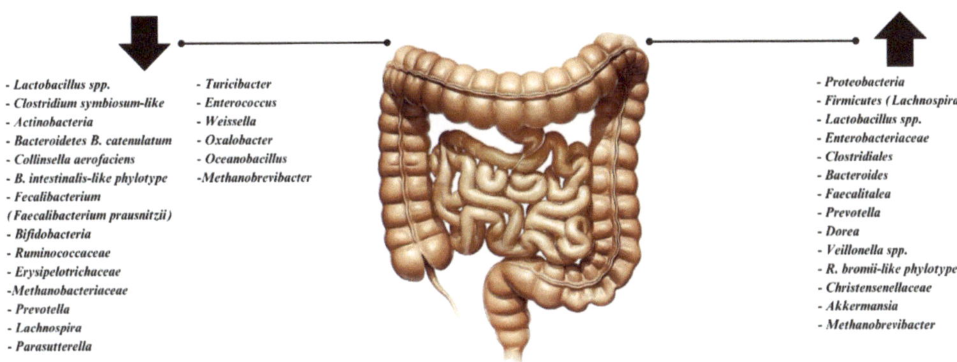

Figure 1. Microbiota alterations in IBS. (The figure is adapted with modifications from Surdea-Blaga et al., 2024, Microbiome in irritable bowel syndrome: advances in the field—A scoping review [25]).

4. The Complex Relationship Between SIBO and IBS

Small intestinal bacterial overgrowth (SIBO), formerly known as "blind loop syndrome", is associated with maldigestion and malabsorption caused by excessive bacterial growth in the small intestine. The symptoms are diarrhea, steatorrhea, and megaloblastic anemia. While jejunal aspirate cultures were formerly the diagnostic gold standard, breath tests like lactulose hydrogen (LHBT) and glucose hydrogen (GHBT) are now widely utilized despite concerns about specificity and false positives [29]. Studies indicate a high incidence of SIBO in IBS patients, while some report symptom relief following antibiotics; however, diagnostic inconsistencies confuse findings [29]. Emerging data have revealed a possible relationship between IBS and SIBO. Gastric achlorhydria, motility disorders, and small bowel stasis are all risk factors for SIBO [30]. While some studies suggest a high SIBO rate in IBS patients and relief of symptoms with antibiotics such as rifaximin, inconsistent evidence and the lack of defined diagnostic criteria confuse the SIBO-IBS link [30]. A case–control study found that 84% of IBS patients tested positive for LHBT, and neomycin therapy alleviated symptoms in these situations [30]. Similarly, the antibiotic rifaximin has shown minor efficacy in IBS-D and is approved by the FDA for this type of condition; however, its advantages do not directly indicate SIBO involvement. Systematic reviews have revealed uneven SIBO prevalence in IBS, with confounding factors such as proton pump inhibitor use aggravating the relationship [30]. Studies have found no obvious symptom differences between SIBO-positive and SIBO-negative people with IBS [30]. Current research does not clearly show SIBO as a causal factor in IBS, emphasizing the need for molecular techniques to investigate the intricate connection between gut microbiota and IBS pathology.

5. Biofilms and IBS

IBS, inflammatory bowel disease (IBD), and colorectal cancer (CRC) are among the gastrointestinal disorders that are influenced by biofilms, which are essential for preserving gut homeostasis. Comprising intricate microbial colonies shielded by a matrix, biofilms display traits including virulence and resistance to antibiotics that help prolong the course of disease [31]. To maintain host–microbiota equilibrium, these biofilms interact with the mucosal microbiota. Polymicrobial and trans-kingdom interactions (encompassing viruses, Prokarya, Eukarya, and Archaea) are key for host–microbiota balance [31]. Disruptions in the integrity of biofilms, frequently brought on by compromised mucus or excessive use of antibiotics, can result in dysbiosis and illness; resistance is increased by biofilm-associated bacterial dispersion and gene transfer, and the makeup of the gut's mucosal microbiota differs from that of the fecal microbiota [31]. Understanding biofilm dynamics is crucial for therapeutic treatments, as biofilm development is especially common in high-density microbial regions like the colon [31].

Biofilms contribute to altered microbiomes and decreased bacterial diversity, which increase the density of bacteria in the gut mucosa and worsen disease pathophysiology [32]. Commensal biofilms can also be beneficial in preventing infections through competitive exclusion even if biofilms in pathogenic conditions like *H. pylori* directly contribute to the development and recurrence of the disease [32]. Moreover, IBS responds differently to therapies that target the gut microbiota, such as FMT, antibiotics, and dietary modifications. Rifaximin, for example, provides short-term symptom alleviation, especially for bloating [33]. However, the impact of these treatments on biofilm-related mechanisms remains poorly understood, and the inconsistent results from FMT trials suggest that biofilms are not fully addressed in current therapies [33]. Future research should focus on biofilm-targeted strategies to better understand and treat GI disorders like IBS.

6. Gut–Brain Axis and IBS Symptoms

The gut microbiota communicates bidirectionally with the autonomic nervous system (ANS), modulating gut motility and secretions through metabolites like serotonin, histamine, and GABA [22,34]. It affects both the central and enteric neural systems, impacting gastrointestinal function and emotional regulation. The microbiota–gut–brain (MGB) axis concept illustrates how these systems interact [3]. Neuroimaging studies utilizing MRI revealed structural and functional abnormalities in the brain associated with IBS, demonstrating that specific microbial signatures correspond to alterations in brain structure and activity [24]. Current research on the MGB axis has produced conflicting results owing to a lack of causal evidence tying changes in the gut microbiota to brain function. Most research have focused on preclinical animal models, indicating immune system interactions, metabolites, and neurotransmitter signaling as important communication pathways [3]. *Eubacterium*, *Bacteroides*, and *Clostridium* (clusters IV, XI, XIII, and XIVa) are key producers of secondary bile acids and SCFAs, which stimulate serotonin synthesis in colonic enterochromaffin cells and regulate gastrointestinal motility [2,8]. Serum serotonin levels differ between IBS subtypes, being lower in IBS-C and higher in IBS-D [3,5]. The serotonin system, particularly 5-HT3 and 5-HT4 receptors, plays a crucial role in gastrointestinal motility and sensory functions [6,8]. 5-HT4 receptors stimulate acetylcholine release, accelerating the peristaltic reflex, whereas 5-HT3 receptors mediate smooth muscle contraction and gut–brain communication [5,8]. Therapeutic interventions targeting these receptors have shown promise; 5-HT3 receptor antagonists alleviate abdominal pain and IBS-D symptoms, while 5-HT4 receptor agonists improve stool frequency, consistency, and abdominal discomfort in IBS-C [3,9]. However, adverse effects, including cardiovascular risks, have led to the withdrawal or restricted use of some medications, driving the development of safer alternatives [6]. Bacterial metabolites such as SCFAs impact neuropeptide production, which controls gastrointestinal motility and sensitivity [8].

Maternal factors, including as food and stress, have a significant impact on the baby microbiome, which affects brain and enteric nervous system development. The early microbiome makeup is essential for long-term gut–brain connection [3]. Research suggests that the timing of microbiota recolonization in germ-free rodents is critical for recovering key brain functions, implying a sensitive period for microbial influence [3]. While human research is scarce, preliminary results point to a link between microbiome composition and cognitive development, particularly in early childhood [3].

The human microbiota has been intensively investigated, with a revised human-to-microbiota cell ratio of 1.3:1, emphasizing microbial cells' considerable genetic contribution [34]. Notably, more than 99% of the genes in the human body are microbial, implying a co-evolutionary relationship that could influence immune responses and epigenetics [34]. The gut microbiota can be therapeutically modified by diet and lifestyle, opening new avenues for treating chronic diseases, notably DGBI.

Communication between the gut and brain occurs via a variety of channels, including neurological, immunological, and metabolic mechanisms, with the vagus nerve playing an important role [13]. The vagus nerve has been identified as the principal pathway for gut microbiota impacts on the central nervous system (CNS), with various bacteria producing neurotransmitters that can influence behavior and brain function [34]. For example, several strains of *Bifidobacteria* have been demonstrated to increase tryptophan levels, a precursor to serotonin [34].

The gut microbiota influences anxiety and stress-related behaviors, with significant differences in microbiome composition found in anorexia nervosa patients compared to healthy controls. (Figure 2 shows an overview of other factors that could be involved in IBS pathogenesis.) Evidence suggests a link between gut microbiota and neurodegenerative illnesses, with certain bacterial species influencing brain health and function [35]. Furthermore, neuropsychological symptoms such as brain fog have been linked to increased SIBO rates, implying a possible gut–brain connection [36].

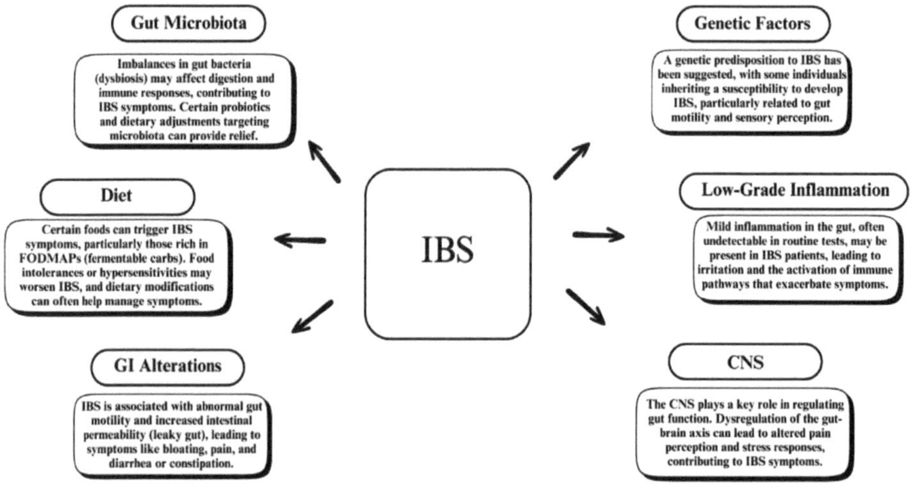

Figure 2. An overview of suggested factors involved in IBS.

7. Microbiome Targeted Treatment in IBS

Probiotics have been demonstrated to boost immunological function, reduce inflammation, and promote gut health in IBS patients by increasing the number of helpful bacteria and promoting the production of SCFAs [20]. Specifically, strains such as *Bifidobacterium* and *Lactobacillus* have shown promise in alleviating IBS symptoms, although their effectiveness varies depending on the strain and patient characteristics [1,37,38]. Probiotics are thought to lower gut inflammation, enhance gut health, and regulate immune responses by modulating pro-inflammatory and anti-inflammatory cytokines [10,39]. Probiotics increase short-chain fatty acids, promote *Lactobacillus* and *Bifidobacterium* colonization, and alleviate colonic hypersensitivity by upregulating μ-opioid and cannabinoid receptor expression. They also enhance gut barrier function, inhibit pathogenic bacteria, produce neurotransmitters, and regulate IL-10/IL-12 levels while decreasing pro-inflammatory cytokines [15,20]. Moreover, certain probiotics may alter gut pain receptors and affect the immune system and brain, contributing to symptom relief, mood improvement, and anxiety reduction [7,39].

While probiotics hold potential, research remains inconsistent. For example, a meta-analysis of 54 randomized controlled trials (RCTs) found that probiotics, particularly *Lactobacillus* and *Bifidobacterium* strains, significantly improved IBS symptoms, especially abdominal pain, but effects varied across strains and formulations [38,40]. In placebo-controlled studies, *B. infantis* significantly improved abdominal pain/discomfort in IBS patients after at least 4 weeks of treatment; *B. lactis* reduced abdominal distension, transit times, pain/discomfort, and global IBS symptoms in female IBS-C patients; *B. animalis* improved bloating within 3 weeks and stool frequency within 6 weeks in IBS-C patients; and *B. bifidum* improved pain, discomfort, bloating, urgency, and quality of life after 4 weeks of treatment [37]. Also, placebo-controlled trials demonstrated that *Lactobacillus* probiotics (*L. plantarum*, *L. rhamnosus*, *L. casei*, and *L. reuteri*) alleviate IBS symptoms, but the outcomes are less consistent compared to *Bifidobacterium* [37]. Another meta-analysis indicated that some probiotics may be useful, and combining probiotics with certain strains relieved symptoms, gastrointestinal discomfort, and bloating [41]. Recent studies have suggested that probiotic combinations are more beneficial than single-strain options, with *Escherichia* and *Streptococcus* combinations resulting in reduced abdominal pain and bloating [7,38]. Nonetheless, the efficacy of probiotics appears to depend on the strain, dosage, and duration of treatment, with some patients experiencing only temporary symptom relief [15,37].

Interestingly, probiotics' ability to promote gut health may extend beyond symptom relief. The composite IBS symptom score decreased significantly in the probiotic group compared to placebo [42]. Probiotic treatment resulted in a 37% reduction in IBS score versus 9% in placebo [42]. They can increase levels of beneficial bacteria like *Bifidobacterium* and *Lactobacillus*, producing SCFAs that are vital for intestinal function [10]. However, the overall effectiveness of probiotics remains inconclusive, with studies highlighting the need for more research to identify the most effective strains and dosages for IBS management [19,43]. Synbiotics, which combine probiotics and prebiotics, have shown promise by providing additional benefits through synergistic effects, improving stool frequency, and reducing bloating [2].

8. Prebiotics in IBS

Prebiotics, non-digestible fibers that promote the growth of beneficial gut bacteria, have shown potential in treating IBS symptoms, particularly bloating and gas [20]. Prebiotic fermentation produces SCFAs, which possess anti-inflammatory properties that may help alleviate IBS symptoms [44]. Clinical trials have indicated that low to moderate doses of prebiotics can relieve symptoms, although large doses may exacerbate bloating and other gastrointestinal discomforts [44,45].

Similar to probiotics, the efficacy of prebiotics in IBS treatment is not uniform across studies. For instance, supplementation with fructooligosaccharides (FOS) and galactooligosaccharides (GOS) promotes the growth of beneficial bacteria, but symptom relief has been inconsistent [36]. Some patients report improvements in bloating and gas with low doses of trans-GOS, but higher doses can increase symptoms due to fermentation [36]. This highlights the importance of individualized approaches when using prebiotics to manage IBS.

9. Synbiotics and Postbiotics

Synbiotics, which combine prebiotics and probiotics, aim to enhance the efficacy of both by improving probiotic survival in the gastrointestinal system [44]. This synergistic approach shows potential in optimizing IBS treatment, with some studies indicating improvements in bowel movement frequency and reductions in bloating [2,45]. However, more research is needed to understand the long-term effects and optimal formulations of synbiotics for IBS patients [5].

In addition to synbiotics, postbiotics, i.e., metabolites produced by probiotics, are emerging as potential therapies for IBS. Although the research on postbiotics is still in its infancy, early findings suggest that they may help reduce inflammation and improve symptoms, especially in diarrhea-predominant IBS [11]. This innovative approach warrants further investigation to determine its effectiveness in clinical settings.

10. Fecal Microbiota Transplantation (FMT) in IBS

FMT has gained attention as a potential therapy for IBS by restoring gut microbiota composition. (Figure 3 summarizes the mechanism of action of different therapeutic approaches for IBS.) Several studies have shown that FMT can result in significant symptom improvements, particularly in patients with severe IBS or high levels of gut dysbiosis [7,46,47]. However, the outcomes of FMT are inconsistent. Some trials report substantial reductions in IBS symptoms, while others show minimal or no effects compared to placebo [9,36,48]. These discrepancies may be influenced by factors such as donor selection, delivery method, and individual patient characteristics [25,48].

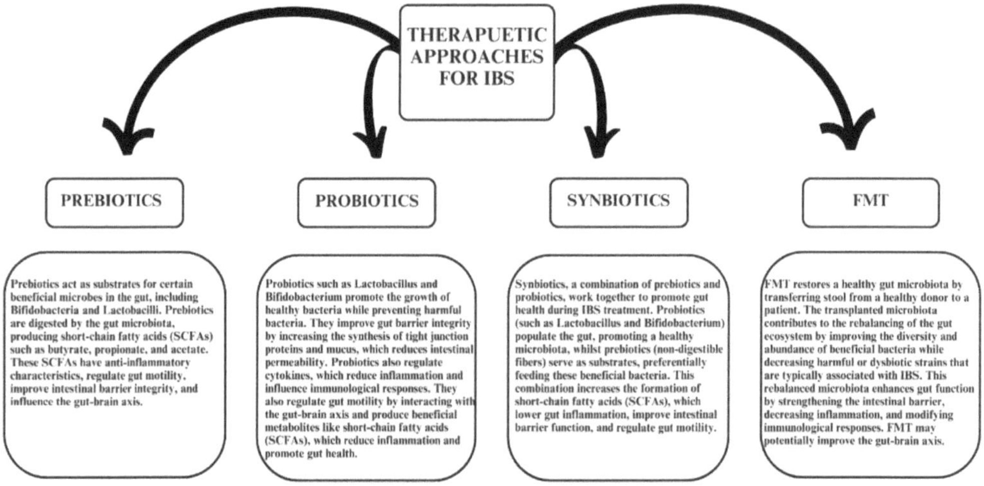

Figure 3. Summary of the mechanisms of action of different therapeutic approaches for IBS.

The mode of FMT administration also appears to play a crucial role in determining its efficacy. For example, older FMT techniques, such as enema or colonoscopy, have been found to be more effective than newer methods like capsule delivery [36,47]. Multiple-donor FMT has shown promise, but its overall effectiveness in treating IBS remains inconclusive [48]. Moreover, the long-term consequences of FMT on gut microbiota and IBS symptom relief are still unknown, necessitating further research to confirm its safety and efficacy [40,43,46].

Despite its potential, FMT's clinical application in IBS management is not without challenges. For instance, male patients have shown reduced response rates to FMT, while patients with severe IBS have reported better outcomes [25]. Additionally, certain bacterial profiles in donors may predict treatment efficacy, highlighting the need for personalized approaches to FMT therapy [25,49].

11. Dietary Interventions for IBS

Dietary interventions, particularly the low-FODMAP diet, are among the most effective first-line treatments for IBS. FODMAPs, or fermentable oligosaccharides, disaccharides, monosaccharides, and polyols, are poorly absorbed carbohydrates that can cause bloating, gas, and discomfort in IBS patients. The low-FODMAP diet is typically implemented in three phases: exclusion, gradual reintroduction, and long-term personalization to meet individual needs [28].

Studies have shown that the low-FODMAP diet can significantly reduce IBS symptoms, with symptom reduction rates ranging from 50% to 76% during the initial elimination phase [15,28]. Long-term adherence to the diet, particularly with the guidance of a dietitian, can provide sustained symptom relief, especially from fructans [15,28]. Dietary factors play an important role in IBS patients, with meal-related symptom aggravation frequently observed, and specific foods such as high-fat meals and poorly digested carbohydrates trigger symptoms through fermentation and modified colonic responses [50]. However, the low-FODMAP diet is not without risks. Long-term use may result in nutrient deficiencies, alterations in gut microbial diversity, and reduced levels of beneficial bacteria such as *Bifidobacterium* [26,28,49].

The efficacy of the low-FODMAP diet varies depending on IBS subtype. For instance, while it significantly alleviates symptoms in diarrhea-predominant IBS patients, it may be less effective in those with constipation-predominant IBS [29,32]. Furthermore, research comparing the low-FODMAP diet to traditional dietary guidance has yielded conflicting results, suggesting that other dietary approaches may be equally beneficial for some patients [25].

In contrast to the low-FODMAP diet, the NICE diet has been proposed as a better long-term solution for IBS patients, benefiting 46% to 54% of patients while avoiding the nutritional risk factors associated with FODMAP restriction [51]. The NICE diet focuses on balanced nutrition, avoiding the stringent restrictions of the low-FODMAP diet, which can lead to calorie restriction and nutritional deficits [51]. Despite the challenges of maintaining the low-FODMAP diet long-term, it remains one of the most effective dietary therapies for IBS symptom management [15].

12. Gluten-Free Diet (GFD) and IBS

Wheat grains contain a variety of components, including proteins such as gluten, which is composed of glutenin and gliadin. Barley, rye, and oats all contain gluten-like proteins known as hordein, secalin, and avenins. Wheat also contains albumins, such as amylase-trypsin inhibitors (ATIs), and starch, which includes fructans, an oligosaccharide classed among the FODMAPs. As a result, people who consume a wheat-based diet are exposed to gluten proteins, ATIs, and FODMAPs, all of which may contribute to gastrointestinal discomfort in IBS patients [27]. This complication makes it difficult to pinpoint the exact components responsible for symptom relief when wheat is removed from the diet [27].

Celiac disease is caused by a particular immune response to gluten, a wheat protein breakdown product that binds to the HLA-DQ2 and HLA-DQ8 receptors on antigen-presenting cells. This interaction results in a mucosal inflammatory response that includes lymphocyte infiltration, crypt enlargement, villous atrophy, and accelerated epithelial cell turnover. IBS symptoms have been connected to an immunological response to gluten, which is comparable to celiac disease; however, the exact mechanism is yet unknown [26]. Prior to starting a gluten-free diet, celiac patients also exhibit key abnormalities seen in IBS, such as increased gut permeability, higher mucosal mast cells, and decreased serotonin transporter expression [26].

Some IBS patients report symptom alleviation from GFD, although the underlying mechanism remains uncertain. Patients with IBS are 3.5 times more likely than controls to report gluten intolerance [26]. Research suggests that reducing fructans rather than gluten may be responsible for the observed symptom relief in non-celiac gluten-sensitivity (NCGS) patients [28]. Despite this, many patients continue to adhere to the GFD long-term, even though it may lead to nutritional deficiencies and potential heavy metal accumulation [28].

While the GFD may provide symptom relief for some IBS patients, its long-term safety and effectiveness are still debated. As with other dietary interventions, individualized approaches are essential to minimize risks and ensure proper nutrition [28,49].

13. Personalized Approaches in IBS Management

The emerging trend in IBS management is the shift towards personalized treatment approaches that consider individual differences in gut microbiota and dietary responses [21,52]. For instance, the gut microbiome may predict a patient's response to the low-FODMAP diet, with higher dysbiosis scores indicating a poor response [9,28]. Similarly, pre-treatment gut flora diversity may serve as a biomarker for predicting the efficacy of FMT in IBS patients [25,46].

Personalized nutrition programs that balance symptom relief with adequate nutrition are vital for long-term IBS management. These programs must be tailored to the patient's unique microbiome composition, symptom profile, and dietary preferences [49]. As research progresses, the integration of microbial and dietary therapies holds promise for improving IBS management and patient outcomes [52].

14. Challenges and Future Research Directions in Microbiome-Targeted Treatments for IBS

The study of microbiome-targeted treatments for IBS has advanced significantly, yet substantial gaps remain in our understanding. These knowledge gaps impede the ability to develop personalized treatment strategies, which could address the diverse subtypes and symptom profiles of IBS. This section highlights key challenges in current research, focusing on the gaps in knowledge and potential avenues for personalized treatments.

15. Gaps in Current Knowledge

One of the major challenges in IBS research is the lack of long-term data on the effects and adverse events related to probiotic interventions. Most existing studies on probiotics for IBS treatment are short-term and do not adequately assess the variability in probiotic formulations or the heterogeneity in patient responses [53]. This is problematic because the effectiveness of probiotics can vary significantly based on the strain used, dosage, and treatment duration, which are not yet optimized for specific IBS subtypes [53]. Additionally, the safety profile of probiotics, particularly with long-term use, remains unclear. More extensive trials with longer follow-up periods are needed to monitor potential side effects and adverse events, which are crucial for refining probiotic therapeutic strategies [53].

Another limitation in the current research is the small sample sizes and short study durations, which restrict the ability to draw robust conclusions, particularly about the efficacy of probiotics across different IBS subtypes. Studies with larger populations and longer durations are essential to evaluate the long-term effectiveness of probiotics and to determine their impact on various IBS subtypes [54]. Furthermore, current trials often focus solely on probiotics, neglecting other potentially beneficial agents such as fiber and prebiotics. The inclusion of these components in future studies could provide a more comprehensive understanding of how various interventions impact IBS symptoms and gut health [54].

In addition to these methodological challenges, there are significant gaps in our understanding of the microbiome–brain interaction, particularly in the context of DGBI, including IBS. The current research on microbiome function and its role in IBS pathophysiology is hampered by inconsistent methodologies for assessing the microbiome, which leads to conflicting results [10]. To address this, large-scale longitudinal studies using multi-omics approaches are needed to investigate the interactions between the microbiome and the host in IBS patients. Future studies should focus on improving the timing and integration of psychological therapies into treatment plans in order to maximize their function alongside medical treatment [55]. Such studies could explain the role of gut–brain communication mechanisms and the potential for microbiome-targeted therapies, such as FMT and psychological interventions, in IBS management [13].

Geographical and population-specific differences also represent a significant gap in IBS research. There is a notable lack of studies focused on IBS patients from Asia, where the microbiota composition may differ due to genetic, dietary, and environmental factors [25]. This underrepresentation limits the generalizability of current findings. Future studies should explore how dietary interventions influence microbiota changes across different IBS subtypes and investigate the efficacy of probiotics and FMT in these populations [25]. Additionally, the role of beneficial bacteria in the pathogenesis of IBS requires further research to determine whether certain strains could be leveraged for targeted therapeutic interventions [25].

16. Potential for Personalized Treatments

Personalized treatments for IBS, which account for individual variations in microbiota composition and host factors, hold great promise but remain an underdeveloped area of research. Current findings often fail to meet clinical needs, highlighting the necessity for further exploration of the specific pathophysiological mechanisms underlying IBS [49]. The complexity of IBS pathophysiology, which can differ significantly between patients, calls for a deeper investigation into how personalized medicine can be integrated into IBS treatment strategies. (Table 1 displays the results of different meta-analyses.) For example, the efficacy of FMT varies across different IBS subtypes, and there is a need to examine how individual patient characteristics, such as microbiome composition, influence treatment outcomes [49]. Similarly, the response to probiotic therapy may differ based on individual genetic and microbial profiles, suggesting that treatments should be tailored to these factors [49].

Table 1. A summary of several meta-analyses evaluating the role of microbiota in IBS.

Reference	Methods	Results
Myneedu et al. 2019 [47]	Authors searched PubMed, Embase, Google Scholar, and abstract books from Digestive Disease Week and United European Gastroenterology Week (2010–2018) for studies on IBS, and they retrieved single-arm and RCTs on FMT for IBS, where the diagnosis was confirmed by a physician or based on ROME I-IV criteria.	Following the SATs, almost 60% of IBS patients reported considerable symptom relief. However, the RCTs revealed varied findings. Some studies indicated improvements in symptoms, while others found no significant difference between the FMT and control groups.
Zhang et al. 2022 [52]	Authors searched for RCTs on the efficacy of probiotics in treating IBS until August 25, 2021. The primary focus was on the rate of symptom reduction as well as changes in overall symptoms. Meta-regression was used to determine whether the length and dose of probiotic treatment had an impact on effectiveness.	*B. coagulans* was found to be the most effective probiotic species for relieving IBS symptoms. *L. plantarum* was shown to confer the highest quality of life (QOL). Meta-regression revealed that probiotic dose had no significant impact on outcomes, whereas treatment time did. Further subgroup analysis found that *B. coagulans* given for 8 weeks was the most effective in relieving these symptoms, outperforming probiotic combinations in the research.
Shang et al. 2022 [39]	Authors conducted a comprehensive search of PubMed, Embase, the Cochrane Library, Web of Science, and China Biology Medicine (CBM). Intervention parameters included probiotic strains, dose, duration, form, and placebo use as well as outcome measures such as symptom reports and scale use.	Three RCTs with 71 patients found that probiotics significantly improved stool consistency compared to placebo. An 8-week therapy period was effective, while 12 weeks provided no benefit. Two RCTs with 74 patients found that probiotics significantly improved fecal *Bifidobacterium* and *Lactobacillus* levels after four weeks. There was no effect after 8 weeks.
Xie et al. 2023 [38]	Reviewers independently gathered crucial information from eligible trials, such as RCT details, participant characteristics, and results. Intention-to-treat analyses were conducted. Transitivity was established by comparing major clinical variables between studies. The network's consistency was verified by node splitting and loop-specific analysis.	The most effective probiotics were *L. acidophilus* (efficacy level A). Other strains, such as *B. bifidum* and *C. butyricum*, provided considerable benefits (efficacy level B). The multistrain group demonstrated the greatest improvement in quality of life. *C. butyricum* also showed significant improvements over placebo. *B. coagulans MTCC 5856* and *S. cerevisiae CNCM I-3856* were found to be most effective at improving stool consistency in IBS-D patients. There were no significant differences between probiotics and placebo for IBS-C in this network analysis.
Jamshidi et al. 2023 [48]	A complete search of the PubMed/Medline and Embase databases was performed to include all relevant publications up to 14 June 2023. To reduce heterogeneity, subgroup analyses were conducted based on FMT preparation, frequency of administration, and route of administration.	Single dosage of FMT administered via colonoscopy significantly decreased patient complaints.Using frozen FMT as an oral capsule significantly increased symptoms compared to the non-FMT placebo.Patients undergoing multiple-donor FMT showed considerable improvement compared to autologous. However, compared to the non-FMT placebo, it had a negative effect on IBS symptoms.
Wu et al. 2024 [40]	Databases were searched, including Medline, Embase and Embase Classic, Cochrane Central Register of Controlled Trials, and Web of Science. Authors also searched for unpublished trial data on ClinicalTrials.gov. Standardized mean differences (SMDs) were utilized to compare the effect sizes of the experimental and placebo groups. The study looked at the effects of several probiotic strains by dividing interventions into subgroups based on the strain.	Probiotics had the greatest treatment benefit over placebo. FMT also showed considerable improvement. Prebiotics did not differ significantly from placebo. Synbiotics also did not reveal significant differences from placebo. Probiotics performed much better than prebiotics and synbiotics. FMT performed much better than prebiotics and synbiotics. Probiotics helped to improve abdominal pain and bloating when compared to placebo.

One of the key challenges in developing personalized treatments is the lack of clarity regarding the optimal probiotic combinations for managing IBS symptoms. While probiotics are commonly used in IBS treatment, it is still unclear which strains, dosages, and combinations are most effective for specific patient subtypes and symptom profiles [40]. This highlights the need for more clinical trials focusing on the efficacy of particular probiotic strains and combinations tailored to different forms of IBS [40]. Such research could benefit from integrating genomic and transcriptomic analyses to better understand the molecular mechanisms of probiotics and their interaction with the host microbiome [40].

Geraniol, an essential oil constituent, has anti-inflammatory, antibacterial, and eubiotic effects on gut microbiota in IBS patients. In a randomized double-blind trial, geraniol reduced overall IBS symptoms and gut microbiota profiles, especially in IBS-M patients [25]. Clinoptilolite, a natural zeolite that has high absorptive ability, was found to reduce diarrhea frequency in a randomized trial of IBS-D patients. A new double-blind RCT investigated the efficacy of GTB1 in 27 IBS-D patients. After four weeks, the GTB1 group's abdominal pain and bloating severity decreased significantly. GTB1 also enhanced the relative abundance of *Lactobacillus* while decreasing *Bacteroides* levels within one week of therapy [25]. Furthermore, randomized trials have suggested that fiber can assist IBS-C patients, with viscous fiber (psyllium) improving stool consistency and frequency by increasing water content [26].

Furthermore, the use of FMT as a treatment for IBS remains in its early stages, with few studies investigating its efficacy across IBS subtypes. Future research should focus on identifying the optimal donor selection criteria, treatment regimens, and bacterial strains for effective FMT therapy in IBS [40]. A more personalized approach to FMT, which considers individual variations in gut microbiota, could significantly improve treatment outcomes for IBS patients [40].

The incorporation of personalized medicine into IBS treatment also requires advancements in understanding how microbiome alterations can affect the broader gut–brain axis. There is still much to learn about how the gut microbiota interacts with the central nervous system and how these interactions influence IBS symptoms. Psychological therapies, including cognitive behavioral therapy (CBT) provided over the phone, have shown long-term efficacy in treating IBS symptoms [55]. Investigating the mechanisms behind these microbiome–brain interactions, particularly in the context of psychological therapies, could open up new avenues for personalized treatment options [13].

17. Conclusions

IBS is a multifactorial disorder influenced by a combination of genetic predispositions, gut microbiota alterations, and psychosocial factors. The growing body of evidence supporting the role of the gut microbiota, particularly its involvement in the gut–brain axis, provides valuable insights into the pathogenesis and management of IBS. While microbiota-targeted therapies such as probiotics, synbiotics, and FMT show promise, the variability in patient responses underscores the need for more personalized treatment approaches. Future research should focus on identifying specific microbial signatures that can serve as biomarkers for IBS and exploring the long-term efficacy and safety of these interventions. A more nuanced understanding of the microbiota's role in different IBS subtypes will be crucial for optimizing individualized therapeutic strategies and improving patient outcomes.

Author Contributions: M.B.A. and A.I. had the idea of the manuscript; M.B.A., M.W. and A.A. drafted the manuscript; A.I., S.-L.P., F.R. and D.L.D. contributed to the writing of the manuscript; A.I. and D.L.D. made substantial contributions to the conception and critically revised the manuscript for important intellectual content. All authors have read and agreed to the published version of the manuscript.

Funding: The authors did not receive any financial support for the research, authorship, and/or publication of this article.

Institutional Review Board Statement: An ethics statement is not applicable because this study is based exclusively on published literature.

Conflicts of Interest: The authors declare that the research was conducted in the absence of any commercial or financial relationships that could be construed as potential conflicts of interest.

References

1. Sperber, A.D.; Bangdiwala, S.I.; Drossman, D.A.; Ghoshal, U.C.; Simren, M.; Tack, J.; Whitehead, W.E.; Dumitrascu, D.L.; Fang, X.; Fukudo, S.; et al. Worldwide Prevalence and Burden of Functional Gastrointestinal Disorders, Results of Rome Foundation Global Study. *Gastroenterology* **2020**, *160*, 99–114.e3. [CrossRef] [PubMed]
2. Chong, P.P.; Chin, V.K.; Looi, C.Y.; Wong, W.F.; Madhavan, P.; Yong, V.C. The microbiome and irritable bowel syndrome—A review on the pathophysiology, current research and future therapy. *Front. Microbiol.* **2019**, *10*, 1136. [CrossRef] [PubMed]
3. Margolis, K.G.; Cryan, J.F.; Mayer, E.A. The microbiota-gut-brain axis: From motility to mood. *Gastroenterology* **2021**, *160*, 1486–1501. [CrossRef] [PubMed]
4. Ghaffari, P.; Shoaie, S.; Nielsen, L.K. Irritable bowel syndrome and microbiome: Switching from conventional diagnosis and therapies to personalized interventions. *J. Transl. Med.* **2022**, *20*, 173. [CrossRef] [PubMed]
5. Iribarren, C.; Maasfeh, L.; Öhman, L.; Simrén, M. Modulating the gut microenvironment as a treatment strategy for irritable bowel syndrome: A narrative review. *Gut Microbiome* **2022**, *3*, e7. [CrossRef]
6. Huang, K.-Y.; Wang, F.-Y.; Lv, M.; Ma, X.-X.; Tang, X.-D.; Lv, L. Irritable bowel syndrome: Epidemiology, overlap disorders, pathophysiology and treatment. *World J. Gastroenterol.* **2023**, *29*, 4120–4135. [CrossRef] [PubMed]
7. Goodoory, V.C.; Ford, A.C. Antibiotics and probiotics for irritable bowel syndrome. *Drugs* **2023**, *83*, 687–699. [CrossRef]
8. Mamieva, Z.; Poluektova, E.; Svistushkin, V.; Sobolev, V.; Shifrin, O.; Guarner, F.; Ivashkin, V. Antibiotics, gut microbiota, and irritable bowel syndrome: What are the relations? *World J. Gastroenterol.* **2022**, *28*, 1204–1219. [CrossRef]
9. Dale, H.F.; Lied, G.A. Gut microbiota and therapeutic approaches for dysbiosis in irritable bowel syndrome: Recent developments and future perspectives. *Turk. J. Med. Sci.* **2020**, *50*, 1632–1641. [CrossRef] [PubMed]
10. Raskov, H.; Burcharth, J.; Pommergaard, H.C.; Rosenberg, J. Irritable bowel syndrome, the microbiota and the gut-brain axis. *Gut Microbes.* **2016**, *7*, 365–383. [CrossRef] [PubMed]
11. Napolitano, M.; Fasulo, E.; Ungaro, F.; Massimino, L.; Sinagra, E.; Danese, S.; Mandarino, F.V. Gut dysbiosis in irritable bowel syndrome: A narrative review on correlation with disease subtypes and novel therapeutic implications. *Microorganisms* **2023**, *11*, 2369. [CrossRef]
12. Porcari, S.; Ingrosso, M.R.; Maida, M.; Eusebi, L.H.; Black, C.; Gasbarrini, A.; Cammarota, G.; Ford, A.C.; Ianiro, G. Prevalence of irritable bowel syndrome and functional dyspepsia after acute gastroenteritis: Systematic review and meta-analysis. *Gut* **2024**, *73*, 1431–1440. [CrossRef]
13. Kraimi, N.; Ross, T.; Pujo, J.; De Palma, G. The gut microbiome in disorders of gut-brain interaction. *Gut Microbes* **2024**, *16*, 2360233. [CrossRef] [PubMed]
14. Marasco, G.; Cremon, C.; Barbaro, M.R.; Stanghellini, V.; Barbara, G. Gut microbiota signatures and modulation in irritable bowel syndrome. *Microbiome Res. Rep.* **2022**, *1*, 11. [CrossRef] [PubMed]
15. Mazzawi, T. Gut microbiota manipulation in irritable bowel syndrome. *Microorganisms* **2022**, *10*, 1332. [CrossRef] [PubMed]
16. Marasco, G.; Cremon, C.; Barbaro, M.R.; Bianco, F.; Stanghellini, V.; Barbara, G. Microbiota modulation in disorders of gut-brain interaction. *Dig. Liver Dis.* **2024**, *56*, 1971–1979. [CrossRef] [PubMed]
17. Deng, X.; Xiao, L.; Luo, M.; Xie, P.; Xiong, L. Intestinal crosstalk between bile acids and microbiota in irritable bowel syndrome. *J. Gastroenterol. Hepatol.* **2023**, *38*, 1072–1082. [CrossRef] [PubMed]
18. Agnello, M.; Carroll, L.N.; Imam, N.; Pino, R.; Palmer, C.; Varas, I.; Greene, C.; Hitschfeld, M.; Gupta, S.; Almonacid, D.E.; et al. Gut microbiome composition and risk factors in a large cross-sectional IBS cohort. *BMJ Open Gastroenterol.* **2020**, *7*, e000345. [CrossRef] [PubMed]
19. Ahlawat, G.M.; Singh, P.K. Methods of determining irritable bowel syndrome and efficiency of probiotics in treatment: A review. *Curr. Ther. Res. Clin. Exp.* **2023**, *99*, 100721. [CrossRef]
20. Shrestha, B.; Shrestha, B.; Patel, D.; Patel, D.; Shah, H.; Shah, H.; Hanna, K.S.; Hanna, K.S.; Kaur, H.; Kaur, H.; et al. The role of gut microbiota in the pathophysiology and therapy of irritable bowel syndrome: A systematic review. *Cureus* **2022**, *14*, e28064. [CrossRef] [PubMed]

21. Ford, A.C.; Sperber, A.D.; Corsetti, M.; Camilleri, M. Irritable bowel syndrome. *Lancet* **2020**, *396*, 1675–1688. [CrossRef] [PubMed]
22. Mishima, Y.; Ishihara, S. Molecular mechanisms of microbiota-mediated pathology in irritable bowel syndrome. *Int. J. Mol. Sci.* **2020**, *21*, 8664. [CrossRef] [PubMed]
23. Takakura, W.; Pimentel, M. Small intestinal bacterial overgrowth and irritable bowel syndrome—An update. *Front. Psychiatry* **2020**, *11*, 664. [CrossRef]
24. Hillestad, E.M.R.; van der Meeren, A.; Nagaraja, B.H.; Bjørsvik, B.R.; Haleem, N.; Benitez-Paez, A.; Sanz, Y.; Hausken, T.; Lied, G.A.; Lundervold, A.; et al. Gut bless you: The microbiota-gut-brain axis in irritable bowel syndrome. *World J. Gastroenterol.* **2022**, *28*, 412–431. [CrossRef]
25. Surdea-Blaga, T.; Ciobanu, L.; Ismaiel, A.; Dumitrascu, D.L. Microbiome in irritable bowel syndrome: Advances in the field—A scoping review. *Microb. Health Dis.* **2024**, *6*, e1017. [CrossRef]
26. Spiller, R. Impact of diet on symptoms of the irritable bowel syndrome. *Nutrients* **2021**, *13*, 575. [CrossRef] [PubMed]
27. Algera, J.; Colomier, E.; Simrén, M. The dietary management of patients with irritable bowel syndrome: A narrative review of the existing and emerging evidence. *Nutrients* **2019**, *11*, 2162. [CrossRef] [PubMed]
28. Rej, A.; Aziz, I.; Tornblom, H.; Sanders, D.S.; Simrén, M. The role of diet in irritable bowel syndrome: Implications for dietary advice. *J. Intern. Med.* **2019**, *286*, 490–502. [CrossRef]
29. Kashyap, P.; Moayyedi, P.; Quigley, E.M.M.; Simren, M.; Vanner, S. Critical appraisal of the SIBO hypothesis and breath testing: A clinical practice update endorsed by the European society of neurogastroenterology and motility (ESNM) and the American neurogastroenterology and motility society (ANMS). *Neurogastroenterol. Motil.* **2024**, *36*, e14817. [CrossRef]
30. Aziz, I.; Törnblom, H.; Simrén, M. Small intestinal bacterial overgrowth as a cause for irritable bowel syndrome. *Curr. Opin. Gastroenterol.* **2017**, *33*, 196–202. [CrossRef]
31. Motta, J.P.; Wallace, J.L.; Buret, A.G.; Deraison, C.; Vergnolle, N. Gastrointestinal biofilms in health and disease. *Nat. Rev. Gastroenterol. Hepatol.* **2021**, *18*, 314–334. [CrossRef]
32. Jandl, B.; Dighe, S.; Baumgartner, M.; Makristathis, A.; Gasche, C.; Muttenthaler, M. Gastrointestinal biofilms—Endoscopic detection, disease relevance and therapeutic strategies. *Gastroenterology* **2024**, *167*, 1098–1112.e5. [CrossRef] [PubMed]
33. Algera, J.P.; Törnblom, H.; Simrén, M. Treatments targeting the luminal gut microbiota in patients with irritable bowel syndrome. *Curr. Opin. Pharmacol.* **2022**, *66*, 102284. [CrossRef] [PubMed]
34. Karakan, T.; Ozkul, C.; Küpeli Akkol, E.; Bilici, S.; Sobarzo-Sánchez, E.; Capasso, R. Gut-brain-Microbiota axis: Antibiotics and functional gastrointestinal disorders. *Nutrients* **2021**, *13*, 389. [CrossRef]
35. Manos, J. The human microbiome in disease and pathology. *APMIS* **2022**, *130*, 690–705. [CrossRef]
36. El-Salhy, M.; Hatlebakk, J.G.; Gilja, O.H.; Bråthen Kristoffersen, A.; Hausken, T. Efficacy of faecal microbiota transplantation for patients with irritable bowel syndrome in a randomised, double-blind, placebo-controlled study. *Gut* **2020**, *69*, 859–867. [CrossRef]
37. Herndon, C.C.; Wang, Y.-P.; Lu, C.-L. Targeting the gut microbiota for the treatment of irritable bowel syndrome. *Kaohsiung J. Med. Sci.* **2020**, *36*, 160–170. [CrossRef] [PubMed]
38. Xie, P.; Luo, M.; Deng, X.; Fan, J.; Xiong, L. Outcome-specific efficacy of different probiotic strains and mixtures in irritable bowel syndrome: A systematic review and network meta-analysis. *Nutrients* **2023**, *15*, 3856. [CrossRef]
39. Shang, X.; E, F.-F.; Guo, K.-L.; Li, Y.-F.; Zhao, H.-L.; Wang, Y.; Chen, N.; Nian, T.; Yang, C.-Q.; Yang, K.-H.; et al. Effectiveness and safety of probiotics for patients with constipation-predominant irritable bowel syndrome: A systematic review and meta-analysis of 10 randomized controlled trials. *Nutrients* **2022**, *14*, 2482. [CrossRef] [PubMed]
40. Wu, Y.; Li, Y.; Zheng, Q.; Li, L. The efficacy of probiotics, prebiotics, synbiotics, and fecal microbiota transplantation in irritable bowel syndrome: A systematic review and network meta-analysis. *Nutrients* **2024**, *16*, 2114. [CrossRef] [PubMed]
41. Goodoory, V.C.; Mais Khasawneh Black, C.J.; Eamonn Martin Quigley Martin, P.; Ford, A.C. Efficacy of Probiotics in Irritable Bowel Syndrome: Systematic Review and Meta-analysis. *Gastroenterology* **2023**, *165*, 1206–1218. [CrossRef]
42. Kajander, K.; Myllyluoma, E.; Rajilić-Stojanović, M.; Kyrönpalo, S.; Rasmussen, M.; Järvenpää, S.; Zoetendal, E.G.; DE Vos, W.M.; Vapaatalo, H.; Korpela, R. Clinical trial: Multispecies probiotic supplementation alleviates the symptoms of irritable bowel syndrome and stabilizes intestinal microbiota. *Aliment. Pharmacol. Ther.* **2007**, *27*, 48–57. [CrossRef] [PubMed]
43. Wollny, T.; Daniluk, T.; Piktel, E.; Wnorowska, U.; Bukłaha, A.; Głuszek, K.; Durnaś, B.; Bucki, R. Targeting the gut microbiota to relieve the symptoms of irritable bowel syndrome. *Pathogens* **2021**, *10*, 1545. [CrossRef] [PubMed]
44. Ford, A.C.; Harris, L.A.; Lacy, B.E.; Quigley, E.M.M.; Moayyedi, P. Systematic review with meta-analysis: The efficacy of prebiotics, probiotics, synbiotics and antibiotics in irritable bowel syndrome. *Aliment. Pharmacol. Ther.* **2018**, *48*, 1044–1060. [CrossRef] [PubMed]
45. Quigley, E.M. The use of probiotics, prebiotics and synbiotics in the management of irritable bowel syndrome. *Eur. Gastroenterol. Hepatol. Rev.* **2012**, *7*, 233–236.
46. Hojo, M.; Nagahara, A. Current perspectives on irritable bowel syndrome: A narrative review. *J. Int. Med. Res.* **2022**, *50*, 3000605221126370. [CrossRef] [PubMed]

47. Myneedu, K.; Deoker, A.; Schmulson, M.J.; Bashashati, M. Fecal microbiota transplantation in irritable bowel syndrome: A systematic review and meta-analysis. *United Eur. Gastroenterol. J.* **2019**, *7*, 1033–1041. [CrossRef] [PubMed]
48. Jamshidi, P.; Farsi, Y.; Nariman, Z.; Hatamnejad, M.R.; Mohammadzadeh, B.; Akbarialiabad, H.; Nasiri, M.J.; Sechi, L.A. Fecal microbiota transplantation in irritable bowel syndrome: A systematic review and meta-analysis of randomized controlled trials. *Int. J. Mol. Sci.* **2023**, *24*, 14562. [CrossRef] [PubMed]
49. Chen, Y.; Feng, S.; Li, Y.; Zhang, C.; Chao, G.; Zhang, S. Gut microbiota and intestinal immunity—A crosstalk in irritable bowel syndrome. *Immunology* **2024**, *172*, 1–20. [CrossRef] [PubMed]
50. Rajilić-Stojanović, M.; Jonkers, D.M.; Salonen, A.; Hanevik, K.; Raes, J.; Jalanka, J.; de Vos, W.M.; Manichanh, C.; Golic, N.; Enck, P.; et al. Intestinal Microbiota and Diet in IBS: Causes, consequences, or epiphenomena? *Am. J. Gastroenterol.* **2015**, *110*, 278–287. [CrossRef] [PubMed]
51. El-Salhy, M.; Patcharatrakul, T.; Gonlachanvit, S. The role of diet in the pathophysiology and management of irritable bowel syndrome. *Indian J. Gastroenterol.* **2021**, *40*, 111–119. [CrossRef] [PubMed]
52. Zhang, T.; Zhang, C.; Zhang, J.; Sun, F.; Duan, L. Efficacy of probiotics for irritable bowel syndrome: A systematic review and network meta-analysis. *Front. Cell. Infect. Microbiol.* **2022**, *12*, 859967. [CrossRef]
53. Umeano, L.; Iftikhar, S.; Alhaddad, S.F.; Paulsingh, C.N.; Riaz, M.F.; Garg, G.; Mohammed, L. Effectiveness of probiotic use in alleviating symptoms of irritable bowel syndrome: A systematic review. *Cureus* **2024**, *16*, e58306. [CrossRef]
54. Ruiz-Sánchez, C.; Escudero-López, B.; Fernández-Pachón, M.-S. Evaluation of the efficacy of probiotics as treatment in irritable bowel syndrome. *Endocrinol. Diabetes Nutr.* **2024**, *71*, 19–30. [CrossRef]
55. Black, C.J.; Thakur, E.R.; Houghton, L.A.; Quigley, E.M.M.; Moayyedi, P.; Ford, A.C. Efficacy of psychological therapies for irritable bowel syndrome: Systematic review and network meta-analysis. *Gut* **2020**, *69*, 1441–1451. [CrossRef] [PubMed]

medicina

MDPI

Review

Hepatic Hemangioma: Review of Imaging and Therapeutic Strategies

Arkadiusz Kacała [1,*], Mateusz Dorochowicz [2], Iwona Matus [2], Michał Puła [3], Adrian Korbecki [3], Michał Sobański [3], Jagoda Jacków-Nowicka [1], Dariusz Patrzałek [4], Dariusz Janczak [4] and Maciej Guziński [1]

1 Department of General, Interventional and Neuroradiology, Wroclaw Medical University, 50-367 Wrocław, Poland; jagoda.jackow-nowicka@umw.edu.pl (J.J.-N.); maciej.guzinski@umw.edu.pl (M.G.)
2 Faculty of Medicine, Wroclaw Medical University, 50-367 Wrocław, Poland; m.dorochowicz97@gmail.com (M.D.); iwona.matus4@gmail.com (I.M.)
3 Department of General, Interventional and Neuroradiology, Wroclaw University Hospital, 50-556 Wrocław, Poland; michal.pula97@gmail.com (M.P.); lek.adriankorbecki@gmail.com (A.K.); michalsobanski11@wp.pl (M.S.)
4 Department of Vascular, General and Transplantation Surgery, Wroclaw Medical University, 50-367 Wrocław, Poland; dariusz.patrzalek@umw.edu.pl (D.P.); dariusz.janczak@umw.edu.pl (D.J.)
* Correspondence: arkadiusz.kacala@gmail.com

Abstract: Hepatic hemangiomas are the most common benign liver tumors. Typically, small- to medium-sized hemangiomas are asymptomatic and discovered incidentally through the widespread use of imaging techniques. Giant hemangiomas (>5 cm) have a higher risk of complications. A variety of imaging methods are used for diagnosis. Cavernous hemangioma is the most frequent type, but radiologists must be aware of other varieties. Conservative management is often adequate, but some cases necessitate targeted interventions. Although surgery was traditionally the main treatment, the evolution of minimally invasive procedures now often recommends transarterial chemoembolization as the treatment of choice.

Keywords: hepatic hemangiomas; atypical hepatic hemangioma; computed tomography (CT); magnetic resonance (MR); ultrasound (US); contrast-enhanced ultrasound (CEUS); transarterial chemoembolization (TACE)

1. Introduction

Hepatic hemangiomas, the most prevalent benign liver tumors, are characterized as slow-flow venous malformations with an incidence rate ranging between 0.4% and 20.0% [1–3]. These tumors are primarily comprised of endothelial cells originating from the hepatic artery [4–6]. Cavernous hemangiomas represent the most frequent pathological subtype. Notably, there is a predilection for women, with reported female-to-male ratios reaching as high as 5:1 [7,8]. The vast majority of hepatic hemangiomas are asymptomatic, maintain a stable size, do not affect liver function, and are incidentally detected during routine abdominal imaging [9–11]. The detection of hepatic hemangiomas has significantly increased in recent years, largely due to advancements in imaging technologies. Hemangiomas that are small to medium in size, defined as less than 4 cm in diameter, typically remain asymptomatic and are managed conservatively.

Giant hemangiomas, particularly those that exhibit progressive growth, pose a higher risk of serious complications, including local compression effects due to the tumor's volume, hemorrhage, Kasabach–Merritt Syndrome, or Budd–Chiari syndrome [12–16]. Such hemangiomas can reach up to 40 cm in diameter and are most commonly found in the right liver lobe, especially in segment IV [17].

This review explores recent developments in the literature on hepatic hemangiomas, with a focus on advancements in surgical and minimally invasive treatment modalities.

The shift towards minimally invasive techniques has expanded the treatment options for hemangiomas previously considered inoperable, reflecting a significant evolution in therapeutic approaches.

2. Materials and Methods

2.1. Literature Search Strategy

A systematic literature search was executed across several electronic databases, including PubMed, MEDLINE, Embase, and Google Scholar, spanning from their inception to 12.01.2024, without imposing language constraints. Search terms employed encompassed "hepatic hemangioma", "liver hemangioma", "imaging", "diagnosis", "therapeutic strategies", "management", "embolization", "surgery," "radiotherapy", "review", and "treatment", along with their various combinations.

2.2. Inclusion and Exclusion Criteria

Inclusion criteria were applied to articles that offered insights into imaging techniques for hepatic hemangiomas, diagnostic methodologies, and therapeutic interventions. Studies that discussed clinical manifestations, histopathological features, epidemiological data, and treatment outcomes were also considered. The selection was limited to peer-reviewed articles, review articles, case reports, and clinical trials. Exclusions were made for animal studies, conference abstracts, and publications without accessible full texts.

2.3. Data Extraction and Analysis

Initial screening involved two independent reviewers assessing the titles and abstracts for relevance. Subsequently, full-text articles were scrutinized to confirm their suitability against the predefined inclusion and exclusion criteria. Any disagreements between the initial reviewers were arbitrated by a third, independent reviewer. The data collation process focused on extracting information regarding the study's design, participant demographics, sample size, imaging results, diagnostic approaches, therapeutic modalities, outcomes, and any reported complications.

2.4. Quality Assessment

The methodological quality of the included studies was evaluated using appropriate tools tailored to the study design: the Cochrane Risk of Bias Tool for randomized controlled trials and the Newcastle–Ottawa Scale for observational studies. Case reports and case series were assessed for their informational clarity and the pertinence of the data provided.

2.5. Synthesis of Results

A narrative synthesis was conducted on the data harvested from the selected studies, aiming to compile a comprehensive summary of the current knowledge on imaging modalities, the diagnostic complexities, and the array of therapeutic interventions available for managing hepatic hemangiomas.

3. Clinical Manifestations of Hepatic Hemangiomas

In the majority of instances, hepatic hemangiomas are asymptomatic and are fortuitously identified during imaging procedures conducted for unrelated medical conditions. This observation holds particularly true for hemangiomas of small to medium size. However, in the case of giant hemangiomas (exceeding 5 cm in diameter) that exhibit rapid growth, various symptoms may manifest, albeit these are typically nonspecific and mimic those associated with a range of other disorders, especially those of gastrointestinal origin. The clinical manifestations commonly observed in patients with giant hepatic hemangiomas include pain in the upper abdominal quadrants, nausea, abdominal distension, dyspepsia, and early satiety. Physical examination seldom reveals a palpable mass.

The complications associated with hepatic hemangiomas are predominantly contingent upon the lesion's size and anatomical location. Mechanical complications can arise, such as spontaneous or trauma-induced rupture, and local compression effects on adjacent anatomical structures. For example, compression of the bile ducts can precipitate jaundice or hemobilia; impingement on the stomach may result in gastric obstruction, leading to symptoms of early satiety and dyspepsia; and compression of the hepatic veins by the tumor mass can obstruct venous outflow, culminating in Budd–Chiari syndrome. [16,18]. Inflammatory complications can manifest as either acute or chronic fever and pain. Hemorrhagic complications encompass intratumoral or intraperitoneal hemorrhage, which may occur with or without associated consumptive coagulopathy. One notable condition in this context is Kasabach–Merritt syndrome, predominantly observed in giant hepatic hemangiomas. This syndrome is characterized by thrombocytopenia, microangiopathic hemolytic anemia, and disseminated intravascular coagulation. Additionally, Klippel–Trenaunay syndrome, a form of congenital hemiatrophy, can lead to the development of nevus flammeus and hemimegalencephaly. Von Hippel–Lindau disease is another condition that results in the formation of hemangiomas in multiple organs, including the brain, retina, pancreas, and liver, further complicating the clinical picture [13–15]. While Kasabach–Merritt syndrome is predominantly reported in pediatric populations, there have been documented instances of its occurrence in adults presenting with giant hepatic hemangiomas [19,20]. It can manifest as a symptom of giant hemangioma, leading to consumption coagulopathy and ensuing thrombocytopenia, prolonged prothrombin time and partial thromboplastin time, and hypofibrinogenemia caused by endothelial defects within the hemangioma. These manifestations may occur alongside or without microangiopathic hemolytic anemia [21]. Degenerative complications include thrombosis, progressive fibrosis, and sclerosis. Noteworthy cases of hepatic hemangiomas include calcified, pedunculated (presents risk of torsion), hepatic hemangioma occurring on the steatotic or cirrhotic liver, and with accompanying arteriovenous shunt or heart failure.

4. Diagnostic Approaches for Hepatic Hemangiomas

Hepatic hemangiomas are commonly identified incidentally through cross-sectional abdominal imaging conducted for routine screening or for purposes unrelated to the investigation of a potential hepatic mass [9–11]. This incidental detection is attributable to the predominantly asymptomatic nature of these lesions, which frequently remain unnoticed during standard physical examinations.

The diagnosis of hepatic hemangiomas utilizes a spectrum of imaging modalities, including conventional ultrasound (US, including B-mode and Doppler), contrast-enhanced ultrasound (CEUS), contrast-enhanced computed tomography (CT), magnetic resonance imaging (MRI), angiography, and nuclear imaging (specifically, scintigraphic studies utilizing Technetium-99m-labeled red blood cells). The specificity and sensitivity of these diagnostic techniques are delineated in Table 1 [22].

Table 1. Sensitivity and specificity of diagnostic methods in hepatic hemangiomas.

Diagnostic Method	Sensitivity (%)	Specificity (%)
Ultrasonography	96.9	60.3
Computed tomography	98.3	55
Magnetic resonance imaging	100	85.7
Tc-99m RBC blood pool scintigraphy	67	100

These diagnostic modalities offer considerable specificity in differentiating hepatic hemangiomas from other vascular neoplasms, benign entities such as adenomas, or malignant lesions including hepatocellular carcinoma (HCC), metastases, and dysplastic nodules. Typically, hepatic hemangiomas are categorized into three histological subtypes: capillary hemangioma, cavernous hemangioma, and sclerosing hemangioma.

4.1. Ultrasound (US)

Ultrasound (US) frequently serves as the initial diagnostic modality for hepatic hemangiomas, favored for its wide availability, non-ionizing nature, and repeatability. However, a significant limitation of US is its dependency on the operator's expertise and the patient's specific characteristics, rendering it highly sensitive to both operator and patient factors. On greyscale ultrasound, hepatic hemangiomas typically present as hyperechoic, well-circumscribed lesions with a uniform appearance, or as hypoechoic masses featuring a hyperechoic rim [23–26] (Figure 1). The hyperechoic pattern observed in ultrasound images of hepatic hemangiomas is linked to their histological makeup, where the echogenicity results from the numerous interfaces between the endothelium-lined sinuses constituting the lesions and the encapsulated blood. Smaller hepatic hemangiomas commonly exhibit this hyperechoic characteristic. In contrast, larger lesions might show heterogeneity, characterized by mixed echogenicity (both hypo- and hyperechoic) arising from potential necrosis, hemorrhage, or fibrosis, leading to classification as atypical hepatic hemangiomas. Doppler ultrasound assessments of most hepatic hemangiomas reveal minimal to absent Doppler flow signals [27].

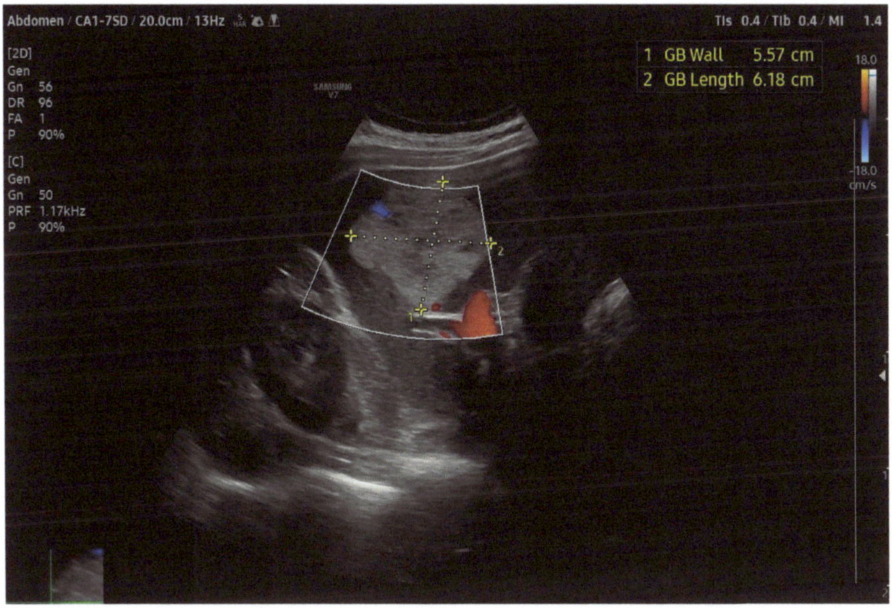

Figure 1. Slightly heterogeneous hyperechoic lesion with absence of flow on color Doppler, characteristic of a large hepatic hemangioma.

Nevertheless, it is imperative to approach the diagnosis of every hyperechoic mass with caution before categorizing it as a hepatic hemangioma. This echogenic pattern may also manifest in a spectrum of other hepatic conditions, encompassing both benign entities (e.g., adenomas) and malignant pathologies (such as hepatocellular carcinoma and metastatic lesions). Consistency in imaging findings across successive examinations constitutes a reliable marker for benign pathology in clinical settings. Ultrasound demonstrates high diagnostic accuracy in distinguishing hepatic hemangiomas from malignant hyperechoic masses, evidencing a sensitivity of 94.1% and specificity of 80.0% for lesions smaller than 3 cm in diameter. The lack of detectable blood flow within a lesion on Doppler ultrasound serves as a robust discriminant for differentiating hepatic hemangioma from hepatocellular carcinoma (HCC), the latter typically exhibiting intra- or peritumoral vascular signals [28].

In the context of hypoechoic lesions, the identification of a peripheral echogenic halo may indicate a hepatic hemangioma. Conversely, the presence of a hypoechoic rim encircling the lesion, often referred to as the "target sign", is infrequently associated with hepatic hemangiomas [27]. Special caution is warranted in the evaluation of hepatic lesions within a steatotic liver, where the altered echotexture may cause a typically hyperechoic hemangioma to appear hypoechoic against the background of an intensely hyperechoic liver parenchyma.

4.2. Contrast-Enhanced Ultrasound (CEUS)

Contrast-enhanced ultrasound (CEUS) represents a more specific diagnostic modality for hepatic hemangiomas (HH) compared to traditional ultrasound techniques. By employing microbubble contrast agents that enhance the visualization of the microvasculature, CEUS facilitates real-time perfusion imaging that mirrors the vascular patterns observable in CT imaging. This feature is exceedingly beneficial for differentiating liver nodules and accurately identifying HH in contrast to adenomas, focal nodular hyperplasia (FNH), hepatocellular carcinoma (HCC), or metastatic lesions. Characteristically, a typical HH demonstrates peripheral nodular enhancement in the arterial phase, followed by complete (occasionally incomplete) centripetal filling in the portal venous and late phases. This enhancement pattern boasts a high sensitivity (98%) for the identification of histologically confirmed HH. Nonetheless, it is critical to acknowledge that HH may, albeit infrequently, exhibit a centrifugal enhancement pattern as well [29,30].

CEUS provides several significant advantages, including the capability for real-time examination and instantaneous results. It allows for the concurrent assessment of multiple lesions, offers the repeatability necessary for follow-up evaluations, and permits the re-injection of contrast agents for enhanced imaging [31]. However, the diagnostic accuracy of CEUS can be compromised in patients with steatosis (fatty liver) or for lesions deeply situated within the body. Moreover, imaging comprehensive views of a large hepatic hemangioma presents a challenge due to the limited penetration depth and field of view of the ultrasound probe [3].

4.3. Endoscopic Ultrasound (EUS)

Endoscopic ultrasound (EUS) offers the ability to visualize and biopsy small, solid liver lesions that may not be detectable through other imaging techniques, or that become evident only during routine staging for gastrointestinal cancers. However, the precise diagnostic utility of EUS in the context of liver diseases remains to be fully elucidated, highlighting the need for comparative studies to define its role more clearly. The therapeutic applications of EUS in hepatic management are expanding [32]. EUS facilitates guided interventions such as fine-needle aspiration (FNA) or biopsies when necessary. Although biopsy is typically not indicated for conventional hepatic hemangiomas, EUS-guided procedures can improve the accuracy and safety of biopsies in atypical cases or when the diagnosis is uncertain [33,34]. It is essential to acknowledge that EUS is a semi-invasive technique, carrying inherent risks of complications.

4.4. Computed Tomography

The characteristic imaging feature of a hepatic hemangioma on computed tomography (CT) scans is a well-circumscribed, hypodense lesion. Upon administration of contrast medium, it demonstrates peripheral nodular enhancement, followed by gradual and homogeneous centripetal fill-in (Figure 2). Nevertheless, this distinct enhancement pattern may not be discernible in lesions smaller than 5 mm, complicating their accurate identification. Atypical hepatic hemangiomas can present with a variety of enhancement patterns on CT imaging [31,35].

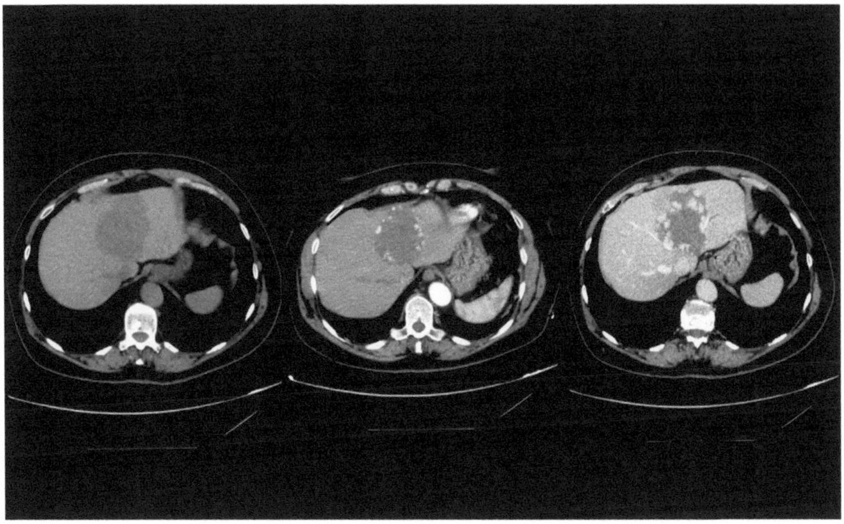

Figure 2. Computed tomography imaging of a giant hepatic hemangioma. The sequence illustrates a precontrast image on the left, an arterial phase image in the center, and a venous phase image on the right. This series effectively demonstrates the delayed contrast filling from the tumor's periphery, characteristic of a hepatic hemangioma.

In the context of hepatic steatosis (fatty liver), particular caution is warranted, as a typical hemangioma might appear hyperdense in comparison to the surrounding hepatic parenchyma. Major limitations of CT imaging include the risk of radiation exposure and the use of iodine-based contrast materials, which carry a potential risk for allergic reactions or contrast-induced nephropathy.

4.5. Magnetic Resonance Imaging

In magnetic resonance imaging (MRI), hepatic hemangiomas are typically characterized by a well-defined, homogenous morphology, manifesting as hypointense on T1-weighted sequences and hyperintense on T2-weighted sequences, a feature often described as the "cotton-wool" appearance [36] (Figure 3). The differentiation between malignancies and hepatic hemangiomas, both of which exhibit hyperintensity on T2-weighted images, can be facilitated by modulating the echo time (TE). While malignant lesions tend to exhibit a reduction in signal intensity, hepatic hemangiomas display an enhanced signal intensity [37].

In MRI diagnostics, the contrast agent employed is gadolinium-based (UCA), rendering it an appropriate option for individuals with allergies to iodinated contrast agents or those with renal insufficiency for whom CT imaging with iodine-based contrast is contraindicated [38].

4.6. Technetium-99m-Labeled Red Blood Cell Imaging

Technetium-99m-labeled red blood cell (Tc-99m RBC) scintigraphy is a noninvasive diagnostic technique offering high specificity for identifying hepatic hemangiomas. In Tc-99m RBC imaging, hepatic hemangiomas exhibit a distinctive perfusion and blood pool mismatch, characterized by diminished perfusion in early dynamic phases with a progressive increment in radiotracer uptake during blood pool phases. Initially, the lesion presents as "cold" or less active, transitioning to intense activity typically within 1–2 h post Tc-99m injection. The sensitivity of this modality is contingent upon the lesion's size: it is 17–20% for lesions under 1 cm, increases to 65–80% for lesions between 1 and 2 cm, and reaches nearly 100% for those exceeding 2 cm in diameter. The specificity of Tc-99m-labeled

RBC scintigraphy, particularly when enhanced with Single Photon Emission Computed Tomography (SPECT), maintains a rate of 100% across all lesion sizes [17].

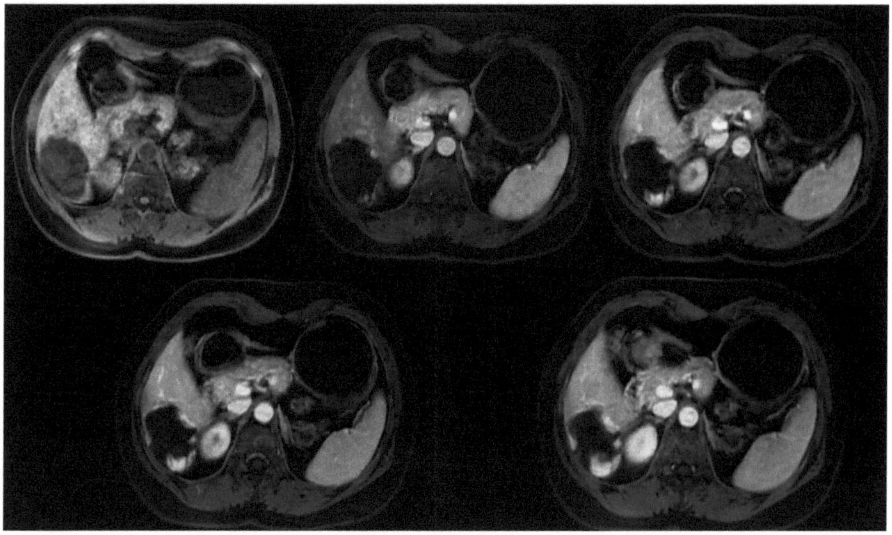

Figure 3. Axial T1-weighted MRI with contrast, multiphase study of a giant hepatic hemangioma. This image sequence demonstrates the gradual peripheral-to-central filling of the hemangioma, highlighting the characteristic enhancement pattern.

However, the sensitivity of this technique is influenced by various factors, including lesion size and anatomical location. Its diagnostic yield significantly improves when integrated with SPECT imaging, yet remains limited for lesions smaller than 1 cm or those positioned in anatomically complex regions [39,40]. The presence of persistent red blood cell activity within the heart, inferior vena cava, and major intrahepatic vessels poses challenges in detecting small hepatic hemangiomas located proximal to these vascular structures in SPECT images [41].

Despite its diagnostic precision, Tc-99m-labeled RBC scintigraphy has been largely superseded as a primary tool for hepatic hemangioma diagnosis due to several drawbacks, such as limited availability, elevated costs, lengthy procedural times, radiation exposure, and the advent of more advanced imaging technologies.

5. Imaging Characterization of Hemangioma Subtypes

Accurate identification of the distinct histological subtypes of hepatic hemangiomas plays a crucial role in the comprehensive diagnostic evaluation. The primary subtypes recognized within hepatic hemangiomas include cavernous hemangioma, capillary hemangioma (also referred to as fast-filling hemangioma), and sclerosing hemangioma [42]. The principal criterion for this classification is the extent of fibrous tissue present within the body of the hemangioma [18]. It is essential to recognize that the unique histological compositions of these lesions may result in imaging appearances that diverge from the conventional semiotics associated with hepatic hemangiomas.

5.1. Cavernous Hemangioma

Cavernous hemangioma is closely aligned with the established radiological profile of hepatic hemangiomas. However, its histological architecture diverges slightly from the classical hepatic hemangioma phenotype. The defining distinction lies in the presence of larger vascular spaces coupled with a reduced quantity of connective tissue. Such a configuration is predominantly observed in lesions smaller than 3 cm in diameter, characterized by well-defined margins and round or lobulated peripheries. Sonographically, this subtype manifests as a hyperechoic lesion with posterior acoustic enhancement, reflecting its histological composition [43]. Analogous to typical hepatic hemangiomas, cavernous hemangiomas seldom generate Doppler signals in both color-coded and spectral Doppler examinations [44]. The CT imaging of cavernous hemangiomas is consistent with the descriptions provided for hepatic hemangiomas in the 'Computed Tomography' section. However, MRI is the preferred imaging modality for cavernous hemangiomas, offering enhanced differentiation between hepatic hemangiomas and malignant hepatic tumors or cysts, contingent upon the inclusion of comprehensive imaging sequences. On T1-weighted MRI, cavernous hemangiomas present as masses with low signal intensity. T2-weighted and diffusion-weighted imaging (DWI) reveal homogeneous hyperintensity, frequently described as the "light bulb sign" [45], attributed to the lesion's cavernous vascular structure facilitating slow blood flow and unrestricted water diffusion. The identification of this sign is particularly crucial in distinguishing flash-filling or sclerosing hemangiomas, which lack nodular enhancement. A T2 relaxation time threshold of 112 ms has demonstrated over 92% accuracy in differentiating hepatic hemangiomas from metastatic lesions [36]. In contrast-enhanced studies, cavernous hemangiomas exhibit enhancement patterns similar to those observed in CT, with early peripheral nodular enhancement followed by delayed, centripetal, and complete enhancement in later phases.

5.2. Capillary Hemangioma

Capillary hemangioma, also known as flash-filling or rapidly-filling hemangioma, represents the second histological subtype of hepatic hemangiomas, accounting for approximately 16% of all hepatic hemangiomas. This subtype is notably more prevalent in hemangiomas measuring less than 1 cm in diameter, comprising 42% of such cases [46]. Flash-filling hemangiomas pose a diagnostic challenge due to their similarity with numerous hypervascular tumors; they exhibit rapid, intense, and uniform contrast enhancement during the arterial phase of contrast-enhanced computed tomography (CT) and T2-weighted magnetic resonance imaging (MRI). A definitive diagnosis can be attained through delayed-phase CT or MRI, where vascular malformations continue to appear significantly attenuated or hyperintense, a characteristic not shared by hypervascular metastases [46]. In contrast to cavernous hemangiomas, capillary hemangiomas present as hypoechoic on ultrasound examinations due to their rapid blood flow through limited vessels and a fibrous stroma. Additionally, the application of color-coded Doppler ultrasound facilitates the detection of intralesional blood flow [42], further aiding in the differentiation of capillary hemangiomas from other vascular abnormalities.

5.3. Sclerosing Hemangioma

Sclerosing hemangioma, occasionally conceptualized as the involutive phase of hemangioma development, is alternatively known as thrombosed or hyalinized hemangioma. This subtype is relatively rare and seldom manifests clinically [46]. The process of hyalinization typically initiates at the lesion's core, leading to the obliteration of vascular channels. Such pronounced alterations significantly modify the lesion's radiological signature, complicating initial diagnostic efforts. Sclerotic transformation tends to produce a heterogenous imaging appearance, characterized by a central fibrous patch surrounded by cystic, fibrotic, and hemorrhagic zones [42]. These areas are distinctly visible as hypoechoic zones on ultrasound and hypodense regions on computed tomography (CT) scans. Furthermore, sclerosing hemangiomas frequently present with irregular contours, capsular retraction, and progressive volume reduction.

Contrary to the imaging profiles of other hepatic hemangioma subtypes, which are marked by pronounced hyperintensity on T2-weighted magnetic resonance (MR) images, hyalinized hemangiomas only demonstrate mild signal elevation. The absence of early enhancement and modest peripheral enhancement during late-phase MR imaging further delineates hyalinized hemangioma from conventional subtypes. Despite these distinctive imaging features, the radiological characteristics of sclerosing hemangioma may not suffice for a conclusive diagnosis, necessitating histopathological evaluation to exclude malignant entities [43].

6. Atypical Hepatic Hemangiomas in Imaging

6.1. Giant Haemangioma

The designation of a hemangioma as "giant" is subject to slight variations across medical literature, but it is commonly defined as a lesion measuring 5 cm in diameter or larger [4,46]. Features such as cystic cavities or central calcifications may be observed, with internal fibrotic septa being a frequent finding.

In ultrasonography (US), giant hemangiomas present a heterogeneous appearance. On non-contrast-enhanced computed tomography (CT) scans, they display a heterogeneous hypodense profile, which may include hypodense central regions [42–45,47–49]. Magnetic resonance imaging (MRI) reveals giant hepatic hemangiomas as hypointense on T1-weighted sequences, with potential alteration in the hyperintensity on T2-weighted images due to hypointense central zones [42–44]. Upon contrast administration, giant hepatic hemangiomas demonstrate peripheral globular enhancement and progressive centripetal filling, a hallmark pattern for these lesions. However, it is noteworthy that complete filling within the lesion is typically not achieved [50].

6.2. Hemangioma with Arterioportal Shunt

The co-occurrence of a hemangioma and an arterioportal shunt has been documented with an incidence rate of up to 26% [42,46,51,52]. This phenomenon is particularly prevalent in small capillary hemangiomas (<2 cm), where the high flow within the compact vascular spaces likely facilitates shunting through potential connections between the hepatic artery and the portal vein [43,52]. Arterioportal shunts are generally diminutive and frequently present as wedge-shaped or irregularly contoured transient enhancements in the arterial phase, indicative of a transient hepatic attenuation difference on computed tomography (CT) or a transient hepatic intensity difference on magnetic resonance imaging (MRI) [51,53–55]. Early opacification of adjacent portal vein branches post-contrast injection may signal the existence of an arterioportal shunt [52].

6.3. Hemangiomatosis

Hepatic hemangiomatosis is an uncommon disorder characterized by the proliferation of numerous hepatic hemangiomas dispersed throughout the liver parenchyma. Although typically asymptomatic in adults, hemangiomatosis is more commonly observed in newborns, where it can be associated with congestive heart failure [56,57]. Unlike the presentation of multiple discrete hepatic hemangiomas, hemangiomatosis features lesions that are ill-defined, extensive, and confluent, potentially encompassing the majority of the hepatic parenchyma. On sonography, these lesions may present as hypo- or hyperechoic, without the distinct peripheral globular enhancement seen in contrast-enhanced computed tomography (CT) and magnetic resonance imaging (MRI), complicating their differentiation from adjacent hepatic tissue. However, MRI retains its diagnostic utility in these cases, as T1- and T2-weighted sequences reveal distinctive signal patterns. Despite heterogeneous enhancement in the arterial phase, these lesions demonstrate progressive enhancement in dynamic late phases, maintaining typical signal intensities on T1- and T2-weighted MR sequences [43,46,56,57].

6.4. Pedunculated Haemangioma

The pedunculated hemangioma represents an exceptionally rare type of lesion that projects from the liver. It is characterized by a clear, encapsulated form, and is connected to the liver via a slender pedicle. Nonetheless, this pedicle may not always be discernible in axial imaging planes, making CT or MRI multiplanar reconstructions beneficial for confirming the hepatic origin of the mass. A specific complication associated with pedunculated hemangioma (PH) is volvulus, which occurs when the lesion twists around its pedicle. This can present as an acute abdominal condition, further complicated by necrosis or hemorrhage [43,58].

6.5. Hepatic Steatosis

Severe hepatic steatosis can influence the apparent enhancement patterns observed in focal hepatic lesions. Even lesions that are typically hypovascular, such as metastases, may demonstrate relatively high attenuation on computed tomography (CT), potentially mimicking hemangiomas with their persistent enhancement pattern. On ultrasound (US), hepatic hemangiomas generally appear as isoechoic or, more commonly, hypoechoic relative to the surrounding hyperintense steatotic liver tissue, frequently exhibiting posterior acoustic enhancement. Additionally, hemangiomas may present with a hypoechoic perilesional halo, a characteristic also observed in malignant tumors within a steatotic liver context. This atypical presentation often necessitates further evaluation through CT or magnetic resonance imaging (MRI). Although hemangiomas in a fatty liver might display a distinct halo on CT or MRI, accurate diagnosis is typically straightforward, facilitated by the lesions' characteristic dynamic enhancement pattern. Magnetic resonance imaging is particularly advantageous for assessing hepatic hemangiomas (HHs) in the context of fatty liver, as the lesions' hyperintensity on T2-weighted images is not affected by hepatic steatosis, maintaining its diagnostic utility [59–61].

6.6. Liver Cirrhosis

The detection and characterization of hepatic hemangiomas in the context of liver cirrhosis can be challenging, as these lesions tend to become more fibrous and decrease in size [59,62]. The prevalence of hepatic hemangiomas is lower in cirrhotic livers than in non-cirrhotic ones [63]. While hepatic hemangiomas may retain their characteristic imaging features, in advanced stages of cirrhosis, they often lose these distinctive traits, complicating diagnostic efforts [63,64]. Magnetic resonance imaging (MRI) is regarded as the preferred modality for evaluating hepatic hemangiomas due to its superior contrast resolution. The diagnostic sensitivity of MRI is further enhanced by T2-weighted sequences. In ultrasonography (US), both dysplastic nodules and hepatocellular carcinoma may appear as hyperechoic nodules, which can mimic the sonographic appearance of hepatic hemangiomas [64].

7. Histology Sampling

Given the vascular nature of hepatic hemangiomas, biopsy procedures involving histological sampling carry a considerable risk of hemorrhage, especially in the context of large, subcapsular lesions. Such procedures may result in severe complications, including mortality [65]. Additionally, the diagnostic yield of biopsy in this context is relatively low [66] leading to the recommendation that biopsy be reserved for lesions exhibiting atypical features.

8. Treatment

The vast majority of hepatic hemangiomas are characterized as small, asymptomatic, and exhibit stable dimensions, with patients generally maintaining normal liver function. These lesions are often incidentally identified during routine abdominal cross-sectional imaging studies [9–11]. Conservative management, encompassing periodic observation and surveillance via imaging at intervals of 6 or 12 months, is typically recommended as a suitable treatment strategy for these lesions. Notably, no cases of malignant transformation within hepatic hemangiomas have been documented [67,68]. Individuals presenting with new-onset pain, showing unresponsiveness to analgesics, undergoing estrogen therapy, experiencing pregnancy, or possessing large hepatic hemangiomas are advised to undergo extended observation as a component of their clinical management.

In instances where patients present with large lesions (exceeding 5 cm in diameter) that demonstrate progressive enlargement and are associated with symptomatology attributable to the lesions, specific therapeutic interventions become imperative [69]. It is essential to exclude all alternative etiologies for the symptoms, such as gastroesophageal reflux disease, peptic ulcer disease, or cholelithiasis, before contemplating interventional procedures.

Historically, surgical interventions, including resection, lobectomy, or enucleation, executed via open surgery or laparoscopy, were the preferred modalities for managing symptomatic cases [70,71]. However, with advancements in minimally invasive interventional techniques, surgical methods are no longer the first-line treatment for patients with multiple or extensive lesions. Alternative therapeutic options, such as radiofrequency ablation, microwave ablation, and arterial embolization, are now available for the treatment of symptomatic hepatic hemangiomas [72].

8.1. Surgical Approach

Surgical interventions for hepatic hemangiomas encompass segmental resection, lobectomy, or enucleation, with the selection of the technique being contingent upon the hemangioma's size and location, alongside the surgeon's expertise and preference. Enucleation is particularly advantageous for hemangiomas that are superficially located with a discernible plane on the liver surface, a zone characterized by compressed hepatic tissue with minimal vasculature due to the hemangioma's expansion. This plane facilitates the hemangioma's removal with negligible blood loss by separating the tumor capsule from the liver parenchyma, thus preserving the majority of the surrounding healthy liver tissue.

Conversely, hemangiomas that are deeply embedded within the liver parenchyma, lack an accessible surface from the Glisson capsule, or span an entire lobe necessitate hepatic resection as the optimal treatment strategy [73]. Surgical treatments, however, carry heightened risks of complications such as hemorrhage, infection, and increased financial costs, particularly for lesions exceeding 10 cm in diameter [74]. Furthermore, in instances of multiple hemangiomas or those proximal to the hepatic portal vessels, surgical resection can lead to less favorable outcomes and a pronounced risk of hemorrhage [75]. The feasibility of surgical intervention may also be limited by the lesion's substantial size, its unfavorable positioning, or specific patient-related factors.

8.2. Radiofrequency Ablation

Radiofrequency ablation (RFA) represents a minimally invasive, efficacious modality for the treatment of both primary and metastatic hepatic neoplasms, with the procedure being executable via percutaneous or laparoscopic approaches [14,76]. Recent applications of percutaneous RFA have also demonstrated success in the management of liver hemangiomas [77–79]. The mechanism of RFA is postulated to entail the induction of localized thermal damage to the flat endothelial cells lining the extensively dilated, non-anastomotic vascular spaces characteristic of these tumors. The utility of RFA in this context is underpinned by the benign and hypervascular nature of liver hemangiomas.

This technique offers several advantages. Primarily, the benign character of the tumor obviates the need for excising a margin of healthy liver tissue surrounding the lesion. Additionally, the tumor's composition, predominantly blood-filled cavities, facilitates the conspicuous collapse of the tumor tissue adjacent to the ablation zone upon application of radiofrequency energy. Moreover, the benign nature of any residual tumor post initial treatment negates the urgency for immediate follow-up intervention, given its non-progressive and non-metastasizing behavior [80].

Conversely, a notable drawback of RFA is the likelihood of hemolysis attributable to the vascular supply of the tumor. The magnitude and risk of this complication escalate with the tumor size, potentially culminating in a spectrum of conditions including hemoglobinuria, hemolytic jaundice, anemia, or renal impairment, contingent on the complication's severity. Complication rates for tumors exceeding 10 cm in diameter have been reported between

34% and 100%, rendering RFA less suitable for managing giant hepatic hemangiomas [81, 82].

8.3. Transarterial Embolization and Chemoembolization

In recent years, transarterial embolization (TAE) has been recognized as an effective strategy for the management of hepatic hemangiomas, functioning by obstructing the blood supply to the lesion without the use of chemotherapeutic agents. This technique is executed via an endovascular route, establishing itself as a cornerstone procedure for vascular occlusion. Similarly, transcatheter arterial chemoembolization (TACE) also aims to occlude the blood supply to the target lesion but incorporates an active biological agent alongside the embolization material. During TACE, various chemotherapeutic agents such as bleomycin, pingyangmycin, or ethanol, mixed with lipiodol, are utilized.

Despite the growing application of these methods, consensus regarding the efficacy of TAE in treating hemangiomas and the spectrum of potential complications remains elusive [74,83–88]. Liu et al. reported that TACE employing pingyangmycin yielded unsatisfactory outcomes in liver hemangioma treatments, highlighting a considerable risk of severe complications [85]. In contrast, Torkian et al., through a systematic review and meta-analysis, posited that TACE, when combined with agents like bleomycin, pingyangmycin, or ethanol mixed with lipiodol, was both safe and efficacious [74]. The popularity of TACE as a primary treatment for giant hepatic hemangiomas has surged. Li et al. documented a multi-center study involving 836 cases, where patients with giant hepatic hemangiomas underwent TACE using a pingyangmycin–lipiodol emulsion. The study reported no mortalities and only two instances of hepatic abscess as severe complications, alongside a notable decrease in lesion size, with the mean diameter reducing from 9.6 $\pm$ 0.8 cm to 3.6 $\pm$ 0.5 cm [6]. Yuan et al. evaluated the medium and long-term outcomes of TACE with a lipiodol–bleomycin emulsion in 241 patients, observing no mortalities or serious complications post-procedure. Patients experienced significant symptomatic relief, with no recurrence of symptoms during follow-up. A satisfactory tumor reduction rate, defined as a decrease in the lesion's maximum diameter by more than 50%, was achieved in 88.1% of cases at the 6-month post-procedure mark [89].

The nature of the blood supply to hemangiomas significantly influences treatment outcomes and the incidence of complications. Therefore, in formulating treatment plans, clinicians should prioritize the evaluation of the hemangioma's blood supply characteristics and dimensions. The current guidelines proposed by Ouyang et al. [90] and Zeng et al. [91] suggest specific considerations for assessing the blood supply type to hemangiomas (Table 2), indicating that in instances where the portal vein supplies blood, no abnormalities are detected in arterial and parenchymal phases, but portal venograms reveal abnormal blood-filled sinuses.

Table 2. Characteristics of blood supply to hepatic hemangioma.

Type of Blood Supply	Artery Characteristics	Arterial Phase	Parenchymal Phase
Rich	Mild to moderate thickening of the arteries	Abnormal blood sinusoids	Dilatation of most blood sinusoids
Moderate	Mild thickening of the arteries	Abnormal blood sinusoids	Dilatation of some blood sinusoids
Poor	No thickening of the arteries	Very few abnormal blood sinusoids	No visible dilatation of blood sinusoids

Postembolization syndrome (PES) represents the most frequent complication following transcatheter arterial chemoembolization (TACE), manifesting as influenza-like symptoms shortly after the intervention. The syndrome is predominantly characterized by abdominal pain, fever, nausea, and vomiting [92,93]. Kacała et al. observed PES in 45.7% of patients post-TACE, with the severity of symptoms varying across cases. In the majority of these instances, the administration of paracetamol was effective for pain management, and PES resolved spontaneously in all cases [93]. Basile et al. have posited that PES should be considered an expected outcome of TACE [94].

8.4. Liver Transplantation

Liver transplantation has been identified as a feasible therapeutic option for managing extensive hepatic hemangiomas associated with life-threatening coagulopathies, such as Kasabach–Merritt syndrome [95], among other indications [96]. Nevertheless, liver transplantation is regarded as a treatment of last resort due to its significant risks and the limited circumstances under which it is deemed appropriate.

9. Future Prospects

With the enhancement of imaging techniques such as MRI, CT, and ultrasound, diagnosing and monitoring hepatic hemangiomas will become significantly more straightforward. These advancements will facilitate the determination of tumor size, location, and characteristics, thereby guiding treatment decisions. Furthermore, the improvement in imaging methods is likely to lead to an increase in incidental findings. Although surgical management was once the preferred method for symptomatic hemangiomas, the rapid development of minimally invasive treatment options has led to a worldwide shift towards these alternatives. Future improvements in these techniques are expected to result in safer and more effective treatments, with lower risks and shorter recovery periods. Additional research is necessary to deepen our understanding of the natural progression of hepatic hemangiomas and to identify risk factors linked to their growth or complications. This knowledge will enable healthcare providers to make more informed decisions about monitoring and treatment. In summary, the outlook for patients with hepatic hemangiomas is optimistic, thanks to progress in diagnostic methods, treatment approaches, and research, all of which contribute to better patient outcomes for these benign liver tumors.

10. Discussion

Hepatic hemangiomas, representing the predominant benign mesenchymal neoplasms of the hepatic tissue, frequently manifest as asymptomatic entities, identified serendipitously through imaging modalities conducted for unrelated reasons, and seldom compromise hepatic functionality. From a histopathological perspective, these neoplasms are characterized by the presence of cavernous venous spaces, which are delineated by a lining of vascular endothelial cells and interspersed with connective tissue septa. The hemodynamics within these lesions are notably impaired, exhibiting a markedly reduced flow rate, with the hepatic artery serving as the principal source of vascular supply [97].

The lesions frequently exhibit stability in dimensional parameters and do not necessitate intervention beyond conservative management and vigilant observation. The lesions frequently exhibit stability in dimensional parameters and do not necessitate intervention beyond conservative management and vigilant observation [9–11]. There are no documented cases of hepatic hemangiomas undergoing malignant transformation; however, steroid or estrogen therapy, as well as pregnancy, have been observed to contribute to an increase in hemangioma size [67,68]. Giant hepatic hemangiomas may manifest with periodic pain, a sensation of abdominal fullness, or the detection of an upper abdominal mass, potentially leading to severe complications such as local compression, persistent pain, and serious conditions like obstructive jaundice, Kasabach–Merritt syndrome, Budd–Chiari syndrome, or spontaneous rupture resulting in intra-abdominal hemorrhage, with mortality rates reaching up to 70% [16,98,99]. The incidence of spontaneous rupture in hepatic hemangiomas ranges from 1% to 4%, with giant subcapsular lesions considered at higher risk [100–103].

Hepatic hemangiomas are commonly identified incidentally during imaging studies, yet their diagnosis and management continue to be debated among clinicians. Imaging techniques such as ultrasound (US), computed tomography (CT), magnetic resonance imaging (MRI), and angiography are crucial for diagnosing and characterizing hepatic hemangiomas. US is often the initial imaging technique employed due to its accessibility and cost-effectiveness. Despite this, there is no consensus on the definitive standard for diagnosing hepatic hemangioma. Historically, angiography was considered the gold standard for this purpose [104]. However, advancements and increased accessibility of various imaging methods have shifted this paradigm. Some researchers now view MRI as the gold standard for diagnosing hepatic hemangiomas [105,106], while others advocate for IV contrast-enhanced abdominal CT scans [69,107]. Furthermore, the role of biopsy and histopathological examination in evaluating liver lesions, especially atypical ones that may resemble malignancies, is crucial, despite the associated risks of such invasive procedures [108,109].

The management strategy for hepatic hemangiomas depends on various factors, including the size, location, symptoms, and potential complications of the lesion. Asymptomatic and small lesions often do not require intervention and can be monitored conservatively with regular imaging. Symptoms of enlarged hepatic hemangiomas are non-specific and can overlap with other pathological conditions; thus, alternative causes of abdominal symptoms such as gallstones, gastroesophageal reflux disease, or peptic ulcer disease should be excluded before considering interventional procedures. Surgery has traditionally been the preferred treatment for symptomatic patients or those with significant lesion growth [70,71]. When symptoms emerge, the lesion grows rapidly, or there is an increased risk of rupture, alternative therapeutic options become necessary, moving beyond the simple dichotomy of resection versus observation [98,110–112].

Surgical intervention necessitates careful consideration due to the potential for complications. Despite some evidence supporting surgical approaches, the risk of extended hospitalization, significant perioperative blood loss, and complications in lesions larger than 10 cm must be evaluated [74,113]. Alternatives to surgical and conservative treatments, such as radiofrequency ablation (RFA) and microwave ablation (MWA), have been explored but found lacking in efficacy and associated with complications, especially in giant lesions [79,81,82,114,115].

Transcatheter arterial chemoembolization (TACE) has recently gained traction as an effective treatment for hepatic hemangiomas. The efficacy of TACE has been extensively reviewed [6,88,89,93,116]. Bleomycin is known for its cytotoxic, antiangiogenic, and sclerosing properties, leading to DNA degradation and eliciting a generalized inflammatory response in the vicinity of the lesion and the portal area [117]. When combined with lipiodol, bleomycin's embolic effect is enhanced, facilitating improved distribution of the chemotherapeutic agent to the targeted site.

However, it is crucial to recognize the limitations and potential complications inherent to each therapeutic approach. Interventional procedures, while effective in alleviating symptoms and reducing the size of hepatic hemangiomas, are associated with risks, including post-procedural bleeding, infection, and hepatic dysfunction. Consequently, a comprehensive evaluation of the risks and benefits of each treatment option is imperative, taking into account the unique characteristics and preferences of each patient.

In conclusion, the management of hepatic hemangiomas necessitates a multidisciplinary strategy, incorporating the expertise of radiologists, hepatologists, and surgeons to customize treatment plans according to the individual requirements of patients. Ongoing research efforts to enhance imaging methodologies and therapeutic techniques are essential to advance the management of hepatic hemangiomas and elevate patient care outcomes.

11. Conclusions

Hepatic cavernous hemangioma is the most prevalent type of benign liver tumor. When treatment is necessary for liver hemangiomas, surgical approaches like hepatic resection or enucleation, performed through open, laparoscopic, or robotic methods, have been historically deemed the first choice. However, in recent years, alternative therapies such as liver transplantation, radiofrequency ablation, transarterial embolization, and transarterial chemoembolization have also been gaining in importance. When deciding on the best treatment approach, it is essential to conduct a thorough assessment that takes into account various factors such as symptoms, size, location, and the presence of any coexisting medical conditions.

Author Contributions: Conceptualization, A.K. (Arkadiusz Kacała) and M.G.; methodology, A.K. (Arkadiusz Kacała) and M.G.; formal analysis, A.K. (Arkadiusz Kacała) and M.D.; investigation, M.D., M.P. and A.K. (Arkadiusz Kacała); writing—original draft preparation, A.K. (Arkadiusz Kacała), M.D., I.M. and M.S.; writing—review and editing, A.K. (Arkadiusz Kacała), A.K. (Adrian Korbecki) M.S., J.J.-N. and M.G.; supervision, M.G., D.J. and D.P. All authors have read and agreed to the published version of the manuscript.

Funding: This research was funded by Wroclaw Medical University.

Institutional Review Board Statement: This study protocol was reviewed and approved by the Bioethics Committee on Wroclaw Medical University, approval number KB 223/2023N (31.08.2023r).

Informed Consent Statement: Not applicable.

Data Availability Statement: No new data were created or analyzed in this study. Data sharing is not applicable to this article.

Conflicts of Interest: The authors declare no conflicts of interest.

References

1. Sadick, M.; Müller-Wille, R.; Wildgruber, M.; Wohlgemuth, W.A. Vascular anomalies (part I): Classification and diagnostics of vascular anomalies. *Rofo* **2018**, *190*, 825–835. [CrossRef]
2. Belghiti, J.; Cauchy, F.; Paradis, V.; Vilgrain, V. Diagnosis and management of solid benign liver lesions. *Nat. Rev. Gastroenterol. Hepatol.* **2014**, *11*, 737–749. [CrossRef]
3. Maruyama, M.; Isokawa, O.; Hoshiyama, K.; Hoshiyama, A.; Hoshiyama, M.; Hoshiyama, Y. Diagnosis and management of giant hepatic hemangioma: The usefulness of contrast-enhanced ultrasonography. *Int. J. Hepatol.* **2013**, *2013*, 802180. [CrossRef] [PubMed]
4. Grieco, M.B.; Miscall, B.G. Giant hemangiomas of the liver. *Surg. Gynecol. Obstet.* **1978**, *147*, 783–787. [PubMed]
5. Karhunen, P.J. Benign hepatic tumours and tumour like conditions in men. *J. Clin. Pathol.* **1986**, *39*, 183–188. [CrossRef] [PubMed]
6. Li, Y.; Jia, Y.; Li, S.; Wang, W.; Wang, Z.; Wang, Y.; Liu, B.; Wang, W.; Chang, H.; Li, Z. Transarterial Chemoembolization of Giant Liver Haemangioma: A Multi-center Study with 836 Cases. *Cell Biochem. Biophys.* **2015**, *73*, 469–472. [CrossRef] [PubMed]
7. Leon, M.; Chavez, L.; Surani, S. Hepatic hemangioma: What internists need to know. *World J. Gastroenterol.* **2020**, *26*, 11–20. [CrossRef] [PubMed]
8. Dockerty, M.B.; Gray, H.K.; Henson, S.W. Benign tumors of the liver. II. Hemangiomas. *Surg. Gynecol. Obstet.* **1956**, *103*, 327–331. [PubMed]
9. Aziz, H.; Brown, Z.J.; Baghdadi, A.; Kamel, I.R.; Pawlik, T.M. A comprehensive review of hepatic hemangioma management. *J. Gastrointest. Surg.* **2022**, *26*, 1998–2007. [CrossRef]
10. Oldhafer, K.J.; Habbel, V.; Horling, K.; Makridis, G.; Wagner, K.C. Benign Liver Tumors. *Visc. Med.* **2020**, *36*, 292–303. [CrossRef]
11. Farhat, W.; Ammar, H.; Said, M.A.; Mizouni, A.; Ghabry, L.; Hammami, E.; Gupta, R.; Habiba Ben Hamada; Mabrouk, M.B.; Ali, A.B. Surgical management of giant hepatic hemangioma: A 10-year single center experience. *Ann. Med. Surg.* **2021**, *69*, 102542. [CrossRef] [PubMed]
12. Liu, X.; Yang, Z.; Tan, H.; Xu, L.; Sun, Y.; Si, S.; Liu, L.; Zhou, W.; Huang, J. Giant liver hemangioma with adult Kasabach-Merritt syndrome: Case report and literature review. *Medicine* **2017**, *96*, e7688. [CrossRef]
13. Erdogan, D.; Busch, O.R.C.; van Delden, O.M.; Bennink, R.J.; ten Kate, F.J.W.; Gouma, D.J.; van Gulik, T.M. Management of liver hemangiomas according to size and symptoms. *J. Gastroenterol. Hepatol.* **2007**, *22*, 1953–1958. [CrossRef]
14. Gandolfi, L.; Leo, P.; Solmi, L.; Vitelli, E.; Verros, G.; Colecchia, A. Natural history of hepatic haemangiomas: Clinical and ultrasound study. *Gut* **1991**, *32*, 677–680. [CrossRef]
15. Sun, J.-H.; Nie, C.-H.; Zhang, Y.-L.; Zhou, G.-H.; Ai, J.; Zhou, T.-Y.; Zhu, T.-Y.; Zhang, A.-B.; Wang, W.-L.; Zheng, S.-S. Transcatheter arterial embolization alone for giant hepatic hemangioma. *PLoS ONE* **2015**, *10*, e0135158. [CrossRef] [PubMed]
16. Sharma, V.; Aggarwal, A.; Singla, R.; Kalra, N.; Chawla, Y.K. Giant hemangioma causing budd-Chiari syndrome. *J. Clin. Exp. Hepatol.* **2014**, *4*, 380–381. [CrossRef] [PubMed]
17. Nakanuma, Y. Non-neoplastic nodular lesions in the liver. *Pathol. Int.* **1995**, *45*, 703–714. [CrossRef]
18. Bajenaru, N.; Balaban, V.; Săvulescu, F.; Campeanu, I.; Patrascu, T. Hepatic hemangioma—Review. *J. Med. Life* **2015**, *8*, 4–11.
19. Oak, C.Y.; Jun, C.H.; Cho, E.A.; Lee, D.H.; Cho, S.B.; Park, C.H.; Joo, Y.E.; Kim, H.S.; Rew, J.S.; Choi, S.K. Hepatic Hemangioma with Kasabach-Merritt Syndrome in an Adult Patient. *Korean J. Gastroenterol.* **2016**, *67*, 220–223. [CrossRef]
20. Aslan, A.; Meyer Zu Vilsendorf, A.; Kleine, M.; Bredt, M.; Bektas, H. Adult Kasabach-Merritt Syndrome due to Hepatic Giant Hemangioma. *Case Rep. Gastroenterol.* **2009**, *3*, 306–312. [CrossRef]

21. Hall, G.W. Kasabach-Merritt syndrome: Pathogenesis and management. *Br. J. Haematol.* **2001**, *112*, 851–862. [CrossRef]
22. Toro, A.; Mahfouz, A.-E.; Ardiri, A.; Malaguarnera, M.; Malaguarnera, G.; Loria, F.; Bertino, G.; Di Carlo, I. What is changing in indications and treatment of hepatic hemangiomas. *Ann. Rev. Ann. Hepatol.* **2014**, *13*, 327–339. [CrossRef]
23. Huang, M.; Zhao, Q.; Chen, F.; You, Q.; Jiang, T. Atypical appearance of hepatic hemangiomas with contrast-enhanced ultrasound. *Oncotarget* **2018**, *9*, 12662–12670. [CrossRef] [PubMed]
24. Moody, A.R.; Wilson, S.R. Atypical hepatic hemangioma: A suggestive sonographic morphology. *Radiology* **1993**, *188*, 413–417. [CrossRef]
25. Bree, R.L.; Schwab, R.E.; Glazer, G.M.; Fink-Bennett, D. The varied appearances of hepatic cavernous hemangiomas with sonography, computed tomography, magnetic resonance imaging and scintigraphy. *Radiographics* **1987**, *7*, 1153–1175. [CrossRef] [PubMed]
26. Kim, K.W.; Kim, T.K.; Han, J.K.; Kim, A.Y.; Lee, H.J.; Park, S.H.; Kim, Y.H.; Choi, B.I. Hepatic hemangiomas: Spectrum of US appearances on gray-scale, power Doppler, and contrast-enhanced US. *Korean J. Radiol.* **2000**, *1*, 191–197. [CrossRef]
27. Kim, T.K.; Han, J.K.; Kim, A.Y.; Park, S.J.; Choi, B.I. Signal from hepatic hemangiomas on power Doppler US: Real or artefactual? *Ultrasound Med. Biol.* **1999**, *25*, 1055–1061. [CrossRef] [PubMed]
28. Hashemi, J.; Kakhki, V.R.D. Accuracy of Gray-Scale and Color Doppler Sonography in Diagnosis of Hepatic Hemangioma, Hepatocellular Carcinoma and Liver Metastasis. *Iran. J. Radiol.* **2008**, *5*, 129–134.
29. Bartolotta, T.V.; Taibbi, A.; Galia, M.; Lo Re, G.; La Grutta, L.; Grassi, R.; Midiri, M. Centrifugal (inside-out) enhancement of liver hemangiomas: A possible atypical appearance on contrast-enhanced US. *Eur. J. Radiol.* **2007**, *64*, 447–455. [CrossRef]
30. Kim, S.; Chung, J.J.; Kim, M.J.; Park, S.; Lee, J.T.; Yoo, H.S. Atypical inside-out pattern of hepatic hemangiomas. *Am. J. Roentgenol.* **2000**, *174*, 1571–1574. [CrossRef]
31. Westwood, M.; Joore, M.; Grutters, J.; Redekop, K.; Armstrong, N.; Lee, K.; Gloy, V.; Raatz, H.; Misso, K.; Severens, J.; et al. Contrast-enhanced ultrasound using SonoVue® (sulphur hexafluoride microbubbles) compared with contrast-enhanced computed tomography and contrast-enhanced magnetic resonance imaging for the characterisation of focal liver lesions and detection of liver metastases: A systematic review and cost-effectiveness analysis. *Health Technol. Assess.* **2013**, *17*, 1–243. [CrossRef]
32. Srinivasan, I.; Tang, S.-J.; Vilmann, A.S.; Menachery, J.; Vilmann, P. Hepatic applications of endoscopic ultrasound: Current status and future directions. *World J. Gastroenterol.* **2015**, *21*, 12544–12557. [CrossRef]
33. Alvarez-Sánchez, M.V.; Jenssen, C.; Faiss, S.; Napoléon, B. Interventional endoscopic ultrasonography: An overview of safety and complications. *Surg. Endosc.* **2014**, *28*, 712–734. [CrossRef]
34. ASGE Technology Committee; Kaul, V.; Adler, D.G.; Conway, J.D.; Farraye, F.A.; Kantsevoy, S.V.; Kethu, S.R.; Kwon, R.S.; Mamula, P.; Pedrosa, M.C.; et al. Interventional EUS. *Gastrointest. Endosc.* **2010**, *72*, 1–4. [CrossRef]
35. Jang, J.Y.; Kim, M.Y.; Jeong, S.W.; Kim, T.Y.; Kim, S.U.; Lee, S.H.; Suk, K.T.; Park, S.Y.; Woo, H.Y.; Kim, S.G.; et al. Current consensus and guidelines of contrast enhanced ultrasound for the characterization of focal liver lesions. *Clin. Mol. Hepatol.* **2013**, *19*, 1–16. [CrossRef] [PubMed]
36. McFarland, E.G.; Mayo-Smith, W.W.; Saini, S.; Hahn, P.F.; Goldberg, M.A.; Lee, M.J. Hepatic hemangiomas and malignant tumors: Improved differentiation with heavily T2-weighted conventional spin-echo MR imaging. *Radiology* **1994**, *193*, 43–47. [CrossRef] [PubMed]
37. Chan, Y.L.; Lee, S.F.; Yu, S.C.H.; Lai, P.; Ching, A.S.C. Hepatic malignant tumour versus cavernous haemangioma: Differentiation on multiple breath-hold turbo spin-echo MRI sequences with different T2-weighting and T2-relaxation time measurements on a single slice multi-echo sequence. *Clin. Radiol.* **2002**, *57*, 250–257. [CrossRef] [PubMed]
38. Tateyama, A.; Fukukura, Y.; Takumi, K.; Shindo, T.; Kumagae, Y.; Kamimura, K.; Nakajo, M. Gd-EOB-DTPA-enhanced magnetic resonance imaging features of hepatic hemangioma compared with enhanced computed tomography. *World J. Gastroenterol.* **2012**, *18*, 6269–6276. [CrossRef] [PubMed]
39. Ziessman, H.A.; Silverman, P.M.; Patterson, J.; Harkness, B.; Fahey, F.H.; Zeman, R.K.; Keyes, J.W. Improved detection of small cavernous hemangiomas of the liver with high-resolution three-headed SPECT. *J. Nucl. Med.* **1991**, *32*, 2086–2091. [PubMed]
40. Birnbaum, B.A.; Weinreb, J.C.; Megibow, A.J.; Sanger, J.J.; Lubat, E.; Kanamuller, H.; Noz, M.E.; Bosniak, M.A. Definitive diagnosis of hepatic hemangiomas: MR imaging versus Tc-99m-labeled red blood cell SPECT. *Radiology* **1990**, *176*, 95–101. [CrossRef] [PubMed]
41. Schillaci, O.; Danieli, R.; Manni, C.; Capoccetti, F.; Simonetti, G. Technetium-99m-labelled red blood cell imaging in the diagnosis of hepatic haemangiomas: The role of SPECT/CT with a hybrid camera. *Eur. J. Nucl. Med. Mol. Imaging* **2004**, *31*, 1011–1015. [CrossRef]
42. Klotz, T.; Montoriol, P.F.; Da Ines, D.; Petitcolin, V.; Joubert-Zakeyh, J.; Garcier, J.M. Hepatic haemangioma: Common and uncommon imaging features. *Diagn. Interv. Imaging* **2013**, *94*, 849–859. [CrossRef] [PubMed]
43. Mamone, G.; Di Piazza, A.; Carollo, V.; Cannataci, C.; Cortis, K.; Bartolotta, T.V.; Miraglia, R. Imaging of hepatic hemangioma: From A to Z. *Abdom. Radiol.* **2020**, *45*, 672–691. [CrossRef] [PubMed]
44. Caseiro-Alves, F.; Brito, J.; Araujo, A.E.; Belo-Soares, P.; Rodrigues, H.; Cipriano, A.; Sousa, D.; Mathieu, D. Liver haemangioma: Common and uncommon findings and how to improve the differential diagnosis. *Eur. Radiol.* **2007**, *17*, 1544–1554. [CrossRef] [PubMed]
45. Mamone, G.; Miraglia, R. The "light bulb sign" in liver hemangioma. *Abdom. Radiol.* **2019**, *44*, 2327–2328. [CrossRef] [PubMed]

46. Vilgrain, V.; Boulos, L.; Vullierme, M.P.; Denys, A.; Terris, B.; Menu, Y. Imaging of atypical hemangiomas of the liver with pathologic correlation. *Radiographics* **2000**, *20*, 379–397. [CrossRef] [PubMed]
47. Valls, C.; Reñe, M.; Gil, M.; Sanchez, A.; Narvaez, J.A.; Hidalgo, F. Giant cavernous hemangioma of the liver: Atypical CT and MR findings. *Eur. Radiol.* **1996**, *6*, 448–450. [CrossRef]
48. Choi, B.I.; Han, M.C.; Park, J.H.; Kim, S.H.; Han, M.H.; Kim, C.W. Giant cavernous hemangioma of the liver: CT and MR imaging in 10 cases. *Am. J. Roentgenol.* **1989**, *152*, 1221–1226. [CrossRef]
49. Danet, I.M.; Semelka, R.C.; Braga, L.; Armao, D.; Woosley, J.T. Giant hemangioma of the liver: MR imaging characteristics in 24 patients. *Magn. Reson. Imaging* **2003**, *21*, 95–101. [CrossRef]
50. Yu, J.S.; Kim, M.J.; Kim, K.W.; Chang, J.C.; Jo, B.J.; Kim, T.H.; Lee, J.T.; Yoo, H.S. Hepatic cavernous hemangioma: Sonographic patterns and speed of contrast enhancement on multiphase dynamic MR imaging. *Am. J. Roentgenol.* **1998**, *171*, 1021–1025. [CrossRef]
51. Kim, K.W.; Kim, T.K.; Han, J.K.; Kim, A.Y.; Lee, H.J.; Choi, B.I. Hepatic hemangiomas with arterioportal shunt: Findings at two-phase CT. *Radiology* **2001**, *219*, 707–711. [CrossRef]
52. Kim, K.W.; Kim, A.Y.; Kim, T.K.; Kim, S.Y.; Kim, M.-J.; Park, M.-S.; Park, S.H.; Lee, K.H.; Kim, J.K.; Kim, P.-N.; et al. Hepatic hemangiomas with arterioportal shunt: Sonographic appearances with CT and MRI correlation. *Am. J. Roentgenol.* **2006**, *187*, W406–W414. [CrossRef]
53. Shimada, M.; Matsumata, T.; Ikeda, Y.; Urata, K.; Hayashi, H.; Shimizu, M.; Sugimachi, K. Multiple hepatic hemangiomas with significant arterioportal venous shunting. *Cancer* **1994**, *73*, 304–307. [CrossRef] [PubMed]
54. Winograd, J.; Palubinskas, A.J. Arterial-portal venous shunting in cavernous hemangioma of the liver. *Radiology* **1977**, *122*, 331–332. [CrossRef]
55. Sousa, M.S.C.; Ramalho, M.; Herédia, V.; Matos, A.P.; Palas, J.; Jeon, Y.H.; Afonso, D.; Semelka, R.C. Perilesional enhancement of liver cavernous hemangiomas in magnetic resonance imaging. *Abdom. Imaging* **2014**, *39*, 722–730. [CrossRef] [PubMed]
56. Guerra, A.; Infante, A.; Rinninella, E.; Spinelli, I.; Mazziotti, M.A.; De Gaetano, A.M.; Pompili, M.; Bonomo, L. A peculiar case of diffuse hemangiomatosis of the left hepatic lobe in an asymptomatic adult patient: Case report and literature review. *Eur. Rev. Med. Pharmacol. Sci.* **2017**, *21*, 1593–1597. [PubMed]
57. Liu, M.C.; Little, E.C. Isolated hepatic hemangiomatosis in 2 septuagenarians. *Radiol. Case Rep.* **2018**, *13*, 1097–1103. [CrossRef]
58. Blondet, A.; Ridereau-Zins, C.; Michalak, S.; Pessaux, P.; Aubertin, A.; Aubé, C. Multiple pedunculated liver hemangiomas presenting with volvulus. *J. Radiol.* **2007**, *88*, 891–894. [CrossRef]
59. Jang, H.-J.; Kim, T.K.; Lim, H.K.; Park, S.J.; Sim, J.S.; Kim, H.Y.; Lee, J.-H. Hepatic hemangioma: Atypical appearances on CT, MR imaging, and sonography. *Am. J. Roentgenol.* **2003**, *180*, 135–141. [CrossRef]
60. Kim, K.W.; Kim, M.J.; Lee, S.S.; Kim, H.J.; Shin, Y.M.; Kim, P.-N.; Lee, M.-G. Sparing of fatty infiltration around focal hepatic lesions in patients with hepatic steatosis: Sonographic appearance with CT and MRI correlation. *Am. J. Roentgenol.* **2008**, *190*, 1018–1027. [CrossRef]
61. Marsh, J.I.; Gibney, R.G.; Li, D.K. Hepatic hemangioma in the presence of fatty infiltration: An atypical sonographic appearance. *Gastrointest. Radiol.* **1989**, *14*, 262–264. [CrossRef]
62. Brancatelli, G.; Federle, M.P.; Blachar, A.; Grazioli, L. Hemangioma in the cirrhotic liver: Diagnosis and natural history. *Radiology* **2001**, *219*, 69–74. [CrossRef] [PubMed]
63. Mastropasqua, M.; Kanematsu, M.; Leonardou, P.; Braga, L.; Woosley, J.T.; Semelka, R.C. Cavernous hemangiomas in patients with chronic liver disease: MR imaging findings. *Magn. Reson. Imaging* **2004**, *22*, 15–18. [CrossRef] [PubMed]
64. Vernuccio, F.; Ronot, M.; Dioguardi Burgio, M.; Lebigot, J.; Allaham, W.; Aubé, C.; Brancatelli, G.; Vilgrain, V. Uncommon evolutions and complications of common benign liver lesions. *Abdom. Radiol.* **2018**, *43*, 2075–2096. [CrossRef] [PubMed]
65. Davies, R. Haemorrhage after fine-needle aspiration biopsy of an hepatic haemangioma. *Med. J. Aust.* **1993**, *158*, 364. [CrossRef] [PubMed]
66. Taavitsainen, M.; Airaksinen, T.; Kreula, J.; Päivänsalo, M. Fine-needle aspiration biopsy of liver hemangioma. *Acta Radiol.* **1990**, *31*, 69–71. [CrossRef] [PubMed]
67. Glinkova, V.; Shevah, O.; Boaz, M.; Levine, A.; Shirin, H. Hepatic haemangiomas: Possible association with female sex hormones. *Gut* **2004**, *53*, 1352–1355. [CrossRef] [PubMed]
68. Mungovan, J.A.; Cronan, J.J.; Vacarro, J. Hepatic cavernous hemangiomas: Lack of enlargement over time. *Radiology* **1994**, *191*, 111–113. [CrossRef] [PubMed]
69. Hoekstra, L.T.; Bieze, M.; Erdogan, D.; Roelofs, J.J.T.H.; Beuers, U.H.W.; van Gulik, T.M. Management of giant liver hemangiomas: An update. *Expert Rev. Gastroenterol. Hepatol.* **2013**, *7*, 263–268. [CrossRef]
70. Abdel Wahab, M.; El Nakeeb, A.; Ali, M.A.; Mahdy, Y.; Shehta, A.; Abdulrazek, M.; El Desoky, M.; Abdel Wahab, R. Surgical Management of Giant Hepatic Hemangioma: Single Center's Experience with 144 Patients. *J. Gastrointest. Surg.* **2018**, *22*, 849–858. [CrossRef]
71. Xie, Q.-S.; Chen, Z.-X.; Zhao, Y.-J.; Gu, H.; Geng, X.-P.; Liu, F.-B. Outcomes of surgery for giant hepatic hemangioma. *BMC Surg.* **2021**, *21*, 186. [CrossRef] [PubMed]
72. Dong, W.; Qiu, B.; Xu, H.; He, L. Invasive management of symptomatic hepatic hemangioma. *Eur. J. Gastroenterol. Hepatol.* **2019**, *31*, 1079–1084. [CrossRef]

73. Fu, X.-H.; Lai, E.C.H.; Yao, X.-P.; Chu, K.-J.; Cheng, S.-Q.; Shen, F.; Wu, M.-C.; Lau, W.Y. Enucleation of liver hemangiomas: Is there a difference in surgical outcomes for centrally or peripherally located lesions? *Am. J. Surg.* **2009**, *198*, 184–187. [CrossRef]
74. Torkian, P.; Li, J.; Kaufman, J.A.; Jahangiri, Y. Effectiveness of Transarterial Embolization in Treatment of Symptomatic Hepatic Hemangiomas: Systematic Review and Meta-analysis. *Cardiovasc. Intervent. Radiol.* **2021**, *44*, 80–91. [CrossRef] [PubMed]
75. Di Carlo, I.; Toro, A. Limiting the surgical indications for liver hemangiomas may help surgeons and patients. *J. Am. Coll. Surg.* **2011**, *212*, 1098–1099. [CrossRef]
76. Pietrabissa, A.; Giulianotti, P.; Campatelli, A.; Di Candio, G.; Farina, F.; Signori, S.; Mosca, F. Management and follow-up of 78 giant haemangiomas of the liver. *Br. J. Surg.* **1996**, *83*, 915–918. [CrossRef] [PubMed]
77. Park, S.Y.; Tak, W.Y.; Jung, M.K.; Jeon, S.W.; Cho, C.M.; Kweon, Y.O.; Kim, K.C. Symptomatic-enlarging hepatic hemangiomas are effectively treated by percutaneous ultrasonography-guided radiofrequency ablation. *J. Hepatol.* **2011**, *54*, 559–565. [CrossRef] [PubMed]
78. Sharpe, E.E.; Dodd, G.D. Percutaneous radiofrequency ablation of symptomatic giant hepatic cavernous hemangiomas: Report of two cases and review of literature. *J. Vasc. Interv. Radiol.* **2012**, *23*, 971–975. [CrossRef]
79. Kong, J.; Gao, R.; Wu, S.; Shi, Y.; Yin, T.; Guo, S.; Xin, Z.; Li, A.; Kong, X.; Ma, D.; et al. Safety and efficacy of microwave versus radiofrequency ablation for large hepatic hemangioma: A multicenter retrospective study with propensity score matching. *Eur. Radiol.* **2022**, *32*, 3309–3318. [CrossRef]
80. Gao, J.; Ke, S.; Ding, X.; Zhou, Y.; Qian, X.; Sun, W. Radiofrequency ablation for large hepatic hemangiomas: Initial experience and lessons. *Surgery* **2013**, *153*, 78–85. [CrossRef]
81. Wu, S.; Gao, R.; Yin, T.; Zhu, R.; Guo, S.; Xin, Z.; Li, A.; Kong, X.; Gao, J.; Sun, W. Complications of radiofrequency ablation for hepatic hemangioma: A multicenter retrospective analysis on 291 cases. *Front. Oncol.* **2021**, *11*, 706619. [CrossRef] [PubMed]
82. Wang, S.; Yang, M.; Yang, X.; Xu, L.; Ke, S.; Ding, X.; Sun, W.; Gao, J. Endothelial pyroptosis underlies systemic inflammatory response following radiofrequency ablation of hepatic hemangiomas. *Scand. J. Clin. Lab. Investig.* **2019**, *79*, 619–628. [CrossRef] [PubMed]
83. Furumaya, A.; van Rosmalen, B.V.; Takkenberg, R.B.; van Delden, O.M.; Dejong, C.H.C.; Verheij, J.; van Gulik, T.M. Transarterial (Chemo-)Embolization and Lipiodolization for Hepatic Haemangioma. *Cardiovasc. Intervent. Radiol.* **2019**, *42*, 800–811. [CrossRef]
84. Özgür, Ö.; Sindel, H.T. Giant hepatic hemangioma treatment with transcatheter arterial embolisation and transcatheter arterial chemoembolisation; Comparative results. *Turk. J. Med. Sci.* **2021**, *51*, 2943–2950. [CrossRef]
85. Liu, X.; Yang, Z.; Tan, H.; Huang, J.; Xu, L.; Liu, L.; Si, S.; Sun, Y. Long-term result of transcatheter arterial embolization for liver hemangioma. *Medicine* **2017**, *96*, e9029. [CrossRef]
86. Özden, İ.; Poyanlı, A.; Önal, Y.; Demir, A.A.; Hoş, G.; Acunaş, B. Superselective transarterial chemoembolization as an alternative to surgery in symptomatic/enlarging liver hemangiomas. *World J. Surg.* **2017**, *41*, 2796–2803. [CrossRef]
87. Della Corte, A.; Marino, R.; Ratti, F.; Palumbo, D.; Guazzarotti, G.; Gusmini, S.; Augello, L.; Cipriani, F.; Fiorentini, G.; Venturini, M.; et al. The Two-Step Treatment for Giant Hepatic Hemangiomas. *J. Clin. Med.* **2021**, *10*, 4381. [CrossRef]
88. Kacała, A.; Dorochowicz, M.; Korbecki, A.; Sobański, M.; Puła, M.; Patrzałek, D.; Janczak, D.; Guziński, M. Transarterial Bleomycin-Lipiodol Chemoembolization for the Treatment of Giant Hepatic Hemangiomas: An Assessment of Effectiveness. *Cancers* **2024**, *16*, 380. [CrossRef]
89. Yuan, B.; Zhang, J.-L.; Duan, F.; Wang, M.-Q. Medium and Long-Term Outcome of Superselective Transcatheter Arterial Embolization with Lipiodol-Bleomycin Emulsion for Giant Hepatic Hemangiomas: Results in 241 Patients. *J. Clin. Med.* **2022**, *11*, 4762. [CrossRef] [PubMed]
90. Ouyang, Y.; Ouyang, X.H.; Gu, S.B. DSA examination and diagnosis of adult hepatic cavernous hemangioma with arteriovenous short circuit. *Zhonghua Fang She Xian Yi Xue Za Zhi* **2000**, *34*, 523–527.
91. Qingle, Z.; Yong, C.; Jianbo, Z.; Kewei, Z.; Yanhao, L.I. Intra-arterial embolization with pingyangmycin-lipiodol emulsion for the treatment of hepatic cavernous hemangioma: An analysis of factors affecting therapeutic results. *J. Interv. Radiol.* **2009**, *12*, 656–660.
92. Sato, M.; Tateishi, R.; Yasunaga, H.; Horiguchi, H.; Yoshida, H.; Matsuda, S.; Koike, K. Mortality and morbidity of hepatectomy, radiofrequency ablation, and embolization for hepatocellular carcinoma: A national survey of 54,145 patients. *J. Gastroenterol.* **2012**, *47*, 1125–1133. [CrossRef]
93. Kacała, A.; Dorochowicz, M.; Patrzałek, D.; Janczak, D.; Guziński, M. Safety and Feasibility of Transarterial Bleomycin-Lipiodol Embolization in Patients with Giant Hepatic Hemangiomas. *Medicina* **2023**, *59*, 1358. [CrossRef] [PubMed]
94. Basile, A.; Carrafiello, G.; Ierardi, A.M.; Tsetis, D.; Brountzos, E. Quality-improvement guidelines for hepatic transarterial chemoembolization. *Cardiovasc. Intervent. Radiol.* **2012**, *35*, 765–774. [CrossRef] [PubMed]
95. Longeville, J.H.; de la Hall, P.; Dolan, P.; Holt, A.W.; Lillie, P.E.; Williams, J.A.; Padbury, R.T. Treatment of a giant haemangioma of the liver with Kasabach-Merritt syndrome by orthotopic liver transplant a case report. *HPB Surg.* **1997**, *10*, 159–162. [CrossRef] [PubMed]
96. Unal, E.; Francis, F.; Aquino, A.; Xu, R.; Morgan, G.; Teperman, L. Liver transplant for mixed capillary-cavernous hemangioma masquerading as hepatocellular carcinoma in a patient with hepatocellular carcinoma. *Exp. Clin. Transplant.* **2011**, *9*, 344–348. [PubMed]
97. Yamashita, Y.; Ogata, I.; Urata, J.; Takahashi, M. Cavernous hemangioma of the liver: Pathologic correlation with dynamic CT findings. *Radiology* **1997**, *203*, 121–125. [CrossRef] [PubMed]

98. Duxbury, M.S.; Garden, O.J. Giant haemangioma of the liver: Observation or resection? *Dig. Surg.* **2010**, *27*, 7–11. [CrossRef] [PubMed]
99. Kim, G.E.; Thung, S.N.; Tsui, W.M.S.; Ferrell, L.D. Hepatic cavernous hemangioma: Underrecognized associated histologic features. *Liver Int.* **2006**, *26*, 334–338. [CrossRef]
100. Ribeiro, M.A.; Papaiordanou, F.; Gonçalves, J.M.; Chaib, E. Spontaneous rupture of hepatic hemangiomas: A review of the literature. *World J. Hepatol.* **2010**, *2*, 428–433. [CrossRef]
101. Yamamoto, T.; Kawarada, Y.; Yano, T.; Noguchi, T.; Mizumoto, R. Spontaneous rupture of hemangioma of the liver: Treatment with transcatheter hepatic arterial embolization. *Am. J. Gastroenterol.* **1991**, *86*, 1645–1649.
102. Chen, Z.-Y.; Qi, Q.-H.; Dong, Z.-L. Etiology and management of hemmorrhage in spontaneous liver rupture: A report of 70 cases. *World J. Gastroenterol.* **2002**, *8*, 1063–1066. [CrossRef]
103. Jain, V.; Ramachandran, V.; Garg, R.; Pal, S.; Gamanagatti, S.R.; Srivastava, D.N. Spontaneous rupture of a giant hepatic hemangioma—Sequential management with transcatheter arterial embolization and resection. *Saudi J. Gastroenterol.* **2010**, *16*, 116–119. [CrossRef]
104. Moinuddin, M.; Allison, J.R.; Montgomery, J.H.; Rockett, J.F.; McMurray, J.M. Scintigraphic diagnosis of hepatic hemangioma: Its role in the management of hepatic mass lesions. *Am. J. Roentgenol.* **1985**, *145*, 223–228. [CrossRef]
105. Seitz, K.; Bernatik, T.; Strobel, D.; Blank, W.; Friedrich-Rust, M.; Strunk, H.; Greis, C.; Kratzer, W.; Schuler, A. Contrast-enhanced ultrasound (CEUS) for the characterization of focal liver lesions in clinical practice (DEGUM Multicenter Trial): CEUS vs. MRI—A prospective comparison in 269 patients. *Ultraschall Med.* **2010**, *31*, 492–499. [CrossRef]
106. Ginting, K.; Tailor, A.; Braverman, T.; Agarwal, A.; Allamaneni, S. Imaging Characteristics and Management of Infected Hepatic Hemangioma: Case-in Discussion. *J. Gastrointest. Abdom. Radiol.* **2021**, *4*, 236–239. [CrossRef]
107. Bailey, J.; Di Carlo, S.; Blackwell, J.; Gomez, D. Same day arterial embolisation followed by hepatic resection for treatment of giant haemangioma. *BMJ Case Rep.* **2016**, *2016*, bcr2015213259. [CrossRef] [PubMed]
108. Yin, X.; Zhang, B.-H.; Qiu, S.-J.; Ren, Z.-G.; Zhou, J.; Chen, X.-H.; Zhou, Y.; Fan, J. Combined hepatocellular carcinoma and cholangiocarcinoma: Clinical features, treatment modalities, and prognosis. *Ann. Surg. Oncol.* **2012**, *19*, 2869–2876. [CrossRef] [PubMed]
109. Lin, S.; Zhang, L.; Li, M.; Cheng, Q.; Zhang, L.; Zheng, S. Atypical hemangioma mimicking mixed hepatocellular cholangiocarcinoma: Case report. *Medicine* **2017**, *96*, e9192. [CrossRef] [PubMed]
110. Yedibela, S.; Alibek, S.; Müller, V.; Aydin, U.; Langheinrich, M.; Lohmüller, C.; Hohenberger, W.; Perrakis, A. Management of hemangioma of the liver: Surgical therapy or observation? *World J. Surg.* **2013**, *37*, 1303–1312. [CrossRef] [PubMed]
111. Schnelldorfer, T.; Ware, A.L.; Smoot, R.; Schleck, C.D.; Harmsen, W.S.; Nagorney, D.M. Management of giant hemangioma of the liver: Resection versus observation. *J. Am. Coll. Surg.* **2010**, *211*, 724–730. [CrossRef]
112. Yamagata, M.; Kanematsu, T.; Matsumata, T.; Utsunomiya, T.; Ikeda, Y.; Sugimachi, K. Management of haemangioma of the liver: Comparison of results between surgery and observation. *Br. J. Surg.* **1991**, *78*, 1223–1225. [CrossRef]
113. Ho, H.-Y.; Wu, T.-H.; Yu, M.-C.; Lee, W.-C.; Chao, T.-C.; Chen, M.-F. Surgical management of giant hepatic hemangiomas: Complications and review of the literature. *Chang Gung Med. J.* **2012**, *35*, 70–78. [CrossRef]
114. Lin, Z.; Zhu, X.; Zhou, J. Ultrasound-guided percutaneous sclerotherapy versus surgical resection in the treatment of large hepatic hemangiomas: A retrospective study. *BMC Surg.* **2022**, *22*, 130. [CrossRef]
115. Ayoobi Yazdi, N.; Mehrabinejad, M.-M.; Dashti, H.; Pourghorban, R.; Nassiri Toosi, M.; Rokni Yazdi, H. Percutaneous Sclerotherapy with Bleomycin and Ethiodized Oil: A Promising Treatment in Symptomatic Giant Liver Hemangioma. *Radiology* **2021**, *301*, 464–471. [CrossRef] [PubMed]
116. Bozkaya, H.; Cinar, C.; Besir, F.H.; Parıldar, M.; Oran, I. Minimally invasive treatment of giant haemangiomas of the liver: Embolisation with bleomycin. *Cardiovasc. Intervent. Radiol.* **2014**, *37*, 101–107. [CrossRef] [PubMed]
117. Oikawa, T.; Hirotani, K.; Ogasawara, H.; Katayama, T.; Ashino-Fuse, H.; Shimamura, M.; Iwaguchi, T.; Nakamura, O. Inhibition of angiogenesis by bleomycin and its copper complex. *Chem. Pharm. Bull.* **1990**, *38*, 1790–1792. [CrossRef] [PubMed]

medicina

MDPI

Review

Current Approach to Risk Factors and Biomarkers of Intestinal Fibrosis in Inflammatory Bowel Disease

Patrycja Dudek and Renata Talar-Wojnarowska *

Department of Digestive Tract Diseases, Medical University of Lodz, 90-153 Lodz, Poland; dudek.patrycja.m@gmail.com

* Correspondence: renata.talar-wojnarowska@umed.lodz.pl; Tel.: +48-42-677-66-67

Abstract: Inflammatory bowel disease (IBD), especially Crohn's disease (CD), characterized by a chronic inflammatory process and progressive intestinal tissue damage, leads to the unrestrained proliferation of mesenchymal cells and the development of bowel strictures. Complications induced by fibrosis are related to high rates of morbidity and mortality and lead to a substantial number of hospitalizations and surgical procedures, generating high healthcare costs. The development of easily obtained, reliable fibrogenesis biomarkers is essential to provide an important complementary tool to existing diagnostic and prognostic methods in IBD management, guiding decisions on the intensification of pharmacotherapy, proceeding to surgical methods of treatment and monitoring the efficacy of anti-fibrotic therapy in the future. The most promising potential markers of fibrosis include cartilage oligomeric matrix protein (COMP), hepatocyte growth factor activator (HGFA), and fibronectin isoform- extra domain A (ED-A), as well as antibodies against granulocyte macrophage colony-stimulating factor (GM-CSF Ab), cathelicidin (LL-37), or circulatory miRNAs: miR-19a-3p and miR-19b-3p. This review summarizes the role of genetic predisposition, and risk factors and serological markers potentially contributing to the pathophysiology of fibrotic strictures in the course of IBD.

Keywords: IBD; strictures; fibrosis; biomarkers

1. Introduction

Inflammatory bowel disease (IBD), such as Crohn's disease (CD) and ulcerative colitis (UC), is characterized by a persistent state of inflammation and progressive intestinal tissue damage, which may lead to uncontrollable mesenchymal cells proliferation and the accumulation of an excessive amount of extracellular matrix (ECM) ingredients. These processes contribute to bowel wall thickening, the development of strictures, and subsequently obstruction—one of the most common complications in the course of IBD, especially CD. The behavior and natural course of CD is highly heterogenous, while the location of the disease remains relatively stable [1–3]. According to the Vienna classification, at the moment of diagnosis, 77% of CD patients were categorized as having the pure inflammatory phenotype of the disease, whereas the development of strictures and fistulae was noticed in 11% and 16% of patients, respectively [3]. This pattern changes dramatically over time, and 5 years after diagnosis, complication rates in patients with CD were reported to range between 48 and 52%. Moreover, 10 years after diagnosis, complications occurred in up to 70% of CD patients, with approximately half of them developing strictures [2–4]. The risk of needing surgical treatment among CD patients is estimated to be between 40 and 71% in the 10-year period after diagnosis [5,6]. The main indications for surgical proceeding include strictures, abscesses, and fistulae. Most often, stricturing CD is treated with strictureplasty or surgical resection. However, recrudescence of the disease at an anastomosis site is frequent, with up to 73% of patients developing recurrent strictures 10 years after strictureplasty [7].

The localization with the greatest likelihood of forming de novo strictures is the ileum and the ileocolonic region. Probably, it is caused by the relatively smaller diameter of this part of the gastrointestinal (GI) tract. Nevertheless, stenotic complications may appear in any region of the GI tract affected by CD: the upper part of the GI tract, the colon, or the rectum. The frequency of stricture formation indicates the most common inflammation sites in the GI tract, with a number of 40–55% stenotic complications occurring in the terminal ileum and colon, 15–25% only in the colon, 25–40% in the ileum alone, and up to 10% affecting the upper part of the GI tract [8,9].

UC, the second main type of IBD, is manifested by continuous inflammatory lesions affecting the inner lining of the large bowel. In the past, UC has not been related to the process of fibrosis. However, recent studies have shown the presence of submucosal fibrosis in up to 100% of colectomy samples from UC patients qualified for surgical treatment due to dysplasia, cancer, or refractory disease [10]. The fibrosis rate is relative to the degree of chronic, but not active inflammation [10,11]. Compared to CD, strictures in UC are much less frequent due to the location of the disease being limited to the large bowel and a wide lumen of the colon. In this form of IBD, stricture formation ranges from 1 to 11.2% of UC patients [12]. Individuals developing strictures should always undergo oncological screening, as a significant proportion of these complications may be related to colorectal cancer. In order to prevent malignant transformation and fibrostenotic complications in UC patients, it is recommended to introduce early colonoscopy surveillance and active anti-inflammatory treatment for better control of the course of the disease [13].

Stricture formation among IBD patients leads to a growing number of hospitalizations, often including surgical treatment, generating high healthcare costs and considerably reducing the quality of life of affected individuals. Easily obtained, reliable biomarkers, such as blood-based markers, would be an essential, complementary instrument in diagnosis, therapy, and monitoring the course of IBD.

This paper aims to summarize the risk factors and biomarkers potentially contributing to the pathophysiology of fibrotic strictures in the course of IBD. The role of genetic predisposition in the development of stenotic complications will also be discussed.

2. Pathophysiology of Intestinal Fibrosis

Intestinal fibrogenesis is a complex, multifactorial process affected by multiple elements, such as genetic factors, gut barrier integrity, microbiota, the immune system, or the regulation of cytokine expression (Figure 1). Two parallel processes are responsible for fibrogenesis in IBD: the expansion of smooth muscle cells and the extensive accumulation of ECM in layers of the bowel wall [14]. In the situation of intestinal tissue damage, the process of mesenchymal cells' accumulation starts in order to secrete ECM components together with growth factors and repair the defect. Mesenchymal cells, characterized by a high motility and versatility, may be gained in the process of the proliferation of existing local mesenchymal cells, cell migration from adherent structures, or differentiation from other types of intestinal cells, like epithelial or endothelial [15]. Intestinal microorganisms and their metabolites, together with growth factors, cytokines secreted by immune and non-immune cells, or even ECM products themselves, are the main factors inducing the processes of mesenchymal cells' activation and differentiation [16]. One of the potential targets of several triggering factors, especially microbial components, are toll-like receptors (TLRs), mainly TLR-4, the activity of which affects epithelial–mesenchymal transition, collagen production, or myofibroblast function [17]. However, in a situation of chronic, severe inflammation, like in IBD, mechanisms of tissue self-regeneration become upregulated, resulting in the accumulation of excessive amounts of ECM products, reducing the intestinal lumen in the place of previous injury, developing stenosis, and subsequently GI tract obstruction. The chronically stimulated mechanisms of tissue regeneration lead to an imbalance between the matrix metalloproteinases (MMP) and cathepsins involved in tissue degradation and the tissue inhibitors of metalloproteinases (TIMPs), impeding their activity [18,19]. The progression of fibrotic changes in the bowel wall can continue independently from the activity of inflammation, which seems to be only a triggering factor for the onset of fibrosis, proceeding in its next steps in a self-perpetuating manner, activated by integrin-mediated mechanisms [19,20].

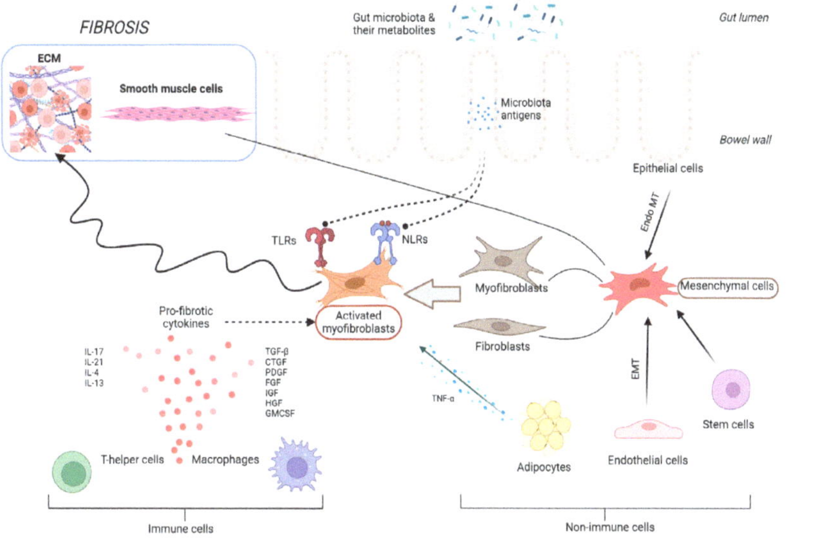

Figure 1. Pathophysiology of intestinal fibrosis. CTGF, connective tissue growth factor; ECM, extracellular matrix; EMT, epithelial–mesenchymal transition; Endo MT, endothelial–mesenchymal transition; FGF, fibroblast growth factor; GMCSF, granulocyte-macrophage colony-stimulating factor; HGF, hepatocyte growth factor; IGF, insulin-like growth factor; IL, interleukin; NLRs, NOD-like receptors; PDGF, platelet derived growth factor; TGF-β, transforming growth factor β; and TNF-α, tumor necrosis factor α.

One of the latest findings in the field of ileal fibrosis pathogenesis applies to the role of gut microbiota reactive antigen-specific T helper (Th) 17 cells. The study by Zhao et al., performed on a mouse and human model, showed that Th17 cells induce intestinal fibrosis via the expression of the epidermal growth factor receptor ligand amphiregulin (AREG) [21]. The intestinal CD4+ T cells of CD patients presented augmented AREG expression in fibrotic sites compared with nonfibrotic bowel segments. Hence, AREG may serve as a new potential biomarker of fibrosis and target for anti-fibrotic treatment in the future. Furthermore, the study proved that, despite multiple analyses of ileal fibrosis pathomechanisms, our knowledge about these processes is still deficient.

A more detailed depiction of the mechanisms involved in the process of intestinal fibrogenesis is beyond the scope of this review and has been raised in other publications.

3. Risk Factors of Fibrogenesis

3.1. Clinical and Environmental Risk Factors

The most commonly studied risk factors of the fibrostenotic CD course include clinical and endoscopic parameters. Nevertheless, it has to be noted that most of the discussed factors are not specific predictors of the fibrostenotising phenotype of the disease, but should be rather considered as parameters showing a tendency towards developing a more serious IBD course, including stricture formation. The risk of the stricturing CD course seems to be independent of sex [22,23]. Its clinical parameters, mostly used for predicting a more complicated phenotype of CD, are: small bowel disease location, perianal disease at diagnosis, and an initial requirement for steroids use [22,24,25]. Most studies also mention a young age at diagnosis as being a risk factor for a complicated CD course, generally defined as the onset of the disease <40 years of age [22]. However, a retrospective cohort study conducted on 1936 IBD patients showed contradictory results. The risk of surgery due to stricturing complications was increased in patients with CD who were 45–59 years of age at diagnosis (p = 0.0023) compared to those aged 15–29 years at diagnosis [23]. Some studies also reported the need for early azathioprine (AZA) therapy as a predictor of disease behavior changes in CD patients. A study conducted on a cohort of 340 CD patients showed that the early use of AZA (p = 0.005), as well as AZA/biological therapy (p = 0.002), was associated with disease behavior changes from B1 (inflammatory phenotype) to B2 (stricturing phenotype)/B3 (penetrating phenotype) [24].

A history of smoking is an environmental risk factor also considered to be of great importance for a more complicated CD course and more rapid progression from diagnosis to the formation of the first stricture. Some previous studies have suggested smoking to be associated with a greater probability of progression to a complicated phenotype of the disease, meaning the development of strictures or fistulae [25–28]. According to other authors, the risk of surgical treatment and further resections during the disease course tends to be higher among smoking individuals [29]. Cosnes et al. found steroids and immunosuppressants requirement to be higher in smokers compared with non-smokers [30]. The mechanisms of the effect of smoking on IBD course are not clear. Most data come from past studies performed in the 1980s and 1990s. Components of tobacco smoke, like nicotine or carbon monoxide, lead to an immunosuppressive effect of smoking, influencing both cellular and humoral immunity. They alter immunoglobulin (Ig) levels, reducing the concentration of serum IgG [31]. They may also change the proportion of immunoregulatory T cells, inducing a reduction in the ratio of T-helper to T-suppressor cells [32]. Smoking has also been connected with altering mucus secretion and the composition in the bowel lumen, what may influence the gut barrier integrity [33]. Furthermore, it may enhance the dysfunction of ileal microvascular perfusion [34].

3.2. Endoscopic Risk Factors

Endoscopy techniques are sensitive methods for the investigation of changes in the superficial layers of the GI tract. However, these procedures are able only to detect severe narrowing of the lumen by visualization or an inability to pass the endoscope and are not appropriate for assessing transmural changes. Endoscopy results can be only partially related to the prediction of a stricturing phenotype, as they rather reflect the activity of the disease and show the IBD behavior well after some complications have occurred. However, some endoscopic findings, such as disease location or mucosal lesions, are considered as predictors of an aggressive disease course. The risk of surgical intervention tends to be higher among patients with extensive, deep, and active mucosal ulcerations [35]. A retrospective study performed by Allez et al. suggested that CD patients with a higher risk of surgery and penetrating complications have a more aggressive course of the disease, with severe lesions in the ileocolon being visualized during endoscopy at symptomatic phases. In a group of 102 patients included in the study, 53 were identified with severe lesions at index colonoscopy, defined as extensive, deep ulcerations affecting more than 10% of the mucosa of a minimum of one colonic segment. During the median 52 months of follow-up, 37 individuals underwent colectomy. The authors observed that the colectomy rate was significantly higher among patients with severe endoscopic lesions compared with those without severe lesions [36].

Disease site is also associated with a complicated course of CD and the need for surgery [37]. The small bowel location of inflammatory changes, rather than the colon, has been defined as being predictive of progression towards stricturing disease and a higher rate of surgery [25]. According to Louis et al., the ileal location of CD is linked with a stricturing phenotype, whereas frequent exacerbations are associated with a penetrating phenotype. The study was performed on a total of 163 CD patients with a non-penetrating, non-stricturing pattern at diagnosis [25]. These conclusions were confirmed in a study by Lakatos et al. performed on 344 CD patients. The results suggest that disease location in the small intestine ($p = 0.001$) and the recognition of perianal disease ($p < 0.001$) are independent predictors of disease behavior changes in CD patients [24]. It becomes evident that groupings of the disease, in particular the Montreal classification, merely identify fibrotic changes after they have become clinically significant. Using this classification to assess risk factors seems to have substantial limitations [38].

On the contrary to CD, the relative risk of stricture formation in UC is much lower. One of the known risk factors of fibrostenosis in UC is the duration of the disease [39,40]. In the study published by Yamagata et al., disease duration was identified as being significantly longer in UC patients with stricturing disease (15.6 years) compared to those without strictures (8.6 years). In this cohort, the incidence of benign stenosis was rated 1.5% over 23 years [40]. Gordon et al. observed a significant association between submucosal fibrosis and the severity of intestinal inflammation ($p < 0.001$), as well as histopathological changes in chronic mucosal injury. No correlation with active inflammation was found. Furthermore, there were no features found on endoscopic mucosal biopsies able to assess the size of the underlying fibrosis or the thickness of the muscularis mucosae [10]. A study conducted on a pediatric population with UC confirmed this hypothesis. In the pediatric UC patients, colorectal submucosal fibrosis and the thickening of the muscularis mucosa were correlated with the presence, chronicity, and degree of inflammation of the mucosa [41]. However, a significant proportion of stenotic complications have been related to the presence of cancer [12,40]. In the study conducted on 1156 patients with UC, 59 of them had colon stenosis, with 24% of these patients being diagnosed with colorectal cancer [12]. The risk of developing colorectal cancer was associated with the duration of the disease (>20 years), location of the disease proximal to the splenic flexure of the colon, and the symptomatic course of stenosis formation. Additionally, the risk of malignant stenotic changes is increased in patients with extensive, active inflammation involving a large part of the intestine, with primary sclerosing cholangitis, or a family history of colorectal cancer < 50 years of age [13].

3.3. *Imaging Techniques in Fibrostenosis Evaluation*

Apart from endoscopy techniques, there are several radiological modalities used for the assessment of IBD complications, including fibrostenosis. In the face of a lack of reliable, clinically useful laboratory markers, radiological techniques are still crucial in the process of assessing fibrostenotic changes in the intestinal tract. For many years, the main problem in using imaging modalities in the diagnosis of fibrostenotic complications referred to a lack of standardized definitions for GI tract strictures. A group of international IBD experts—the CrOhN's disease anti-fibrotic STRICTure therapies (CONSTRICT) group—has provided some defined radiological criteria for ileal stenosis. Due to consensus, a naïve small bowel stricture may be defined as a combination of three features found in cross-sectional imaging: localized luminal narrowing (reduction in luminal diameter of at least 50%, compared to the adjacent normal bowel tract), bowel wall thickening (25% increase in wall thickness compared to the adjacent healthy bowel loop), and pre-stricture dilation (luminal diameter more than 3 cm) [42]. All available cross-sectional imaging techniques today, like Computed Tomography (CT), Magnetic Resonance (MR), and Intestinal Ultrasound (IUS), have the ability to detect strictures, varying in terms of accuracy, availability, exposure to radiation, or cost effectiveness. One of the greatest concerns remains distinguishing between inflammatory-predominant strictures and the fibrotic type of bowel stenosis, as none of the currently available imaging techniques are able to accurately assess the amount of accumulated fibrosis.

CT techniques, including CT enterography (CTE), are characterized by a high sensitivity and specificity in bowel stenosis identification, reaching 85–100% and up to 100%, respectively [43–45]. The main limitation of this type of testing refers to radiation exposure, which may exclude from its use a significant group of CD patients—the pediatric population or young adults, who are especially susceptible to the long-term effects of radiation. Another limitation includes the questionable clinical usefulness of CTE in distinguishing different types of strictures and identifying those associated with fibrostenosis. A retrospective study conducted on a group of 22 CD patients, who underwent a surgical resection of a small bowel stricture, showed that CTE was a sensitive tool for identifying inflammatory changes (p = 0.002), such as mesenteric hypervascularity, mesenteric fat stranding, and mucosal hyperenhancement; however, it did not predict the presence of tissue fibrosis [46].

MR modalities, especially MR enterography (MRE), have gained popularity in CD management in the last years. MRE is characterized by being comparable to CTE sensitivity and specificity in stenosis identification, estimated at 75–92% and 90–95%, respectively [47,48]. However, the main advantage of MR modalities is the lack of exposure to radiation, altogether making MR an optimal technique for the diagnosis of intestinal strictures and assessment of anti-fibrotic therapy response [42]. Limitations of this technique include restricted availability, a long examination time, and higher costs, in comparison to CT. MR findings predictive for stenosis include T1 and T2 isointensity or hypointensity, delayed mural hyperenhancement relative to the normal bowel, and an elevated magnetization transfer ratio [49]. Recently, some novel modalities of MR imaging have been tested, such as Type I Collagen Targeted MR Imaging Probe. A study held on a rat model showed a correlation with the severity of bowel fibrosis (p = 0.021), presenting this technique as a promising method for predicting the progression of fibrotic changes and monitoring the therapeutic response [50].

There are several ultrasound (US) techniques with high diagnostic potential in CD management and fibrosis detection: B-Mode IUS (B-IUS), strain elastography (SE), shear wave elastography (SWE), colour Doppler imaging (CDI) and contrast-enhanced ultrasound (CEUS). Markers of fibrostenosis in US include a thickened bowel wall with a lack of vascularity or contrast enhancement, prestenotic lumen dilation with an increased fluid content, and the presence of stratification in contrast with a loss of stratification typical for inflammatory changes with a low degree of fibrosis [49,51].

Due to its high availability, good tolerance among patients, and low costs, IUS with its modalities seems to be an ideal diagnostic tool for CD patients. However, recently published data have shown that US's diagnostic value remains unclear [52]. The main limitation of IUS may be its high dependence on the skills and experience of the operator, which leads to significant variability in results. Furthermore, US techniques have a low ability to obtain some segments of GI tract, like the duodenum and rectum, as well presenting limited visualization among obese patients. Despite the great potential of these techniques, more studies are needed to understand the precise significance of each radiological parameter and assess cut-off values in different US modes. There is a high discrepancy in evaluating the diagnostic accuracy of US techniques in stenosis detection, with sensitivity varying from 74% to 100% and specificity ranging from 89% to 91% [53–55]. The detection of stenosis improves significantly when using a US modality with contrast enhancement. In a study comparing the accuracy of conventional US and contrast-enhanced techniques in assessing CD complications, i.a., intestinal stenosis, the sensitivity in stricture detection was 74% and 89% for conventional and contrast US, respectively [56]. In another study, the diagnostic value of transabdominal US and contrast US in small bowel lesion detection was evaluated on a group of 28 CD patients. The sensitivity of at least one stricture detection was 76% for conventional US and 94% for the contrast-enhanced technique [57].

A great challenge in CD stricture diagnosis rises from distinguishing inflammatory from fibrostenotic lesions, what may be a matter of great importance in CD management therapy choice, treatment modification, or shifting to surgical procedures. US techniques, such as CDI or CEUS, seem to be a promising tool for the differentiation of such lesions, as they are able to assess the parameters of the bowel wall, appropriate for evaluating the grade of fibrosis in the thickened wall, vascularity, perfusion, neoangiogenesis, and the presence of piercing vessels. However, available data on the clinical usefulness of US modalities in assessing fibrosis in bowel strictures are scarce. A recently published meta-analysis including 14 studies showed that US techniques were inaccurate in differentiating inflammatory from fibrotic stenosis [52]. Another US technique—elastography—also seems to be a promising tool in assessing fibrostenosis in CD patients. Its clinical usefulness is based on changes in the mechanical and elastic properties of the bowel wall due to ECM products' deposition and smooth muscle proliferation in the process of fibrosis, making the measurement of ileal tissue stiffness a marker of fibrosis [58]. Elastography includes two modalities: strain elastography (SE), which measures the bowel stiffness in response to external tissue compression, and shear wave elastography (SWE), the function of which is based on the speed of acoustic wave propagation in tissues differing in stiffness. In the study by Fraquelli et al. conducted on 23 CD patients qualified for terminal ileum resection, SE strain ratio measurements correlated significantly with the severity of fibrotic bowel lesions in a histological image analysis ($p < 0.0001$) [59]. In another study, authors assessed the use of real-time elastography (RTE) in bowel fibrosis detection. Affected and unaffected by stenotic changes, the ileal segments of 10 CD patients were examined pre-, intra-, and postoperatively with different techniques, with a correlation found between RTE, direct tensiometry, and the histological examination results [60]. Further studies evaluating the clinical usefulness of elastography are needed due to small study groups, the high heterogeneity of the used modalities, and no established cut-off values, which hampers defining the role of these techniques in distinguishing different types of bowel stenosis.

3.4. Biomarkers of Fibrosis in IBD

Multiple studies have tried to identify laboratory parameters and biomarkers which would be able to estimate the risk of the fibrostenotic course of IBD, detect the initial stages of fibrosis prior to symptoms, and assess the outcome of a patient's therapy. Table 1 summarizes the identified potential biomarkers—serologic, genetic, and histologic—associated with stenotic complications.

Table 1. Potential biomarkers of fibrogenesis in inflammatory bowel disease.

Category	Biomarkers
Extracellular matrix proteins	Collagen I Collagen III Collagen IV Collagen degradation products (fragments of type I (C1M), III (PRO-C3, C3M), IV (PRO-C4, C4M, C4G), and VI (C6Ma3)) Fibronectin isoform ED-A Cartilage oligomeric matrix protein (COMP)
Growth factors	Transforming growth factor β (TGF-β) Hepatocyte growth factor activator (HGFA)
Cytokine antibodies	Anti TGF-β antibodies Anti interleukin 10 (IL-10) antibodies Anti granulocyte-macrophage colony-stimulating factor (GM-CSF) antibodies
Antimicrobial antibodies	Anti-Saccharomyces cerevisiae antibody (ASCA) Anti-zymogen granule membrane glycoprotein 2 (GP2) antibodies Anti-flagellins: A4-Fla2, anti Fla-X, anti-CBir1 Anti-Escherichia coli outer membrane porin C (anti-OmpC) Anti-CD associated bacterial sequence (I2) Cathelicidin (LL-37)
Genetic variants	Caspase activation recruitment domain (NOD2/CARD15) L-selectin (CD62L) Micro-RNA (miR-19a-3p, miR-19b-3p) Genetic variation of cytokines: IL-12B, IL 10 Tumor necrosis factor (TNF) ligand superfamily member 15
Histhopathology/Tissue based markers	Mast cell density TGF-β activated kinase 1 (TAK1) Ovarian cancer G-protein coupled receptor 1(OGR1) mRNA Cholesterol 25 hydroxylase (CH25H) mRNA

Table 1. *Cont.*

Category	Biomarkers
Other	Fecal calprotectin (FC) Fecal lactoferrin (FL) a2-Heremans-Schmid glycoprotein (AHSG/fetuin A) Elafin Mannan-binding lectin (MBL) C-reactive protein (CRP)

Comparably, little attention has been focused on routinely used and widely available tests, such as C-reactive protein (CRP) concentration. Multiple studies have evaluated the use of CRP in IBD, especially in CD, for establishing a diagnosis, monitoring disease activity, or assessing the response to treatment. The role of CRP as a predictive biomarker in GI tract stricture formation is unclear and study results remain inconsistent. In a cross-sectional study using proteomics to identify potential biomarkers of stricturing CD, no significant correlation with CRP, leukocyte, platelet, and hemoglobin concentration was found [61]. A newly published study demonstrated that a higher serum erythrocyte sedimentation rate and platelet counts, but not CRP, were associated with CD patients' strictures [62].

3.5. Extracellular Matrix Proteins

Using ECM proteins as potential biomarkers of intestinal fibrogenesis intuitively seems to be an expected proceeding, as an excessive accumulation of ECM products is related to the process of remodeling and stricture formation in the intestinal wall. The predominant matrix molecules are collagens, with two major types—collagen I and collagen III—being involved in fibrogenesis [63]. Several studies have evaluated the circulating metabolites of connective tissue, but the results were not consistent and collagens and their properties did not receive the status of being a reliable biomarker of intestinal fibrostenosis [63,64]. More promising outcomes were achieved in a study based on the measurement of the serum levels of molecules involved in collagen turnover and degradation (fragments of collagen type I (C1M), III (PRO-C3, C3M), IV (PRO-C4, C4M, C4G), and VI (C6Ma3)) in a group of CD patients in comparison to healthy individuals. A high level of degradation of collagen type I, III, and IV and excessive formation of collagen type IV were associated with the stricturing phenotype of CD [65].

Recently, some interesting findings were also reported concerning fibronectin. Fibronectin can occur in up to 20 different isoforms due to the alternative splicing of the primary transcript, with every isoform having a different function. Splicing variant ED-A is connected to cell proliferation and the differentiation of fibroblasts into myofibroblasts. Tissue stiffness is one of the known factors which affects the alternative splicing of fibronectin. In the study by de Bruyn et al., increased expression of fibronectin isoform ED-A was observed in an immunohistochemical examination of intestinal samples obtained from CD patients unresponsive to infliximab (IFX) therapy, who underwent ileocecal resection. According to this study, the tissue of the IFX failure patients was characterized by increased stiffness because of higher levels of collagen and fibronectin. The thickness of the muscularis mucosa of those individuals was substantially greater than the mucosa of subjects naïve to IFX [66].

Cartilage oligomeric matrix protein (COMP) is a glycoprotein from the thrombospondin family, which takes part in ECM production and tissue remodeling in response to damage [67]. COMP interacts with other ECM components, including different types of collagen (I, II, IX, XII, and XIV), matrillin-3, aggrecan, fibronectin, and proteases (MMP-3,-12,-13), directly linked to ECM formation [68]. Other roles of COMP include ECM protein export and the correct integration of ECM. Disorders of these functions cause skeletal dysplasias, wound healing abnormalities, and fibrosis in multiple organ systems [69,70]. The function of COMP is highly integrated with transforming growth factor β (TGF-β), which plays an important role in regulating myofibroblast activity and ECM characteristics. In the process of fibrosis, COMP and TGF-β interact mutually, affecting the activity and expression of each other, in a self-perpetuating cycle [71,72]. A dysregulated expression of COMP has been found in numerous pathologies connected to cartilage destruction and fibrosis, like rheumathoid arthritis, idiopathic pulmonary fibrosis, or scleroderma [72–74]. The clinical potential of COMP as a biomarker is associated with the secretion of high levels of this protein into the bloodstream, which enables an indication of COMP serum concentration using conventional methods. In a study conducted by Stidham et al., subjects with fibrostenotic and inflammation-predominant CD phenotypes underwent a comparison of their quantitative serum glycoproteome profiles [61]. The COMP serum levels were elevated in the fibrostenotic vs. inlammatory CD group of patients ($p = 0.012$). Increased concentrations of COMP among subjects with fibrostenosis persisted even after the resection of the affected parts of the intestine. The constantly elevated COMP expression may exhibit a susceptibility for fibrotic changes in response to tissue damage and inflammation.

3.6. Growth Factors

Another group of interest as potential biomarkers of fibrostenosis are growth factors. Among these, transforming growth factor β (TGF-β) plays a predominant role, regulating the process of fibrosis in many organs, including the intestine, contributing to disorders such as diabetic nephropathy, rheumathoid arthritis, radiation-induced fibrosis, or myocarditis [75–79]. TGF-β belongs to a large superfamily of activins/bone morphogenetic proteins. Produced by various types of cells, TGF-β is characterized by pleiotropic activity, including the regulation of the immune response, cell proliferation, and oncogenesis. The association between TGF-β level and intestinal strictures in CD patients was investigated, and it was proved that the expression of TGF-β was increased in the intestinal mucosa covering strictures compared to non-strictured parts of the intestines of patients with fibrostenosing CD [80]. An elevated level of expression of TGF-β1 and active TGF-β1 was also found in the muscle cells of intestinal strictures, obtained from surgically resected ileal segments of CD patients [81].

Hepatocyte growth factor activator (HGFA) is a protease secreted into the blood by the liver in order to activate hepatocyte growth factor (HGF) as a response to tissue damage. HGF is a multipotent molecule produced by various types of cells, including fibroblasts, taking part in crucial processes such as the regeneration and protection of tissues, epithelial to mesenchymal cell transformation, the apoptosis of myofibroblasts or protection from chronic inflammation, and fibrosis [82]. HGF has an antagonistic relationship with TGF-β, inhibiting fibrotic remodeling [83]. The administration of HGF or HGF gene therapy contributes to anti-fibrotic effects in lung, liver, renal, cardiac, and brain injuries, which was confirmed in animal models [84–88]. In the aforementioned study by Stidham et al., HGFA serum levels were significantly elevated in a fibrostenotic group of CD patients compared to subjects with the inflammatory phenotype ($p = 0.031$). Within the group with the fibrosis-predominant phenotype, HGFA levels significantly declined following the resection of the fibrostenotic intestine ($p = 0.015$). Elevated serum HGFA levels in fibrostenotic subjects, with a significant decline after surgical resection, suggest the usefulness of this enzyme as a marker of accumulated fibrotic bowel damage [61].

3.7. Cytokine Antibodies

Endogenous autoantibodies to cytokines are able to modulate inflammation by creating a state of relative immunodeficiency in IBD patients, predisposing them to chronic inflammatory processes in the intestinal mucosa. In the study by Ebert et al., the concentration of antibodies recognizing TGF-β was significantly higher in UC patients ($p < 0.01$), compared with normal sera. In the same study, anti-IL-10 antibody levels were found to be greater in CD ($p < 0.05$) patients than among healthy individuals [89]. In a subsequent study, an increased concentration of neutralizing autoantibodies against granulocyte macrophage colony-stimulating factor (GM-CSF Ab) was observed in a population of adult and pediatric CD patients, however, GM-CSF Ab level was found to be especially elevated among subjects with ileal disease involvement and the stricturing CD phenotype ($p < 0.001$). Another important finding in this research, performed additionally on an animal model, was the loss of the barrier function of the intestinal mucosa and the development of transmural ileitis after exposure to non-steroidal anti-inflammatory drugs (NSAID) in GM-CSF-null mice and NOD2-null mice, in which GM-CSF was neutralized [90]. Parallel results were obtained in another study. The authors found that an elevated concentration of GM-CSF Ab, disease duration greater than 3 years, and ileal location of the disease were independent risk factors of stricturing/penetrating CD behavior and intestinal resection [91]. GM-CSF, which is produced by the immune cells of the lamina propria, plays an important role in regulating intestinal inflammatory processes by supporting epithelial barrier integrity or stimulating crypt cell proliferation in acute tissue injury. Deficiency of GM-CSF can contribute to a relative immunodeficiency and disorder in ileal homeostasis [92].

3.8. Antimicrobial Antibodies

Searching for biomarkers associated with the gut microbiota, such as antimicrobial antibodies or antimicrobial proteins, seems to be promising, as molecules connected to enteric flora might be unique markers specific for intestinal fibrosis, distinguishing ileal from other types of organ fibrosis. Dysbiosis is related to IBD in general, and in addition, changes in enteric microbiota composition may be characteristic for different types of disease phenotypes [93]. Alterations in the gut microbiota and their metabolites, together with a loss of ileal barrier integrity, lead to the translocation of microbial antigens to the bowel mucosa or portal circulation and indirectly stimulate the production of fibrotic agents by immune and non-immune cells [94]. The process of antigen recognition by immune and non-immune cells takes place with a contribution from pattern recognition receptors (PRRs), like Toll-like receptors (TLRs) and nucleotide-binding and oligomerization domain (NOD)-like receptors (NLRs) [95]. Due to the role of dysregulated intestinal immune response in the pathogenesis of IBD, multiple studies have been conducted evaluating the clinical usefulness of antimicrobial antibodies in UC and, in particular, CD management [96]. An association between antimicrobial antibodies level and IBD behavior or phenotype was the most prominent regarding anti-Saccharomyces cerevisiae antibodies (ASCA). A prospective cohort study evaluated their prevalence and relationship with IBD. Positive ASCA was found to occur more frequently in CD patients with stricturing ($p = 0.003$) or penetrating ($p = 0.012$) complications compared to subjects with the pure inflammatory phenotype of CD at diagnosis. Furthermore, patients with ASCA presence had at least a twice higher risk of the evolution of their disease course to being more severe during follow-up ($p < 0.001$) [97]. This association was confirmed in several other studies [96–98]. In the study by Degenhardt et al., ASCA IgG and IgA were qualitatively and quantitatively associated with CD, CD complications (fistula and stenosis), and the need for surgical treatment [98]. This research has also shown link between another antibody type—serum anti-zymogen granule membrane glycoprotein 2 (GP2) antibodies. GP2 is thought to play an important role in immunomodulatory processes. The expression of GP2 in human enterocytes suggests that the pathogenesis of Crohn's disease is, apart from multiple other factors, associated with anti-GP2 response [99]. Anti-GP2 IgA and IgG levels were found to be exclusively connected to the stricturing CD course and the need for surgical intervention, independently of disease location. No significant association with the fistulizing phenotype of CD, early disease onset, or disease activity was found [98]. In another study, the results showed that CD patients with the presence of IgA and/or IgG ASCA antibodies and anti-GP2 IgG antibodies, compared to seronegative individuals, had an early disease onset ($p < 0.0001$) and greater risk of both ileal and colonic disease ($p < 0.0001$), as well as forming strictures ($p < 0.0001$) [100].

Other antibodies which are associated with the fibrostenotic CD phenotype include both antimicrobial molecules: anti-flagellins A4-Fla2, anti Fla-X, anti-CBir1, and anti- Escherichia coli outer membrane porin C (anti-OmpC), and those that are nonantimicrobial, such as antineutrophil cytoplasmic antibody (ANCA)/perinuclear ANCA (pANCA) [101–105]. Another antimicrobial peptide which gave promising results is cathelicidin (LL-37, also known as hCAP18). Cathelicidin expression was found in multiple tissues, like the mucosa of the colon, breast, salivary glands, or some types of immune cells. Cathelicidin in the intestinal epithelium is responsible for ensuring epithelial barrier integrity or bacterial adhesion. An increased expression of cathelicidin was found in the mucosa of UC patients [106]. A study performed by Tran et al. showed that serum cathelicidin levels were inversely correlated with the activity of the disease (Partial Mayo Score) in UC patients, which is consistent with its known anti-inflammatory effect. Cathelicidin concentration combined with CRP level indicated the activity of UC more accurately than using either of these parameters independently. The study also demonstrated that low LL-37 levels among CD patients indicated a higher risk of developing intestinal strictures ($p = 0.0485$); however, it was not determined whether LL-37 levels were associated with the development of other complications like fistulae. The study was performed on two cohorts of IBD patients—80 UC patients and 95 CD patients. The serum levels of LL-37 were assessed using ELISA tests [107].

3.9. Fecal Biomarkers

Fecal calprotectin (FC) and fecal lactoferrin (FL)-neutrophil-derived proteins are the two most commonly used fecal biomarkers in clinical trials. The role of FC is well-established, with a significant correlation with intestinal inflammation, serving as a useful tool in CD evaluation. The use of FL testing has been limited mainly to research, probably due to the low stability of lactoferrin at room temperature.

Although fecal biomarkers seem to have a great potential to serve as easily obtained, non-invasive indicators of structuring CD, available data concerning this type of markers are limited. Only a few studies have evaluated fecal markers in the context of GI tract strictures. In a recent study, FC and FL levels were assessed to predict disease recurrence in CD patients with anastomotic strictures who underwent surgical treatment. The patients included in the study were evaluated by postoperative colonoscopy. Endoscopic balloon dilation was performed in subjects with strictures at the site of anastomosis, unable to pass by the colonoscope, regardless of the patients' symptoms. Stool samples for FC and FL were collected on the day preceding bowel cleaning. Both FC and FL levels were significantly associated with the endoscopic recurrence of anastomotic strictures ($p < 0.001$), with an optimal cut-off value of 90.85 μg/g for FC and 5.6 μg/g for FL [108]. The use of FC as a potential biomarker of stricturing CD was also discussed in a recently published article. The authors assessed the management of stricturing CD in two pregnant patients using FC levels and intestinal ultrasound, proving that FC can serve as a complementary tool to ultrasound findings in confirming therapeutic response. Moreover, an increased FC level during pregnancy is associated with later exacerbation and a higher risk of adverse fetal and maternal outcomes [109].

3.10. Tissue-Based Biomarkers

Histopathologic analyses of intestinal fibrosis may provide some critical information about the pathogenesis of stricture formation, leading to the development of antifibrotic therapies. The most remarkable changes in strictured intestinal tissue include chronic inflammation, hypertrophia of muscularis propria, and hyperplasia of the smooth muscle layer in the submucosa. The 'inflammation–smooth muscle hyperplasia axis' appears to be the crucial patomechanism of stricture formation in the course of CD [110]. However, no standardized scoring system to grade the severity of histological fibrosis is currently available, which hampers further investigations and comparisons of study results [111]. The main limitation to the clinical use of tissue-based biomarkers is their low availability, requiring endoscopic procedures, which makes them less significant in IBD management compared to easily obtained serum biomarkers. The second major objection is the limited value of endoscopic mucosa biopsy samples for diagnosis of intestinal fibrosis, as stricture formation is a transmural process. This may be the cause of lacking studies confirming histopathological biomarkers.

Potential markers include transforming growth factor beta activated kinase 1 (TAK1). A study conducted on 26 IBD patients evaluated surgical ileal samples obtained from individuals with the stricturing phenotype of CD. The concentrations of TAK1 and its phosphorylated form—pTAK1—were elevated in the ileal specimens of CD patients compared with healthy subjects and correlated with the level of intestinal fibrosis ($p < 0.01$) [112]. Another study evaluated the expression of fibrosis markers and pH-sensing receptors in ileal samples from CD patients who had undergone ileocaecal resection because of fibrostenotic complications. The expression of pH-sensing ovarian cancer G-protein coupled receptor-1 [OGR1/GPR68] was found to be elevated in the ileal samples of fibrostenotic patients and positively correlated with the expression of pro-fibrotic cytokines and pro-collagens ($p = 0.016$) [113]. In a different study, the authors observed a gradual increase in cholesterol 25 hydroxylase (CH25H) expression in samples, comparing, as follows: healthy control ileal tissue, non-fibrotic ileal tissue of CD patients, and fibrotic ileal tissues from the same CD patients ($p < 0.05$). Samples were obtained from subjects who underwent ileocaecal resection because of stenotic complications. The expression of CH25H was strongly correlated with the expressions of various fibrosis mediators (COL-1, COL-3, SMA, and TGF-β) [114]. As all the experiments were conducted on a small group of subjects, further studies are required to elucidate the exact significance of tissue-based markers.

3.11. Genetic Variants

Genetic variants have also been considered as markers of stricture formation in IBD. The first gene proven to be linked with CD was the nucleotide binding and oligomerization domain, named later the caspase activation recruitment domain (NOD2/CARD15). Later research proved the association of this gene with susceptibility to the stricturing phenotype of CD. A meta-analysis including CD patients showed that owners of at least one high-risk variant of NOD2/CARD15 had a slightly increased risk of familiar disease, modestly elevated risk of the stricturing CD phenotype, and significantly higher risk of small bowel disease [115]. Another study revealed that the presence of a single NOD2 mutation was associated with an 8% increase in the risk of a complicated CD course (stricturing or fistulizing) and a 41% increase in the risk among subjects owning two mutations. Furthermore, individuals with any NOD2 mutation presented a 58% elevated risk of needing surgery, whereas the risk of perianal disease remained unchanged. The authors of the study assumed that CD patients with two mutations of NOD2/CARD15, due to a high risk of a complicated course of the disease, may benefit from the early intensification of therapy [116]. On the contrary, some studies have not confirmed the association of NOD2/CARD15 variants with the stricturing course of CD [25,101,102]. It is still unclear whether the observed relationship between NOD2/CARD15 gene variants and stricturing CD is a real association, or only a reflection of a high proportion of CD patients who develop complications. A significant limitation of using gene variants as biomarkers is the fact that this does not take into account the impact of environmental factors on disease course, such as microbiome or nutrition.

Another group of interest as candidate biomarkers of stricturing CD are circulatory micro-RNAs (miR)—short noncoding RNA fragments regulating the gene expression in epigenetic mechanisms. Aberrant miRNA expression is related to the pathogenesis of fibrosis. Suppression of miR-29 has been linked to liver or renal fibrosis [117,118]. A study by Lewis et al. identified miR-19-3p to be a potential marker of the stenotic phenotype of CD. Patients with stricturing CD, compared to control CD patients, presented reduced serum concentrations of miR-19a-3p and miR-19b-3p ($p = 0.007$ and $p = 0.008$, respectively). The association between miR-19-3p and stenotic CD seemed to be independent of clinical factors, such as disease duration, disease activity, location, gender, or age. A 4-year patient follow-up supported this hypothesis [119]. Other variants of miR-19 have been also linked to fibrotic processes. A lower concentration of miR-19a-5p in the peripheral blood was found in interstitial lung fibrosis, as well as cardiac and liver fibrosis [120–122]. The usefulness of miRNA in IBD management was also confirmed in a prospective study conducted on 77 IBD patients. The authors stated that miR-320a blood levels were strongly correlated with the exacerbation of CD and reflected endoscopic and clinical disease activity, as well as reaching a response to treatment [123]. The limitations to the potential utility of miRNA as a biomarker include difficulties with isolation and purification.

4. Conclusions

Despite the increasing number of studies on the pathogenesis of fibrosis, our understanding of the patomechanisms of the stricturing CD phenotype and association between biomarkers and strictures remains limited. Currently, there is still a lack of clinically approved biomarkers of intestinal fibrostenosis. High hopes were raised for microbial biomarkers due to their specificity for gut microbiota and, thus, ileal fibrosis. However, ASCA antibodies or NOD2/CARD15-related markers seem to show a tendency towards a more severe CD course rather than being representative of the IBD stricturing phenotype. Promising results have been achieved according to other types of biomarkers—ECM compounds, such as COMP or growth factors, like HGFA. Despite a high correlation with the stricturing CD phenotype presented in previous studies and potentially easy obtainment, the main objection against their clinical usefulness may be a lack of specificity for ileal fibrosis. Most biomarkers derived from growth factors, cytokines, or ECM compounds have been already found to be associated with fibrosis in multiple other organs, which may be misleading in further studies. The development of imaging techniques has enabled GI tract stricture detection, however, distinguishing between inflammatory and fibrotic types of ileal stenosis is still ineffective. Non-invasive and easily obtained fecal biomarkers seem to have great potential, showing an eventual direction for fibrosis marker development. FC has a well-established position in IBD evaluation, as still no other valuable fecal biomarkers, specific for ileal fibrosis, have been recently found.

Expansion in drug development has led to better control of inflammation in IBD course, however, the available anti-inflammatory therapies still have little impact on the reduction in or reversibility of GI tract fibrosis, remaining a great medical challenge. The progression of intestinal fibrosis is partially independent of the inflammatory process and indicates an urgent need for the identification of reliable, noninvasive biomarkers, which could be useful in the management of IBD patients, especially those with CD. Further studies on the pathogenesis underlying the stricturing CD phenotype and its associated biomarkers may contribute to the optimalization of IBD patients' management and better long-term outcomes. Advances may be hampered by a lack of validated endpoints, which would enable scientists to compare the results of clinical trials.

Author Contributions: P.D. and R.T.-W. provided the overall concept and framework of the manuscript; P.D. researched and identified appropriate articles, and wrote the manuscript; P.D. and R.T.-W. revised the manuscript. All authors have read and agreed to the published version of the manuscript.

Funding: This research received no external funding.

Institutional Review Board Statement: Not applicable.

Informed Consent Statement: Not applicable.

Data Availability Statement: No new data were created.

Conflicts of Interest: The authors declare no conflicts of interest.

References

1. Louis, E.; Reenaers, C.; Belaiche, J. Does the phenotype at diagnosis (e.g., fibrostenosing, inflammatory, perforating) predict the course of Crohn's disease? *Inflamm. Bowel Dis.* **2008**, *14* (Suppl. S2), 59–60. [CrossRef]
2. Cosnes, J.; Cattan, S.; Blain, A.; Beaugerie, L.; Carbonnel, F.; Parc, R.; Gendre, J.-P. Long-term evolution of disease behavior of Crohn's disease. *Inflamm. Bowel Dis.* **2002**, *8*, 244–250. [CrossRef] [PubMed]
3. Louis, E.J.; Collard, A.; Oger, A.F.; Groote, E.; De Belaiche, J. Location and behavior of Crohn's disease according to Vienna classification evolution over the course of the disease. *Gastroenterology* **2001**, *120*, A141. [CrossRef]

4. Tarrant, K.M.; Barclay, M.L.; Frampton, C.M.A.; Gearry, R.B. Perianal disease predicts changes in Crohn's disease phenotype—Results of a population-based study of inflammatory bowel disease phenotype. *Am. J. Gastroenterol.* **2008**, *103*, 3082–3093. [CrossRef] [PubMed]
5. Wolters, F.L.; Russel, M.G.; Sijbrandij, J.; Schouten, L.J.; Odes, S.; Riis, L.; Munkholm, P.; Langholz, E.; Bodini, P.; O'Morain, C.; et al. Disease outcome of inflammatory bowel disease patients: General outline of a Europe-wide population-based 10-year clinical follow-up study. *Scand. J. Gastroenterol.* **2006**, *41*, 46–54. [CrossRef] [PubMed]
6. Peyrin-Biroulet, L.; Loftus, E.V.J.; Colombel, J.F.; Sandborn, W.J. The Natural History of Adult Crohn's Disease in Population-Based Cohorts. *Off. J. Am. Coll. Gastroenterol. ACG* **2010**, *105*, 501–523. [CrossRef] [PubMed]
7. Lightner, A.L.; Vogel, J.D.; Carmichael, J.C.; Keller, D.S.M.; Shah, S.A.; Mahadevan, U.; Kane, S.V.M.; Paquette, I.M.; Steele, S.R.M.; Feingold, D.L. The American Society of Colon and Rectal Surgeons Clinical Practice Guidelines for the Surgical Management of Crohn's Disease. *Dis. Colon Rectum* **2020**, *63*, 1028–1052. [CrossRef] [PubMed]
8. Freeman, H.J. Natural history and clinical behavior of Crohn's disease extending beyond two decades. *J. Clin. Gastroenterol.* **2003**, *37*, 216–219. [CrossRef]
9. Nikolaus, S.; Schreiber, S. Diagnostics of Inflammatory Bowel Disease. *Gastroenterology* **2007**, *133*, 1670–1689. [CrossRef]
10. Gordon, I.O.; Agrawal, N.; Goldblum, J.R.; Fiocchi, C.; Rieder, F. Fibrosis in ulcerative colitis: Mechanisms, features, and consequences of a neglected problem. *Inflamm. Bowel Dis.* **2014**, *20*, 2198–2206. [CrossRef]
11. Gordon, I.O.; Agrawal, N.; Willis, E.; Goldblum, J.R.; Lopez, R.; Allende, D.; Liu, X.; Patil, D.Y.; Yerian, L.; El-Khider, F.; et al. Fibrosis in ulcerative colitis is directly linked to severity and chronicity of mucosal inflammation. *Aliment. Pharmacol. Ther.* **2018**, *47*, 922–939. [CrossRef]
12. Gumaste, V.; Sachar, D.B.; Greenstein, A.J. Benign and malignant colorectal strictures in ulcerative colitis. *Gut* **1992**, *33*, 938–941. [CrossRef]
13. Park, S.C.; Jeen, Y.T. The clinical significance and risk factors of colorectal stricture in ulcerative colitis. *Gut Liver* **2020**, *14*, 535–536. [CrossRef] [PubMed]
14. Solitano, V.; Dal Buono, A.; Gabbiadini, R.; Wozny, M.; Repici, A.; Spinelli, A.; Vetrano, S.; Armuzzi, A. Fibro-Stenosing Crohn's Disease: What Is New and What Is Next? *J. Clin. Med.* **2023**, *12*, 3052. [CrossRef] [PubMed]
15. Flier, S.N.; Tanjore, H.; Kokkotou, E.G.; Sugimoto, H.; Zeisberg, M.; Kalluri, R. Identification of epithelial to mesenchymal transition as a novel source of fibroblasts in intestinal fibrosis. *J. Biol. Chem.* **2010**, *285*, 20202–20212. [CrossRef] [PubMed]
16. D'Haens, G.; Rieder, F.; Feagan, B.G.; Higgins, P.D.R.; Panés, J.; Maaser, C.; Rogler, G.; Löwenberg, M.; Van Der Voort, R.; Pinzani, M.; et al. Challenges in the Pathophysiology, Diagnosis, and Management of Intestinal Fibrosis in Inflammatory Bowel Disease. *Gastroenterology* **2022**, *162*, 26–31. [CrossRef] [PubMed]
17. Jun, Y.K.; Kwon, S.H.; Yoon, H.T.; Park, H.; Soh, H.; Lee, H.J.; Im, J.P.; Kim, J.S.; Kim, J.W.; Koh, S.-J. Toll-like receptor 4 regulates intestinal fibrosis via cytokine expression and epithelial-mesenchymal transition. *Sci. Rep.* **2020**, *10*, 19867. [CrossRef] [PubMed]
18. Mckaig, B.C.; Mcwilliams, D.; Watson, S.A.; Mahida, Y.R. Expression and Regulation of Tissue Inhibitor of Metalloproteinase-1 and Matrix Metalloproteinases by Intestinal Myofibroblasts in Inflammatory Bowel Disease. *Am. J. Pathol.* **2003**, *162*, 1355–1360. [CrossRef] [PubMed]
19. Speca, S.; Giusti, I.; Rieder, F.; Latella, G. Cellular and molecular mechanisms of intestinal fibrosis. *World J. Gastroenterol.* **2012**, *18*, 3635–3661. [CrossRef] [PubMed]
20. Wells, R.G. The role of matrix stiffness in regulating cell behavior. *Hepatology* **2008**, *47*, 1394–1400. [CrossRef]
21. Zhao, X.; Yang, W.; Yu, T.; Yu, Y.; Cui, X.; Zhou, Z.; Yang, H.; Yu, Y.; Bilotta, A.J.; Yao, S.; et al. Th17 Cell-Derived Amphiregulin Promotes Colitis-Associated Intestinal Fibrosis Through Activation of mTOR and MEK in Intestinal Myofibroblasts. *Gastroenterology* **2023**, *164*, 89–102. [CrossRef]
22. Beaugerie, L.; Seksik, P.; Nion-Larmurier, I.; Gendre, J.P.; Cosnes, J. Predictors of crohn's disease. *Gastroenterology* **2006**, *130*, 650–656. [CrossRef] [PubMed]
23. Bernell, O.; Lapidus, A.; Hellers, G. Risk factors for surgery and postoperative recurrence in Crohn's disease. *Ann. Surg.* **2000**, *231*, 38–45. [CrossRef] [PubMed]
24. Lakatos, P.L.; Czegledi, Z.; Szamosi, T.; Banai, J.; David, G.; Zsigmond, F.; Pandur, T.; Erdelyi, Z.; Gemela, O.; Papp, J.; et al. Perianal disease, small bowel disease, smoking, prior steroid or early azathioprine/biological therapy are predictors ofdisease behavior change in patients with Crohn's disease. *World J. Gastroenterol.* **2009**, *15*, 3504–3510. [CrossRef] [PubMed]
25. Louis, E.; Michel, V.; Hugot, J.P.; Reenaers, C.; Fontaine, F.; Delforge, M.; El Yafi, F.; Colombel, J.F.; Belaiche, J. Early development of stricturing or penetrating pattern in Crohn's disease is influenced by disease location, number of flares, and smoking but not by NOD2/CARD15 genotype. *Gut* **2003**, *52*, 552–557. [CrossRef]
26. Westhovens, R. Clinical efficacy of new JAK inhibitors under development. Just more of the same? *Rheumatology* **2019**, *58*, i27–i33. [CrossRef] [PubMed]
27. Mahid, S.S.; Minor, K.S.; Stevens, P.L.; Galandiuk, S. The role of smoking in Crohn's disease as defined by clinical variables. *Dig. Dis. Sci.* **2007**, *52*, 2897–2903. [CrossRef]
28. Picco, M. Tobacco consumption and disease duration are associated with fistulizing and stricturing behaviors in the first 8 years of Crohn's disease. *Am. J. Gastroenterol.* **2003**, *98*, 363–368. [CrossRef]
29. Reese, G.E.; Nanidis, T.; Borysiewicz, C.; Yamamoto, T.; Orchard, T.; Tekkis, P.P. The effect of smoking after surgery for Crohn's disease: A meta-analysis of observational studies. *Int. J. Colorectal. Dis.* **2008**, *23*, 1213–1221. [CrossRef]

30. Cosnes, J. Tobacco and IBD: Relevance in the understanding of disease mechanisms and clinical practice. *Best Pract. Res. Clin. Gastroenterol.* **2004**, *18*, 481–496. [CrossRef]
31. Tarbiah, N.; Todd, I.; Tighe, P.J.; Fairclough, L.C. Cigarette smoking differentially affects immunoglobulin class levels in serum and saliva: An investigation and review. *Basic Clin. Pharmacol. Toxicol.* **2019**, *125*, 474–483. [CrossRef]
32. Miller, L.G.; Goldstein, G.; Murphy, M.; Ginns, L.C. Reversible alterations in immunoregulatory T cells in smoking. Analysis by monoclonal antibodies and flow cytometry. *Chest* **1982**, *82*, 526–529. [CrossRef]
33. Zijlstra, F.J.; Srivastava, E.D.; Rhodes, M.; van Dijk, A.P.; Fogg, F.; Samson, H.J.; Copeman, M.; Russell, M.A.; Feyerabend, C.; Williams, G.T. Effect of nicotine on rectal mucus and mucosal eicosanoids. *Gut* **1994**, *35*, 247. [CrossRef]
34. Hatoum, O.A.; Binion, D.G.; Otterson, M.F.; Gutterman, D.D. Acquired microvascular dysfunction in inflammatory bowel disease: Loss of nitric oxide-mediated vasodilation. *Gastroenterology* **2003**, *125*, 58–69. [CrossRef] [PubMed]
35. Allez, M.; Lémann, M. Role of endoscopy in predicting the disease course in inflammatory bowel disease. *World J. Gastroenterol.* **2010**, *16*, 2626–2632. [CrossRef] [PubMed]
36. Allez, M.; Lemann, M.; Bonnet, J.; Cattan, P.; Jian, R.; Modigliani, R. Long term outcome of patients with active Crohn's disease exhibiting extensive and deep ulcerations at colonoscopy. *Am. J. Gastroenterol.* **2002**, *97*, 947–953. [CrossRef] [PubMed]
37. Lazarev, M.; Huang, C.; Bitton, A.; Cho, J.H.; Duerr, R.H.; McGovern, D.P.; Proctor, D.D.; Regueiro, M.; Rioux, J.D.; Schumm, P.P.; et al. Relationship between proximal Crohn's disease location and disease behavior and surgery: A cross-sectional study of the IBD genetics consortium. *Am. J. Gastroenterol.* **2013**, *108*, 106–112. [CrossRef] [PubMed]
38. Silverberg, M.S.; Satsangi, J.; Ahmad, T.; Arnott, I.D.R.; Bernstein, C.N.; Brant, S.R.; Caprilli, R.; Colombel, J.-F.; Gasche, C.; Geboes, K.; et al. Toward an integrated clinical, molecular and serological classification of inflammatory bowel disease: Report of a Working Party of the 2005 Montreal World Congress of Gastroenterology. *Can. J. Gastroenterol.* **2005**, *19* (Suppl. A), 5A–36A. [CrossRef] [PubMed]
39. Rieder, F.; Fiocchi, C.; Rogler, G.; Disease, D.; Clinic, C.; Foundation, C.C. Mechanisms, management, and treatment of fibrosis in patients with inflammatory bowel diseases. *HHS Public Access* **2018**, *152*, 340–350. [CrossRef] [PubMed]
40. Yamagata, M.; Mikami, T.; Tsuruta, T.; Yokoyama, K.; Sada, M.; Kobayashi, K.; Katsumata, T.; Koizumi, W.; Saigenji, K.; Okayasu, I. Submucosal fibrosis and basic-fibroblast growth factor-positive neutrophils correlate with colonic stenosis in cases of ulcerative colitis. *Digestion* **2011**, *84*, 12–21. [CrossRef]
41. Gordon, I.O.; Abushamma, S.; Kurowski, J.A.; Holubar, S.D.; Kou, L.; Lyu, R.; Rieder, F. Paediatric Ulcerative Colitis Is a Fibrotic Disease and Is Linked with Chronicity of Inflammation. *J. Crohn's Colitis* **2022**, *16*, 804–821. [CrossRef]
42. Rieder, F.; Bettenworth, D.; Ma, C.; Parker, C.E.; Williamson, L.A.; Nelson, S.A.; van Assche, G.; Di Sabatino, A.; Bouhnik, Y.; Stidham, R.W.; et al. An expert consensus to standardise definitions, diagnosis and treatment targets for anti-fibrotic stricture therapies in Crohn's disease. *Aliment. Pharmacol. Ther.* **2018**, *48*, 347–357. [CrossRef] [PubMed]
43. Chiorean, M.V.; Sandrasegaran, K.; Saxena, R.; Maglinte, D.D.; Nakeeb, A.; Johnson, C.S. Correlation of CT enteroclysis with surgical pathology in Crohn's disease. *Am. J. Gastroenterol.* **2007**, *102*, 2541–2550. [CrossRef] [PubMed]
44. Vogel, J.; Da Luz Moreira, A.; Baker, M.; Hammel, J.; Einstein, D.; Stocchi, L.; Fazio, V. CT enterography for Crohn's disease: Accurate preoperative diagnostic imaging. *Dis. Colon Rectum* **2007**, *50*, 1761–1769. [CrossRef] [PubMed]
45. Solem, C.A.; Loftus, E.V.; Fletcher, J.G.; Baron, T.H.; Gostout, C.J.; Petersen, B.T.; Tremaine, W.J.; Egan, L.J.; Faubion, W.A.; Schroeder, K.W.; et al. Small-bowel imaging in Crohn's disease: A prospective, blinded, 4-way comparison trial. *Gastrointest. Endosc.* **2008**, *68*, 255–266. [CrossRef] [PubMed]
46. Adler, J.; Punglia, D.R.; Dillman, J.R.; Polydorides, A.D.; Dave, M.; Al-Hawary, M.M.; Platt, J.F.; McKenna, B.J.; Zimmermann, E.M. Computed tomography enterography findings correlate with tissue inflammation, not fibrosis in resected small bowel Crohn's disease. *Inflamm. Bowel Dis.* **2012**, *18*, 849–856. [CrossRef] [PubMed]
47. Pous-Serrano, S.; Frasson, M.; Palasí Giménez, R.; Sanchez-Jordá, G.; Pamies-Guilabert, J.; Llavador Ros, M.; Mateu, P.N.; Garcia-Granero, E. Accuracy of magnetic resonance enterography in the preoperative assessment of patients with Crohn's disease of the small bowel. *Color. Dis.* **2017**, *19*, O126–O133. [CrossRef] [PubMed]
48. Fiorino, G.; Bonifacio, C.; Peyrin-Biroulet, L.; Minuti, F.; Repici, A.; Spinelli, A.; Fries, W.; Balzarini, L.; Montorsi, M.; Malesci, A.; et al. Prospective comparison of computed tomography enterography and magnetic resonance enterography for assessment of disease activity and complications in ileocolonic Crohn's disease. *Inflamm. Bowel Dis.* **2011**, *17*, 1073–1080. [CrossRef] [PubMed]
49. Sleiman, J.; Chirra, P.; Gandhi, N.S.; Baker, M.E.; Lu, C.; Gordon, I.O.; Viswanath, S.E.; Rieder, F. Crohn's disease related strictures in cross-sectional imaging: More than meets the eye? *United Eur. Gastroenterol. J.* **2022**, *10*, 1167–1178. [CrossRef]
50. Li, Z.; Lu, B.; Lin, J.; He, S.; Huang, L.; Wang, Y.; Meng, J.; Li, Z.; Feng, S.-T.; Lin, S.; et al. A Type I Collagen-Targeted MR Imaging Probe for Staging Fibrosis in Crohn's Disease. *Front. Mol. Biosci.* **2021**, *8*, 762355. [CrossRef]
51. Coelho, R.; Ribeiro, H.; Maconi, G. Bowel thickening in Crohn's disease: Fibrosis or inflammation? Diagnostic ultrasound imaging tools. *Inflamm. Bowel Dis.* **2017**, *23*, 23–34. [CrossRef]
52. Xu, C.; Jiang, W.; Wang, L.; Mao, X.; Ye, Z.; Zhang, H. Intestinal Ultrasound for Differentiating Fibrotic or Inflammatory Stenosis in Crohn's Disease: A Systematic Review and Meta-analysis. *J. Crohn's Colitis* **2022**, *16*, 1493–1504. [CrossRef] [PubMed]
53. Maconi, G.; Bollani, S.; Bianchi Porro, G. Ultrasonographic Detection of Intestinal Complications in Crohn's Disease. *Dig. Dis. Sci.* **1996**, *41*, 1643–1648. [CrossRef] [PubMed]

54. Kohn, A.; Cerro, T.; Milite, G.; De Angelis, E.; Prantera, C. Prospective evaluation of transabdominal bowel sonography in the diagnosis of intestinal obstruction in Crohn's disease: Comparison with plain abdominal film and small bowel enteroclysis. *Inflamm. Bowel Dis.* **1999**, *5*, 153–157. [CrossRef] [PubMed]
55. Gasche, C.; Moser, G.; Turetschek, K.; Schober, E.; Moeschl, P.; Oberhuber, G. Transabdominal bowel sonography for the detection of intestinal complications in Crohn's disease. *Gut* **1999**, *44*, 112. [CrossRef] [PubMed]
56. Parente, F.; Greco, S.; Molteni, M.; Anderloni, A.; Sampietro, G.M.; Danelli, P.G.; Bianco, R.; Gallus, S.; Bianchi Porro, G. Oral contrast enhanced bowel ultrasonography in the assessment of small intestine Crohn's disease. A prospective comparison with conventional ultrasound, x ray studies, and ileocolonoscopy. *Gut* **2004**, *53*, 1652–1657. [CrossRef] [PubMed]
57. Calabrese, E.; Seta, F.; La Buccellato, A.; Virdone, R.; Pallotta, N.; Corazziari, E.; Cottone, M. Crohn's Disease: A Comparative Prospective Study of Transabdominal Ultrasonography, Small Intestine Contrast Ultrasonography, and Small Bowel Enema. *Inflamm. Bowel Dis.* **2005**, *11*, 139–145. [CrossRef] [PubMed]
58. Stidham, R.W.; Higgins, P.D.R. Imaging of intestinal fibrosis: Current challenges and future methods. *United Eur. Gastroenterol. J.* **2016**, *4*, 515–522. [CrossRef] [PubMed]
59. Fraquelli, M.; Branchi, F.; Cribiù, F.M.; Orlando, S.; Casazza, G.; Magarotto, A.; Massironi, S.; Botti, F.; Contessini-Avesani, E.; Conte, D.; et al. The role of ultrasound elasticity imaging in predicting ileal fibrosis in Crohn's disease patients. *Inflamm. Bowel Dis.* **2015**, *21*, 2605–2612. [CrossRef]
60. Baumgart, D.C.; Müller, H.P.; Grittner, U.; Metzke, D.; Fischer, A.; Guckelberger, O.; Pascher, A.; Sack, I.; Vieth, M.; Rudolph, B. US-based real-time elastography for the detection of fibrotic gut tissue in patients with stricturing Crohn disease. *Radiology* **2015**, *275*, 889–899. [CrossRef]
61. Stidham, R.W.; Wu, J.; Shi, J.; Lubman, D.M.; Higgins, P.D.R. Serum glycoproteome profiles for distinguishing intestinal fibrosis from inflammation in Crohn's disease. *PLoS ONE* **2017**, *12*, e0170506. [CrossRef] [PubMed]
62. Li, T.; Qian, Y.; Bai, T.; Li, J. Prediction of complications in inflammatory bowel disease using routine blood parameters at diagnosis. *Ann. Transl. Med.* **2022**, *10*, 185. [CrossRef] [PubMed]
63. Kjeldsen, J.; Schaffalitzky De Muckadell, O.B.; Junker, P. Seromarkers of collagen I and III metabolism in active Crohn's disease. Relation to disease activity and response to therapy. *Gut* **1995**, *37*, 805–810. [CrossRef] [PubMed]
64. De Simone, M.; Cioffi, U.; Contessini-Avesani, E.; Oreggia, B.; Paliotti, R.; Pierini, A.; Bolla, G.; Oggiano, E.; Ferrero, S.; Magrini, F.; et al. Elevated serum procollagen type III peptide in splanchnic and peripheral circulation of patients with inflammatory bowel disease submitted to surgery. *BMC Gastroenterol.* **2004**, *4*, 29. [CrossRef] [PubMed]
65. Bourgonje, A.R.; Alexdottir, M.S.; Otten, A.T.; Loveikyte, R.; Bay-Jensen, A.C.; Pehrsson, M.; van Dullemen, H.M.; Visschedijk, M.C.; Festen, E.A.M.; Weersma, R.K.; et al. Serological biomarkers of type I, III and IV collagen turnover are associated with the presence and future progression of stricturing and penetrating Crohn's disease. *Aliment. Pharmacol. Ther.* **2022**, *56*, 675–693. [CrossRef]
66. De Bruyn, J.R.; Becker, M.A.; Steenkamer, J.; Wildenberg, M.E.; Meijer, S.L.; Buskens, C.J.; Bemelman, W.A.; Löwenberg, M.; Ponsioen, C.Y.; van den Brink, G.R.; et al. Intestinal fibrosis is associated with lack of response to infliximab therapy in Crohn's disease. *PLoS ONE* **2018**, *13*, e0190999. [CrossRef]
67. Tan, K.; Lawler, J. The interaction of Thrombospondins with extracellular matrix proteins. *J. Cell Commun. Signal.* **2009**, *3*, 177–187. [CrossRef]
68. Di Cesare, P.E.; Chen, F.S.; Moergelin, M.; Carlson, C.S.; Leslie, M.P.; Perris, R.; Fang, C. Matrix-matrix interaction of cartilage oligomeric matrix protein and fibronectin. *Matrix Biol.* **2002**, *21*, 461–470. [CrossRef]
69. Holden, P.; Meadows, R.S.; Chapman, K.L.; Grant, M.E.; Kadler, K.E.; Briggs, M.D. Cartilage Oligomeric Matrix Protein Interacts with Type IX Collagen, and Disruptions to These Interactions Identify a Pathogenetic Mechanism in a Bone Dysplasia Family. *J. Biol. Chem.* **2001**, *276*, 6046–6055. [CrossRef]
70. Mann, H.H.; Özbek, S.; Engel, J.; Paulsson, M.; Wagener, R. Interactions between the cartilage oligomeric matrix protein and matrilins: Implications for matrix assembly and the pathogenesis of chondrodysplasias. *J. Biol. Chem.* **2004**, *279*, 25294–25298. [CrossRef]
71. Farina, G.; Lemaire, R.; Pancari, P.; Bayle, J.; Widom, R.L.; Lafyatis, R. Cartilage oligomeric matrix protein expression in systemic sclerosis reveals heterogeneity of dermal fibroblast responses to transforming growth factor β. *Ann. Rheum. Dis.* **2009**, *68*, 435–441. [CrossRef]
72. Farina, G.; Lemaire, R.; Korn, J.H.; Widom, R.L. Cartilage oligomeric matrix protein is overexpressed by scleroderma dermal fibroblasts. *Matrix Biol.* **2006**, *25*, 213–222. [CrossRef] [PubMed]
73. Wisłowska, M.; Jabłońska, B. Serum cartilage oligomeric matrix protein (COMP) in rheumatoid arthritis and knee osteoarthritis. *Clin. Rheumatol.* **2005**, *24*, 278–284. [CrossRef] [PubMed]
74. Martinez, F.J.; Collard, H.R.; Pardo, A.; Raghu, G.; Richeldi, L.; Selman, M.; Swigris, J.J.; Taniguchi, H.; Wells, A.U. Idiopathic pulmonary fibrosis. *Nat. Rev. Dis. Prim.* **2017**, *3*, 1949–1961. [CrossRef] [PubMed]
75. Sato, M.; Muragaki, Y.; Saika, S.; Roberts, A.B.; Ooshima, A. Targeted disruption of TGF-β1/Smad3 signaling protects against renal tubulointerstitial fibrosis induced by unilateral ureteral obstruction. *J. Clin. Investig.* **2003**, *112*, 1486–1494. [CrossRef] [PubMed]
76. Rieder, F.; Fiocchi, C. Intestinal fibrosis in inflammatory bowel disease—Current knowledge and future perspectives. *J. Crohn's Colitis* **2008**, *2*, 279–290. [CrossRef] [PubMed]

77. Wahl, S.M.; Chen, W. Transforming growth factor β-induced regulatory T cells referee inflammatory and autoimmune diseases. *Arthritis Res. Ther.* **2005**, *7*, 62–68. [CrossRef] [PubMed]
78. Pohlers, D.; Brenmoehl, J.; Löffler, I.; Müller, C.K.; Leipner, C.; Schultze-Mosgau, S.; Stallmach, A.; Kinne, R.W.; Wolf, G. TGF-β and fibrosis in different organs—Molecular pathway imprints. *Biochim. Biophys. Acta Mol. Basis Dis.* **2009**, *1792*, 746–756. [CrossRef]
79. Gervaz, P.; Morel, P.; Vozenin-Brotons, M.C. Molecular Aspects of Intestinal Radiation-Induced Fibrosis. *Curr. Mol. Med.* **2009**, *9*, 273–280. [CrossRef]
80. Di Sabatino, A.; Jackson, C.L.; Pickard, K.M.; Buckley, M.; Rovedatti, L.; Leakey, N.A.B.; Picariello, L.; Cazzola, P.; Monteleone, G.; Tonelli, F.; et al. Transforming growth factor β signalling and matrix metalloproteinases in the mucosa overlying Crohn's disease strictures. *Gut* **2009**, *58*, 777–789. [CrossRef]
81. Li, C.; Flynn, R.S.; Grider, J.R.; Murthy, K.S.; Kellum, J.M.; Akbari, H.; Kuemmerle, J.F. Increased activation of latent TGF-β1 by αVβ3 in human Crohn's disease and fibrosis in TNBS colitis can be prevented by cilengitide. *Inflamm. Bowel Dis.* **2013**, *19*, 2829–2839. [CrossRef] [PubMed]
82. Nakamura, T.; Sakai, K.; Nakamura, T.; Matsumoto, K. Hepatocyte growth factor twenty years on: Much more than a growth factor. *J. Gastroenterol. Hepatol.* **2011**, *26* (Suppl. S1), 188–202. [CrossRef]
83. Mungunsukh, O.; Day, R.M. Transforming growth factor-β1 selectively inhibits hepatocyte growth factor expression via a micro-RNA-199-dependent posttranscriptional mechanism. *Mol. Biol. Cell* **2013**, *24*, 2088–2097. [CrossRef] [PubMed]
84. Crestani, B.; Marchand-Adam, S.; Quesnel, C.; Plantier, L.; Borensztajn, K.; Marchal, J.; Mailleux, A.; Soler, P.; Dehoux, M. Hepatocyte growth factor and lung fibrosis. *Proc. Am. Thorac. Soc.* **2012**, *9*, 158–163. [CrossRef] [PubMed]
85. Azuma, J.; Taniyama, Y.; Takeya, Y.; Iekushi, K.; Aoki, M.; Dosaka, N.; Matsumoto, K.; Nakamura, T.; Ogihara, T.; Morishita, R. Angiogenic and antifibrotic actions of hepatocyte growth factor improve cardiac dysfunction in porcine ischemic cardiomyopathy. *Gene Ther.* **2006**, *13*, 1206–1213. [CrossRef]
86. Mizuno, S.; Matsumoto, K.; Nakamura, T. Hepatocyte growth factor suppresses interstitial fibrosis in a mouse model of obstructive nephropathy. *Kidney Int.* **2001**, *59*, 1304–1314. [CrossRef]
87. Umeda, Y.; Marui, T.; Matsuno, Y.; Shirahashi, K.; Iwata, H.; Takagi, H.; Matsumoto, K.; Nakamura, T.; Kosugi, A.; Mori, Y.; et al. Skeletal muscle targeting in vivo electroporation-mediated HGF gene therapy of bleomycin-induced pulmonary fibrosis in mice. *Lab. Investig.* **2004**, *84*, 836–844. [CrossRef]
88. Ueki, T.; Kaneda, Y.; Tsutsui, H.; Nakanishi, K.; Sawa, Y.; Morishita, R.; Matsumoto, K.; Nakamura, T.; Takahashi, H.; Okamoto, E.; et al. Hepatocyte growth factor gene therapy of liver cirrhosis in rats. *Nat. Med.* **1999**, *5*, 226–230. [CrossRef]
89. Ebert, E.C.; Panja, A.; Das, K.M.; Praveen, R.; Geng, X.; Rezac, C.; Bajpai, M. Patients with inflammatory bowel disease may have a transforming growth factor-β-, interleukin (IL)-2- or IL-10-deficient state induced by intrinsic neutralizing antibodies. *Clin. Exp. Immunol.* **2009**, *155*, 65–71. [CrossRef]
90. Han, X.; Uchida, K.; Jurickova, I.; Koch, D.; Willson, T.; Samson, C.; Bonkowski, E.; Trauernicht, A.; Kim, M.-O.; Tomer, G.; et al. Granulocyte-Macrophage Colony-Stimulating Factor Autoantibodies in Murine Ileitis and Progressive Ileal Crohn's Disease. *Gastroenterology* **2009**, *136*, 1261–1271.e3. [CrossRef]
91. Gathungu, G.; Kim, M.O.; Ferguson, J.P.; Sharma, Y.; Zhang, W.; Ng, S.M.E.; Bonkowski, E.; Ning, K.; Simms, L.A.; Croft, A.R.; et al. Granulocyte-macrophage colony-stimulating factor autoantibodies: A marker of aggressive Crohn's disease. *Inflamm. Bowel Dis.* **2013**, *19*, 1671–1680. [CrossRef] [PubMed]
92. Shi, Y.; Liu, C.H.; Roberts, A.I.; Das, J.; Xu, G.; Ren, G.; Zhang, Y.; Zhang, L.; Yuan, Z.R.; Tan, H.S.; et al. Granulocyte-macrophage colony-stimulating factor (GM-CSF) and T-cell responses: What we do and don't know. *Cell Res.* **2006**, *16*, 126–133. [CrossRef] [PubMed]
93. Frank, D.N.; Robertson, C.E.; Hamm, C.M.; Kpadeh, Z.; Zhang, T.; Chen, H.; Zhu, W.; Sartor, R.B.; Boedeker, E.C.; Harpaz, N.; et al. Disease phenotype and genotype are associated with shifts in intestinal-associated microbiota in inflammatory bowel diseases. *Inflamm. Bowel Dis.* **2011**, *17*, 179–184. [CrossRef]
94. Rieder, F. The gut microbiome in intestinal fibrosis: Environmental protector or provocateur? *Sci. Transl. Med.* **2013**, *5*, 190ps10. [CrossRef]
95. Takeuchi, O.; Akira, S. Pattern Recognition Receptors and Inflammation. *Cell* **2010**, *140*, 805–820. [CrossRef]
96. Steiner, C.A.; Berinstein, J.A.; Louissaint, J.; Higgins, P.D.R.; Spence, J.R.; Shannon, C.; Lu, C.; Stidham, R.W.; Fletcher, J.G.; Bruining, D.H.; et al. Biomarkers for the Prediction and Diagnosis of Fibrostenosing Crohn's Disease: A Systematic Review. *Clin. Gastroenterol. Hepatol.* **2022**, *20*, 817–846.e10. [CrossRef] [PubMed]
97. Solberg, I.C.; Lygren, I.; Cvancarova, M.; Jahnsen, J.; Stray, N.; Sauar, J.; Schreiber, S.; Moum, B.; Vatn, M.H. Predictive value of serologic markers in a population-based Norwegian cohort with inflammatory bowel disease. *Inflamm. Bowel Dis.* **2009**, *15*, 406–414. [CrossRef]
98. Degenhardt, F.; Dirmeier, A.; Lopez, R.; Lang, S.; Kunst, C.; Roggenbuck, D.; Reinhold, D.; Szymczak, S.; Rogler, G.; Klebl, F.; et al. Serologic Anti-GP2 Antibodies Are Associated with Genetic Polymorphisms, Fibrostenosis, and Need for Surgical Resection in Crohn's Disease. *Inflamm. Bowel Dis.* **2016**, *22*, 2648–2657. [CrossRef]
99. Roggenbuck, D.; Hausdorf, G.; Martinez-Gamboa, L.; Reinhold, D.; Büttner, T.; Jungblut, P.R.; Porstmann, T.; Laass, M.W.; Henker, J.; Büning, C.; et al. Identification of GP2, the major zymogen granule membrane glycoprotein, as the autoantigen of pancreatic antibodies in Crohn's disease. *Gut* **2009**, *58*, 1620–1628. [CrossRef]

100. Pavlidis, P.; Komorowski, L.; Teegen, B.; Liaskos, C.; Koutsoumpas, A.L.; Smyk, D.S.; Perricone, C.; Mytilinaiou, M.G.; Stocker, W.; Forbes, A.; et al. Diagnostic and clinical significance of Crohn's disease-specific pancreatic anti-GP2 and anti-CUZD1 antibodies. *Clin. Chem. Lab Med.* **2016**, *54*, 249–256. [CrossRef]
101. Mow, W.S.; Vasiliauskas, E.A.; Lin, Y.C.; Fleshner, P.R.; Papadakis, K.A.; Taylor, K.D.; Landers, C.J.; Abreu-Martin, M.T.; Rotter, J.I.; Yang, H.; et al. Association of Antibody Responses to Microbial Antigens and Complications of Small Bowel Crohn's Disease. *Gastroenterology* **2004**, *126*, 414–424. [CrossRef]
102. Schoepfer, A.M.; Schaffer, T.; Mueller, S.; Flogerzi, B.; Vassella, E.; Seibold-Schmid, B.; Seibold, F. Phenotypic associations of Crohn's disease with antibodies to flagellins A4-Fla2 and Fla-X, ASCA, p-ANCA, PAB, and NOD2 mutations in a Swiss cohort. *Inflamm. Bowel Dis.* **2009**, *15*, 1358–1367. [CrossRef] [PubMed]
103. Vasiliauskas, E.A.; Kam, L.Y.; Karp, L.C.; Gaiennie, J.; Yang, H.; Targan, S.R. Marker antibody expression stratifies Crohn's disease into immunologically homogeneous subgroups with distinct clinical characteristics. *Gut* **2000**, *47*, 487–496. [CrossRef] [PubMed]
104. Papadakis, K.A.; Yang, H.; Ippoliti, A.; Mei, L.; Elson, C.O.; Hershberg, R.M.; Vasiliauskas, E.A.; Fleshner, P.R.; Abreu, M.T.; Taylor, K.; et al. Anti-flagellin (CBir1) phenotypic and genetic Crohn's disease associations. *Inflamm. Bowel Dis.* **2007**, *13*, 524–530. [CrossRef]
105. Targan, S.R.; Landers, C.J.; Yang, H.; Lodes, M.J.; Cong, Y.; Papadakis, K.A.; Vasiliauskas, E.; Elson, C.O.; Hershberg, R.M. Antibodies to CBir1 flagellin define a unique response that is associated independently with complicated Crohn's disease. *Gastroenterology* **2005**, *128*, 2020–2028. [CrossRef]
106. Schauber, J.; Rieger, D.; Weiler, F.; Wehkamp, J.; Eck, M.; Fellermann, K.; Scheppach, W.; Gallo, R.L.; Stange, E.F. Heterogeneous expression of human cathelicidin hCAP18/LL-37 in inflammatory bowel diseases. *Eur. J. Gastroenterol. Hepatol.* **2006**, *18*, 615–621. [CrossRef] [PubMed]
107. Hoang-Ngoc Tran, D.; Wang, J.; Ha, C.; Ho, W.; Mattai, S.A.; Oikonomopoulos, A.; Weiss, G.; Lacey, P.; Cheng, M.; Shieh, C.; et al. Circulating cathelicidin levels correlate with mucosal disease activity in ulcerative colitis, risk of intestinal stricture in Crohn's disease, and clinical prognosis in inflammatory bowel disease. *BMC Gastroenterol.* **2017**, *17*, 63.
108. Lopes, S.; Andrade, P.; Rodrigues-Pinto, E.; Afonso, J.; Macedo, G.; Magro, F. Fecal marker levels as predictors of need for endoscopic balloon dilation in Crohn's disease patients with anastomotic strictures. *World J. Gastroenterol.* **2017**, *23*, 6482–6490. [CrossRef]
109. Prentice, R.; Wright, E.K.; Flanagan, E.; Ross, A.L.; Bell, S.J. The Use of Fecal Calprotectin and Intestinal Ultrasound in the Evaluation and Management of Stricturing Crohn's Disease in Pregnancy. *Inflamm. Bowel Dis.* **2022**, *28*, E13–E16. [CrossRef]
110. Chen, W.; Lu, C.; Hirota, C.; Iacucci, M.; Ghosh, S.; Gui, X. Smooth muscle hyperplasia/hypertrophy is the most prominent histological change in Crohn's fibrostenosing bowel strictures: A semiquantitative analysis by using a novel histological grading scheme. *J. Crohn's Colitis* **2017**, *11*, 92–104. [CrossRef]
111. Gordon, I.O.; Bettenworth, D.; Bokemeyer, A.; Srivastava, A.; Rosty, C.; de Hertogh, G.; Robert, M.E.; Valasek, M.A.; Mao, R.; Kurada, S.; et al. Histopathology Scoring Systems of Stenosis Associated With Small Bowel Crohn's Disease: A Systematic Review. *Gastroenterology* **2020**, *158*, 137–150.e1. [CrossRef]
112. Grillo, A.R.; Scarpa, M.; D'Incà, R.; Brun, P.; Scarpa, M.; Porzionato, A.; De Caro, R.; Martines, D.; Buda, A.; Angriman, I.; et al. TAK1 is a key modulator of the profibrogenic phenotype of human ileal myofibroblasts in Crohn's disease. *Am. J. Physiol. Gastrointest Liver Physiol.* **2015**, *309*, G443–G454. [CrossRef]
113. Hutter, S.; van Haaften, W.T.; Hünerwadel, A.; Baebler, K.; Herfarth, N.; Raselli, T.; Mamie, C.; Misselwitz, B.; Rogler, G.; Weder, B.; et al. Intestinal Activation of pH-Sensing Receptor OGR1 [GPR68] Contributes to Fibrogenesis. *J. Crohn's Colitis* **2018**, *12*, 1348–1358. [CrossRef]
114. Raselli, T.; Wyss, A.; Gonzalez Alvarado, M.N.; Weder, B.; Mamie, C.; Spalinger, M.R.; Van Haaften, W.T.; Dijkstra, G.; Sailer, A.W.; Silva, P.H.I.; et al. The oxysterol synthesising enzyme CH25H contributes to the development of intestinal fibrosis. *J. Crohns Colitis* **2019**, *13*, 1186–1200. [CrossRef]
115. Economou, M.; Trikalinos, T.A.; Loizou, K.T.; Tsianos, E.V.; Ioannidis, J.P.A. Differential effects of NOD2 variants on Crohn's disease risk and phenotype in diverse populations: A metaanalysis. *Am. J. Gastroenterol.* **2004**, *99*, 2393–2404. [CrossRef] [PubMed]
116. Adler, J.; Rangwalla, S.C.; Dwamena, B.A.; Higgins, P.D.R. The prognostic power of the nod2 genotype for complicated crohn's disease: A meta-analysis. *Am. J. Gastroenterol.* **2011**, *106*, 699–712. [CrossRef] [PubMed]
117. Roderburg, C.; Urban, G.W.; Bettermann, K.; Vucur, M.; Zimmermann, H.; Schmidt, S.; Janssen, J.; Koppe, C.; Knolle, P.; Castoldi, M.; et al. Micro-RNA profiling reveals a role for miR-29 in human and murine liver fibrosis. *Hepatology* **2011**, *53*, 209–218. [CrossRef] [PubMed]
118. Qin, W.; Chung, A.C.K.; Huang, X.R.; Meng, X.M.; Hui, D.S.C.; Yu, C.M.; Sung, J.J.; Lan, H.Y. TGF-β/Smad3 signaling promotes renal fibrosis by inhibiting miR-29. *J. Am. Soc. Nephrol.* **2011**, *22*, 1462–1474. [CrossRef] [PubMed]
119. Lewis, A.; Mehta, S.; Hanna, L.N.; Rogalski, L.A.; Jeffery, R.; Nijhuis, A.; Kumagai, T.; Biancheri, P.; Bundy, J.G.; Bishop, C.L.; et al. Low Serum Levels of MicroRNA-19 Are Associated with a Stricturing Crohn's Disease Phenotype. *Inflamm. Bowel Dis.* **2015**, *21*, 1926–1934. [CrossRef]
120. Huang, Y.; Dai, Y.; Zhang, J.; Cheng, J.; Lu, Y.; Li, D.; Wang, C.; Ma, K.; Liao, G.; Xue, F.; et al. Circulating miRNAs might be promising biomarkers to reflect the dynamic pathological changes in smoking-related interstitial fibrosis. *Toxicol. Ind. Health* **2014**, *30*, 182–191.

121. van Almen, G.C.; Verhesen, W.; van Leeuwen, R.E.W.; van de Vrie, M.; Eurlings, C.; Schellings, M.W.M.; Swinnen, M.; Cleutjens, J.P.M.; van Zandvoort, M.A.M.J.; Heymans, S.; et al. MicroRNA-18 and microRNA-19 regulate CTGF and TSP-1 expression in age-related heart failure. *Aging Cell* **2011**, *10*, 769–779. [CrossRef] [PubMed]
122. Lakner, A.M.; Steuerwald, N.M.; Walling, T.L.; Ghosh, S.; Li, T.; Mckillop, I.H.; Russo, M.W.; Bonkovsky, H.L.; Schrum, L.W. Inhibitory effects of microRNA 19b in hepatic stellate cell-mediated fibrogenesis. *Hepatology* **2012**, *56*, 300–310. [CrossRef] [PubMed]
123. Cordes, F.; Demmig, C.; Bokemeyer, A.; Brückner, M.; Lenze, F.; Lenz, P.; Nowacki, T.; Tepasse, P.; Schmidt, H.H.; Schmidt, M.A.; et al. MicroRNA-320a Monitors Intestinal Disease Activity in Patients with Inflammatory Bowel Disease. *Clin. Transl. Gastroenterol.* **2020**, *11*, e00134. [CrossRef]

medicina

MDPI

Review

Idiopathic Slow Transit Constipation: Pathophysiology, Diagnosis, and Management

Luke J. Vlismas [1,2,*], William Wu [1,3] and Vincent Ho [1,3]

1 Deptartment of Gastroenterology, Campbelltown Hospital, Campbelltown, NSW 2560, Australia; williamwenhao.wu@health.nsw.gov.au (W.W.); v.ho@westernsydney.edu.au (V.H.)
2 School of Medicine and Public Health, University of Newcastle, Newcastle, NSW 2308, Australia
3 School of Medicine, Western Sydney University, Campbelltown, NSW 2560, Australia
* Correspondence: lukevlismas@gmail.com

Abstract: Slow transit constipation (STC) has an estimated prevalence of 2–4% of the general population, and although it is the least prevalent of the chronic constipation phenotypes, it more commonly causes refractory symptoms and is associated with significant psychosocial stress, poor quality of life, and high healthcare costs. This review provides an overview of the pathophysiology, diagnosis, and management options in STC. STC occurs due to colonic dysmotility and is thought to be a neuromuscular disorder of the colon. Several pathophysiologic features have been observed in STC, including reduced contractions on manometry, delayed emptying on transit studies, reduced numbers of interstitial cells of Cajal on histology, and reduced amounts of excitatory neurotransmitters within myenteric plexuses. The underlying aetiology is uncertain, but autoimmune and hormonal mechanisms have been hypothesised. Diagnosing STC may be challenging, and there is substantial overlap with the other clinical constipation phenotypes. Prior to making a diagnosis of STC, other primary constipation phenotypes and secondary causes of constipation need to be ruled out. An assessment of colonic transit time is required for the diagnosis and can be performed by a number of different methods. There are several different management options for constipation, including lifestyle, dietary, pharmacologic, interventional, and surgical. The effectiveness of the available therapies in STC differs from that of the other constipation phenotypes, and prokinetics often make up the mainstay for those who fail standard laxatives. There are few available management options for patients with medically refractory STC, but patients may respond well to surgical intervention. STC is a common condition associated with a significant burden of disease. It can present a clinical challenge, but a structured approach to the diagnosis and management can be of great value to the clinician. There are many therapeutic options available, with some having more benefits than others.

Keywords: slow transit constipation; constipation; colon; dysmotility; enteric nervous system; manometry; pathophysiology; diagnosis; management; prokinetcs

1. Introduction

Constipation is the symptom of unsatisfactory defecation and can occur either in association with identifiable triggers or as a primary chronic condition. Chronic idiopathic constipation (CIC) is a common condition affecting a significant proportion of adults worldwide. A 2011 meta-analysis by Suares and Ford found a worldwide prevalence of 14%, with variations geographically [1]. CIC is one of the most common gastrointestinal complaints and reasons for an ambulatory review, and frequently impacts on quality of life [2–5]. Several known risk factors for CIC exist, most notably a female gender and an increased age, particularly age over 65. A higher prevalence of low socioeconomic status has also been observed [1,4,6].

Disease phenotypes of CIC include impaired evacuation due to dyssynergic defecation (DD), colonic dysmotility resulting in slow transit constipation (STC), or constipation without evidence of abnormal defecation or delayed colonic transit (normal transit constipation;

NTC) [3,4,7,8]. Normal transit constipation is the most common phenotype and frequently overlaps with constipation-predominant irritable bowel syndrome (IBS-C) [7].

STC is the least common of the CIC phenotypes, however, variations in prevalence occur depending on the setting, and the true population prevalence is difficult to determine as the majority of patients with NTC are successfully managed in the primary care setting and many patients with chronic constipation do not require advanced investigation to define STC [4,7]. The prevalence of patients with STC within a population of patients with CIC has been reported to range between 15–30% [9], giving an estimated STC prevalence of 2–4% in the general population, based on the abovementioned worldwide prevalence of CIC.

STC is sometimes classified as a functional gut disorder, as in the Rome Foundation's classification of disorders of gut-brain interaction [10]; however, there is objective evidence of the disease in these patients based on motility studies and pathologic examination of colectomy specimens [2,4,7,11,12], and the condition is likely to be a neuromuscular disease of the colon [7]. Although the aetiology of STC remains unclear, our understanding is evolving, and hormonal and autoimmune mechanisms have been proposed [7,13,14]. The microbiome may also play a role in the aetiology of some patients, but its overall contribution to the pathophysiology remains unclear [15].

STC may be challenging to manage and, at its extreme, may require surgical intervention. It frequently results in poor quality of life and significant psychosocial stress, and also commonly results in high health care burden with frequent presentations to health care [6]. It can be difficult to distinguish between the phenotypes of constipation clinically, and although management strategies for each are similar initially, the management of those who fail standard first-line therapies differs greatly. Therefore, having an algorithmic approach to the diagnosis and management is of vital importance to the clinician.

This review article summarises the findings of a literature review on the topic of slow transit constipation in adults, providing readers with an overview of the pathophysiology, diagnostic modalities, and management options for STC. It also proposes a framework for approaching the diagnosis and management.

2. Definitions and Classification

Constipation is generally defined as unsatisfactory defecation, characterised by increased stool firmness, reduced frequency of bowel movements, and/or difficult evacuation [3,7,8,16,17]. The term chronic generally refers to abnormalities that are present for three months or longer, and the development of STC is generally insidious, other than in certain secondary causes such as spinal cord injury. The aetiologies of constipation are numerous but can be classified as either primary or secondary. Primary constipation, often synonymous with idiopathic constipation, relates to intrinsic colonic or anorectal dysfunction, whereas secondary constipation occurs as a result of structural abnormalities, systemic disease, or medications [7,17].

The American Gastroenterological Association (AGA) classifies chronic constipation into three phenotypes: DD; STC; and NTC [3]. DD results in an impaired rectal evacuation and may or may not have a secondary delayed colonic transit due to rectal outlet obstruction. NTC is constipation without evidence of DD and with a normal colonic transit time. Some patients with CIC, particularly those with NTC, have an overlap with IBS-C, which is predominantly characterised by abdominal pain in addition to bowel disturbances [3,18].

STC occurs due to colonic dysmotility, resulting in delayed colonic transit times not due to DD. A proportion of these patients have a co-existing upper gastrointestinal dysmotility, with one study reporting a delayed gastric emptying in 34%, a delayed small bowel transit in 10%, and both in 8% [8]. The term colonic inertia refers to a state of severely impaired colonic motility with an absence of post-prandial increased motor activity or a lack of response to stimulant laxatives [2,7,17,18].

The conditions defined by AGA's classification appear in the Rome IV criteria as functional constipation and functional defecation disorders, with IBS again being defined separately but may co-exist [10]. However, the Rome IV Criteria are based on symptoms alone and are not as useful when discussing STC as there is no requirement for colonic transit studies for the diagnosis in this classification system, and the majority of patients with functional constipation have normal transit times [4,10].

3. Pathophysiology

3.1. Normal Physiology of the Colon

The primary function of the colon is water reabsorption and waste transportation towards the rectum where it is excreted as stool via the anus [7,12]. These functions rely on complex interactions between the endocrine, nervous, and muscular systems.

3.1.1. Control of Colonic Function

The majority of lower gastrointestinal function is under involuntary control; however, the process of defecation has voluntary and involuntary mechanisms. The colonic function is maintained primarily by neural and hormonal input [7].

The motor function is coordinated by input from the enteric nervous system, which contains both sympathetic and parasympathetic nerves. The enteric nervous system interfaces with the colonic smooth muscle via the colonic myenteric plexuses and the interstitial cells of Cajal (ICC) [2,7,12]. The ICCs act as the pace-making cells of the colon, mediating the signals of the enteric nervous system and the colonic smooth muscle, and are essential in the generation and propagation of electrical slow waves [4]. Both stimulating (e.g., serotonin [5-hydroxytryptamine, 5-HT], and acetylcholine) and inhibitory (e.g., nitric oxide) neurotransmitters are released by the enteric nerves to produce peristaltic waves [7].

The endocrine system contributes to both the motility and the fluid/electrolyte function of the colon. Hormones such as cholecystokinin and motilin contribute to the postprandial increase in colonic motor activity (the gastrocolic reflex), and hormones, such as the thyroid hormone, interact with the enteric nervous system to regulate intestinal motility. Similar to its action in the kidneys, aldosterone also helps to regulate sodium and water reabsorption in the colon [4,12].

3.1.2. Fluid and Electrolyte Homeostasis

The colon contributes to fluid and electrolyte homeostasis, reabsorbing 1–2 L of fluid per day [4,7]. The amount of water reabsorbed is a time-dependent process, and hence, the states that result in a delayed evacuation of faecal material result in harder, smaller stools [7].

Sodium is actively reabsorbed through multiple active transport channels. Countering this, chloride, and subsequently sodium, are secreted through chloride channels, though this function is largely inactive in the normal state, resulting in a net reabsorption of fluid and electrolytes [7]. Water is passively reabsorbed or secreted in response to osmotic gradients created by these processes in balance with the osmotic pressure of the intestinal contents [7].

3.1.3. Motor Function

The normal colonic transit times in adults range from 20–72 h [7]. Multiple different types of motor patterns occur in the colon and anorectum and can be propagating or non-propagating [4,12]. Non-propagating motor patterns serve as segmentation and mixing functions and aid in fluid and electrolyte reabsorption [4,7,12,18]. Non-propagating motor patterns are low-amplitude and occur as random contractions, as well as short-length peristaltic contractions, both in the antegrade and retrograde directions. These short peristaltic contractions are the result of the spontaneous myogenic slow waves created by ICCs. Retrograde peristaltic contractions act as a normal physiologic brake: in the right colon, it delays ileocaecal emptying and increases nutrient absorption in the small bowel; and in the left colon, it increases colonic transit time and subsequently water reabsorption, as well as assisting with the control of continence [12].

Propagating motor patterns result in powerful contractions which propel contents from the right to left colon towards the anus, resulting in mass movements, which are the main type of propulsive motility of the colon [4,7,12,18]. High-amplitude propagating contractions (HAPCs) can be seen with high-resolution manometry and are the manometric description of mass movements [4,12]. HAPCs occur spontaneously a few times each day, typically in the morning, and can be triggered or augmented by certain triggers, such as eating (the gastrocolic reflex) [4,7]. Pan-colonic pressurisations are simultaneous pressure increases across the length of the colon, which occur in unison with internal anal sphincter relaxation, resulting in the urge to defecate and facilitating the evacuation of bowel motions [12].

Defecation is the process of the evacuation of stool from the rectum via the anus. The process begins with rectal filling, followed by the coordination of relaxation of the muscles of the pelvic floor and anal sphincter and contraction of the abdominal wall and rectum [7].

3.2. Pathophysiology of Constipation

The disruptions of physiologic mechanisms leading to constipation vary greatly between the different phenotypes. The pathophysiology of NTC is unclear but is likely multifactorial. DD results from an impaired coordination of the muscles of defecation, leading to an impaired relaxation or paradoxical contraction of the anus, and/or an inadequate rectal and abdominal propulsive force. In some patients, DD may result in delayed colon transit time due to rectal outlet obstruction [4,7].

STC is thought to be a neuromuscular disorder of the colon, and dysmotility can be demonstrated by various means. Manometric studies have displayed a reduction in the number or complete absence of HAPCs, an impaired or absent gastrocolic reflex, and an overall reduced motor activity of both propagating and non-propagating patterns [4,7,12]. Ambulatory 24 h colonic manometry has demonstrated a similar nocturnal colonic pressure activity in STC compared with the controls but with an attenuation or absence of the normal increase in motor activity on waking [15]. An increase in retrograde peristaltic contractions has also been demonstrated during manometry, resulting in an exaggerated colonic break function [19]. Transit studies have shown delayed emptying, particularly of the proximal colon, and some patients may also have co-existing dysmotility of the stomach or small bowel [4,8].

Although the aetiology of STC remains unclear, several pathophysiologic features have been observed in these patients, and therefore, our understanding is evolving [7,13]. There is a strong female predominance and hormonal contributions to the aetiology have been hypothesised. Colectomy specimens have demonstrated increased progesterone receptors, which correlate with alterations to the contractile and inhibitory G-proteins [13]. Autoantibodies have been demonstrated in the pathology specimens of a small proportion of patients with gastrointestinal dysmotility, including STC, suggesting a possible autoimmune aetiology in some patients [14].

Patients may also have abnormal or reduced numbers of interstitial cells of Cajal [4,7,15,18], and in the majority of cases of patients who have undergone colectomy for refractory STC, histological examination shows an abnormal or reduced number of ICCs [11]. Additionally, reduced amounts of excitatory neurotransmitters within myenteric plexus neurons have been demonstrated [2,7].

Differences in the gut microbiome and metabolites have been observed in patients with STC, including an increased prevalence of methanogenic flora [15,20]. Although the overall contribution to the pathophysiology of STC and aetiological mechanisms remain unclear, methane gas, a product of the fermentation of dietary fibre by intestinal bacteria, has been shown to delay gastrointestinal transit and impair motility in animal models [15]. An examination of the microbiome may also act as a potential biomarker in the diagnosis of STC [20].

4. Diagnosis

Making a confident diagnosis of STC can be challenging, as symptoms overlap substantially with other phenotypes of CIC and secondary causes of constipation. Therefore, having an algorithmic approach to the diagnosis and management can be of great use to the clinician.

4.1. Differential Diagnoses

Prior to making a diagnosis of STC, it is important to consider and exclude the differential diagnoses of chronic constipation.

4.1.1. Other Phenotypes of Primary Chronic Constipation

As mentioned above, the phenotypes of CIC can be classified as one of DD, STC, or NTC [3]. A fourth phenotype is sometimes described, where patients have overlapped DD and STC [8], though these patients would be classified as having DD with delayed colonic transit in the AGA's classification system. For most of these patients, their delay in transit is due to rectal outlet obstruction and can be overcome with management of the DD; however, some patients may have true colonic dysmotility.

IBS-C is also a differential diagnosis to consider, and if a patient describes abdominal pain as their predominant symptom, then this may be the more appropriate diagnosis, particularly if there is no evidence of DD or delayed colonic transit in the investigations. The phenotypes of CIC can co-exist with a diagnosis of IBS-C, most commonly NTC [18]. Table 1 lists the phenotypes of primary chronic constipation.

Table 1. Phenotypes of primary chronic constipation.

Chronic Idiopathic Constipation
Dyssynergic defecation (with or without delayed colonic transit) Slow transit constipation Normal transit constipation †
Constipation predominant irritable bowel syndrome †

† IBS-C and NTC often co-exist.

4.1.2. Secondary Causes of Chronic Constipation

Prior to making a diagnosis of idiopathic STC, secondary causes should be considered, and the reversible risk factors addressed. The secondary causes of constipation may cause constipation either by inducing colonic dysmotility or by other pathophysiologic mechanisms.

Because colonic motility occurs through an interaction between hormonal, neuronal, and muscular systems, most of the secondary causes of STC are metabolic or neuromuscular disorders, as well as from medications that interact with these systems. Neurologic disorders are common secondary causes of STC and may be conditions that affect the central nervous system, such as Parkinsons disease, multiple sclerosis, or stroke, the peripheral nervous system, as in diabetic enteric neuropathy, or a combination of the two, as can occur in spinal cord injury. Additionally, some neurologic conditions may cause an overlap with STC and DD; this commonly occurs from spinal cord injury [2–4,7,18,21]. Table 2 lists the secondary causes of STC.

Other than processes that result in colonic dysmotility, other secondary causes of constipation should also be considered in the initial assessment. For example, a mechanical obstruction, such as from malignancy, stricture, or rectocoele, can obstruct the passage of faeces and cause constipation [3,7,18]. Other conditions associated with constipation, including psychiatric disorders, such as depression and eating disorders, cognitive impairment, immobility, cardiac disease, and non-coeliac gluten sensitivity, should also be considered [3,18,22].

Table 2. Secondary causes of slow transit constipation.

Neurologic Disorders
Parkinson's disease
Multiple sclerosis
Stroke
Spinal cord injury
Diabetic enteric neuropathy
Myopathies
Systemic sclerosis
Amyloidosis
Metabolic disorders
Hypothyroidism
Hypercalcaemia
Uraemia
Diabetes mellitus
Medications
Opiates
Anticholinergics (e.g., antidepressants, antispasmodics, antipsychotics)
Dopaminergics (e.g., levodopa, dopamine agonists, antipsychotics)
Calcium channel blockers
5-HT3 antagonists

4.2. Clinical Assessment

The purpose of the initial assessment is to exclude secondary causes, elicit any red flags, and characterise the nature and severity of the patient's constipation to allow for a correct classification of their constipation phenotype [3,4,6]. A review of a patient's medical history and medication list is an effective way to screen for secondary causes of constipation and may be addressed to improve symptoms without the need for advanced investigations. Red flags should be screened for, and if present, should prompt an investigation with a colonoscopy and/or cross-sectional imaging, primarily to exclude colorectal cancer [3,4].

A characterisation to determine the timing of onset, associated features, frequency of bowel motions, and description of stool form can be of use. The Bristol Stool Form Scale (BSFS) values < 3 correlate with delayed colonic transit time, whereas the frequency of defecation may not correlate well with the colonic transit time, particularly if not in the extreme [3,4,23,24]. In addition to the core features of hard, infrequent, and/or difficult-to-pass stools, patients with CIC may have a range of symptoms, including a sensation of an anorectal blockade, a feeling of incomplete evacuation, painful defecation, a need for digitation, abdominal pain, bloating, nausea, and vomiting [4,8,17]. Patients with STC typically experience a reduced urge to defecate and may have associated abdominal pain, nausea, and vomiting; however, it is difficult to distinguish STC from the other phenotypes of CIC by history alone [6,7]. Additionally, because some patients have extra-colonic gastrointestinal dysmotility, some associated symptoms relate more to upper gastrointestinal conditions, such as gastroparesis [8]. Because STC frequently causes psychosocial stress and impacts quality of life, it is important to assess the impact that the disease is having on a patient [6].

A clinical assessment can be useful to exclude DD. Some features of the history are more suggestive of this phenotype, including a sensation of an anorectal blockade, a feeling of incomplete evacuation, or a need for digitation [3]. The rectal examination findings of increased anal tone, impaired anal sphincter relaxation or paradoxical contraction, and/or decreased perineal descent have been shown to be an effective diagnostic tool for DD, with a sensitivity and specificity of 75% and 87%, respectively [3,18,25].

4.3. Investigations and Diagnostic Workup

Further investigation may not be required after initial assessment if there are no red flags present and the patient responds to first line management.

4.3.1. Investigations to Rule out Secondary Causes

A colonoscopy is frequently performed in patients for the investigation of constipation, though in the absence of red flag features, a colonoscopy is often of low yield and may not be required [4]. However, a colonoscopy should be performed to exclude colorectal cancer if the patient has any red flags, is required as part of a bowel cancer screening program, or is refractory to medical management and is being considered for surgery [3]. Similarly, an investigation with cross-sectional imaging, such as an abdominal CT, may be appropriate if structural causes, such as intra-abdominal malignancy, are suspected from the initial assessment. Laboratory investigations can add value in a subset of patients, including screening for hypercalcaemia and hypothyroidism, though these are uncommon causes in those whose primary complaints are constipation [6].

4.3.2. Investigations for Primary Constipation

Further investigation for suspected idiopathic constipation is generally only required for those who have failed simple laxative therapy. In this situation, an evaluation for DD or STC is important as these phenotypes are more commonly difficult to manage [17]. If patients with suspected CIC have not responded adequately to simple laxatives, a localisation to either the colon or the anorectum allows for the initiation of appropriate management [7].

If available, an assessment of anorectal function should be performed prior to colon transit studies to identify if DD is present, particularly if suspicion is high based on the clinical assessment. High-resolution anorectal manometry is the gold standard for diagnosing DD [4]. Other available tests of anorectal function include rectal balloon expulsion test, anal electromyography, and defecography [3,4,8,18].

Assessment of Colonic Motility

After excluding a rectal outlet obstruction and reversible secondary causes of constipation, an assessment of colonic transit is the next step in the workup of suspected STC and is essential to make the diagnosis [6,18]. In order to perform the testing of colonic motility, medications that alter transit times should be ceased prior in order to assess the true intrinsic colonic motility [3]. There are a number of methods of assessing colonic motility available in clinical practice, as well as those which are generally only performed in a research setting.

The radio-opaque marker test is often the standard diagnostic test used, which is widely available and simple to perform [3,4,7]. In this test, a capsule containing 20 radio-opaque markers is swallowed, and a plain film abdominal radiograph is taken 5 days later, with retention of 5 or more markers indicating slow transit. Some capsules contain a different number of markers, so a cut-off of 20% of the original number of markers is generally used. Although this test performs well in identifying STC, the number of retained markers does not correlate well with the severity or quality of life [4]. The Metcalf method is an alternative method for performing the radio-opaque marker test, which is able to approximate the total and segmental colonic transit times, which involves taking capsules on consecutive days. The number of retained markers on an X-ray the day after, both totally and segmentally, are counted [18].

Colonic scintigraphy is an alternative method that provides the total and segmental colonic transit times but is frequently less available than the radio-opaque marker test in clinical practice [3,4,7]. For this test, the patients consume a radio-isotope-labelled meal, and the transit time is calculated by making timed measurements of the residual radioactivity [4]. The scintigraphic images at 24 and 48 h are able to define the delayed colonic transit, and the results are given as a percentage of the radioactivity remaining in each colonic segment. By 48 h, a separation between patients with and without STC can be demonstrated, with the upper limit of normal being defined as the mean +/− 2 SD in the healthy controls [26]. Whole gut scintigraphy can also be used to assess for co-existing extra-colonic dysmotility.

In clinical practice, wireless motility capsules are the third most commonly used but are generally only available in limited settings, such as research centres. These capsules measure the chemical properties of the intestinal contents along the gastrointestinal tract to determine the transit time. They are able to provide information on transit through the stomach and small bowel as well but are unable to provide information on segmental colonic transit [3,4,7].

These three techniques used to measure colonic transit times are comparable in accuracy and have correlated well when performed on patients with constipation [3]. High-resolution colonic manometry is generally only performed in a research setting for clinical trials and studies on physiology but provides additional detail about the motor function of the colon [27].

Figure 1 provides an approach to the diagnosis of STC and the other constipation phenotypes.

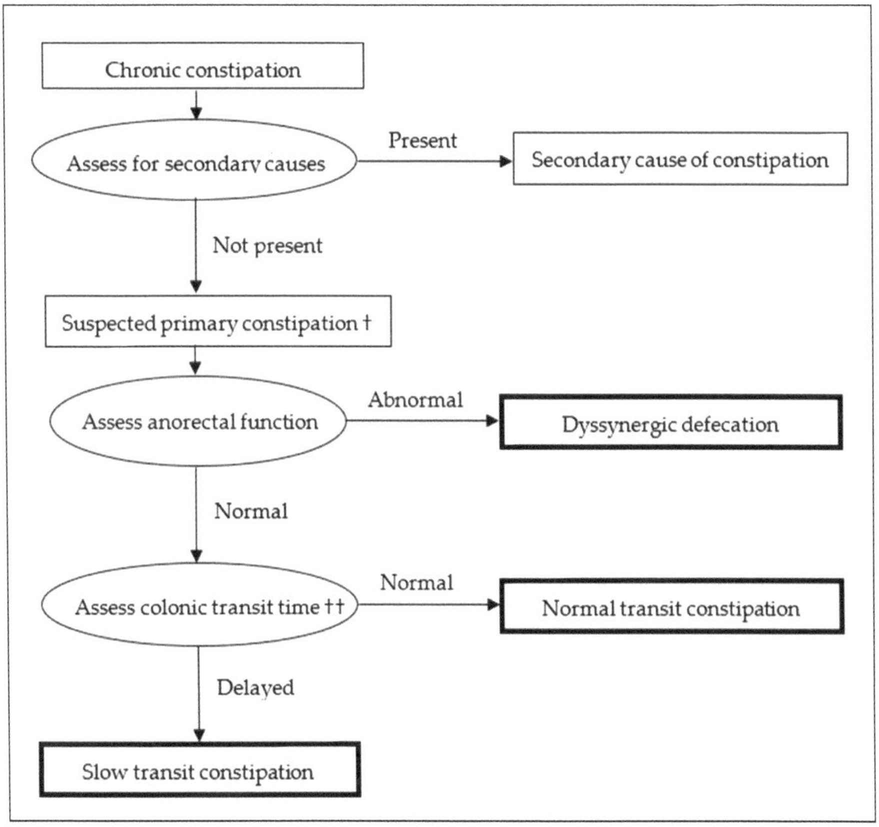

Figure 1. Diagnostic algorithm for slow transit constipation. † Simple laxatives should be trialled, and further investigations only performed in those who do not respond. †† If anorectal function testing is not available, it may be reasonable to proceed with colonic transit studies if suspicion of DD is not high based on clinical assessment, but testing should be pursued if there is persisting difficulty with management.

5. Management

There are several different management options for constipation, including dietary, pharmacologic, interventional, and surgical. A large proportion of people with constipation are managed by simple measures, such as fibre supplementation and standard laxatives, and these treatments should precede the use of the advanced investigations of anorectal and colonic functions listed above.

There is much overlap between the constipation phenotypes in the treatments available, with the exception of DD, which is best managed non-pharmacologically. If DD is identified by the testing of anorectal function, anorectal biofeedback, and pelvic floor physiotherapy are the most effective treatment, and these patients are commonly refractory to pharmacologic therapy [4,7].

One of the challenges in managing patients with STC is a lack of evidence specific to those with confirmed delayed colonic transit times, particularly for the pharmacologic trials. The inclusion criteria for most of the trials do not require an assessment of colonic transit; instead, patients are defined as having either functional constipation or CIC, and thus the trials would include patients with different constipation phenotypes. Given that patients with STC are more likely to be refractory to therapy compared to those with NTC, it is likely that the trials for more advanced therapies, such as prokinetics, do include a significant proportion of patients with STC, and the trials that have performed an assessment of transit time reflect this. However, it may be unclear how effective the therapies are in those with confirmed STC.

Each of the different management options, including their effectiveness in STC, will be discussed.

5.1. Lifestyle, Dietary and Fibre Supplementation

Increasing oral hydration is often recommended, but in the absence of dehydration, this has not been beneficial [3,4]; however, many fibre supplements and laxatives are recommended alongside increased oral hydration. Exercise has been shown to improve gastrointestinal symptoms and the quality of life in patients with IBS [3,4], but studies have shown mixed results regarding the effects of exercise on colonic transit time, and there is limited data on its effect in those with STC. However, increasing physical activity, and particularly addressing inactivity, may increase gut transit times [4,28–31].

Fibre supplementation may alter the water content and consistency of the stools, as well as affect the gastrointestinal microbiota by their prebiotic effect. Although soluble fibre supplements can be an effective treatment for many patients with CIC [3,4,32], they may have a limited benefit in slow transit constipation and may worsen the patient's symptoms, such as bloating and abdominal pain [18]. This lack of efficacy is demonstrated by delayed transit times in the colon transit studies which define STC, a method that requires the consumption of a high volume of fibre to perform [33].

Probiotics can be recommended; however, the role of probiotics in the management of CIC is unclear [4]. The proposed mechanisms of benefit in constipation include the restoration of non-pathogenic gastrointestinal microbiota and the increased bacterial production of lactate and short-chain fatty acids. For STC, their effectiveness is similarly unclear, though a 2014 meta-analysis by Dimidi et al., investigating the effects of probiotics in patients with functional constipation, showed a significantly improved whole gut transit time, stool frequency, and stool consistency; however, there was significant heterogeneity between the studies and the high risk of bias, and the outcomes in patients with STC were not observed [34].

5.2. Pharmacologic

There are multiple pharmacologic targets for the treatment of constipation, including gut motility, secretory function of the colon, and faecal fluid composition [4,7].

5.2.1. Osmotic Laxatives

Osmotic laxatives passively draw water into the intestinal lumen by osmotic gradients, which increases stool water content and facilitates colon propulsion [3,7,18]. Polyethylene glycol (PEG) containing osmotic laxatives is commonly used as first-line pharmacotherapy for CIC [4,35]. Lactulose, a non-absorbable carbohydrate, is also commonly used; however, PEG was shown to be more effective than lactulose for CIC in a 2010 meta-analysis by Lee-Robichaud et al., and its use may be limited by its common side effects of bloating and flatulence [35,36]. Magnesium oxide is an alternative that has been shown to improve the frequency of bowel movements and quality of life when compared with the placebo, but patients should be monitored for hypermagnesaemia, particularly those with renal impairment [18,35,37]. The goal of osmotic laxative therapy is to produce soft but not liquid stools, with doses being titrated to achieve this [3]. Patients with STC may or may not respond to osmotic laxatives, but these should be trialled in all patients with CIC, preferably prior to undertaking advanced investigations.

5.2.2. Stimulant Laxatives

Stimulant laxatives are irritant substances that directly stimulate the afferent nerves or the gastrointestinal smooth muscle to induce gut motility, including colonic HAPCs [7]. Several stimulant laxatives, including bisocodyl, sodium picosulfate, and senna, were shown to improve constipation and quality of life in CIC [4,35–39]. The side effects commonly experienced include abdominal pain and cramping, and diarrhoea [35]. Similarly, stimulant suppositories, such as bisacodyl and glycerin, can be used to improve stool consistency and the ease of defecation in chronic constipation [3].

The long-term safety of stimulant laxatives is commonly questioned in clinical practice. However, there is no evidence that long-term use has any negative impact on colonic motility or that it induces physiologic dependence [3,4,33,40,41].

Although effective in other forms of constipation, this class may have limited effectiveness in those with STC, as studies have demonstrated a reduced colonic motor response to these agents, and colonic inertia is defined by a lack of response to these agents [33]. However, a trial of stimulant laxatives, typically in combination with other classes such as osmotic laxatives, should be attempted.

5.2.3. Stool Softeners

Stool softeners are surfactants that reduce the surface tension of faecal material and promote water retention within the stool. The common agents in this class are docusate and liquid paraffin. They may provide some benefit to patients with constipation but often provide little improvement to patients with CIC when used in isolation and are shown to be inferior to psyllium in improving stool frequency. Their effectiveness in treating STC is unclear [3,18].

5.2.4. Secretagogues

Secretagogues target the chloride channels and induce electrolyte and fluid secretion, thereby increasing the faecal water content [4,7,35]. The increase in fluid content both accelerates colonic transit and improves the ease of defecation [3]. Lubiprostone, linaclotide, and plecanatide are the available agents in this class and can be effective in CIC, though their availability varies between regions, and their use may be limited by the side effects, particularly diarrhoea [16,35]. A 2023 meta-analysis by Chang et al., comparing lubiprostone to a placebo in patients with CIC, demonstrated an increased number of spontaneous bowel movements by 2/week. However, there was no subgroup analysis performed on the patients with STC. Although the increased intestinal fluid content induced by secretagogues may accelerate gastrointestinal transit times [35], their effectiveness in the management of STC have not been studied in depth.

5.2.5. Bile Acid Transporter Inhibitors

Elobixibat is a new treatment currently under development. It is an inhibitor of ileal bile acid transport, which induces a state of bile acid malabsorption, increasing colonic fluid secretion and promoting colonic motility [4]. It has shown promise in patients with CIC, and a 2019 post-hoc analysis of two phase-three trials by Nakajima et al. showed efficacy in patients with severe constipation and implied a benefit in those with STC. Using the criteria of <2 bowel movements per week and BSFS <3, which can be independent predictors for STC, suggested it is effective in those with both STC and NTC [42,43]. However, further studies are required to better define its effectiveness in those with STC.

5.2.6. Prokinetics

Prokinetics stimulate gastrointestinal motility, inducing intestinal peristalsis and augmenting propagating contractions [7], and are generally the most effective medical therapies for patients with STC. There are several different classes of prokinetic agents in use.

5-HT4 receptor agonists facilitate acetylcholine release from enteric neurons and include multiple agents in the class, such as prucalopride, cisapride, tegaserod, mosapride, and itopride [3,4,7,16,35]. Prucalopride has a potent colonic prokinetic effect and has the strongest evidence to support its use in STC. It is also the most commonly used of the 5-HT4 agonists for CIC in clinical practice. Unlike cisapride, tegaserod, and itopride, prucalopride has not shown any relevant electrocardiographic changes, nor has it been associated with adverse cardiovascular effects [16,18,44,45]. A 2011 meta-analysis by Ford and Suares reviewed the efficacy of prucalopride in CIC, analysing seven RCTs that compared prucalopride with a placebo in 2639 participants with CIC. This meta-analysis showed a clinical response of 28.3% vs. 13.3% in those treated with prucalopride vs. a placebo, respectively, corresponding to a NNT of six [46]. One 2002 RCT by Emmanuel et al., which was included in the above meta-analysis, performed whole-gut transit studies using the radio-opaque marker test on all the participants before and after the treatment period and also performed sub-group analyses on those with STC vs. NTC. Of the total 74 participants, a majority (58%) were classified as STC. Prucalopride at a dose of 1 mg daily reduced the number of retained markers in all the patients when compared with a placebo. A significant reduction in the number of retained markers in those with STC, but not those with NTC, was also demonstrated. 22% of the prucalopride-treated patients with delayed transit at the baseline improved to normal transit times, compared with only 5% in the placebo group [47].

Other 5-HT4 receptor agonists have also been used in STC. Cisapride has both cholinergic and serotonergic effects. It has a pan-gastrointestinal prokinetic effect, with more of an effect on upper gastrointestinal motility than colonic, and may have a greater role in patients with co-existing gastroparesis [48]. Mosapride has shown effectiveness in patients with secondary causes of STC, such as parkinsonism and diabetes [49,50]. Tegaserod was previously used for the management of constipation but has been removed from the market and is no longer available [51]. Velusetrag and naronapride are also 5-HT4 receptor agonists which are currently undergoing clinical trials [4].

Colchicine is an anti-inflammatory medication commonly used for the treatment of gout, which has a dose-dependent side effect of inducing diarrhoea and can be used in the management of CIC. The exact mechanism by which it results in diarrhoea is unclear, but it ultimately induces intestinal secretions and colonic motility [33,52]. A 2010 RCT by Taghavi et al. compared colchicine 1 mg daily to a placebo in patients with confirmed STC and showed significantly improved symptom scores and increased frequency of spontaneous bowel movements in the treatment group, with 26/30 participants treated with colchicine having an acceptable symptomatic response [52].

Misoprostol is a synthetic prostaglandin-E_1 analogue used to treat and prevent non-steroidal anti-inflammatory drug-related peptic ulcers that has the common side effect of diarrhoea. In addition to its effect on gastric acid production, it also increases gastrointestinal fluid production and motility [33,53]. A 1997 open-label trial by Roarty et al. observed the effect of oral misoprostol at a starting dose of 200 μg TDS in 18 patients with refractory constipation. An intolerance to the medication due to abdominal discomfort was common; 6/18 patients dropped out prior to the completion of the study period, but 10/12 participants who tolerated the medication had an improved frequency of bowel movements [53].

The motilin receptor agonist erythromycin has prokinetic properties, which are more pronounced in the upper gastrointestinal tract, but can also stimulate distal colonic motility in a patient with reduced plasma motilin [6]. However, a 1998 trial by Bassotti et al., investigating the effect of intravenous erythromycin on colonic motility in 18 participants with STC, concluded that it had little prokinetic effects in the colon, although some increased activity in the distal colon was demonstrated at a low dose [54]. Anecdotally, some experts have seen the benefit of erythromycin in STC, and a trial may be reasonable in patients, particularly those with co-existing upper gastrointestinal involvement [6].

Cholinesterase inhibitors stimulate upper and lower gastrointestinal motility, with parenteral neostigmine commonly used in the treatment of acute intestinal pseudo-obstructions and oral pyridostigmine having shown a benefit in the management of chronic and recurrent intestinal pseudo-obstruction [55–57]. Pyridostigmine reduces colonic transit times and improves the symptoms in patients with chronic constipation and those with secondary causes of slow transit constipation [58–62]. However, studies on those with idiopathic STC are lacking. A 2010 study by O'Dea et al. investigated the efficacy of pyridostigmine in patients with severe constipation or recurrent pseudo-obstruction, which included six patients with STC. This study showed a benefit in only one patient, with the remaining five ceasing the medication, four of which ultimately required colectomy for refractory STC [57]. Its use may be limited because of the cholinergic side effects, but serious adverse events are rare [57,60]. Although larger randomised trials are required to better assess its effectiveness in patients with idiopathic STC, based on its efficacy in similar conditions, physiologic plausibility, and safety profile, it may be reasonable to trial pyridostigmine for patients with idiopathic STC who have failed other prokinetic medications.

Some prokinetic agents that have an effect on the upper gut, like metoclopramide and domperidone, have no effect on colonic motility and are not useful in STC.

Table 3 summarises the above advanced pharmacological therapies used in STC, including the recommended doses and regimens.

Table 3. Summary of advanced pharmacotherapies used in slow transit constipation.

Medication	Mechanism	Recommended Regimen	Comments
Prokinetics			
Prucalopride	5-HT4 agonist	1–2 mg daily, oral Maximum 4 mg/day	Typical first line prokinetic in STC.
Cisapride	Cholinergic; 5-HT4 agonist	10 mg QID, oral	May be preferred in patients with co-existing gastroparesis.
Mosapride	5-HT4 agonist	5 mg TDS, oral	Evidence for use in secondary causes of STC but limited in idiopathic STC.
Colchicine	Uncertain	1 mg daily, oral	Limited evidence in STC, but available evidence suggests benefit.
Misoprostol	Prostaglandin analogue	200 μg TDS, oral Maximum 2400 μg/day	May be limited by abdominal discomfort. Limited evidence in STC.
Erythromycin	Motilin receptor agonist	40 mg TDS, oral or IV Maximum 2 g/day	Conflicting data for benefit in STC.
Pyridostigmine	Cholinesterase inhibitor	60 mg TDS, oral Maximum 720 mg/day	Physiologically plausible and beneficial in similar conditions (pseudo-obstruction and secondary STC), but limited evidence in idiopathic STC.

Table 3. *Cont.*

Medication	Mechanism	Recommended Regimen	Comments
Bile acid transporter inhibitors			
Elobixibat	Bile acid transporter antagonist	5–15 mg daily, oral	Limited evidence in STC, but available evidence suggests benefit.
Secretagogues			
Lubiorostone	Chloride channel agonist	24 µg BD, oral	Limited evidence in STC, but effective in severe CIC.
Linaclotide	CFTR agonist	72–145 µg daily, oral Maximum 290 µg/day	Limited evidence in STC, but effective in severe CIC.
Plecanatide	CFTR agonist	3 mg daily, oral	Limited evidence in STC, but effective in severe CIC.

5.3. *Interventional and Surgical*

There are a number of interventional and surgical methods that have been used for the treatment of medically refractory STC, and the choice depends on a patient's profile, disease phenotype, and severity.

5.3.1. Faecal Microbiota Transplant

Faecal microbiota transplantation (FMT) involves the delivery of donor faecal matter to the recipient's gastrointestinal lumen and has been beneficial for a number of different gastrointestinal conditions. FMT can be delivered by a number of different techniques, including a nasointestinal tube, a colonoscopy, or an enema. A 2017 RCT by Tian et al. investigated the use of FMT in patients with STC who showed significantly improved symptoms with FMT compared with conventional treatment, with a clinical cure rate of 36.7% and 13.3% [63], respectively. Unfortunately, the 2018 long-term follow-up study of this cohort showed a loss of efficacy over time in some patients [64].

5.3.2. Electrical Stimulation

Sacral nerve stimulation has been used for the treatment of STC to induce colonic propagating contractions. Earlier uncontrolled studies suggested a benefit in the patients with STC; however, subsequent higher-quality studies have shown that it was not associated with improved symptoms of constipation or an increase in colonic transit times, as well as high rates of patient dissatisfaction and risks of complications, such as infection, and haematoma [65–68].

Transcutaneous electrical stimulation has been used to improve the symptoms associated with STC, with more experience in the paediatric population than in adults. A 2016 Cochrane Review on its use in children with STC was unable to draw any conclusions due to the low quality of evidence and the high risk of bias in the included studies [69]. There are a few studies in the adult population; however, a 2017 RCT by Yang et al. compared transcutaneous electrical stimulation to sham intervention in 28 women with STC which showed a significant improvement in the symptoms and defecation frequency [70]. Overall, the effectiveness of transcutaneous electrical stimulation in the treatment of STC remains unclear, but it may be beneficial to some patients and is a safe therapy with no serious adverse effects.

Colonic pacing with intramuscular electrode placement is an experimental treatment for STC, which has shown some promise in animal models and a limited number of humans, but more research is required before its use can be recommended [71].

5.3.3. Acupuncture

Acupuncture can be a safe treatment option in the management of CIC and can improve symptoms of constipation, however there is a high degree of heterogeneity in the studies investigating its use [72]. There is limited evidence in its use in patients with STC; however, a 2013 RCT by Peng et al. showed significant improvement in stool frequency with deep puncture acupuncture therapy when compared with shallow puncture and western medication groups at the six month follow up visit, but the outcomes at the earlier assessments were not significantly different [73].

5.3.4. Transanal Irrigation

Transanal irrigation can be beneficial to patients with CIC, as well as those with secondary constipation, including from a spinal cord injury. A 2015 meta-analysis by Emmett et al. investigated the effectiveness of transanal irrigation in patients with functional constipation and demonstrated a 50.4% response rate across the seven uncontrolled studies (254 participants), although substantial heterogeneity was present [74]. Despite not being well-investigated in patients with idiopathic STC, it is a safe and well-tolerated adjunct and may be reasonable to trial in agreeable patients.

5.3.5. Antegrade Colonic Enemas

The creation of cecostomy or appendicostomy, either through percutaneous endoscopic cecostomy or appendiceal conduits, respectively, allows for the use of antegrade colonic enemas to promote colonic emptying [3,17,18]. The choice between these two modalities depends on the patient's profile and the surgeon's preference, but appendiceal conduits are generally preferred in the paediatric population [33]. These interventions are less invasive than colectomy, particularly endoscopic cecostomy, which can be performed under local anaesthetic and conscious sedation and can improve symptoms of constipation in the majority of patients [18]. There is a larger pool of evidence in the paediatric population than in adults, but two uncontrolled cohort studies in adults have demonstrated a benefit. A 2004 retrospective study by Lees et al. reported on 32 patients with refractory constipation caused by STC, DD, or mixed STC/DD who underwent cecostomy/appendicostomy conduit creation, with satisfactory function achieved in 47% [75]. A 2001 prospective study by Rongen et al. observed 12 patients with medically refractory STC who underwent cecostomy/appendicostomy conduit creation and showed an improved median defecation frequency from 1/week to 1/day; there were no major complications, but 4/12 ultimately required colectomy due to persisting constipation [76]. Although there is limited evidence available in adults, the results of the above studies suggest a benefit in a population of patients who may otherwise require colectomy, and the creation of cecostomy/appendicostomy does not appear to affect their suitability for further surgeries.

A number of different irrigation solutions are used for antegrade colonic enemas, including tap water, saline, PEG, glycerin, and mineral oil [77].

5.3.6. Surgery

Various forms of surgery have been used for medically refractory STC, but the most common and most effective is a total colectomy, either with ileorectal anastomosis or ileostomy formation [4,7,78]. Ileostomy without colectomy can be considered in patients who are at high operative risk [2]. Segmental colectomy has been used for treatment but may be ineffective if the remaining colon is also disordered and does not perform better than ileorectal anastomosis in trials, and so total colectomy is generally the preferred surgery [2,18].

Surgical intervention is rarely indicated in patients with constipation because the optimal patient selection is of vital importance, but in the correct circumstances, outcomes can be good, and the patients' symptoms may respond well [4,7]. Prior to the consideration of surgery, reversible secondary causes need to be excluded, and the patients should have medically refractory STC and have exhausted pharmacologic options. A 2017 systematic review by Knowles et al. found an 86% average satisfaction rate, with rates from individual studies ranging from 81–89% [78]. However, the adverse event rate is not insignificant, with a total complication rate of 24%, comprising a mortality of 0.4%, a re-operation rate of 13%, and a small bowel obstruction rate of 15%. Additionally, patients commonly have long-term symptoms following surgery, including abdominal pain in 30–50%, bloating in 10–40%, recurrence of constipation in 10–30%, and diarrhoea in 5–15% [2,78].

Colectomy is only suitable for patients with proven STC and is not suitable for those with NTC [3,4]. Surgery is rarely indicated in patients whose phenotype is DD unless their symptoms are refractory to biofeedback and pelvic floor physiotherapy. When DD and STC co-exist, DD should be treated prior to the consideration of surgery. If surgery is to be considered despite addressing DD, an ileostomy is preferred over an ileorectal anastomosis [4,33].

For patients with both STC and extra-colonic gastrointestinal dysmotility, it can be hypothesised that a patient's upper gastrointestinal dysmotility may improve with colectomy for the management of STC; however, a 2001 cohort study by Mollen et al., investigating the effects of colectomy on gastric emptying in patients with STC, showed no difference before and after surgery on the gastric emptying time [79]. Therefore, the use of colectomy should undergo careful consideration in those with both colonic and extra-colonic dysmotility, as these patients have lower satisfaction rates [18]. Similarly, patients with isolated STC, whose predominant symptoms are abdominal pain or bloating, are more likely to have persisting symptoms. In both of these circumstances, a trial with a loop ileostomy may be performed to determine the suitability to proceed with colectomy [18].

Figure 2 provides an algorithm for the management of patients with STC, and Table 4 summarises the therapeutic trials which have reported on patients with confirmed STC.

Table 4. Summary of therapeutic trials reporting on slow transit constipation.

Author, Year, Article Type	Treatment	Population	Study Characteristics	Outcomes
Emmanuel et al. [47] 2002 RCT	Prucalopride 1 mg	Females aged over 18 with functional constipation. Whole gut transit was performed on all participants, and subgroup analysis on those with STC was performed.	74 (all female) participants, 43 classified with STC. Overall, 37 treatment, 37 placebo. Of those with STC, 22 treatment, 21 placebo.	Prucalopride reduced the number of retained markers in all patients when compared with placebo by 11.2 vs. 1.1 ($p < 0.05$), respectively. Prucalopride significantly reduced the number of retained markers in those with STC by 17.3 ($p < 0.05$), but the change in baseline by 1.6 in NTC was not significant.
Taghavi et al. [52] 2010 RCT	Colchicine 1 mg daily	Patients with chronic constipation who had STC confirmed with colon transit time.	60 participants (47 female). 30 treatment, 30 placebo.	Colchicine significantly improved symptom scores and increased frequency of spontaneous bowel movements. 26/30 participants treated with colchicine had an acceptable symptomatic response.

Table 4. *Cont.*

Author, Year, Article Type	Treatment	Population	Study Characteristics	Outcomes
Roarty et al. [53] 1997 Open-label trial	Misoprostol 200 μg TDS. Dose titration based on response and tolerance was allowed, with a range of 400–2400μg/day.	Adults with chronic constipation refractory to available medical therapy, who had STC confirmed with colonic transit time.	18 participants (15 females). All received treatment.	Intolerance to misoprostol due to abdominal discomfort was common, with 6/18 patients dropping out prior to completion of the study period. 10/12 participants who tolerated misoprostol had improved frequency of bowel movements. Of the patients who tolerated the medication, mean bowel movement frequency improved from 11.25 to 4.8 days ($p = 0.0004$).
Bassotti et al. [54] 1998 Open-label	Erythromycin 50, 200, and 500 mg IV	Females with severe constipation with confirmed STC with colonoscopically positioned manometric probe, and effects of treatment on motility were assessed.	18 participants (all female). All received placebo infusion followed by treatment.	Erythromycin had little prokinetic effects in the colon, although some increased activity in the distal colon was demonstrated at a low dose.
Bharucha et al. [60] 2013 RCT	Pyridostigmine 60 mg TDS initially. Increased every three days to a maximum of 120 mg TDS, based on effect and tolerance.	Diabetic patients with CIC. All patients had scintigraphy to determine colonic transit time, and 13/30 participants had confirmed slow transit.	30 patients (22 female) 16 received treatment, and 14 received placebo. Of the 13 participants with STC, eight received pyridostigmine and five placebo.	Significantly increased colonic transit overall ($p < 0.01$), as well as improved stool frequency and consistency ($p = 0.04$). 7/8 vs. 2/5 patients with STC had normalisation of colonic transit times with pyridostigmine vs. placebo, respectively.
O'Dea et al. [57] 2010 Open-label	Pyridostigmine 10 mg BD initially, increased if required.	Adults with refractory STC or recurrent pseudo-obstruction who were being considered for colectomy.	13 overall, six with STC. All patients received treatment.	Of those with STC, 1/6 participants had improved symptoms. 4/5 who had no benefit ultimately underwent colectomy.
Tian et al. [63] 2017 RCT	FMT 100 mL by nasointestinal tube daily for six days, in addition to conventional therapy. Compared unblinded to conventional therapy alone.	Adults with refractory STC.	60 participants (40 female). 30 received FMT plus conventional therapy, 30 received conventional therapy.	FMT plus conventional therapy resulted in a clinical cure rate of 36.7% vs. 13.3% ($p = 0.04$) compared with conventional therapy alone. Treatment compared with control was also associated with an increased number of CSBMs per week (3.2 vs. 2.1, $p = 0.001$) and colonic transit time (58.5 vs. 73.6 h, $p < 0.00001$).

Table 4. *Cont.*

Author, Year, Article Type	Treatment	Population	Study Characteristics	Outcomes
Dinning et al. [65] 2015 RCT	SNS	Adults with medically refractory STC confirmed by scintigraphy.	Of 59 participants who underwent peripheral nerve evaluation to assess for suitability for permanent SNS, 55 participants (51 females) proceeded with permanent SNS insertion and were included. All patients received both actual and sham stimulations in a cross-over design.	There was no significant difference with either supraseonsory or subsensory stimulation compared with sham stimulations in any of the outcome measures.
Zerbib et al. [67] 2017 RCT	SNS	Adults with medically refractory CIC. All patients underwent assessment of colonic transit times using radio-opaque marker test. 28/36 of the initial participants, and 16/20 of those who progressed to permanent SNS, were classified as STC.	Of 36 participants (34 female) who underwent peripheral nerve evaluation to assess for suitability for permanent SNS, 20 responded and received a permanent SNS and were included. All patients received both actual and sham stimulations in a cross-over design.	There was no significant difference between on- and off- periods of stimulation in any of the outcomes measured.
Yiannakou et al. [68] 2019 RCT	SNS	Adults with medically refractory CIC. All patients underwent assessment of colonic transit times. 30/45 of initial participants were classified as STC.	Of the 45 participants (43 female) who underwent peripheral nerve evaluation to assess for suitability for permanent SNS, 29 were responders, 2/29 did not proceed, and 27 ultimately received a permanent SNS and were included. All patients received both actual and sham stimulations in a cross-over design.	There was no significant difference between on- and off- periods of stimulation in any of the outcomes measured. Additionally, there was no difference between those who were discriminate and indiscriminate responders during the peripheral nerve evaluation.
Ng et al. [69] 2016 Systematic review	TES	Children with STC confirmed by scintigraphy.	10 studies reporting on a single RCT cohort of 42 children (18 girls) aged 8–18 years, with additional data from their subsequent long-term studies. 21 received TES, 21 received sham stimulation.	TES was associated with a significantly reduced colonic transit time compared with sham stimulation (mean difference 1.05, 95%CI 0.36–1.74). There was no statistical difference between TES and sham stimulation in terms of CSBM/week, soiling or QOL.

Table 4. *Cont.*

Author, Year, Article Type	Treatment	Population	Study Characteristics	Outcomes
Yang et al. [70] 2017 RCT	TES	Women with STC.	28 participants (all female). 14 received TES, 14 received sham stimulation.	TES improved symptoms scores and frequency of SBMs compared with sham stimulation ($p < 0.05$).
Martellucci and Valeri. [71] 2013 Pilot study	Colonic pacing	Adults with medically refractory STC.	Two participants (both female). Both underwent intramuscular electrode placement for colonic pacing.	Number of SBM/week improved from 0.3 to 3.5 in one patient, and 0.5 to 2.5 in the other. Both patients were able to subsequently cease all conventional therapy for constipation and there were no complications.
Peng et al. [73] 2013 RCT	Acupuncture		128 participants. 64 received deep puncture, 33 shallow puncture, and 31 western medication.	Defecation frequency improved from 1.8 to 3.9 SBMs/week with deep puncture acupuncture but did not meet statistical significance ($p > 0.05$). Deep puncture acupuncture was significantly associated with improved defecation frequency to 3.5 SBMs/week at the six month follow up visit ($p < 0.05$).
Lees et al. [75] 2004 Cohort	ACE	Medically refractory CIC (combination of STC, DD, mixed STC/DD patients)	32 participants (26 female) Median age 35. All received ACE.	28/32 required further conduit procedure (19/32 reversed). Satisfactory ACE function achieved in 47%. 12 ultimately went on to surgery (colectomy/ileostomy). Further surgical interventions not affected by prior caecostomy.
Rongen et al. [76] 2001 Cohort	ACE	Medically refractory STC	12 participants (8 female) Mean age 43. All received ACE.	Median defecation frequency improved from 1/week to 1/day. 4/12 ultimately required colectomy. Further surgery not compromised by preceding caecostomy.

Table 4. *Cont.*

Author, Year, Article Type	Treatment	Population	Study Characteristics	Outcomes
Knowles et al. [78] 2017 Systematic review	Surgery	Patients undergoing colectomy for medically refractory STC.	40 studies including 2045 participants. All patients received surgery.	Colectomy resulted in a global satisfaction rate of 86% (range 81–89%). Peri-operative complications occurred in 24.4% (range 17.8–31.7%), with a mortality rate of 0.4%. Abdominal pain and bloating present in 20–50%. Persistent constipation present in 10–30%. Diarrhoea and/or incontinence in 5–15%.

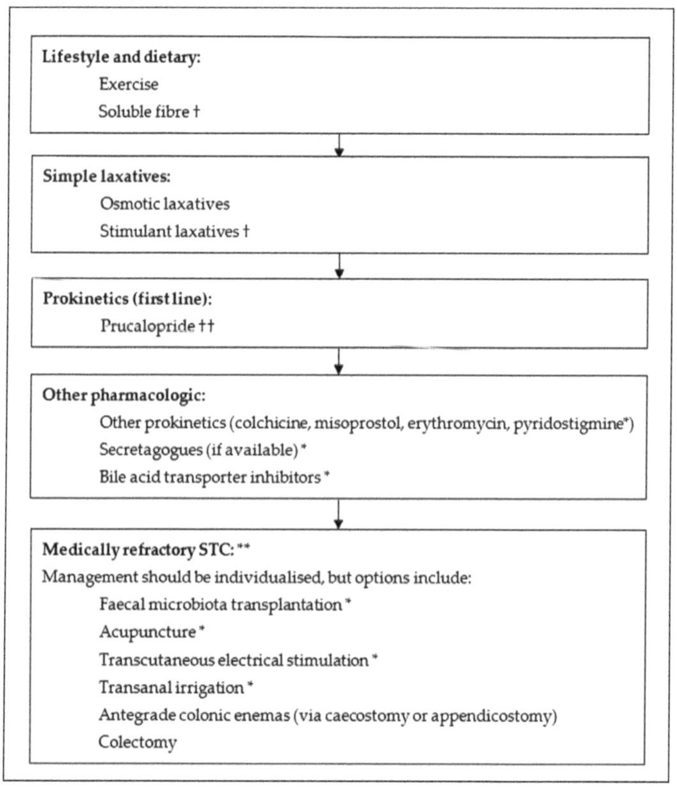

Figure 2. Management algorithm for slow transit constipation. † If there is no improvement with these therapies, then they should not be emphasised and discontinuation should be considered due to minimal benefit in STC. †† If there is no response to prucalopride, the other 5-HT4 agonists may be trialled, but may be similarly ineffective. The use of prokinetic agents of other classes may have more benefit in this circumstance. * Limited evidence in STC to guide management but may be beneficial in some patients. ** Assure that DD has been excluded and STC confirmed.

5.4. Future Directions

Future studies should continue to investigate the pathophysiology and therapeutic options for patients affected by STC. Continued advancements in our understanding of the underlying pathophysiology leading to STC will help to guide future studies, with the ultimate aim of identifying therapeutic targets. These areas include the neuromuscular function of the colon, as well as the microbiome.

Given the relative paucity of evidence for pharmacotherapy in patients with confirmed STC, it would be beneficial if future trials could be performed to address this, with inclusion/exclusion criteria designed to exclude the other constipation phenotypes.

Pharmacologic agents currently under investigation include the 5-HT4 receptor agonists velusetrag and naropride, which are currently undergoing clinical trials for use in CIC [4]. It would be beneficial if the available secretagogues, including lubiprostone, linaclotide, and plecanatide, underwent further studies into their effectiveness in the management of patients with STC, given their benefit in patients with severe CIC. Similarly, the bile acid transporter class shows promise with elobixibat, but further research is required to evaluate its efficacy in patients with STC.

Although sacral nerve stimulation appears to provide no benefit, transcutaneous electrical stimulation may hold some promise in patients with STC but requires further evaluation with larger randomized controlled trials. Additionally, although colonic pacing with intramuscular electrodes is currently experimental, its use may prove to be a useful therapy to avoid surgery in otherwise refractory cases but requires further evaluation.

6. Conclusions

STC is a significant condition that has an estimated prevalence of 2–4% in the general population. It frequently impacts quality of life and is associated with significant psychosocial stress and high healthcare costs. Our understanding of the pathophysiology is evolving, but it is likely to be a neuromuscular disorder of the colon. Observed abnormalities include reduced motor activity on manometry; delayed emptying on transit studies; hormonal changes, abnormal neurotransmitter activity, and reduced ICCs on histology; and alterations of the microbiome. The underlying aetiology is uncertain, but autoimmune and hormonal mechanisms have been hypothesised. It can be a challenging condition to manage, but a structured approach to the diagnosis and management can be of great value to the clinician. Therapeutic options include lifestyle and dietary changes, laxatives, pharmacotherapy, and interventional therapies, with prokinetic agents generally providing the most effective medical therapy for these patients. Though it is rarely required, medically refractory STC may respond well to colectomy.

Author Contributions: Conceptualization, L.J.V. and V.H.; methodology, L.J.V. and V.H.; formal analysis, L.J.V.; investigation, L.J.V.; resources, L.J.V. and W.W.; data curation, L.J.V.; writing—original draft preparation, L.J.V.; writing—review and editing, W.W. and V.H.; visualization, L.J.V.; supervision, V.H.; project administration, L.J.V. All authors have read and agreed to the published version of the manuscript.

Funding: This research received no external funding.

Institutional Review Board Statement: Not applicable.

Informed Consent Statement: Not applicable.

Data Availability Statement: Data sharing not applicable.

Conflicts of Interest: The authors declare no conflict of interest.

References

1. Suares, N.C.; Ford, A.C. Prevalence of, and risk factors for, chronic idiopathic constipation in the community: Systematic review and meta-analysis. *Am. J. Gastroenterol.* **2011**, *106*, 1582–1591; quiz 1581, 1592. [CrossRef] [PubMed]
2. Andromanakos, N.P.; Pinis, S.I.; Kostakis, A.I. Chronic severe constipation: Current pathophysiological aspects, new diagnostic approaches, and therapeutic options. *Eur. J. Gastroenterol. Hepatol.* **2015**, *27*, 204–214. [CrossRef]

3. Bharucha, A.E.; Pemberton, J.H.; Locke, G.R., 3rd. American Gastroenterological Association technical review on constipation. *Gastroenterology* **2013**, *144*, 218–238. [CrossRef] [PubMed]
4. Black, C.J.; Ford, A.C. Chronic idiopathic constipation in adults: Epidemiology, pathophysiology, diagnosis and clinical management. *Med. J. Aust.* **2018**, *209*, 86–91. [CrossRef] [PubMed]
5. Peery, A.F.; Crockett, S.D.; Barritt, A.S.; Dellon, E.S.; Eluri, S.; Gangarosa, L.M.; Jensen, E.T.; Lund, J.L.; Pasricha, S.; Runge, T.; et al. Burden of Gastrointestinal, Liver, and Pancreatic Diseases in the United States. *Gastroenterology* **2015**, *149*, 1731–1741.e3. [CrossRef]
6. El-Salhy, M. Chronic idiopathic slow transit constipation: Pathophysiology and management. *Colorectal Dis.* **2003**, *5*, 288–296. [CrossRef]
7. Andrews, C.N.; Storr, M. The pathophysiology of chronic constipation. *Can. J. Gastroenterol.* **2011**, *25* (Suppl. B), 16B–21B. [CrossRef]
8. Shahid, S.; Ramzan, Z.; Maurer, A.H.; Parkman, H.P.; Fisher, R.S. Chronic idiopathic constipation: More than a simple colonic transit disorder. *J. Clin. Gastroenterol.* **2012**, *46*, 150–154. [CrossRef]
9. Bassotti, G.; Roberto, G.D.; Sediari, L.; Morelli, A. Toward a definition of colonic inertia. *World J. Gastroenterol.* **2004**, *10*, 2465–2467. [CrossRef]
10. Drossman, D.A.; Hasler, W.L. Rome IV-Functional GI Disorders: Disorders of Gut-Brain Interaction. *Gastroenterology* **2016**, *150*, 1257–1261. [CrossRef]
11. Huizinga, J.D.; Hussain, A.; Chen, J.H. Interstitial cells of Cajal and human colon motility in health and disease. *Am. J. Physiol. Gastrointest. Liver Physiol.* **2021**, *321*, G552–G575. [CrossRef] [PubMed]
12. Feldman, M.; Friedman, L.S.; Brandt, L.J. *Sleisenger and Fordtran's Gastrointestinal and Liver Disease: Pathophysiology, Diagnosis, Management*, 11th ed.; Elsevier: Philadelphia, PA, USA, 2020.
13. Sharma, A.; Rao, S. Constipation: Pathophysiology and Current Therapeutic Approaches. *Handb. Exp. Pharmacol.* **2017**, *239*, 59–74. [CrossRef] [PubMed]
14. Tornblom, H.; Lang, B.; Clover, L.; Knowles, C.H.; Vincent, A.; Lindberg, G. Autoantibodies in patients with gut motility disorders and enteric neuropathy. *Scand. J. Gastroenterol.* **2007**, *42*, 1289–1293. [CrossRef] [PubMed]
15. Rao, S.S.; Sadeghi, P.; Beaty, J.; Kavlock, R. Ambulatory 24-hour colonic manometry in slow-transit constipation. *Am. J. Gastroenterol.* **2004**, *99*, 2405–2416. [CrossRef] [PubMed]
16. Thayalasekeran, S.; Ali, H.; Tsai, H.H. Novel therapies for constipation. *World J. Gastroenterol.* **2013**, *19*, 8247–8251. [CrossRef]
17. Bharucha, A.E.; Wald, A. Chronic Constipation. *Mayo Clin. Proc.* **2019**, *94*, 2340–2357. [CrossRef]
18. Bharucha, A.E.; Lacy, B.E. Mechanisms, Evaluation, and Management of Chronic Constipation. *Gastroenterology* **2020**, *158*, 1232–1249.e3. [CrossRef]
19. Wlodarczyk, J.; Wasniewska, A.; Fichna, J.; Dziki, A.; Dziki, L.; Wlodarczyk, M. Current Overview on Clinical Management of Chronic Constipation. *J. Clin. Med.* **2021**, *10*, 1738. [CrossRef]
20. Tian, H.; Chen, Q.; Yang, B.; Qin, H.; Li, N. Analysis of Gut Microbiome and Metabolite Characteristics in Patients with Slow Transit Constipation. *Dig. Dis. Sci.* **2021**, *66*, 3026–3035. [CrossRef]
21. Bharucha, A.E.; Philips, S.F. Slow-transit Constipation. *Curr. Treat Options Gastroenterol.* **2001**, *4*, 309–315. [CrossRef]
22. Roszkowska, A.; Pawlicka, M.; Mroczek, A.; Balabuszek, K.; Nieradko-Iwanicka, B. Non-Celiac Gluten Sensitivity: A Review. *Medicina* **2019**, *55*, 222. [CrossRef] [PubMed]
23. Lewis, S.J.; Heaton, K.W. Stool form scale as a useful guide to intestinal transit time. *Scand. J. Gastroenterol.* **1997**, *32*, 920–924. [CrossRef] [PubMed]
24. Saad, R.J.; Rao, S.S.; Koch, K.L.; Kuo, B.; Parkman, H.P.; McCallum, R.W.; Sitrin, M.D.; Wilding, G.E.; Semler, J.R.; Chey, W.D. Do stool form and frequency correlate with whole-gut and colonic transit? Results from a multicenter study in constipated individuals and healthy controls. *Am. J. Gastroenterol.* **2010**, *105*, 403–411. [CrossRef] [PubMed]
25. Tantiphlachiva, K.; Rao, P.; Attaluri, A.; Rao, S.S. Digital rectal examination is a useful tool for identifying patients with dyssynergia. *Clin. Gastroenterol. Hepatol.* **2010**, *8*, 955–960. [CrossRef] [PubMed]
26. Southwell, B.R.; Clarke, M.C.; Sutcliffe, J.; Hutson, J.M. Colonic transit studies: Normal values for adults and children with comparison of radiological and scintigraphic methods. *Pediatr. Surg. Int.* **2009**, *25*, 559–572. [CrossRef] [PubMed]
27. Li, Y.W.; Yu, Y.J.; Fei, F.; Zheng, M.Y.; Zhang, S.W. High-resolution colonic manometry and its clinical application in patients with colonic dysmotility: A review. *World J. Clin. Cases* **2019**, *7*, 2675–2686. [CrossRef] [PubMed]
28. Jensen, M.M.; Pedersen, H.E.; Clemmensen, K.K.B.; Ekblond, T.S.; Ried-Larsen, M.; Faerch, K.; Brock, C.; Quist, J.S. Associations Between Physical Activity and Gastrointestinal Transit Times in People with Normal Weight, Overweight, and Obesity. *J. Nutr.* 2023, *in press*. [CrossRef]
29. Song, B.K.; Cho, K.O.; Jo, Y.; Oh, J.W.; Kim, Y.S. Colon transit time according to physical activity level in adults. *J. Neurogastroenterol. Motil.* **2012**, *18*, 64–69. [CrossRef]
30. Oettle, G.J. Effect of moderate exercise on bowel habit. *Gut* **1991**, *32*, 941–944. [CrossRef]
31. Robertson, G.; Meshkinpour, H.; Vandenberg, K.; James, N.; Cohen, A.; Wilson, A. Effects of exercise on total and segmental colon transit. *J. Clin. Gastroenterol.* **1993**, *16*, 300–303. [CrossRef]
32. Suares, N.C.; Ford, A.C. Systematic review: The effects of fibre in the management of chronic idiopathic constipation. *Aliment. Pharmacol. Ther.* **2011**, *33*, 895–901. [CrossRef] [PubMed]

33. Wald, A. Slow Transit Constipation. *Curr. Treat Options Gastroenterol.* **2002**, *5*, 279–283. [CrossRef] [PubMed]
34. Dimidi, E.; Christodoulides, S.; Fragkos, K.C.; Scott, S.M.; Whelan, K. The effect of probiotics on functional constipation in adults: A systematic review and meta-analysis of randomized controlled trials. *Am. J. Clin. Nutr.* **2014**, *100*, 1075–1084. [CrossRef]
35. Chang, L.; Chey, W.D.; Imdad, A.; Almario, C.V.; Bharucha, A.E.; Diem, S.; Greer, K.B.; Hanson, B.; Harris, L.A.; Ko, C.; et al. American Gastroenterological Association-American College of Gastroenterology Clinical Practice Guideline: Pharmacological Management of Chronic Idiopathic Constipation. *Gastroenterology* **2023**, *164*, 1086–1106. [CrossRef] [PubMed]
36. Lee-Robichaud, H.; Thomas, K.; Morgan, J.; Nelson, R.L. Lactulose versus Polyethylene Glycol for Chronic Constipation. *Cochrane Database Syst. Rev.* **2010**, CD007570. [CrossRef]
37. Morishita, D.; Tomita, T.; Mori, S.; Kimura, T.; Oshima, T.; Fukui, H.; Miwa, H. Senna Versus Magnesium Oxide for the Treatment of Chronic Constipation: A Randomized, Placebo-Controlled Trial. *Am. J. Gastroenterol.* **2021**, *116*, 152–161. [CrossRef]
38. Kamm, M.A.; Mueller-Lissner, S.; Wald, A.; Richter, E.; Swallow, R.; Gessner, U. Oral bisacodyl is effective and well-tolerated in patients with chronic constipation. *Clin. Gastroenterol. Hepatol.* **2011**, *9*, 577–583. [CrossRef]
39. Mueller-Lissner, S.; Kamm, M.A.; Wald, A.; Hinkel, U.; Koehler, U.; Richter, E.; Bubeck, J. Multicenter, 4-week, double-blind, randomized, placebo-controlled trial of sodium picosulfate in patients with chronic constipation. *Am. J. Gastroenterol.* **2010**, *105*, 897–903. [CrossRef]
40. Muller-Lissner, S.A.; Kamm, M.A.; Scarpignato, C.; Wald, A. Myths and misconceptions about chronic constipation. *Am. J. Gastroenterol.* **2005**, *100*, 232–242. [CrossRef]
41. Wald, A. Is chronic use of stimulant laxatives harmful to the colon? *J. Clin. Gastroenterol.* **2003**, *36*, 386–389. [CrossRef]
42. Acosta, A.; Camilleri, M. Elobixibat and its potential role in chronic idiopathic constipation. *Ther. Adv. Gastroenterol.* **2014**, *7*, 167–175. [CrossRef] [PubMed]
43. Nakajima, A.; Taniguchi, S.; Kurosu, S.; Gillberg, P.G.; Mattsson, J.P.; Camilleri, M. Efficacy, long-term safety, and impact on quality of life of elobixibat in more severe constipation: Post hoc analyses of two phase 3 trials in Japan. *Neurogastroenterol. Motil.* **2019**, *31*, e13571. [CrossRef] [PubMed]
44. Camilleri, M.; Piessevaux, H.; Yiannakou, Y.; Tack, J.; Kerstens, R.; Quigley, E.M.M.; Ke, M.; Da Silva, S.; Levine, A. Efficacy and Safety of Prucalopride in Chronic Constipation: An Integrated Analysis of Six Randomized, Controlled Clinical Trials. *Dig. Dis. Sci.* **2016**, *61*, 2357–2372. [CrossRef] [PubMed]
45. Huang, X.; Lv, B.; Zhang, S.; Fan, Y.H.; Meng, L.N. Itopride therapy for functional dyspepsia: A meta-analysis. *World J. Gastroenterol.* **2012**, *18*, 7371–7377. [CrossRef] [PubMed]
46. Ford, A.C.; Suares, N.C. Effect of laxatives and pharmacological therapies in chronic idiopathic constipation: Systematic review and meta-analysis. *Gut* **2011**, *60*, 209–218. [CrossRef]
47. Emmanuel, A.V.; Roy, A.J.; Nicholls, T.J.; Kamm, M.A. Prucalopride, a systemic enterokinetic, for the treatment of constipation. *Aliment. Pharmacol. Ther.* **2002**, *16*, 1347–1356. [CrossRef]
48. Tack, J.; Coremans, G.; Janssens, J. A risk-benefit assessment of cisapride in the treatment of gastrointestinal disorders. *Drug Saf.* **1995**, *12*, 384–392. [CrossRef]
49. Ueno, N.; Inui, A.; Satoh, Y. The effect of mosapride citrate on constipation in patients with diabetes. *Diabetes Res. Clin. Pract.* **2010**, *87*, 27–32. [CrossRef]
50. Liu, Z.; Sakakibara, R.; Odaka, T.; Uchiyama, T.; Uchiyama, T.; Yamamoto, T.; Ito, T.; Asahina, M.; Yamaguchi, K.; Yamaguchi, T.; et al. Mosapride citrate, a novel 5-HT4 agonist and partial 5-HT3 antagonist, ameliorates constipation in parkinsonian patients. *Mov. Disord.* **2005**, *20*, 680–686. [CrossRef]
51. Morrow, A. ZELNORM® (Tegaserod) Notice of Withdrawal from Market. Alfasigma USA, Inc. 2022. Available online: https://www.myzelnorm.com/assets/pdfs/Press%20Release%20on%20Notice%20of%20Withdrawal.pdf (accessed on 2 November 2023).
52. Taghavi, S.A.; Shabani, S.; Mehramiri, A.; Eshraghian, A.; Kazemi, S.M.; Moeini, M.; Hosseini-Asl, S.M.; Saberifiroozi, M.; Alizade-Naeeni, M.; Mostaghni, A.A. Colchicine is effective for short-term treatment of slow transit constipation: A double-blind placebo-controlled clinical trial. *Int. J. Colorectal Dis.* **2010**, *25*, 389–394. [CrossRef]
53. Roarty, T.P.; Weber, F.; Soykan, I.; McCallum, R.W. Misoprostol in the treatment of chronic refractory constipation: Results of a long-term open label trial. *Aliment. Pharmacol. Ther.* **1997**, *11*, 1059–1066. [CrossRef] [PubMed]
54. Bassotti, G.; Chiarioni, G.; Vantini, I.; Morelli, A.; Whitehead, W.E. Effect of different doses of erythromycin on colonic motility in patients with slow transit constipation. *Z. Gastroenterol.* **1998**, *36*, 209–213. [PubMed]
55. Sen, A.; Chokshi, R. Update on the Diagnosis and Management of Acute Colonic Pseudo-obstruction (ACPO). *Curr. Gastroenterol. Rep.* **2023**, *25*, 191–197. [CrossRef] [PubMed]
56. Wilkie, B.D.; Noori, J.; Johnston, M.; Woods, R.; Keck, J.O.; Behrenbruch, C. Pyridostigmine in chronic intestinal pseudo-obstruction—A systematic review. *ANZ J. Surg.* **2023**, *93*, 2086–2091. [CrossRef] [PubMed]
57. O'Dea, C.J.; Brookes, J.H.; Wattchow, D.A. The efficacy of treatment of patients with severe constipation or recurrent pseudo-obstruction with pyridostigmine. *Colorectal Dis.* **2010**, *12*, 540–548. [CrossRef] [PubMed]
58. Law, N.M.; Bharucha, A.E.; Undale, A.S.; Zinsmeister, A.R. Cholinergic stimulation enhances colonic motor activity, transit, and sensation in humans. *Am. J. Physiol. Gastrointest. Liver Physiol.* **2001**, *281*, G1228–G1237. [CrossRef] [PubMed]
59. Soufi-Afshar, I.; Moghadamnia, A.; Bijani, A.; Kazemi, S.; Shokri-Shirvani, J. Comparison of pyridostigmine and bisacodyl in the treatment of refractory chronic constipation. *Casp. J. Intern Med.* **2016**, *7*, 19–24.

60. Bharucha, A.E.; Low, P.; Camilleri, M.; Veil, E.; Burton, D.; Kudva, Y.; Shah, P.; Gehrking, T.; Zinsmeister, A.R. A randomised controlled study of the effect of cholinesterase inhibition on colon function in patients with diabetes mellitus and constipation. *Gut* **2013**, *62*, 708–715. [CrossRef]
61. Ahuja, N.K.; Mische, L.; Clarke, J.O.; Wigley, F.M.; McMahan, Z.H. Pyridostigmine for the treatment of gastrointestinal symptoms in systemic sclerosis. *Semin. Arthritis Rheum.* **2018**, *48*, 111–116. [CrossRef]
62. Ly, A.; Rahman, M.; Song, D. Pyridostigmine as an Effective Treatment for Atonic Colon in Parkinson's Disease (P13-11.007). *Neurology* **2022**, *98*, 1539. [CrossRef]
63. Tian, H.; Ge, X.; Nie, Y.; Yang, L.; Ding, C.; McFarland, L.V.; Zhang, X.; Chen, Q.; Gong, J.; Li, N. Fecal microbiota transplantation in patients with slow-transit constipation: A randomized, clinical trial. *PLoS ONE* **2017**, *12*, e0171308. [CrossRef] [PubMed]
64. Ding, C.; Fan, W.; Gu, L.; Tian, H.; Ge, X.; Gong, J.; Nie, Y.; Li, N. Outcomes and prognostic factors of fecal microbiota transplantation in patients with slow transit constipation: Results from a prospective study with long-term follow-up. *Gastroenterol. Rep.* **2018**, *6*, 101–107. [CrossRef] [PubMed]
65. Dinning, P.G.; Hunt, L.; Patton, V.; Zhang, T.; Szczesniak, M.; Gebski, V.; Jones, M.; Stewart, P.; Lubowski, D.Z.; Cook, I.J. Treatment efficacy of sacral nerve stimulation in slow transit constipation: A two-phase, double-blind randomized controlled crossover study. *Am. J. Gastroenterol.* **2015**, *110*, 733–740. [CrossRef] [PubMed]
66. Patton, V.; Stewart, P.; Lubowski, D.Z.; Cook, I.J.; Dinning, P.G. Sacral Nerve Stimulation Fails to Offer Long-term Benefit in Patients with Slow-Transit Constipation. *Dis. Colon Rectum* **2016**, *59*, 878–885. [CrossRef] [PubMed]
67. Zerbib, F.; Siproudhis, L.; Lehur, P.A.; Germain, C.; Mion, F.; Leroi, A.M.; Coffin, B.; Le Sidaner, A.; Vitton, V.; Bouyssou-Cellier, C.; et al. Randomized clinical trial of sacral nerve stimulation for refractory constipation. *Br. J. Surg.* **2017**, *104*, 205–213. [CrossRef]
68. Yiannakou, Y.; Etherson, K.; Close, H.; Kasim, A.; Mercer-Jones, M.; Plusa, S.; Maier, R.; Green, S.; Cundall, J.; Knowles, C.; et al. A randomized double-blinded sham-controlled cross-over trial of tined-lead sacral nerve stimulation testing for chronic constipation. *Eur. J. Gastroenterol. Hepatol.* **2019**, *31*, 653–660. [CrossRef]
69. Ng, R.T.; Lee, W.S.; Ang, H.L.; Teo, K.M.; Yik, Y.I.; Lai, N.M. Transcutaneous electrical stimulation (TES) for treatment of constipation in children. *Cochrane Database Syst. Rev.* **2016**, *11*, CD010873. [CrossRef]
70. Yang, Y.; Yim, J.; Choi, W.; Lee, S. Improving slow-transit constipation with transcutaneous electrical stimulation in women: A randomized, comparative study. *Women Health* **2017**, *57*, 494–507. [CrossRef]
71. Martellucci, J.; Valeri, A. Colonic electrical stimulation for the treatment of slow-transit constipation: A preliminary pilot study. *Surg. Endosc.* **2014**, *28*, 691–697. [CrossRef]
72. Wang, L.; Xu, M.; Zheng, Q.; Zhang, W.; Li, Y. The Effectiveness of Acupuncture in Management of Functional Constipation: A Systematic Review and Meta-Analysis. *Evid. Based Complement. Alternat. Med.* **2020**, *2020*, 6137450. [CrossRef]
73. Peng, W.N.; Wang, L.; Liu, Z.S.; Guo, J.; Cai, H.J.; Ni, J.N.; Duan, J.X.; Yang, D.L. Analysis on follow-up efficacy and safety of slow transit constipation treated with individualized deep puncture at Tianshu (ST 25): A multi-central randomized controlled trial. *Zhongguo Zhen Jiu* **2013**, *33*, 865–869. [PubMed]
74. Emmett, C.D.; Close, H.J.; Yiannakou, Y.; Mason, J.M. Trans-anal irrigation therapy to treat adult chronic functional constipation: Systematic review and meta-analysis. *BMC Gastroenterol.* **2015**, *15*, 139. [CrossRef] [PubMed]
75. Lees, N.P.; Hodson, P.; Hill, J.; Pearson, R.C.; MacLennan, I. Long-term results of the antegrade continent enema procedure for constipation in adults. *Colorectal Dis.* **2004**, *6*, 362–368. [CrossRef] [PubMed]
76. Rongen, M.J.; van der Hoop, A.G.; Baeten, C.G. Cecal access for antegrade colon enemas in medically refractory slow-transit constipation: A prospective study. *Dis. Colon Rectum* **2001**, *44*, 1644–1649. [CrossRef]
77. Chu, D.I.; Balsara, Z.R.; Routh, J.C.; Ross, S.S.; Wiener, J.S. Experience with glycerin for antegrade continence enema in patients with neurogenic bowel. *J. Urol.* **2013**, *189*, 690–693. [CrossRef]
78. Knowles, C.H.; Grossi, U.; Horrocks, E.J.; Pares, D.; Vollebregt, P.F.; Chapman, M.; Brown, S.; Mercer-Jones, M.; Williams, A.B.; Yiannakou, Y.; et al. Surgery for constipation: Systematic review and practice recommendations: Graded practice and future research recommendations. *Colorectal Dis.* **2017**, *19* (Suppl. S3), 101–113. [CrossRef]
79. Mollen, R.M.; Hopman, W.P.; Oyen, W.J.; Kuijpers, H.H.; Edelbroek, M.A.; Jansen, J.B. Effect of subtotal colectomy on gastric emptying of a solid meal in slow-transit constipation. *Dis. Colon Rectum* **2001**, *44*, 1189–1195. [CrossRef]

medicina

MDPI

Opinion

Joint Group and Multi Institutional Position Opinion: Cirrhotic Cardiomyopathy—From Fundamentals to Applied Tactics

Ivan Rankovic [1,2,*,†], Ivana Babic [3,*,†], Jelena Martinov Nestorov [4,5], Jelena Bogdanovic [3,5], Maja Stojanovic [5,6], Jovanka Trifunovic [7], Nikola Panic [5,8], Mihailo Bezmarevic [9], Jelena Jevtovic [4], Dusan Micic [5,10], Vladimir Dedovic [5,11], Nemanja Djuricic [11], Filip Pilipovic [12], Elena Curakova Ristovska [13], Tijana Glisic [4,5], Sanja Kostic [14], Nemanja Stojkovic [15], Nata Joksimovic [3], Mileva Bascarevic [6], Aleksandra Bozovic [3], Lewis Elvin [1,2], Ajibola Onifade [1,2], Keith Siau [1,2], Elizaveta Koriakovskaia [16] and Vladimir Milivojevic [4,5]

1 Gastroenterology and Liver Unit, Royal Cornwall Hospitals NHS Trust, London TR1 3LJ, UK; ajibolaonifade453@gmail.com (A.O.); keithsiau@nhs.net (K.S.)
2 Medical School, University of Exeter, Exeter TR10 9FE, UK
3 Clinic for Endocrinology, Diabetes and Metabolic Diseases, University Clinical Centre of Serbia, 11 000 Belgrade, Serbia; jeca.bogdanovic@yahoo.com (J.B.); natajoksimovic9@gmail.com (N.J.); alex91@hotmail.rs (A.B.)
4 Clinic for Gastroenterology and Hepatology, University Clinical Centre of Serbia, 11 000 Belgrade, Serbia; jelenamartinov@yahoo.com (J.M.N.); jelenajev89@gmail.com (J.J.); tijana.glisic78@gmail.com (T.G.); dotorevlada@gmail.com (V.M.)
5 Faculty of Medicine, University of Belgrade, 11 000 Belgrade, Serbia; dr.maja.stojanovic@gmail.com (M.S.); nikola.panicmail@gmail.com (N.P.); ducamicic@yahoo.com (D.M.); vladeda@gmail.com (V.D.)
6 Clinic for Allergy and Immunology, University Clinical Centre of Serbia, 11 000 Belgrade, Serbia; milevabascarevic333@gmail.com
7 Faculty of Dentistry Pancevo, University of Business Academy in Novi Sad, 21 000 Novi Sad, Serbia; drjotrifunovic@aol.com
8 Center for Digestive Endoscopy, University Clinic "Dr Dragisa Misovic", 11 000 Belgrade, Serbia
9 Clinic for General Surgery, Military Medical Academy, Military Medical Academy Medical Faculty, University of Defense, 11 000 Belgrade, Serbia; bezmarevicm@gmail.com
10 Clinic for Emergency Surgery, Emergency Centre, University Clinical Centre of Serbia, 11 000 Belgrade, Serbia
11 Clinic for Cardiology, University Clinical Centre of Serbia, 11 000 Belgrade, Serbia; nemanjabg83@yahoo.com
12 Institute for Orthopedic Surgery "Banjica", 11 000 Belgrade, Serbia; filip011@gmail.com
13 University Clinic for Gastroenterohepatology, 1000 Skopje, North Macedonia; elenacurakova@yahoo.com
14 Clinic for Gynecology and Obstetrics, University Clinical Centre of Serbia, 11 000 Belgrade, Serbia; cara.kostic@gmail.com
15 Department of Cardiology, University Clinic "Dr Dragisa Misovic", 11 000 Belgrade, Serbia; neckos94@gmail.com
16 Department of Cardiology, Moscow State University of Medicine and Dentistry, 127473 Moscow, Russia; koriakovskaia.liza@gmail.com
* Correspondence: doctorranke@gmail.com (I.R.); babicivana993@gmail.com (I.B.)
† These authors contributed equally to this work.

Abstract: Cirrhotic cardiomyopathy (CCM) is a diagnostic entity defined as cardiac dysfunction (diastolic and/or systolic) in patients with liver cirrhosis, in the absence of overt cardiac disorder. Pathogenically, CCM stems from a combination of systemic and local hepatic factors that, through hemodynamic and neurohormonal changes, affect the balance of cardiac function and lead to its remodeling. Vascular changes in cirrhosis, mostly driven by portal hypertension, splanchnic vasodilatation, and increased cardiac output alongside maladaptively upregulated feedback systems, lead to fluid accumulation, venostasis, and cardiac dysfunction. Autocrine and endocrine proinflammatory cytokines (TNF-alpha, IL-6), as well as systemic endotoxemia stemming from impaired intestinal permeability, contribute to myocardial remodeling and fibrosis, which further compromise the contractility and relaxation of the heart. Additionally, relative adrenal insufficiency is often present in cirrhosis, further potentiating cardiac dysfunction, ultimately leading to the development of CCM. Considering its subclinical course, CCM diagnosis remains challenging. It relies mostly on stress echocardiography or advanced imaging techniques such as speckle-tracking echocardiography. Currently, there is no specific treatment for CCM, as it vastly overlaps with the treatment of heart failure. Diuretics play a central role. The role of non-selective beta-blockers in treating portal hypertension is established; however, their role in CCM remains somewhat controversial as their effect on prognosis is unclear. However, our group still advocates them as essential tools in optimizing the neurohumoral pathologic axis that perpetuates CCM. Other targeted therapies with direct anti-inflammatory and antioxidative effects still lack sufficient evidence for wide approval. This is not only a review but also a comprehensive distillation of the insights from practicing clinical hepatologists and other specialties engaged in advanced approaches to treating liver disease and its sequelae.

Keywords: cirrhotic cardiomyopathy; cirrhosis; pathogenesis; treatment

1. Introduction

Chronic liver disease (CLD) is a progressive loss of liver function as well as destruction of liver tissue, resulting in cirrhosis as its final manifestation [1]. Cardiac dysfunction is a well-known complication of cirrhosis. It includes structural and functional changes in the heart muscle, collectively referred to as cirrhotic cardiomyopathy (CCM). CCM is a significant yet often underdiagnosed complication in patients with advanced liver disease. Its epidemiology, though less studied than other cirrhotic complications, reveals a growing concern as cirrhosis-related mortality rises globally [2]. Therefore, it is imperative that we address this issue with an approach that accounts for the complex pathodynamics at play.

2. Definition and Epidemiology

CCM is a diagnostic entity defined as cardiac dysfunction in patients with cirrhosis in the absence of other known cardiac disorders. CCM is typified by impaired contractile responsiveness to physiological stress and altered diastolic relaxation, with electrophysiological abnormalities, in the absence of other known cardiac disorders [3,4]. Clinically, it is characterized by suboptimal ventricular response of the heart muscle to stress, which leads to the inadequate perfusion of organs. Considering its usual subclinical nature, currently there is no universally accepted and clinically implemented diagnostic criteria regarding CCM, despite modern imaging techniques [5]. Available epidemiological data is very heterogeneous: studies indicate that the prevalence of CCM in cirrhotic patients ranges from 40% to 70%, varying depending on diagnostic criteria and population studied [2]. In 2005, an initial diagnostic criteria for CCM was proposed at the World Congress of Gastroenterology following a consensus conference [6]. This has recently been superseded by the criteria proposed by the Cirrhotic Cardiomyopathy Consortium based on echocardiographic imaging parameters (Table 1) [2].

Table 1. Direct comparison of diagnostic criteria: Montreal Consortium and Cirrhotic Cardiomyopathy Consortium.

Criteria	Montreal Consortium (2005)	Cirrhotic Cardiomyopathy Consortium (2019)
	Absence of Clinically Significant Cardiac Disease	
Systolic dysfunction	LVEF (resting) < 55%; Abnormal systolic contractile response to stress	LVEF (resting) < 50%; Absolute GLS < −18%

Table 1. *Cont.*

Criteria	Montreal Consortium (2005)	Cirrhotic Cardiomyopathy Consortium (2019)
Diastolic dysfunction	E/A ratio < 1; TDE > 200 ms; IVRT > 80 ms	Septal E′ velocity < 7 cm/s; E/E′ ratio ≥ 15; LAVI > 34 mL/m^2; TR velocity > 2.8 m/s
Supportive criteria	ECG: prolonged QT interval; ECHO: chronotropic incompetence to stress; electromechanical uncoupling/dyssynchrony; left atrial enlargement; left ventricular hypertrophy; Biochemical: elevated BNP, NT-proBNP, troponin	Diastolic dysfunction grading—modified ASE criteria Indeterminate grade—additional criteria: 1. IVRT 2. PV 3. Valsalva 4. atrial strain

LVEF—left ventricular ejection fraction; GLS—global longitudinal strain; E/A—peak early velocity/peak atrial velocity; TDE—tissue Doppler velocity; IVRT—isovolumetric relaxation time; LAVI—left atrial volume index; TR—tricuspid regurgitation; Septal E′—early diastolic mitral annulus velocity; E/E′—diastolic LV filling pressure; PV—pulmonary valve.

3. Pathogenesis

Pathogenetically, CCM appears to stem from a combination of systemic and local hepatic factors that collectively affect the balance of cardiac function and lead to its remodeling, as presented in Figure 1.

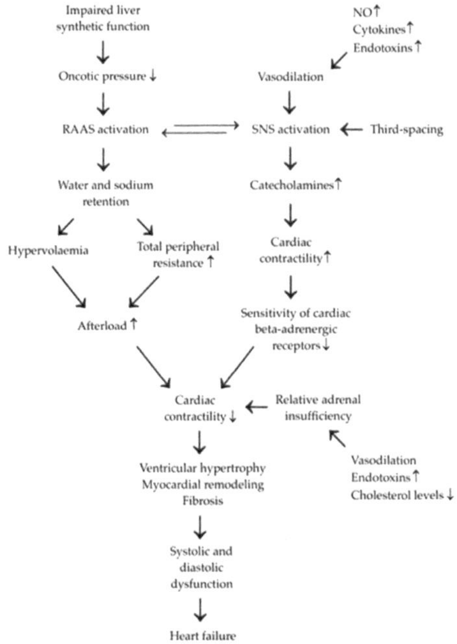

Figure 1. CCM pathogenesis. RAAS—renin–angiotensin–aldosterone system; SNS—sympathetic nervous system; NO—nitric oxide.

3.1. Hyperdynamic Changes

Cirrhosis is a condition characterized by systemic vasodilation, which is a prerequisite for the occurrence of various hemodynamic disorders in the body. Due to the hyperdynamic response of the cardiovascular system in the state of cirrhosis, mostly driven by portal hypertension, splanchnic vasodilation, and increased cardiac output, there are changes in the heart muscle function. This results in ventricular hypertrophy and eventually, systolic and diastolic dysfunction, as well as other compensatory abnormalities in the heart muscle that arise from maladaptive homeostatic mechanisms [7].

Mechanisms of cirrhosis that contribute to heart failure are based primarily upon failure of the synthetic function of the liver [8]. As fibrosis progresses, the synthetic production decreases. Pertinent to heart failure and the propagation of CCM is the drop in oncotic pressure and the intimate relationship this has with the kidneys. A drop in intravascular oncotic pressure yields intravascular volume depletion and thus activation of the renin–angiotensin–aldosterone system (RAAS) [9]. This is regulated by renal perfusion pressure, sodium levels in the ultrafiltrate of the distal convoluted tubule, intrinsic sympathetic nervous system (SNS) activity, and the negative feedback of angiotensin II (AII) on the juxtaglomerular cells [9]. In order to compensate for the decrease in vascular resistance, SNS activation occurs, i.e., heart muscle contractility increases, but water and sodium are retained through the RAAS. In cirrhosis, these feedback systems become maladaptively upregulated, culminating in increased sodium retention (causing further exacerbation of hypervolemia) and increased total peripheral resistance (further increasing afterload) [5,9]. This is especially evident during exertion and stress. Accumulation of fluid outside the intravascular and intracellular spaces (third-spacing), venostasis, and low arterial pressure upregulates the SNS primarily through augmenting the release of catecholamines. This in turn acts on cardiomyocytes and leads to a decrease in the expression of beta-adrenergic receptors, a hallmark of CCM [10]. Additionally, an increased production of nitric oxide (NO) by the failing liver, in the presence of circulating endotoxins and cytokines, and a relative state of adrenal insufficiency have further depressive effects on inotropy and chronotropy [8,10].

3.2. Inflammation

Cirrhosis with impaired liver function leads to the accumulation of toxins, which can have a direct effect on the cardiac muscle cells, leading to functional and structural changes [8]. Chronic low-grade inflammation is associated with compensated cirrhosis. Chronic exposure to elevated levels of circulating cytokines, such as tumor necrosis factor-alpha (TNF-α) and interleukin-6 (IL-6), plays a pivotal role in the inflammatory milieu that characterizes cirrhosis [11]. Decompensated cirrhosis further potentiates the systemic inflammation. These cytokines contribute to myocardial remodeling and fibrosis, which further compromise the contractility and relaxation of the heart due to altered cardiac metabolism as a result of impaired energy production and utilization [11]. It has been proven that myocardial fibrosis and the increase in myocardial mass that occurs during cirrhosis causes a decrease in compliance of the myocardial wall with subsequent impairment of ventricular filling [12]. Additionally, due to venous congestion, third-spacing, and intestinal dysmotility, there is an increased intestinal permeability, with bacteria and endotoxins translocating into the systemic circulation. This subsequently increases vasodilation and thus worsens the burden on the heart and the course of the disease [13]. Therefore, these patients experience the cardinal symptoms of heart failure, predominantly in the form of lethargy, dyspnea, fluid retention, and a diminished exercise tolerance. In addition, these patients also have an increased susceptibility to infections.

3.3. Hormonal Alterations

As a consequence of cirrhosis, various hemodynamic abnormalities develop, such as hyperdynamic vascular insufficiency, reduced peripheral vascular resistance, reduced arterial pressure, increased cardiac output, hyporesponsiveness to vasopressors, increased levels of proinflammatory cytokines, and accordingly, studies have established that endocrine insufficiency is a common cause. Adrenal insufficiency (AI) is a common comorbidity in cirrhosis [14]. It can also play a significant role in the pathogenesis and progression of CCM. Research has proven that pituitary dysfunction can be a consequence of cirrhosis [15]. More recent research has shown that liver transplantation removes these abnormalities, which supports previous studies and attests to the liver's role in maintaining normal endocrine function. The prevalence of adrenal insufficiency in cirrhotic patients ranges from 30% to 60%, depending on the diagnostic criteria and severity of liver disease [14,15]. In this context, adrenal insufficiency is often relative, meaning that the adrenal glands fail to produce adequate levels of cortisol in response to the increased physiological demands imposed by cirrhosis [14]. This condition, known as relative adrenal insufficiency (RAI), has profound effects on cardiovascular function, including reduced myocardial contractility, hypotension, and increased susceptibility to shock.

Cortisol is the main glucocorticoid secreted by the adrenal cortex under the control of adrenocorticotropic hormone (ACTH) released from the pituitary gland. Among the factors of diurnal cortisol secretion, stress plays the greatest role [16]. During stress and serious illnesses, the activation of the hypothalamic–pituitary–adrenal axis by the action of cytokines and other factors leads to an increased secretion of corticotropin, which stimulates the production of ACTH and consequently increases the release of cortisol into the circulatory system [16]. Cortisol is a major component of cellular adaptation to stress, so an adequate level of cortisol is necessary to increase cardiac output and vascular tone and to reduce the release of proinflammatory cytokines [16].

The mechanism of adrenal insufficiency may include impaired function of total cholesterol, high-density lipoprotein cholesterol (HDL), low-density lipoprotein cholesterol (LDL), and increased levels of proinflammatory cytokines and circulating endotoxins [16]. Since the adrenal gland does not store cortisol, it must be synthesized from its precursor, cholesterol, under conditions of stress. In liver failure there is a low level of cholesterol for cortisol synthesis, thus favoring adrenal insufficiency under stress conditions [14].

Animal and human studies show that liver disease is associated with a decrease in cortisol levels and an increase in circulating glucocorticoids. In situations where the function of the pituitary gland is disturbed, and therefore the secretion of cortisol, a disturbed response of the cardiovascular system to stress results is observed [14,15]. Additionally, cortisol elimination is also impaired in cirrhosis.

4. Diagnosis

The clinical diagnosis of CCM remains challenging, primarily due to the lack of universally accepted diagnostic criteria and the subtle nature of its manifestations.

According to the Montreal Criteria, introduced in 2005 at the World Congress of Gastroenterology, CCM was characterized by the presence of systolic or diastolic dysfunction, along with additional indicators such as increased left ventricular mass, electrophysiological disturbances (most notably prolonged QT intervals), elevated natriuretic peptide values, and an abnormal response to stress [17].

Systolic dysfunction, in general, refers to the left ventricular relaxation impairment, described by decreased ejection fraction, usually as a result of hampered myocardial contractility [18,19]. Systolic function in patients with cirrhosis is normal or increased at rest, as a result of hyperdynamic circulation that masks the true state of the heart muscle and left ventricle. Damage to beta-adrenergic receptors and the presence of cardio-depressant substances are listed as some of the factors causing systolic dysfunction.

The characteristic of this clinical condition is impaired electrical conduction, which is reflected in electrocardiographic abnormalities, the most noticeable of which is a prolonged QT interval, which corresponds to ventricular depolarization and repolarization of the heart muscle. This finding is present in 30–50% of patients with cirrhosis, regardless of the underlying etiology and its prevalence parallels the severity of cirrhosis as assessed by the Child–Pugh scoring system [20]. Biomarkers such as B-type natriuretic peptide (BNP) and troponins have shown promise in identifying subclinical cardiac involvement; however, given their relatively low specificity due to the large number of confounding factors that can contribute to troponin or BNP elevation, their role in CCM diagnosis is still limited [21].

Considering that cardiac dysfunction may not be visible during the resting state, exercise or pharmacologic stress echocardiography is a critical tool in this regard, as it can reveal the inadequate increase in cardiac output that is characteristic of CCM [22]. Additionally, the role of advanced imaging modalities, such as cardiac magnetic resonance imaging (MRI), speckle-tracking echocardiography, and nuclear myocardial perfusion scanning, is increasingly recognized for their ability to detect early myocardial changes, including subtle fibrosis and impaired strain patterns [21].

In that sense, a revised criteria for CCM diagnosis were proposed in 2019. According to the 2019 Cirrhotic Cardiomyopathy Consortium Criteria, CCM is diagnosed when either systolic or diastolic dysfunction is detected in an echocardiography study at rest [23]. Systolic dysfunction is defined as a left ventricular ejection fraction (LVEF) < 50% or an absolute global longitudinal strain (GLS) value below 18%. The initial criteria included patients with GLS higher than 22%, but this was later withdrawn. Diastolic dysfunction criteria have been modified for patients with cirrhosis.

5. Treatment

When it comes to therapy, to date there is no strict treatment for CCM. When CCM progresses and produces symptoms of heart failure, patients are treated as those diagnosed with a form of heart failure (Figure 2). The heart, liver, and kidneys share an inextricable relationship with regards to the autoregulation of cardiac output. Dysfunction in any of these organs will propagate as a cycle. Considering this from the perspective of CCM, we can better understand the targeted therapies. Perturbation of the RAAS is a likely key driver in the etiology of CCM and probably underpins the morbidity and mortality benefits conferred by therapies such as angiotensin-converting enzyme inhibitors (ACE-I), angiotensin receptor blockers (ARBs), beta-blockers, and mineralocorticoid antagonists (MRAs) [24].

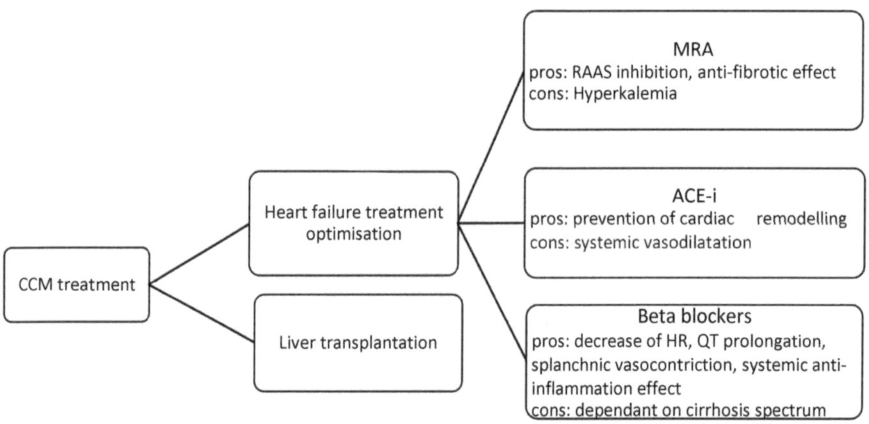

Figure 2. Current recommendations for CCM treatment. MRA—mineralocorticoid receptor antagonist; ACE-I—angiotensin-converting enzyme inhibitor; HR—heart rate.

5.1. Heart Failure Treatment Optimisation

Treatment choice relies heavily on the stage of heart failure and systolic dysfunction. MRAs are indicated in patients with severe heart failure and decrease associated morbidity and mortality [25]. This is particularly pertinent considering the other advantages that these medications confer in cirrhotic patients. Hyperdynamic changes, RAAS activation, and sodium and water retention are strongly mitigated by MRAs. Additionally, it is well established that angiotensin II exerts intrahepatic vasoconstrictive and profibrotic effects; thus, the antifibrotic effect of novel MRAs can further contribute to the prevention of cirrhosis progression and decompensation [26]. Our group has found the use of eplerenone helpful due to its higher avidity with fewer adverse effects when compared to other drugs in this class.

The implementation of ACE-I remains controversial. If initiated during the early stages of CCM, they can prevent and delay cardiac remodeling. However, due to its subclinical nature, the diagnosis of CCM is usually delayed and only identified towards the latter stages. This greatly limits the usefulness of ACE-I within the physician's armamentarium, as traditionally, ACE-Is are avoided in the decompensated cirrhotic patient due to the risk of hepatorenal syndrome and hypotension [26].

The beneficial role of beta-blockers in the treatment of CCM is multifactorial. Non-selective beta-blockers (NSBBs) reduce heart rate and cardiac output via β1-adrenergic receptor blockage, while via β2-adrenergic blockage they exert splanchnic vasoconstriction [26]. Further to this, NSBBs downregulate the RAAS by antagonizing beta-adrenoreceptors that are expressed at the juxtaglomerular apparatus. Additionally, due to the unopposed adrenergic tone, NSBBs cause a mild increase in peripheral resistance [26,27]. NSBBs have been shown to reduce the prolonged QT interval and the hyperdynamic load in patients with cirrhosis, but whether the correction of the QT interval has a positive effect on prognosis is doubtful [28]. Current guidelines indicate the use of NSBBs in primary and secondary prevention of variceal bleeding, considering their effect is predominantly exerted through the decrease in portal hypertension [27]. Our group feels that in certain settings, when QT interval responds to beta-blockade, the use of selective or semi-selective beta-blockers (e.g., carvedilol) may be of benefit. Carvedilol acts by blocking α1 adrenergic receptors and lowers intrahepatic vascular resistance. An additional beneficial effect of beta-blockers is mirrored in their ability to, via β2 receptor blockage, stimulate intestinal motility, improve gut dysbiosis, and prevent systemic bacteriemia and endotoxemia, thus ultimately exerting a systemic anti-inflammatory effect [27]. Additionally, effective beta-blockade requires appropriate dosing because of receptor upregulation and adaptation, and it is crucial to emphasize the importance of personalized dosing strategies. The "windows" of the longstanding, somewhat controversial "window theory" regarding the timing of beta-blocker implementation have now been opened to patients with compensated cirrhosis and those with small varices and uncomplicated ascites. This "expansion" allows for the broader beta-blocker implementation in preventing and treating portal hypertension-related complications, including CCM [27].

The implementation of glucocorticoid therapy in treating CCM remains controversial since large studies evaluating its effect on CCM are lacking. A limited number of small-scale studies have mostly included patients with decompensated cirrhosis in whom CCM has not been previously diagnosed. Empirical glucocorticoid therapy in asymptomatic patients is not recommended. However, temporary glucocorticoid therapy in the state of RAI has been shown to reduce hypotension and improve vital signs, ultimately leading to a better prognosis [29]. We feel that the use of glucocorticoids is particularly helpful when liver deterioration is precipitated by infection and would suggest considering using them when heart failure in cirrhosis is coupled with sepsis.

5.2. Liver Transplantation

Liver transplantation remains the cornerstone treatment for cirrhotic patients. It has been shown that successful liver transplantation improves all complications and organ-related hemodynamic dysfunctions, including CCM [6]. Normalization of cardiac structure and function, as well as normalization of QT prolongation, was observed as early as one year following transplantation. However, cardiac dysfunction remains a major risk factor for perioperative management as well as post-transplantation clinical course. In that sense, preoperative clinical assessment, monitoring of the cardiovascular system during and after the operation, and proper postoperative management are mandatory in order to improve post-transplantation outcome [6].

6. Current Controversies and Future Perspective

Considering the multifaceted mechanisms of cardiac dysfunction in CCM and other non-cirrhotic heart disease, therapeutic possibilities are vastly limited and may not be applicable in CCM. In that sense, understanding the pathophysiological mechanism behind CCM may lead to more targeted therapies, such as antioxidants, anti-inflammatory, and anti-apoptotic agents [30,31]. We reviewed the literature for potential anti-renin therapies, specifically aliskiren, but found no positive results. Currently, research in this field is mostly limited to animal studies, rendering the therapeutic progress in this area rather slow [32,33]. The role of statins in treating CCM remains controversial, considering that, due to its subclinical nature, CCM usually becomes evident in the state of cirrhotic decompensation. However, it has been shown that atorvastatin exhibits not only lipid-lowering capabilities but also has an effect on the reduction in the inflammatory cytokines in plasma, such as TNF-α and IL-6 [34]. It also decreases the cardiac dysfunction marker N-terminal pro-brain natriuretic peptide (BNP) concentration, and it seemingly decreased cardiac chronotropic hyperresponsiveness in animal models [31,35]. Given its acceptable safety profile, it may be a good option in patients with cirrhosis except those with severely decompensated liver function. We feel statins are an attractive approach and have used them as a continued therapy for cirrhotic patients at high cardiovascular risk. Atorvastatin, being one of the most potent statins, is a primary choice for us; however, we also consider alternatives such as lovastatin, pravastatin, and rosuvastatin. In our experience, low-dose long-term statin therapy may stop the deleterious mechanisms in CCM formation. However, well-designed clinical studies are needed to fully clarify the potential benefits of the pleiotropic effects of statins.

7. Conclusions

In conclusion, the diagnosis of CCM, though challenging, is of paramount importance in the management of cirrhotic patients. Advances in diagnostic modalities and a deeper understanding of their clinical implications are essential steps toward improving outcomes in this population. As research progresses, CCM may transition from being a largely unrecognized complication to a central focus in the care of patients with advanced liver disease.

Author Contributions: Conceptualization, I.R., I.B., E.K. and V.M.; writing—original draft preparation, I.R., V.M., I.B., J.M.N., J.B., M.S., J.T., N.P., M.B. (Mihailo Bezmarevic), J.J., D.M., N.D., V.D., F.P., E.C.R., S.K., T.G., N.S., N.J., M.B. (Mileva Bascarevic), A.B., L.E., A.O., K.S. and E.K.; writing—review and editing, I.R., V.M., L.E., K.S. and I.B.; resources, I.R., A.O., K.S. and V.M.; visualization, I.B. and E.K.; supervision, I.R., I.B., M.S., N.P., M.B. (Mihailo Bezmarevic), L.E., A.O., K.S. and V.M. All authors have read and agreed to the published version of the manuscript.

Funding: This research received no external funding.

Acknowledgments: The authors wish to thank the Gastroenterology and Liver unit i.e., Hepatology department of the Royal Cornwall Hospitals (RCHT) NHS trust and University of Exeter—NHS England Education SouthWest for fostering research opportunities. The authors are especially grateful to team members of the Liver Unit of the RCHT: Hyder Hussaini, Liz Farrington, William David Stableforth, Mary C McKenna, Mark Holmes, Kim Hicks, Toni Harvey and all other staff members for their diligence and support.

Conflicts of Interest: The authors declare no conflicts of interest.

References

1. Sharma, A.; Nagalli, S. Chronic Liver Disease. [Updated 2023 July 3]. In *StatPearls [Internet]*; StatPearls Publishing: Treasure Island, FL, USA, 2024. Available online: https://www.ncbi.nlm.nih.gov/books/NBK554597/ (accessed on 24 December 2024).
2. Liu, H.; Naser, J.A.; Lin, G.; Lee, S.S. Cardiomyopathy in cirrhosis: From pathophysiology to clinical care. *JHEP Rep.* **2023**, *6*, 100911. [CrossRef] [PubMed]
3. Moller, S.; Dumcke, C.W.; Krag, A. The heart and the liver. *Expert. Rev. Gastroenterol. Hepatol.* **2009**, *3*, 51–64. [CrossRef]
4. Wiese, S.; Hove, J.D.; Bendtsen, F.; Møller, S. Cirrhotic cardiomyopathy: Pathogenesis and clinical relevance. *Nat. Rev. Gastroenterol. Hepatol.* **2014**, *11*, 177–186. [CrossRef] [PubMed]
5. Izzy, M.; VanWagner, L.B.; Lin, G.; Altieri, M.; Findlay, J.Y.; Oh, J.K.; Watt, K.D.; Lee, S.S. Cirrhotic Cardiomyopathy Consortium. Redefining Cirrhotic Cardiomyopathy for the Modern Era. *Hepatology* **2020**, *71*, 334–345. [CrossRef]
6. Liu, H.; Jayakumar, S.; Traboulsi, M.; Lee, S.S. Cirrhotic cardiomyopathy: Implications for liver transplantation. *Liver Transpl.* **2017**, *23*, 826–835. [CrossRef]
7. Yumusak, O.; Doulberis, M. Update on cirrhotic cardiomyopathy: From etiopathogenesis to treatment. *Ann. Gastroenterol.* **2024**, *37*, 381–391. [CrossRef]
8. Liu, H.; Nguyen, H.H.; Yoon, K.T.; Lee, S.S. Pathogenic Mechanisms Underlying Cirrhotic Cardiomyopathy. *Front. Netw. Physiol.* **2022**, *2*, 849253. [CrossRef]
9. McGrath, M.S.; Wentworth, B.J. The Renin–Angiotensin System in Liver Disease. *Int. J. Mol. Sci.* **2024**, *25*, 5807. [CrossRef]
10. Carvalho, M.V.H.; Kroll, P.C.; Kroll, R.T.M.; Carvalho, V.N. Cirrhotic cardiomyopathy: The liver affects the heart. *Braz. J. Med. Biol. Res.* **2019**, *52*, e7809. [CrossRef] [PubMed]
11. Dirchwolf, M.; Ruf, A.E. Role of systemic inflammation in cirrhosis: From pathogenesis to prognosis. *World J. Hepatol.* **2015**, *7*, 1974–1981. [CrossRef] [PubMed]
12. Isaak, A.; Praktiknjo, M.; Jansen, C.; Faron, A.; Sprinkart, A.M.; Pieper, C.C.; Chang, J.; Luetkens, J.A. Myocardial Fibrosis and Inflammation in Liver Cirrhosis: MRI Study of the Liver-Heart Axis. *Radiology* **2020**, *297*, 51–61. [CrossRef] [PubMed]
13. Fukui, H. Leaky Gut and Gut-Liver Axis in Liver Cirrhosis: Clinical Studies Update. *Gut Liver.* **2021**, *15*, 666–676. [CrossRef] [PubMed]
14. Wentworth, B.J.; Siragy, H.M. Adrenal Insufficiency in Cirrhosis. *J. Endocr. Soc.* **2022**, *6*, bvac115. [CrossRef] [PubMed]
15. Hartl, L.; Simbrunner, B.; Jachs, M.; Wolf, P.; Bauer, D.J.M.; Scheiner, B.; Balcar, L.; Reiberger, T. An impaired pituitary-adrenal signalling axis in stable cirrhosis is linked to worse prognosis. *JHEP Rep.* **2023**, *5*, 100789. [CrossRef] [PubMed]
16. Lightman, S.L.; Birnie, M.T.; Conway-Campbell, B.L. Dynamics of ACTH and Cortisol Secretion and Implications for Disease. *Endocr. Rev.* **2020**, *41*, bnaa002. [CrossRef] [PubMed]
17. Farr, M.; Schulze, P.C. Recent advances in the diagnosis and management of cirrhosis-associated cardiomyopathy in liver transplant candidates: Advanced echo imaging, cardiac biomarkers, and advanced heart failure therapies. *Clin. Med. Insights Cardiol.* **2015**, *8* (Suppl. S1), 67–74. [CrossRef]
18. Dourakis, S.P.; Geladari, E.; Geladari, C.; Vallianou, N. Cirrhotic Cardiomyopathy: The Interplay Between Liver and Cardiac Muscle. How Does the Cardiovascular System React When the Liver is Diseased? *Curr. Cardiol. Rev.* **2021**, *17*, 78–84. [CrossRef]
19. Ruiz-del-Árbol, L.; Serradilla, R. Cirrhotic cardiomyopathy. *World J. Gastroenterol.* **2015**, *21*, 11502–11521. [CrossRef]
20. Bernardi, M.; Maggioli, C.; Dibra, V.; Zaccherini, G. QT interval prolongation in liver cirrhosis: Innocent bystander or serious threat? *Expert. Rev. Gastroenterol. Hepatol.* **2012**, *6*, 57–66. [CrossRef]
21. Šimić, S.; Svaguša, T.; Grgurević, I.; Mustapić, S.; Žarak, M.; Prkačin, I. Markers of cardiac injury in patients with liver cirrhosis. *Croat. Med. J.* **2023**, *64*, 362–373. [CrossRef] [PubMed]
22. Dimitroglou, Y.; Aggeli, C.; Alexopoulou, A.; Tsartsalis, D.; Patsourakos, D.; Koukos, M.; Tousoulis, D.; Tsioufis, K. The Contemporary Role of Speckle Tracking Echocardiography in Cirrhotic Cardiomyopathy. *Life* **2024**, *14*, 179. [CrossRef] [PubMed]
23. Luo, Y.; Yin, S.; Chen, Q.; Liu, J.; Chong, Y.; Zhong, J. Comparison of the 2005 Montreal Criteria and the 2019 Cirrhotic Cardiomyopathy Consortium Criteria for the Diagnosis of Cirrhotic Cardiomyopathy. *Am. J. Cardiol.* **2023**, *208*, 180–189. [CrossRef]

24. Bodys-Pełka, A.; Kusztal, M.; Raszeja-Wyszomirska, J.; Główczyńska, R.; Grabowski, M. What's New in Cirrhotic Cardiomyopathy?—Review Article. *J. Pers. Med.* **2021**, *11*, 1285. [CrossRef]
25. Pozzi, M.; Grassi, G.; Ratti, L.; Favini, G.; Dell'Oro, R.; Redaelli, E.; Calchera, I.; Mancia, G. Cardiac, neuroadrenergic, and portal hemodynamic effects of prolonged aldosterone blockade in postviral child A cirrhosis. *Am. J. Gastroenterol.* **2005**, *100*, 1110–1116. [CrossRef] [PubMed]
26. Felli, E.; Nulan, Y.; Selicean, S.; Wang, C.; Gracia-Sancho, J.; Bosch, J. Emerging Therapeutic Targets for Portal Hypertension. *Curr. Hepatol. Rep.* **2023**, *22*, 51–66. [CrossRef]
27. Albillos, A.; Krag, A. Beta-blockers in the era of precision medicine in patients with cirrhosis. *J. Hepatol.* **2023**, *78*, 866–872. [CrossRef]
28. Zambruni, A.; Trevisani, F.; Di Micoli, A.; Savelli, F.; Berzigotti, A.; Bracci, E.; Caraceni, P.; Bernardi, M. Effect of chronic beta-blockade on QT interval in patients with liver cirrhosis. *J. Hepatol.* **2008**, *48*, 415–421. [CrossRef] [PubMed]
29. Wu, C.H.; Guo, L.; Hao, D.; Wang, Q.; Ye, X.; Ito, M.; Huang, B.; Li, X.A. Relative adrenal insufficiency is a risk factor and endotype of sepsis—A proof-of-concept study to support a precision medicine approach to guide glucocorticoid therapy for sepsis. *Front. Immunol.* **2023**, *13*, 1110516. [CrossRef] [PubMed]
30. Liu, H.; Nguyen, H.H.; Hwang, S.Y.; Lee, S.S. Oxidative Mechanisms and Cardiovascular Abnormalities of Cirrhosis and Portal Hypertension. *Int. J. Mol. Sci.* **2023**, *24*, 16805. [CrossRef]
31. Liu, H.; Ryu, D.; Hwang, S.; Lee, S.S. Therapies for Cirrhotic Cardiomyopathy: Current Perspectives and Future Possibilities. *Int. J. Mol. Sci.* **2024**, *25*, 5849. [CrossRef]
32. Nam, S.W.; Liu, H.; Wong, J.Z.; Feng, A.Y.; Chu, G.; Merchant, N.; Lee, S.S. Cardiomyocyte apoptosis contributes to pathogenesis of cirrhotic cardiomyopathy in bile duct-ligated mice. *Clin. Sci.* **2014**, *127*, 519–526. [CrossRef]
33. Yang, Y.Y.; Liu, H.; Nam, S.W.; Kunos, G.; Lee, S.S. Mechanisms of TNFalpha-induced cardiac dysfunction in cholestatic bile duct-ligated mice: Interaction between TNFalpha and endocannabinoids. *J. Hepatol.* **2010**, *53*, 298–306. [CrossRef] [PubMed]
34. Bielecka-Dabrowa, A.; Mikhailidis, D.P.; Rizzo, M.; von Haehling, S.; Rysz, J.; Banach, M. The influence of atorvastatin on parameters of inflammation left ventricular function, hospitalizations and mortality in patients with dilated cardiomyopathy—5-year follow-up. *Lipids Health Dis.* **2013**, *12*, 47. [CrossRef] [PubMed]
35. Liao, J.K. Statin therapy for cardiac hypertrophy and heart failure. *J. Investig. Med.* **2004**, *52*, 248–253. [CrossRef] [PubMed]

medicina MDPI

Brief Report

Ultrasound Prevalence and Clinical Features of Nonalcoholic Fatty Liver Disease in Patients with Inflammatory Bowel Diseases: A Real-Life Cross-Sectional Study

Ludovico Abenavoli [1,*], Rocco Spagnuolo [1], Giuseppe Guido Maria Scarlata [1], Emidio Scarpellini [2], Luigi Boccuto [3] and Francesco Luzza [1]

1 Department of Health Sciences, University "Magna Graecia", Viale Europa, 88100 Catanzaro, Italy; spagnuolo@unicz.it (R.S.); giuseppeguidomaria.scarlata@unicz.it (G.G.M.S.); luzza@unicz.it (F.L.)
2 Translationeel Onderzoek van Gastroenterologische Aandoeningen (T.A.R.G.I.D.), Gasthuisberg University Hospital, KU Leuven, Herestraat 49, 3000 Leuven, Belgium; emidio.scarpellini@med.kuleuven.be
3 Healthcare Genetics and Genomics Doctoral Program, School of Nursing, College of Behavioral, Social and Health Sciences, Clemson University, Clemson, SC 29631, USA; lboccut@clemson.edu
* Correspondence: l.abenavoli@unicz.it; Tel.: +39-0961-3694387

Abstract: *Background and Objectives*: Inflammatory bowel disease (IBD) is a condition characterized by chronic intestinal inflammation. We can identify two major forms: Crohn's disease (CD) and ulcerative colitis (UC). One of the extraintestinal manifestations of IBD is nonalcoholic fatty liver disease (NAFLD). IBD and NAFLD share common pathogenetic mechanisms. Ultrasound (US) examination is the most commonly used imaging method for the diagnosis of NAFLD. This cross-sectional observational retrospective study aimed to evaluate the US prevalence of NAFLD in IBD patients and their clinical features. *Materials and Methods*: A total of 143 patients with IBD underwent hepatic US and were divided into two different groups according to the presence or absence of NAFLD. Subsequently, new exclusion criteria for dysmetabolic comorbidities (defined as plus) were applied. *Results*: The US prevalence of NAFLD was 23% (21% in CD and 24% in UC, respectively). Most IBD–NAFLD patients were male and older and showed significantly higher values for body mass index, waist circumference, disease duration, and age at onset than those without NAFLD. IBD–NAFLD patients showed a significantly higher percentage of stenosing phenotype and left-side colitis. Regarding metabolic features, IBD–NAFLD patients showed a significantly higher percentage of hypertension and IBD plus dysmetabolic criteria. Also, higher values of alanine aminotransferase and triglycerides and lower levels of high-density lipoproteins are reported in these patients. *Conclusions*: We suggest performing liver US screening in subjects affected by IBD to detect NAFLD earlier. Also, patients with NAFLD present several metabolic comorbidities that would fall within the new definition of metabolic-associated fatty liver disease. Finally, we encourage larger longitudinal studies, including healthy controls, to provide further confirmation of our preliminary data.

Keywords: Crohn's disease; ulcerative colitis; liver steatosis; hepatic ultrasound; metabolism

1. Introduction

1.1. Crosstalk between IBD and NAFLD

Inflammatory bowel disease (IBD) is an inflammatory condition encompassing two major forms: Crohn's disease (CD) and ulcerative colitis (UC). They are characterized by an unregulated and abnormal immune response induced by environmental stimuli in genetically predisposed subjects [1]. In about 5–50% of patients with IBD, there are several extraintestinal manifestations such as musculoskeletal, ocular, cutaneous, and hepatobiliary. Hepatobiliary manifestations include primary sclerosing cholangitis, autoimmune/granulomatous hepatitis, and in particular, nonalcoholic fatty liver disease (NAFLD) [2,3]. NAFLD is currently the main cause of chronic liver disease in the general population worldwide and ranges from simple fatty liver to steatohepatitis to advanced fibrosis and finally cirrhosis [4,5]. It can be considered a manifestation of metabolic syndrome often associated with obesity, insulin resistance, dyslipidemia, and hypertension [6–8]. The prevalence of NAFLD in IBD patients is broadly variable due to different diagnostic methodologies and ranges from 20–30% of patients identified using hepatic ultrasound (US) to 24% of individuals diagnosed by magnetic resonance enterography to 71% of cases with transient elastography [9–11]. The overall prevalence is approximately 32% and, thus, considerably higher than the general population rate (25.2%) [12]. Despite a large number of studies, the pathogenetic mechanisms related to the onset of steatosis and the development of liver damage in patients with IBD are not entirely understood. Also, other risk factors can be involved in this association, such as chronic inflammation, drug-induced liver injury, prolonged steroid exposure, malnutrition, and gut dysbiosis [13,14]. In the genetic field, a previous study has shown how patients with IBD carrying the p.I148M missense variant in the patatin-like phospholipase domain-containing protein 3 (PNPLA3) gene, an important common genetic determinant of liver fat content and progression to chronic liver disease, have higher susceptibility to hepatic steatosis and liver damage [15]. A more recent cross-sectional study by Rodriguez-Duque et al. on 838 IBD patients compared with 1718 controls showed that these patients are at higher risk of developing fatty liver, not only for their weight or the presence of hypertension, diabetes, or high cholesterol but also for variables related to intestinal disease, such as IBD duration, activity, and prior surgery, that can be considered major predictors of incident NAFLD [16–19].

1.2. Diagnostic Approaches in NAFLD

Liver involvement associated with NAFLD in IBD patients complicates therapeutic management and increases the risk of hospitalization and mortality [20]. Thus, it is essential to adopt an appropriate diagnostic approach aimed at identifying and staging early NAFLD in IBD patients. Specifically, Hamaguchi's score operates with an abdominal US scoring system to provide accurate indications of hepatic steatosis, visceral obesity, and metabolic syndrome [21]. The diagnosis of NAFLD requires hepatic fat assessment by imaging techniques or histology, excluding other causes of secondary fat accumulation (e.g., use of alcohol or steatogenic drugs) [22]. The gold standard in the diagnosis of NAFLD is liver biopsy, but it is an invasive and not very reproducible as well as expensive technique [23]. At the same time, the use of transient elastography makes it possible to determine liver stiffness and quantify steatosis using controlled attenuation parameters with high accuracy, but it is not accessible worldwide. Furthermore, it requires technical expertise and is unreliable in patients with severe obesity and ascites [24,25]. Therefore, US examination is the most common item performed in clinical practice for the diagnosis of NAFLD [26]. However, although US is an easily reproducible and inexpensive technique, it has high interindividual variability [27]. Currently, US data on NAFLD in IBD patients are quite heterogeneous. A recent meta-analysis showed a prevalence with different imaging techniques of 30% and that the risk of NAFLD was two times higher in IBD patients versus healthy subjects [28]. Another study showed the US prevalence of NAFLD in IBD patients treated with biological therapy at 54% [29]. Similar results were obtained by Shintaku et al., with a US prevalence of NAFLD of 45% among 71 enrolled IBD patients [30]. In addition, due to the newly proposed nomenclature of metabolic-associated fatty liver disease (MAFLD), there is a need for continuous evaluation of the clinical features of these patients, especially from a metabolic perspective [7].

1.3. Aims

This cross-sectional study aimed to evaluate the US prevalence of NAFLD in patients with IBD and to evaluate their clinical features.

2. Materials and Methods

2.1. Patients

We retrospectively enrolled 143 patients with clinical, endoscopic, and radiological diagnoses of IBD [1]. According to specific inclusion criteria: (i) patients of age $\geq$18, (ii) patients subjected to hepatic US at hospital admission; and exclusion criteria: (i) patients with a history of alcohol or drug abuse, (ii) patients with previous or current viral hepatitis infection, (iii) patients with autoimmune liver disease, (iv) cirrhotic patients, (v) patients with malignancies, (vi) pregnant and/or lactating women. From each patient were collected (i) demographic and anthropometric data, (ii) disease characteristics, (iii) disease location and phenotype, (iv) dysmetabolic comorbidities, (v) laboratory parameters, and (vi) medications.

2.2. NAFLD Diagnosis

All patients underwent liver evaluation by US, according to a previous study by Mancina et al. [15]. Briefly, abdominal US was performed by the same experienced operator with a grayscale scanner device (LOGIQ S8 XDclear 2.0+, GE HealthCare, Milan, Italy) using a 3.5-MHz convex transducer with B-mode image evaluation. Individuals were fasting at least 4 h before the procedure. Before the procedure, the subjects followed a fiber-free diet and took 80 mg of simethicone thrice daily for 3 days. Hepatic steatosis was graded as mild (steatosis grade 1 or S1), moderate (steatosis grade 2 or S2), or severe (steatosis grade 3 or S3). Mild liver steatosis (S1) features were defined as a slight increase in liver echogenicity with a slight exaggeration of liver and kidney echo discrepancy. Moderate liver steatosis (S2) features were defined as an increase in liver echogenicity and loss of echoes from the wall of the portal vein with a greater posterior beam attenuation and greater discrepancy between hepatic and renal echoes. The features of severe liver steatosis (S3) were defined as a greater reduction in beam penetration, loss of echoes from most of the portal vein wall, and an even larger discrepancy between hepatic and renal echoes. Hepatic steatosis was defined as a steatosis grade of $\geq$S1 [27,31,32].

Anamnestic, laboratory, and endoscopic data were also collected. If laboratory and endoscopic data were not available, data resulting from investigations carried out on another date, ranging from 15 days before or after the date of the hepatic US, were used.

2.3. Study Design

Patients were stratified according to the presence or absence of hepatic fat accumulation at US examination. We chose to adopt additional exclusion criteria (self-defined plus) to identify which patients had liver steatosis not attributable to dysmetabolic comorbidities: obesity, high waist circumference, insulin resistance, type 2 diabetes mellitus (T2DM), hypertension, and dyslipidemia. Application of these criteria allows the evaluation of hepatic steatosis independently from factors attributable to metabolic syndrome. This approach was similarly applied in a previous study by Angelico et al. [33].

2.4. Statistical Analysis

We reported quantitative variables as mean $\pm$ standard deviation (SD) and nominal variables as percentages and absolute numbers. Comparisons of continuous variables were performed by the Student's *t*-test or the Mann–Whitney U test, considering each quantitative trait after testing it for normality using the Shapiro–Wilk test. Differences between categorical variables were assessed by the chi-square (χ^2) test. A p-value < 0.05 was considered statistically significant. The data were analyzed using SPSS 26.0 software (IBM Corp., Armonk, NY, USA).

2.5. Ethics

This study was approved by the local ethics committee of Magna Graecia University (protocol number 2014/49). This study was conducted in compliance with the principles outlined in the Declaration of Helsinki. Informed written consent was obtained from each participating patient.

3. Results

3.1. Characteristics of the Patients Enrolled

All 143 IBD patients were subjected to hepatic US to assess the presence of steatosis. Among them, 33 patients (11 with CD and 22 with UC, respectively) showed hepatic steatosis, while 110 patients (41 with CD and 69 with UC, respectively) did not show hepatic steatosis. Subsequently, self-defined plus exclusion criteria were applied, obtaining 81 IBD patients, divided into 35 patients with CD and 46 with UC (Figure 1).

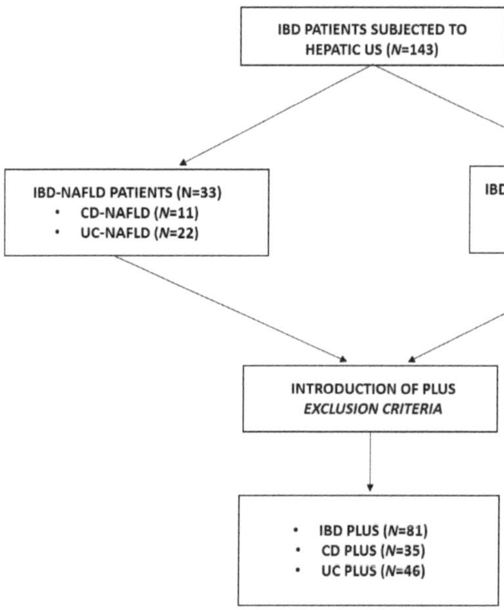

Figure 1. Workflow of study design. Patients were enrolled in the study and divided into different groups.

3.2. Comparison between IBD Patients

The main clinical and laboratory features of the subjects enrolled in our study are summarized in Table 1. Most IBD patients under investigation were males (n = 82, 57%), with a mean age of 45 ± 16 years and a body mass index (BMI) and waist circumference of 25 ± 4 kg/m^2 and 91 ± 12 cm, respectively. Most UC patients showed mild or severe liver steatosis (n = 18, 20% and n = 2, 2%, respectively). On the other hand, CD showed a higher percentage of moderate liver steatosis (n = 4, 8%). Ninety-one (63%) patients had UC, and most of them (n = 48, 54%) extended to the entire colon. Fifty-two (37%) had CD, and most of them with an ileal and ileocolonic extension: 41% and 42%, respectively. About 40% of CD patients (n = 21) had a stenosing disease phenotype. Twenty-four (17%) had previously undergone surgery. UC patients showed more dysmetabolic comorbidities than CD patients but similar levels on laboratory parameters. IBD patients were treated with salicylate (n = 75, 52%), azathioprine (n = 47, 33%), and biological therapy (n = 86, 60%).

3.3. US Prevalence of NAFLD among IBD Patients

The US prevalence of NAFLD was 23%, considering the unified sample (IBD): 21% and 24% in CD and UC, respectively (Figure 2).

Table 1. Clinical and laboratory features of patients.

	IBD (N = 143)	CD (N = 52)	UC (N = 91)
Demographic and Anthropometric			
Age (years)	45 ± 16	44 ± 17	45 ± 15
Male gender, n (%)	82 (57)	31 (60)	51 (56)
Active smoker, n (%)	4 (3)	3 (6)	1 (1)
BMI (kg/m^2)	25 ± 4	24 ± 4	25 ± 5
Waist circumference (cm)	91 ± 12	89 ± 11	91 ± 13
Disease characteristic			
Disease duration (years)	12 ± 9	13 ± 9	11 ± 10
Age at onset (years)	33 ± 15	34 ± 13	32 ± 14
CD (Harvey–Bradshaw index)	-	7 ± 3	-
UC (full Mayo Score)	-	-	2 ± 0.7
Relapse/year	1.3 ± 0.7	1.3 ± 0.9	1.2 ± 0.7
Active disease, n (%)	47 (33)	27 (52)	20 (22)
Extraintestinal manifestations, n (%)	26 (18)	13 (25)	13 (14)
NAFLD, n (%)	33 (23)	11 (21)	22 (24)
Mild steatosis, n (%)	24 (17)	6 (11)	18 (20)
Moderate steatosis, n (%)	6 (4)	4 (8)	2 (2)
Severe steatosis, n (%)	3 (2)	1 (2)	2 (2)
Surgery, n (%)	24 (17)	17 (33)	7 (8)
CD disease location and phenotype, n (%)			
Ileal	-	21 (41)	-
Colonic	-	8 (15)	-
Ileo–colonic	-	22 (42)	-
Upper GI	-	1 (2)	-
Inflammatory	-	16 (31)	-
Fistulizing	-	15 (29)	-
Stenosing	-	21 (40)	-
UC disease location, n (%)			
Proctitis	-	-	8 (8)
Proctosigmoiditis	-	-	19 (21)
Left side	-	-	16 (17)
Pancolitis	-	-	48 (54)
Dysmetabolic comorbidities, n (%)			
T2DM	11 (8)	1 (2)	10 (11)
Hypertension	24 (17)	8 (15)	16 (18)
Dyslipidemia	18 (13)	5 (10)	13 (14)
IBD plus dysmetabolic criteria	81 (57)	35 (67)	46 (50)
Laboratory parameters			
ALT (UI/L)	19 ± 10	18 ± 9	20 ± 11
AST (UI/L)	20 ± 9	22 ± 9	21 ± 10
Total cholesterol (mg/dL)	168 ± 42	167 ± 41	168 ± 43
LDL (mg/dL)	104 ± 35	103 ± 34	105 ± 35
HDL (mg/dL)	56 ± 17	55 ± 17	56 ± 16
Triglycerides (mg/dL)	99 ± 48	99 ± 49	100 ± 48
Fasting blood glucose (mg/dL)	88 ± 20	87 ± 19	89 ± 21
Fasting insulinemia (mg/dL)	10 ± 7	10 ± 8	9 ± 8
HOMA-IR	2 ± 2	2 ± 1	2 ± 3
Fecal calprotectin (mcg/gr)	501 ± 797	492 ± 802	509 ± 804

Table 1. *Cont.*

	IBD (*N* = 143)	CD (*N* = 52)	UC (*N* = 91)
Medication, *n* (%)			
Salicylates, *n* (%)	75 (52)	24 (46)	51 (56)
Azathioprine, *n* (%)	47 (33)	16 (31)	31 (34)
>3 cycles of steroids, *n* (%)	34 (24)	9 (17)	25 (27)
Biological therapy, *n* (%)	86 (60)	33 (63)	53 (58)
Anti-TNF-α, *n* (%)	61 (71)	25 (76)	36 (68)
Vedolizumab, *n* (%)	16 (19)	2 (6)	14 (26)
Ustekinumab, *n* (%)	9 (10)	6 (18)	3 (6)
>1 Biological drug, *n* (%)	23 (16)	4 (8)	19 (21)
Current biological therapy duration (years)	3 ± 2	4 ± 2	2 ± 2
Total biological therapy duration (years)	5 ± 4	4 ± 3	4 ± 2

Legend: IBD, inflammatory bowel disease; CD, Crohn's disease; UC, ulcerative Colitis; BMI, body mass index; NAFLD, nonalcoholic fatty liver disease; T2DM, type 2 diabetes mellitus; ALT, alanine aminotransferase; AST, aspartate aminotransferase; LDL, low-density lipoprotein; HDL, high-density lipoprotein; HOMA-IR, homeostasis model assessment of insulin resistance; TNF-α, tumor necrosis factor-alfa.

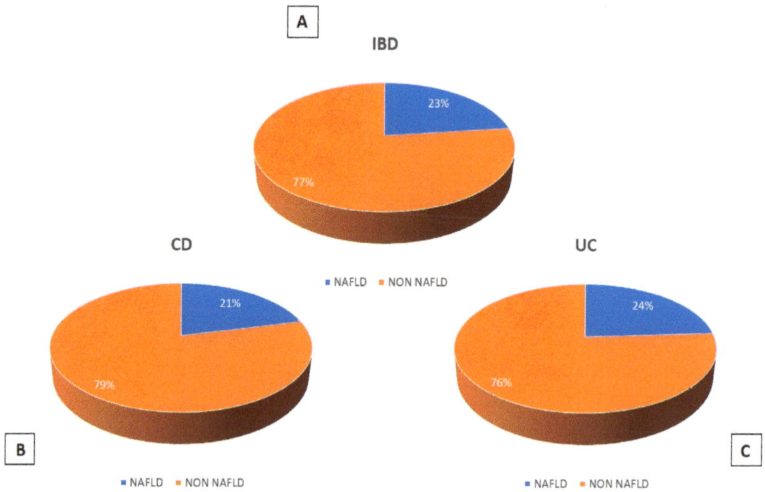

Figure 2. (**A**) US prevalence of NAFLD in IBD patients; (**B**) US prevalence of NAFLD in patients with CD; (**C**) US prevalence of NAFLD in patients with UC.

3.4. Comparison between IBD Patients with and without NAFLD

As shown in Table 2, subjects with IBD were stratified according to the ultrasonographic of NAFLD.

Most IBD–NAFLD patients were males (*n* = 24 (73%) vs. *n* = 58 (53%), *p* = 0.047) and showed significantly higher values than IBD non-NAFLD patients for age (53 ± 13 vs. 43 ± 17 years, *p* = 0.03), BMI (27 ± 5 vs. 24 ± 4 kg/m^2, *p* < 0.001), and waist circumference (100 ± 11 vs. 88 ± 11 cm, *p* < 0.001). None of the IBD–NAFLD patients was an active smoker (*n* = 0 vs. *n* = 4 (4%), *p* = 0.266). Furthermore, a significantly higher percentage of IBD–NAFLD patients reported hypertension (*n* = 13 (39%) vs. *n* = 11 (10%), *p* < 0.001) and IBD plus dysmetabolic criteria (*n* = 26 (78%) vs. *n* = 36 (33%), *p* < 0.001). No significant differences between the two groups for type 2 diabetes mellitus (T2DM; *n* = 4 (12%) vs. *n* = 7 (7%), *p* = 0.278) and dyslipidemia (*n* = 5 (15%) vs. *n* = 13 (12%), *p* = 0.564) were found. Regarding laboratory parameters, IBD–NAFLD patients showed significantly higher values of alanine aminotransferase (ALT; 22 ± 10 vs. 18 ± 9 UI/L, *p* = 0.034) and triglycerides (123 ± 63 vs. 93 ± 40 mg/dL, *p* = 0.002) but significantly lower values of high-density lipoproteins (HDL; 48 ± 16 vs. 58 ± 17 mg/dL, p = 0.005). No significant differences were found between the two groups for the other laboratory parameters. IBD–NAFLD patients showed a higher disease duration (15 ± 10 vs. 11 ± 9 years, p = 0.044) and age at onset (38 ± 16 vs. 32 ± 15 years, p = 0.047) than IBD non-NAFLD patients. Most IBD–NAFLD patients had UC (n = 22 (66%) vs. n = 69 (63%), p = 0.830), while in the other group, there was a higher percentage of CD patients (n = 41 (37%) vs. n = 11 (33%), p = 0.837). Among IBD patients, there was a significant difference between groups for the stenosing phenotype (n = 7 (64%) vs. n = 12 (29%), p = 0.035) and left-side colitis (n = 7 (32%) vs. n = 9 (13%), p = 0.044). There was no significant difference between groups for the other disease locations and phenotypes. Otherwise, IBD non-NAFLD patients showed a higher Harvey–Bradshaw Index (7 ± 3 vs. 5 ± 2, p = 0.033) than IBD patients with NAFLD. In addition, none of IBD patients treated with vedolizumab showed NAFLD (n = 0 vs. n = 16 (23%), p = 0.023), while most IBD–NAFLD patients were treated with antitumor necrosis factor-alfa (TNF-α; n = 15 (88%) vs. n = 46 (67%), p = 0.841) and ustekinumab (n = 2 (12%) vs. n = 7 (10%), p = 1.000) but with no statistically significant difference between the two groups. Finally, regarding the other medications, most IBD–NAFLD patients were treated with salicylate (n = 20 (61%) vs. n = 55 (50%), p = 0.324), azathioprine (n = 10 (30%) vs. n = 37 (33%), p = 0.834), or were undergoing surgery (n = 7 (21%) vs. n = 17 (15%), p = 0.435), with no significant differences between groups.

Table 2. Comparison between IBD patients stratified for NAFLD and non-NAFLD.

	IBD–NAFLD (*N* = 33)	**IBD Non-NAFLD (*N* = 110)**	***p*-Value**
Demographic and Anthropometric			
Age (years)	53 ± 13	43 ± 17	0.03
Male gender, *n* (%)	24 (73)	58 (53)	0.047
Active smoker, *n* (%)	0	4 (4)	0.266
BMI (kg/m^2)	27 ± 5	24 ± 4	<0.001
Waist circumference (cm)	100 ± 11	88 ± 11	<0.001
Disease characteristic			
Disease duration (years)	15 ± 10	11 ± 9	0.044
Age at onset (years)	38 ± 16	32 ± 15	0.047
CD, *n* (%)	11 (33)	41 (37)	0.837
UC, *n* (%)	22 (66)	69 (63)	0.830
CD (Harvey–Bradshaw index)	5 ± 2	7 ± 3	0.033
UC (full Mayo Score)	2 ± 0.6	2 ± 0.8	0.612
Relapse/year	1.3 ± 0.7	1.3 ± 0.8	1.000
Active disease, *n* (%)	12 (36)	35 (32)	0.675
Extraintestinal manifestations, *n* (%)	7 (21)	19 (17)	0.612
Surgery, *n* (%)	7 (21)	17 (15)	0.435
CD disease location and phenotype, *n* (%)			
Ileal *	3 (27)	18 (44)	0.30
Colonic *	1 (9)	7 (17)	0.51
Ileo–colonic *	7 (64)	15 (37)	0.106
Upper GI *	0	1 (2)	0.60
Inflammatory *	1 (9)	16 (39)	0.06
Fistulizing *	3 (27)	13 (32)	0.70
Stenosing *	7 (64)	12 (29)	0.035
UC disease location, *n* (%)			
Proctitis *	0	8 (12)	0.09
Proctosigmoiditis *	2 (9)	17 (25)	0.11
Left side *	7 (32)	9 (13)	0.044
Pancolitis *	13 (59)	35 (50)	0.42

Table 2. *Cont.*

	IBD–NAFLD (*N* = 33)	IBD Non-NAFLD (*N* = 110)	*p*-Value
Dysmetabolic comorbidities, *n* (%)			
T2DM	4 (12)	7 (7)	0.278
Hypertension	13 (39)	11 (10)	<0.001
Dyslipidemia	5 (15)	13 (12)	0.564
IBD plus dysmetabolic criteria	26 (78)	36 (33)	<0.001
Laboratory parameter			
ALT (UI/L)	22 ± 10	18 ± 9	0.034
AST (UI/L)	22 ± 10	20 ± 9	0.187
Total cholesterol (mg/dL)	168 ± 47	167 ± 40	0.994
LDL (mg/dL)	107 ± 38	103 ± 33	0.591
HDL (mg/dL)	48 ± 16	58 ± 17	0.005
Triglycerides (mg/dL)	123 ± 63	93 ± 40	0.002
Fasting blood glucose (mg/dL)	92 ± 25	87 ± 18	0.156
Fasting insulinemia (mg/dL)	10 ± 8	10 ± 7	0.106
HOMA-IR	3 ± 2	2 ± 2	0.078
Fecal calprotectin (mcg/gr)	439 ± 911	519 ± 764	0.613
Medication, *n* (%)			
Salicylates, *n* (%)	20 (61)	55 (50)	0.324
Azathioprine, *n* (%)	10 (30)	37 (33)	0.834
>3 cycles of steroids, *n* (%)	9 (27)	25 (23)	0.643
Biological therapy, *n* (%)	17 (51)	69 (63)	0.311
Anti-TNF-α, *n* (%)	15 (88)	46 (67)	0.841
Vedolizumab, *n* (%)	0	16 (23)	0.023
Ustekinumab, *n* (%)	2 (12)	7 (10)	1.000
>1 Biological drug, *n* (%)	5 (15)	18 (16)	1.000
Current biological therapy duration (years)	4 ± 3	3 ± 2	0.188
Total biological therapy duration (years)	5 ± 3	4 ± 3	0.251

Legend: IBD, inflammatory bowel disease; CD, Crohn's disease; UC, ulcerative colitis; BMI, body mass index; NAFLD, nonalcoholic fatty liver disease; T2DM, type 2 diabetes mellitus; ALT, alanine aminotransferase; AST, aspartate aminotransferase; LDL, low-density lipoprotein; HDL, high-density lipoprotein; HOMA-IR, homeostasis model assessment of insulin resistance; TNF-α, tumor necrosis factor-alfa. * The *p*-value was evaluated with regard to CD and UC patients, respectively.

4. Discussion

NAFLD is frequently associated with IBD: both metabolic features and intestinal inflammation are involved in the pathogenesis of IBD-associated NAFLD. In this context, our study showed a US prevalence of NAFLD of 23% in IBD patients. These data are in line with recent epidemiological investigations that showed a US prevalence of 20–50% [12,29,30,34]. Given that IBD is a risk factor for NAFLD, this result underlines the importance of performing hepatic US in at-risk patients. Moreover, this approach should be applied in clinical practice not only to patients with IBD but also to other risk categories. In addition, due to its low cost, it should be used to follow disease progress over time [8]. In this complex interplay between genetic, metabolic, inflammatory, and pharmacological factors, the existing causative relationship and the underlying pathogenic mechanisms that might recognize the gut microbiota as a key link remain unclear [10]. Most IBD–NAFLD patients were male with an older age and age at onset than IBD patients without NAFLD. These data are explained by longer disease duration in patients with liver steatosis and were confirmed by Sourianarayanane et al., who described NAFLD patients as older (46.0 ± 13.3 vs. 42.0 ± 14.1 years; $p = 0.018$) and with a later onset of IBD compared with the control group (37.2 ± 15.3 vs. 28.7 ± 23.8 years; $p = 0.002$) [35]. These data were confirmed by Glassner et al., who showed that IBD–NAFLD patients had significantly longer disease duration than IBD-only patients (20 ± 12.2 vs. 10 ± 7.7 years, $p = 0.004$) [6]. Longer disease duration may lead to several risk factors for NAFLD, such as chronic inflammation and dysbiosis of the gut microbiota. It is plausible that gut dysbiosis may play a pivotal role in the biochemical and metabolic pathways that correlate with the onset and progression of IBD-associated NAFLD [5]. No significant difference was found for the number of relapses, extraintestinal manifestations, and active disease, according to Scrivo et al. [36].

At the same time, the IBD–NAFLD patients of our cohort showed a significantly higher BMI and waist circumference and a significantly higher percentage of hypertension than IBD non-NAFLD patients. These findings, along with the significantly lower HDL and higher triglycerides levels in patients with liver steatosis, support the use of new MAFLD nomenclature, which includes in the definition of NAFLD the additional dysmetabolic comorbidities we investigated in this study and additional risk factors, such as genetics and environmental factors and gut dysbiosis [7,37]. Our data are in agreement with the study by Magri et al. on a cohort of patients with characteristics similar to the present investigation. Indeed, NAFLD patients showed higher BMI and waist circumference vs. non-NAFLD patients. Furthermore, additional parameters such as visceral and body fat were evaluated. In this regard, the percentage of visceral fat was higher in NAFLD patients [38]. Hoffmann et al. also confirmed this evidence in their monocentric retrospective study performed on 153 IBD patients [39]. In addition, confirming this evidence, Saroli Palumbo et al. indicated how extrahepatic diseases such as chronic kidney disease and cardiovascular diseases are more common among IBD–NAFLD patients [40]. As expected, ALT levels were significantly higher in IBD–NAFLD patients than in the IBD non-NAFLD group. Indeed, liver enzyme levels and BMI are robust predictors of the risk of NAFLD in IBD [9]. Among IBD patients, the percentage of left-side colitis in UC was significantly higher in patients with liver steatosis vs. the non-NAFLD group. This finding is consistent with a previous study showing that a more extensive disease and a higher number of annual relapses and surgeries correlate with more severe steatosis [41]. On the other hand, the stenosing phenotype percentage in CD patients was significantly higher in NAFLD patients than in non-NAFLD. The possible reason is that our cohort is characterized by patients with long-term disease, who are thus more likely to have a more severe phenotype and, consequently, a greater susceptibility to liver steatosis [42]. Furthermore, recent studies have investigated whether CD is a stronger risk factor for developing NAFLD than UC. However, this hypothesis remains to be investigated because of the many genetic, environmental, and metabolic factors that play a major role in the establishment of hepatic steatosis in IBD patients [43]. Regarding biological therapy, vedolizumab was the only biological drug with a significant statistical difference between IBD–NAFLD patients and IBD non-NAFLD patients. In the literature, an antihepatic steatosis effect of anti-TNF-α treatments has been suggested [6], while there is no evidence of this effect for vedolizumab [44,45]. However, given the cross-sectional nature of this study, investigation of the mechanistic role of biological therapy in IBD–NAFLD patients falls beyond the purpose of our aims. Finally, 78% of the IBD population showed NAFLD without additional metabolic features (IBD plus dysmetabolic criteria). This high number is now first described, considering that the "multiple-hit hypothesis" of NAFLD includes as risk factors several comorbidities analyzed in this study, namely, obesity, insulin resistance, T2DM, and dyslipidemia [46]. Our data provide an analysis of the prevalence and clinical features of NAFLD in IBD patients admitted to a real-life hospital setting. This is one of the few observational studies in the literature that describes the clinical features of liver steatosis in IBD patients with and without dysmetabolic comorbidities.

The main limitations of our study are the small number of patients involved and the absence of a healthy control group. The latter is an interesting point as 25–30% of healthy people can have liver steatosis findings at screening abdominal US [6]. Thus, controlled trials in this field are needed to confirm these results.

5. Conclusions

Our data suggest the importance of performing US examinations in patients with IBD to detect NAFLD as early as possible. This clinical strategy can be central in improving the management of subjects affected by both these conditions. In addition, patients with NAFLD present several metabolic comorbidities that would fall within the new definition of MAFLD. Our preliminary results invite further confirmation on larger longitudinal studies including healthy controls.

Author Contributions: Conceptualization: L.A. and R.S.; formal analysis: L.A. and R.S.; data curation: G.G.M.S.; writing—original draft preparation: L.A., R.S. and G.G.M.S.; writing—review and editing: L.A., R.S., E.S. and L.B.; supervision: F.L. All authors have read and agreed to the published version of the manuscript.

Funding: This research received no external funding.

Institutional Review Board Statement: The study was conducted in accordance with the Declaration of Helsinki and approved by the ethics committee of the University of Magna Graecia (protocol number 2014/49, approval date 22 April 2021).

Informed Consent Statement: The present retrospective study was conducted anonymously, according to the principles of good clinical practice. Informed written consent was obtained from each participating patient.

Data Availability Statement: The data presented in this study are available on request from the corresponding author.

Conflicts of Interest: Ludovico Abenavoli is the Guest editor of the Special Issue "Editorial Board Members' Collection Series: Gastrointestinal and Hepatic Diseases".

References

1. Spagnuolo, R.; Larussa, T.; Iannelli, C.; Cosco, C.; Nisticò, E.; Manduci, E.; Bruno, A.; Boccuto, L.; Abenavoli, L.; Luzza, F.; et al. COVID-19 and Inflammatory Bowel Disease: Patient Knowledge and Perceptions in a Single Center Survey. *Medicina* **2020**, *56*, 407. [CrossRef] [PubMed]
2. Malik, T.F.; Aurelio, D.M. Extraintestinal Manifestations of Inflammatory Bowel Disease. In *StatPearls*; StatPearls Publishing: Treasure Island, FL, USA, 2023.
3. Marotto, D.; Atzeni, F.; Ardizzone, S.; Monteleone, G.; Giorgi, V.; Sarzi-Puttini, P. Extra-intestinal manifestations of inflammatory bowel diseases. *Pharmacol. Res.* **2020**, *161*, 105206. [CrossRef] [PubMed]
4. Martínez-Domínguez, S.J.; García-Mateo, S.; Laredo, V.; Gargallo-Puyuelo, C.J.; Gallego Llera, B.; López de la Cruz, J.; Gomollón, F. Liver Fibrosis in Non-Alcoholic Fatty Liver Disease and Progression to Hepatocellular Carcinoma in Patients with Inflammatory Bowel Disease: A Systematic Review. *Cancers* **2023**, *15*, 3367. [CrossRef] [PubMed]
5. Abenavoli, L.; Giubilei, L.; Procopio, A.C.; Spagnuolo, R.; Luzza, F.; Boccuto, L.; Scarpellini, E. Gut Microbiota in Non-Alcoholic Fatty Liver Disease Patients with Inflammatory Bowel Diseases: A Complex Interplay. *Nutrients* **2022**, *14*, 5323. [CrossRef]
6. Glassner, K.; Malaty, H.M.; Abraham, B.P. Epidemiology and Risk Factors of Nonalcoholic Fatty Liver Disease among Patients with Inflammatory Bowel Disease. *Inflamm. Bowel Dis.* **2017**, *23*, 998–1003. [CrossRef]
7. Abenavoli, L.; Scarlata, G.G.M.; Scarpellini, E.; Boccuto, L.; Spagnuolo, R.; Tilocca, B.; Roncada, P.; Luzza, F. Metabolic-Dysfunction-Associated Fatty Liver Disease and Gut Microbiota: From Fatty Liver to Dysmetabolic Syndrome. *Medicina* **2023**, *59*, 594. [CrossRef]
8. Spagnuolo, R.; Montalcini, T.; De Bonis, D.; Ferro, Y.; Cosco, C.; Mazza, E.; Romeo, S.; Doldo, P.; Pujia, A. Weight Gain and Liver Steatosis in Patients with Inflammatory Bowel Diseases. *Nutrients* **2019**, *11*, 303. [CrossRef]
9. Hong, Q.; Shen, J.; Feng, Q.; Zheng, Q.; Qiao, Y. Prevalence and predictors of non-alcoholic liver disease on MRI among patients with Crohn's disease. *BMC Gastroenterol.* **2022**, *22*, 183.
10. Spagnuolo, R.; Abenavoli, L.; Corea, A.; Larussa, T.; Mancina, R.M.; Cosco, C.; Luzza, F.; Doldo, P. Multifaceted pathogenesis of liver steatosis in inflammatory bowel disease: A systematic review. *Eur. Rev. Med. Pharmacol. Sci.* **2021**, *25*, 5818–5825.
11. Principi, M.; Iannone, A.; Losurdo, G.; Mangia, M.; Shahini, E.; Albano, F.; Rizzi, S.F.; La Fortezza, R.F.; Lovero, R.; Contaldo, A.; et al. Nonalcoholic Fatty Liver Disease in Inflammatory Bowel Disease: Prevalence and Risk Factors. *Inflamm. Bowel Dis.* **2018**, *24*, 1589–1596. [CrossRef]
12. Lin, A.; Roth, H.; Anyane-Yeboa, A.; Rubin, D.T.; Paul, S. Prevalence of Nonalcoholic Fatty Liver Disease in Patients with Inflammatory Bowel Disease: A Systematic Review and Meta-analysis. *Inflamm. Bowel Dis.* **2021**, *27*, 947–955. [CrossRef] [PubMed]
13. Chao, C.Y.; Battat, R.; Al Khoury, A.; Restellini, S.; Sebastiani, G.; Bessissow, T. Co-existence of non-alcoholic fatty liver disease and inflammatory bowel disease: A review article. *World J. Gastroenterol.* **2016**, *22*, 7727–7734. [CrossRef] [PubMed]
14. Larussa, T.; Abenavoli, L.; Fabiano, G.; Mancuso, M.A.; Polimeni, N.; Dumitrascu, D.L.; Luzza, F. Gut microbiota in inflammatory bowel disease: A target for therapy not to be missed. *Minerva Gastroenterol.* **2021**, *67*, 357–368. [CrossRef] [PubMed]
15. Mancina, R.M.; Spagnuolo, R.; Milano, M.; Brogneri, S.; Morrone, A.; Cosco, C.; Lazzaro, V.; Russo, C.; Ferro, Y.; Pingitore, P.; et al. PNPLA3 148M Carriers with Inflammatory Bowel Diseases Have Higher Susceptibility to Hepatic Steatosis and Higher Liver Enzymes. *Inflamm. Bowel Dis.* **2016**, *22*, 134–140. [CrossRef]
16. Bessissow, T.; Le, N.H.; Rollet, K.; Afif, W.; Bitton, A.; Sebastiani, G. Incidence and Predictors of Nonalcoholic Fatty Liver Disease by Serum Biomarkers in Patients with Inflammatory Bowel Disease. *Inflamm. Bowel Dis.* **2016**, *22*, 1937–1944. [CrossRef]

17. Kablawi, D.; Aljohani, F.; Palumbo, C.S.; Restellini, S.; Bitton, A.; Wild, G.; Afif, W.; Lakatos, P.L.; Bessissow, T.; Sebastiani, G. Nonalcoholic Fatty Liver Disease Increases Cardiovascular Risk in Inflammatory Bowel Diseases. *Crohns Colitis 360* **2023**, *5*, otad004. [CrossRef]
18. Rodriguez-Duque, J.C.; Calleja, J.L.; Iruzubieta, P.; Hernández-Conde, M.; Rivas-Rivas, C.; Vera, M.I.; Garcia, M.J.; Pascual, M.; Castro, B.; García-Blanco, A.; et al. Increased risk of MAFLD and Liver Fibrosis in Inflammatory Bowel Disease Independent of Classic Metabolic Risk Factors. *Clin. Gastroenterol. Hepatol. Off. Clin. Pract. J. Am. Gastroenterol. Assoc.* **2023**, *21*, 406–414. [CrossRef]
19. Mancina, R.M.; De Bonis, D.; Pagnotta, R.; Cosco, C.; Cosco, V.; Montalcini, T.; Pujia, A.; Doldo, P.; Spagnuolo, R. Ulcerative Colitis as an Independent Risk Factor for Hepatic Steatosis. *Gastroenterol. Nurs. Off. J. Soc. Gastroenterol. Nurses Assoc.* **2020**, *43*, 292–297. [CrossRef]
20. Nguyen, D.L.; Bechtold, M.L.; Jamal, M.M. National trends and inpatient outcomes of inflammatory bowel disease patients with concomitant chronic liver disease. *Scand. J. Gastroenterol.* **2014**, *49*, 1091–1095. [CrossRef]
21. Hamaguchi, M.; Kojima, T.; Itoh, Y.; Harano, Y.; Fujii, K.; Nakajima, T.; Kato, T.; Takeda, N.; Okuda, J.; Ida, K.; et al. The severity of ultrasonographic findings in nonalcoholic fatty liver disease reflects the metabolic syndrome and visceral fat accumulation. *Am. J. Gastroenterol.* **2007**, *102*, 2708–2715. [CrossRef]
22. Chalasani, N.; Younossi, Z.; Lavine, J.E.; Diehl, A.M.; Brunt, E.M.; Cusi, K.; Charlton, M.; Sanyal, A.J. The diagnosis and management of non-alcoholic fatty liver disease: Practice Guideline by the American Association for the Study of Liver Diseases, American College of Gastroenterology, and the American Gastroenterological Association. *Hepatology* **2012**, *55*, 2005–2023. [CrossRef] [PubMed]
23. Kobyliak, N.; Abenavoli, L. The role of liver biopsy to assess non-alcoholic fatty liver disease. *Rev. Recent. Clin. Trials* **2014**, *9*, 159–169. [CrossRef] [PubMed]
24. Abenavoli, L.; Beaugrand, M. Transient elastography in non-alcoholic fatty liver disease. *Ann. Hepatol.* **2012**, *11*, 172–178. [CrossRef] [PubMed]
25. Chang, P.E.; Goh, G.B.; Ngu, J.H.; Tan, H.K.; Tan, C.K. Clinical applications, limitations and future role of transient elastography in the management of liver disease. *World J. Gastrointest. Pharmacol. Ther.* **2016**, *7*, 91–106. [CrossRef]
26. Pouwels, S.; Sakran, N.; Graham, Y.; Leal, A.; Pintar, T.; Yang, W.; Kassir, R.; Singhal, R.; Mahawar, K.; Ramnarain, D. Non-alcoholic fatty liver disease (NAFLD): A review of pathophysiology, clinical management and effects of weight loss. *BMC Endocr. Disord.* **2022**, *22*, 63. [CrossRef]
27. Şendur, H.N.; Cerit, M.N.; Ibrahimkhanli, N.; Şendur, A.B.; Özhan Oktar, S. Interobserver Variability in Ultrasound-Based Liver Fat Quantification. *J. Ultrasound Med.* **2023**, *42*, 833–841. [CrossRef]
28. Zamani, M.; Alizadeh-Tabari, S.; Singh, S.; Loomba, R. Meta-analysis: Prevalence of, and risk factors for, non-alcoholic fatty liver disease in patients with inflammatory bowel disease. *Aliment. Pharmacol. Ther.* **2022**, *55*, 894–907. [CrossRef]
29. Likhitsup, A.; Dundulis, J.; Ansari, S.; Patibandla, S.; Hutton, C.; Kennedy, K.; Helzberg, J.H.; Chhabra, R. High prevalence of non-alcoholic fatty liver disease in patients with inflammatory bowel disease receiving anti-tumor necrosis factor therapy. *Ann. Gastroenterol.* **2019**, *32*, 463–468. [CrossRef]
30. Shintaku, D.; Lopes, M.; de Oliveira, A.C.; Beraldo, R.; Godoi, G.; Castelhano, N.; Pereira, J.; Vulcano, D.; de Oliveira, E.C.; Herrerias, G.; et al. P049 Investigation of Liver Diseases by Ultrasound in Patients with Inflammatory Bowel Disease. *Am. J. Gastroenterol.* **2021**, *116*, S12–S13. [CrossRef]
31. Ferraioli, G.; Soares Monteiro, L.B. Ultrasound-based techniques for the diagnosis of liver steatosis. *World J. Gastroenterol.* **2019**, *25*, 6053–6062. [CrossRef]
32. European Association for the Study of the Liver (EASL); European Association for the Study of Diabetes (EASD); European Association for the Study of Obesity (EASO). EASL-EASD-EASO Clinical Practice Guidelines for the management of non-alcoholic fatty liver disease. *J. Hepatol.* **2016**, *64*, 1388–1402. [CrossRef] [PubMed]
33. Angelico, F.; Pastori, D.; Del Ben, M. Impact of the New Metabolic-Associated Fatty Liver Disease (MAFLD) on NAFLD Patients Classification in Italy. *Clin. Gastroenterol. Hepatol.* **2021**, *19*, 2683–2684. [CrossRef] [PubMed]
34. Sagami, S.; Ueno, Y.; Tanaka, S.; Fujita, A.; Hayashi, R.; Oka, S.; Hyogo, H.; Chayama, K. Significance of non-alcoholic fatty liver disease in Crohn's disease: A retrospective cohort study. *Hepatol. Res.* **2017**, *47*, 872–881. [CrossRef] [PubMed]
35. Sourianarayanane, A.; Garg, G.; Smith, T.H.; Butt, M.I.; McCullough, A.J.; Shen, B. Risk factors of non-alcoholic fatty liver disease in patients with inflammatory bowel disease. *J. Crohn's Colitis* **2013**, *7*, 279–285. [CrossRef]
36. Scrivo, B.; Celsa, C.; Busacca, A.; Giuffrida, E.; Pipitone, R.M.; Grimaudo, S.; Calogero, C.; Petta, S.; Cappello, M. P162 Prevalence of NAFLD (nonalcoholic fatty liver disease) and fibrosis in inflammatory bowel disease: The impact of traditional risk factors, intestinal inflammation and genetic phenotype. *J. Crohn's Colitis* **2020**, *14*, 219–220. [CrossRef]
37. Eslam, M.; Sanyal, A.J.; George, J.; International Consensus Panel. MAFLD: A Consensus-Driven Proposed Nomenclature for Metabolic Associated Fatty Liver Disease. *Gastroenterology* **2020**, *158*, 1999–2014.e1. [CrossRef]
38. Magrì, S.; Paduano, D.; Chicco, F.; Cingolani, A.; Farris, C.; Delogu, G.; Tumbarello, F.; Lai, M.; Melis, A.; Casula, L.; et al. Nonalcoholic fatty liver disease in patients with inflammatory bowel disease: Beyond the natural history. *World J. Gastroenterol.* **2019**, *25*, 5676–5686. [CrossRef]
39. Hoffmann, P.; Jung, V.; Behnisch, R.; Gauss, A. Prevalence and risk factors of nonalcoholic fatty liver disease in patients with inflammatory bowel diseases: A cross-sectional and longitudinal analysis. *World J. Gastroenterol.* **2020**, *26*, 7367–7381. [CrossRef]

40. Saroli Palumbo, C.; Restellini, S.; Chao, C.Y.; Aruljothy, A.; Lemieux, C.; Wild, G.; Afif, W.; Lakatos, P.L.; Bitton, A.; Cocciolillo, S.; et al. Screening for Nonalcoholic Fatty Liver Disease in Inflammatory Bowel Diseases: A Cohort Study Using Transient Elastography. *Inflamm. Bowel Dis.* **2019**, *25*, 124–133. [CrossRef]
41. Sartini, A.; Gitto, S.; Bianchini, M.; Verga, M.C.; Di Girolamo, M.; Bertani, A.; Del Buono, M.; Schepis, F.; Lei, B.; De Maria, N.; et a Non-alcoholic fatty liver disease phenotypes in patients with inflammatory bowel disease. *Cell Death Dis.* **2018**, *9*, 87. [CrossRef
42. Cho, C.W.; You, M.W.; Oh, C.H.; Lee, C.K.; Moon, S.K. Long-term Disease Course of Crohn's Disease: Changes in Diseas Location, Phenotype, Activities, and Predictive Factors. *Gut Liver* **2022**, *16*, 57–170. [CrossRef] [PubMed]
43. Kodali, A.; Okoye, C.; Klein, D.; Mohamoud, I.; Olanisa, O.O.; Parab, P.; Chaudhary, P.; Mukhtar, S.; Moradi, A.; Hamid, I Crohn's Disease is a Greater Risk Factor for Nonalcoholic Fatty Liver Disease Compared to Ulcerative Colitis: A Systemati Review. *Cureus* **2023**, *15*, e42995. [CrossRef] [PubMed]
44. Papaefthymiou, A.; Potamianos, S.; Goulas, A.; Doulberis, M.; Kountouras, J.; Polyzos, S.A. Inflammatory Bowel Disease-associated Fatty Liver Disease: The Potential Effect of Biologic Agents. *J. Crohn's Colitis* **2022**, *16*, 852–862. [CrossRef]
45. Mitrovic, M.; Marković, S.; Kalaba, A.; Zarić, D.; Kralj, D.; Milić, A.; Svorcan, P. The effect of anti-TNF-α and anti-integrin agents on liver steatosis in inflammatory bowel disease patients with non-alcoholic fatty liver disease. *J. Crohn's Colitis* **2023**, *17*, i837. [CrossRef]
46. Zarghamravanbakhsh, P.; Frenkel, M.; Poretsky, L. Metabolic causes and consequences of nonalcoholic fatty liver disease (NAFLD). *Metabol. Open* **2021**, *12*, 100149, Erratum in *Metabol. Open* **2023**, *17*, 100231. [CrossRef] [PubMed]

MDPI AG
Grosspeteranlage 5
4052 Basel
Switzerland
Tel.: +41 61 683 77 34

Medicina Editorial Office
E-mail: medicina@mdpi.com
www.mdpi.com/journal/medicina

www.ingramcontent.com/pod-product-compliance
Lightning Source LLC
Chambersburg PA
CBHW041415210926
51850CB00013B/3279